MCAT®

Behavioral Sciences Review 2025–2026

Online + Book

Edited by Alexander Stone Macnow, MD

ACKNOWLEDGMENTS

Editor-in-Chief, 2025–2026 Edition
M. Dominic Eggert

Contributing Editor, 2025–2026 Edition
Elisabeth Fassas, MD MSc

Prior Edition Editorial Staff: Christopher Durland; Charles Pierce, MD; Jason Selzer

Published by Kaplan North America, LLC dba Kaplan Publishing
1515 West Cypress Creek Road
Fort Lauderdale, Florida 33309

Printed in India.

ISBN: 978-1-5062-9404-9

10 9 8 7 6 5 4 3 2 1

TABLE OF CONTENTS

GO ONLINE

kaptest.com/booksonline

THE *KAPLAN MCAT REVIEW* TEAM

Alexander Stone Macnow, MD
Editor-in-Chief

Áine Lorié, PhD
Editor

Derek Rusnak, MA
Editor

Mikhail Alexeeff
Kaplan MCAT Faculty

Laura L. Ambler
Kaplan MCAT Faculty

Krista L. Buckley, MD
Kaplan MCAT Faculty

Kristen L. Russell, ME
Editor

Pamela Willingham, MSW
Editor

Melinda Contreras, MS
Kaplan MCAT Faculty

Samantha Fallon
Kaplan MCAT Faculty

Jason R. Selzer
Kaplan MCAT Faculty

M. Dominic Eggert
Editor

Faculty Reviewers and Editors: Elmar R. Aliyev; James Burns; Jonathan Cornfield; Alisha Maureen Crowley; Nikolai Dorofeev, MD; Benjamin Downer, MS; Colin Doyle; Christopher Durland; Marilyn Engle; Eleni M. Eren; Raef Ali Fadel; Elizabeth Flagge; Adam Grey; Rohit Gupta, Jonathan Habermacher; Tyra Hall-Pogar, PhD; Justine Harkness, PhD; Scott Huff; Samer T. Ismail; Ae-Ri Kim, PhD; Elizabeth A. Kudlaty; Kelly Kyker-Snowman, MS; Ningfei Li; John P. Mahon; Brandon McKenzie; Matthew A. Meier; Nainika Nanda; Caroline Nkemdilim Opene; Kaitlyn E. Prenger; Uneeb Qureshi; Jason Selzer; Allison St. Clair; Bela G. Starkman, PhD; Chris Sun; Michael Paul Tomani, MS; Bonnie Wang; Ethan Weber; Lauren K. White; Nicholas M. White; Allison Ann Wilkes, MS; Kerranna Williamson, MBA; and Tony Yu

Thanks to Rebecca Anderson; Jeff Batzli; Eric Chiu; Tim Eich; Tyler Fara; Owen Farcy; Dan Frey; Robin Garmise; Rita Garthaffner; Joanna Graham; Allison Gudenau; Allison Harm; Beth Hoffberg; Aaron Lemon-Strauss; Keith Lubeley; Diane McGarvey; Petros Minasi; Beena P V; John Polstein; Deeangelee Pooran-Kublall, MD, MPH; Rochelle Rothstein, MD; Larry Rudman; Srividhya Sankar; Sylvia Tidwell Scheuring; Carly Schnur; Aiswarya Sivanand; Todd Tedesco; Karin Tucker; Lee Weiss; Christina Wheeler; Kristen Workman; Amy Zarkos; and the countless others who made this project possible.

GETTING STARTED CHECKLIST

Getting Started Checklist

- [ ] Register for your free online assets—including full-length tests, Science Review Videos, and additional practice materials—at **www.kaptest.com/booksonline.**
- [ ] Create a study calendar that ensures you complete content review and sufficient practice by Test Day!
- [ ] As you finish a chapter and the online practice for that chapter, check it off on the table of contents.
- [ ] Register to take the MCAT at **www.aamc.org/mcat**.
- [ ] Set aside time during your prep to make sure the rest of your application—personal statement, recommendations, and other materials—is ready to go!
- [ ] Take a moment to admire your completed checklist, then get back to the business of prepping for this exam!

PREFACE

And now it starts: your long, yet fruitful journey toward wearing a white coat. Proudly wearing that white coat, though, is hopefully only part of your motivation. You are reading this book because you want to be a healer.

If you're serious about going to medical school, then you are likely already familiar with the importance of the MCAT in medical school admissions. While the holistic review process puts additional weight on your experiences, extracurricular activities, and personal attributes, the fact remains: along with your GPA, your MCAT score remains one of the two most important components of your application portfolio—at least early in the admissions process. Each additional point you score on the MCAT pushes you in front of thousands of other students and makes you an even more attractive applicant. But the MCAT is not simply an obstacle to overcome; it is an opportunity to show schools that you will be a strong student and a future leader in medicine.

We at Kaplan take our jobs very seriously and aim to help students see success not only on the MCAT, but as future physicians. We work with our learning science experts to ensure that we're using the most up-to-date teaching techniques in our resources. Multiple members of our team hold advanced degrees in medicine or associated biomedical sciences, and are committed to the highest level of medical education. Kaplan has been working with the MCAT for over 50 years and our commitment to premed students is unflagging; in fact, Stanley Kaplan created this company when he had difficulty being accepted to medical school due to unfair quota systems that existed at the time.

We stand now at the beginning of a new era in medical education. As citizens of this 21st-century world of healthcare, we are charged with creating a patient-oriented, culturally competent, cost-conscious, universally available, technically advanced, and research-focused healthcare system, run by compassionate providers. Suffice it to say, this is no easy task. Problem-based learning, integrated curricula, and classes in interpersonal skills are some of the responses to this demand for an excellent workforce—a workforce of which you'll soon be a part.

We're thrilled that you've chosen us to help you on this journey. Please reach out to us to share your challenges, concerns, and successes. Together, we will shape the future of medicine in the United States and abroad; we look forward to helping you become the doctor you deserve to be.

Good luck!

Alexander Stone Macnow, MD
Editor-in-Chief
Department of Pathology and Laboratory Medicine
Hospital of the University of Pennsylvania

BA, Musicology—Boston University, 2008
MD—Perelman School of Medicine at the University of Pennsylvania, 2013

ABOUT THE MCAT

Anatomy of the MCAT

Here is a general overview of the structure of Test Day:

Section	Number of Questions	Time Allotted
Test-Day Certification		4 minutes
Tutorial (optional)		10 minutes
Chemical and Physical Foundations of Biological Systems	59	95 minutes
Break (optional)		10 minutes
Critical Analysis and Reasoning Skills (CARS)	53	90 minutes
Lunch Break (optional)		30 minutes
Biological and Biochemical Foundations of Living Systems	59	95 minutes
Break (optional)		10 minutes
Psychological, Social, and Biological Foundations of Behavior	59	95 minutes
Void Question		3 minutes
Satisfaction Survey (optional)		5 minutes

The structure of the four sections of the MCAT is shown below.

Chemical and Physical Foundations of Biological Systems	
Time	95 minutes
Format	• 59 questions • 10 passages • 44 questions are passage-based, and 15 are discrete (stand-alone) questions. • Score between 118 and 132
What It Tests	• Biochemistry: 25% • Biology: 5% • General Chemistry: 30% • Organic Chemistry: 15% • Physics: 25%

Critical Analysis and Reasoning Skills (CARS)	
Time	90 minutes
Format	• 53 questions • 9 passages • All questions are passage-based. There are no discrete (stand-alone) questions. • Score between 118 and 132
What It Tests	Disciplines: • Humanities: 50% • Social Sciences: 50% Skills: • *Foundations of Comprehension*: 30% • *Reasoning Within the Text*: 30% • *Reasoning Beyond the Text*: 40%
Biological and Biochemical Foundations of Living Systems	
Time	95 minutes
Format	• 59 questions • 10 passages • 44 questions are passage-based, and 15 are discrete (stand-alone) questions. • Score between 118 and 132
What It Tests	• Biochemistry: 25% • Biology: 65% • General Chemistry: 5% • Organic Chemistry: 5%
Psychological, Social, and Biological Foundations of Behavior	
Time	95 minutes
Format	• 59 questions • 10 passages • 44 questions are passage-based, and 15 are discrete (stand-alone) questions. • Score between 118 and 132
What It Tests	• Biology: 5% • Psychology: 65% • Sociology: 30%
Total	
Testing Time	375 minutes (6 hours, 15 minutes)
Total Seat Time	447 minutes (7 hours, 27 minutes)
Questions	230
Score	472 to 528

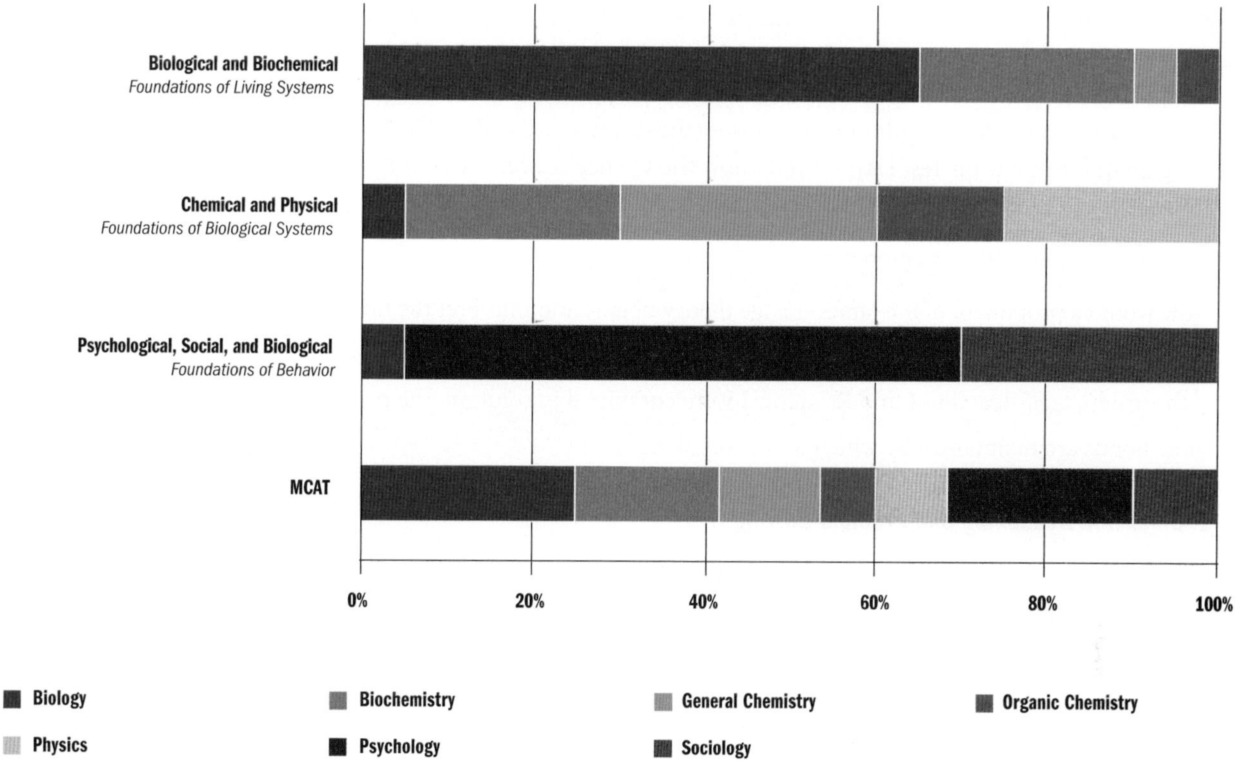

Scientific Inquiry and Reasoning Skills (SIRS)

The AAMC has defined four *Scientific Inquiry and Reasoning Skills* (SIRS) that will be tested in the three science sections of the MCAT:

1. *Knowledge of Scientific Concepts and Principles* (35% of questions)
2. *Scientific Reasoning and Problem-Solving* (45% of questions)
3. *Reasoning About the Design and Execution of Research* (10% of questions)
4. *Data-Based and Statistical Reasoning* (10% of questions)

Let's see how each one breaks down into more specific Test Day behaviors. Note that the bullet points of specific objectives for each of the SIRS are taken directly from the *Official Guide to the MCAT Exam*; the descriptions of what these behaviors mean and sample question stems, however, are written by Kaplan.

Skill 1: *Knowledge of Scientific Concepts and Principles*

This is probably the least surprising of the four SIRS; the testing of science knowledge is, after all, one of the signature qualities of the MCAT. Skill 1 questions will require you to do the following:

- Recognize correct scientific principles
- Identify the relationships among closely related concepts
- Identify the relationships between different representations of concepts (verbal, symbolic, graphic)
- Identify examples of observations that illustrate scientific principles
- Use mathematical equations to solve problems

At Kaplan, we simply call these Science Knowledge or Skill 1 questions. Another way to think of Skill 1 questions is as "one-step" problems. The single step is either to realize which scientific concept the question stem is suggesting or to take the concept stated in the question stem and identify which answer choice is an accurate application of it. Skill 1 questions are particularly prominent among discrete questions (those not associated with a passage). These questions are an opportunity to gain quick points on Test Day—if you know the science concept attached to the question, then that's it! On Test Day, 35% of the questions in each science section will be Skill 1 questions.

Here are some sample Skill 1 question stems:

- How would a proponent of the James–Lange theory of emotion interpret the findings of the study cited in the passage?
- Which of the following most accurately describes the function of FSH in the human menstrual cycle?
- If the products of Reaction 1 and Reaction 2 were combined in solution, the resulting reaction would form:
- Ionic bonds are maintained by which of the following forces?

Skill 2: *Scientific Reasoning and Problem-Solving*

The MCAT science sections do, of course, move beyond testing straightforward science knowledge; Skill 2 questions are the most common way in which it does so. At Kaplan, we also call these Critical Thinking questions. Skill 2 questions will require you to do the following:

- Reason about scientific principles, theories, and models
- Analyze and evaluate scientific explanations and predictions
- Evaluate arguments about causes and consequences
- Bring together theory, observations, and evidence to draw conclusions
- Recognize scientific findings that challenge or invalidate a scientific theory or model
- Determine and use scientific formulas to solve problems

Just as Skill 1 questions can be thought of as "one-step" problems, many Skill 2 questions are "two-step" problems, and more difficult Skill 2 questions may require three or more steps. These questions can require a wide spectrum of reasoning skills, including integration of multiple facts from a passage, combination of multiple science content areas, and prediction of an experiment's results. Skill 2 questions also tend to ask about science content without actually mentioning it by name. For example, a question might describe the results of one experiment and ask you to predict the results of a second experiment without actually telling you what underlying scientific principles are at work—part of the question's difficulty will be figuring out which principles to apply in order to get the correct answer. On Test Day, 45% of the questions in each science section will be Skill 2 questions.

Here are some sample Skill 2 question stems:

- Which of the following experimental conditions would most likely yield results similar to those in Figure 2?
- All of the following conclusions are supported by the information in the passage EXCEPT:
- The most likely cause of the anomalous results found by the experimenter is:
- An impact to a person's chest quickly reduces the volume of one of the lungs to 70% of its initial value while not allowing any air to escape from the mouth. By what percentage is the force of outward air pressure increased on a 2 cm^2 portion of the inner surface of the compressed lung?

Skill 3: *Reasoning About the Design and Execution of Research*

The MCAT is interested in your ability to critically appraise and analyze research, as this is an important day-to-day task of a physician. We call these questions Skill 3 or Experimental and Research Design questions for short. Skill 3 questions will require you to do the following:

- Identify the role of theory, past findings, and observations in scientific questioning
- Identify testable research questions and hypotheses
- Distinguish between samples and populations and distinguish results that support generalizations about populations
- Identify independent and dependent variables
- Reason about the features of research studies that suggest associations between variables or causal relationships between them (such as temporality and random assignment)
- Identify conclusions that are supported by research results
- Determine the implications of results for real-world situations
- Reason about ethical issues in scientific research

Over the years, the AAMC has received input from medical schools to require more practical research skills of MCAT test takers, and Skill 3 questions are the response to these demands. This skill is unique in that the outside knowledge you need to answer Skill 3 questions is not taught in any one undergraduate course; instead, the research design principles needed to answer these questions are learned gradually throughout your science classes and especially through any laboratory work you have completed. It should be noted that Skill 3 comprises 10% of the questions in each science section on Test Day.

Here are some sample Skill 3 question stems:

- What is the dependent variable in the study described in the passage?
- The major flaw in the method used to measure disease susceptibility in Experiment 1 is:
- Which of the following procedures is most important for the experimenters to follow in order for their study to maintain a proper, randomized sample of research subjects?
- A researcher would like to test the hypothesis that individuals who move to an urban area during adulthood are more likely to own a car than are those who have lived in an urban area since birth. Which of the following studies would best test this hypothesis?

Skill 4: *Data-Based and Statistical Reasoning*

Lastly, the science sections of the MCAT test your ability to analyze the visual and numerical results of experiments and studies. We call these Data and Statistical Analysis questions. Skill 4 questions will require you to do the following:

- Use, analyze, and interpret data in figures, graphs, and tables
- Evaluate whether representations make sense for particular scientific observations and data
- Use measures of central tendency (mean, median, and mode) and measures of dispersion (range, interquartile range, and standard deviation) to describe data
- Reason about random and systematic error

- Reason about statistical significance and uncertainty (interpreting statistical significance levels and interpreting a confidence interval)
- Use data to explain relationships between variables or make predictions
- Use data to answer research questions and draw conclusions

Skill 4 is included in the MCAT because physicians and researchers spend much of their time examining the results of their own studies and the studies of others, and it's very important for them to make legitimate conclusions and sound judgments based on that data. The MCAT tests Skill 4 on all three science sections with graphical representations of data (charts and bar graphs), as well as numerical ones (tables, lists, and results summarized in sentence or paragraph form). On Test Day, 10% of the questions in each science section will be Skill 4 questions.

Here are some sample Skill 4 question stems:

- According to the information in the passage, there is an inverse correlation between:
- What conclusion is best supported by the findings displayed in Figure 2?
- A medical test for a rare type of heavy metal poisoning returns a positive result for 98% of affected individuals and 13% of unaffected individuals. Which of the following types of error is most prevalent in this test?
- If a fourth trial of Experiment 1 was run and yielded a result of 54% compliance, which of the following would be true?

SIRS Summary

Discussing the SIRS tested on the MCAT is a daunting prospect given that the very nature of the skills tends to make the conversation rather abstract. Nevertheless, with enough practice, you'll be able to identify each of the four skills quickly, and you'll also be able to apply the proper strategies to solve those problems on Test Day. If you need a quick reference to remind you of the four SIRS, these guidelines may help:

Skill 1 (**Science Knowledge**) questions ask:

- Do you remember this science content?

Skill 2 (**Critical Thinking**) questions ask:

- Do you remember this science content? And if you do, could you please apply it to this novel situation?
- Could you answer this question that cleverly combines multiple content areas at the same time?

Skill 3 (**Experimental and Research Design**) questions ask:

- Let's forget about the science content for a while. Could you give some insight into the experimental or research methods involved in this situation?

Skill 4 (**Data and Statistical Analysis**) questions ask:

- Let's forget about the science content for a while. Could you accurately read some graphs and tables for a moment? Could you make some conclusions or extrapolations based on the information presented?

Critical Analysis and Reasoning Skills (CARS)

The *Critical Analysis and Reasoning Skills* (CARS) section of the MCAT tests three discrete families of textual reasoning skills; each of these families requires a higher level of reasoning than the last. Those three skills are as follows:

1. *Foundations of Comprehension* (30% of questions)
2. *Reasoning Within the Text* (30% of questions)
3. *Reasoning Beyond the Text* (40% of questions)

These three skills are tested through nine humanities- and social sciences-themed passages, with approximately 5 to 7 questions per passage. Let's take a more in-depth look into these three skills. Again, the bullet points of specific objectives for each of the CARS are taken directly from the *Official Guide to the MCAT Exam*; the descriptions of what these behaviors mean and sample question stems, however, are written by Kaplan.

Foundations of Comprehension

Questions in this skill will ask for basic facts and simple inferences about the passage; the questions themselves will be similar to those seen on reading comprehension sections of other standardized exams like the SAT® and ACT®. *Foundations of Comprehension* questions will require you to do the following:

- Understand the basic components of the text
- Infer meaning from rhetorical devices, word choice, and text structure

This admittedly covers a wide range of potential question types including Main Idea, Detail, Inference, and Definition-in-Context questions, but finding the correct answer to all *Foundations of Comprehension* questions will follow from a basic understanding of the passage and the point of view of its author (and occasionally that of other voices in the passage).

Here are some sample *Foundations of Comprehension* question stems:

- **Main Idea**—The author's primary purpose in this passage is:
- **Detail**—Based on the information in the second paragraph, which of the following is the most accurate summary of the opinion held by Schubert's critics?
- **(Scattered) Detail**—According to the passage, which of the following is FALSE about literary reviews in the 1920s?
- **Inference (Implication)**—Which of the following phrases, as used in the passage, is most suggestive that the author has a personal bias toward narrative records of history?
- **Inference (Assumption)**—In putting together the argument in the passage, the author most likely assumes:
- **Definition-in-Context**—The word "obscure" (paragraph 3), when used in reference to the historian's actions, most nearly means:

Reasoning Within the Text

While *Foundations of Comprehension* questions will usually depend on interpreting a single piece of information in the passage or understanding the passage as a whole, *Reasoning Within the Text* questions require more thought because they will ask you to identify the purpose of a particular piece of information in the context of the passage, or ask how one piece of information relates to another. *Reasoning Within the Text* questions will require you to:

- Integrate different components of the text to draw relevant conclusions

In other words, questions in this skill often ask either *How do these two details relate to one another?* or *What else must be true that the author didn't say?* The CARS section will also ask you to judge certain parts of the passage or even judge the author. These questions, which fall under the *Reasoning Within the Text* skill, can ask you to identify authorial bias, evaluate the credibility of cited sources, determine the logical soundness of an argument, identify the importance of a particular fact or statement in the context of the passage, or search for relevant evidence in the passage to support a given conclusion. In all, this category includes Function and Strengthen–Weaken (Within the Passage) questions, as well as a smattering of related—but rare—question types.

Here are some sample *Reasoning Within the Text* question stems:

- **Function**—The author's discussion of the effect of socioeconomic status on social mobility primarily serves which of the following functions?
- **Strengthen–Weaken (Within the Passage)**—Which of the following facts is used in the passage as the most prominent piece of evidence in favor of the author's conclusions?
- **Strengthen–Weaken (Within the Passage)**—Based on the role it plays in the author's argument, *The Possessed* can be considered:

Reasoning Beyond the Text

The distinguishing factor of *Reasoning Beyond the Text* questions is in the title of the skill: the word *Beyond*. Questions that test this skill, which make up a larger share of the CARS section than questions from either of the other two skills, will always introduce a completely new situation that was not present in the passage itself; these questions will ask you to determine how one influences the other. *Reasoning Beyond the Text* questions will require you to:

- Apply or extrapolate ideas from the passage to new contexts
- Assess the impact of introducing new factors, information, or conditions to ideas from the passage

The *Reasoning Beyond the Text* skill is further divided into Apply and Strengthen–Weaken (Beyond the Passage) questions, and a few other rarely appearing question types.

Here are some sample *Reasoning Beyond the Text* question stems:

- **Apply**—If a document were located that demonstrated Berlioz intended to include a chorus of at least 700 in his *Grande Messe des Morts*, how would the author likely respond?
- **Apply**—Which of the following is the best example of a "virtuous rebellion," as it is defined in the passage?
- **Strengthen–Weaken (Beyond the Passage)**—Suppose Jane Austen had written in a letter to her sister, "My strongest characters were those forced by circumstance to confront basic questions about the society in which they lived." What relevance would this have to the passage?
- **Strengthen–Weaken (Beyond the Passage)**—Which of the following sentences, if added to the end of the passage, would most WEAKEN the author's conclusions in the last paragraph?

CARS Summary

Through the *Foundations of Comprehension* skill, the CARS section tests many of the reading skills you have been building on since grade school, albeit in the context of very challenging doctorate-level passages. But through the two other skills (*Reasoning Within the Text* and *Reasoning Beyond the Text*), the MCAT demands that you understand the deep structure of passages and the arguments within them at a very advanced level. And, of course, all of this is tested under very tight timing restrictions: only 102 seconds per question—and that doesn't even include the time spent reading the passages.

Here's a quick reference guide to the three CARS skills:

Foundations of Comprehension questions ask:

- Did you understand the passage and its main ideas?
- What does the passage have to say about this particular detail?
- What must be true that the author did not say?

Reasoning Within the Text questions ask:

- What's the logical relationship between these two ideas from the passage?
- How well argued is the author's thesis?

Reasoning Beyond the Text questions ask:

- How does this principle from the passage apply to this new situation?
- How does this new piece of information influence the arguments in the passage?

Scoring

Each of the four sections of the MCAT is scored between 118 and 132, with the median at approximately 125. This means the total score ranges from 472 to 528, with the median at about 500. Why such peculiar numbers? The AAMC stresses that this scale emphasizes the importance of the central portion of the score distribution, where most students score (around 125 per section, or 500 total), rather than putting undue focus on the high end of the scale.

Note that there is no wrong answer penalty on the MCAT, so you should select an answer for every question—even if it is only a guess.

The AAMC has released the 2020–2022 correlation between scaled score and percentile, as shown on the following page. It should be noted that the percentile scale is adjusted and renormalized over time and thus can shift slightly from year to year. Percentile rank updates are released by the AAMC around May 1 of each year.

Total Score	Percentile	Total Score	Percentile
528	100	499	43
527	100	498	39
526	100	497	36
525	100	496	33
524	100	495	31
523	99	494	28
522	99	493	25
521	98	492	23
520	97	491	20
519	96	490	18
518	95	489	16
517	94	488	14
516	92	487	12
515	90	486	11
514	88	485	9
513	86	484	8
512	83	483	6
511	81	482	5
510	78	481	4
509	75	480	3
508	72	479	3
507	69	478	2
506	66	477	1
505	62	476	1
504	59	475	1
503	56	474	<1
502	52	473	<1
501	49	472	<1
500	46		

Source: AAMC. 2023. Summary of MCAT Total and Section Scores. Accessed October 2023.
https://students-residents.aamc.org/mcat-research-and-data/percentile-ranks-mcat-exam

Further information on score reporting is included at the end of the next section (see *After Your Test*).

MCAT Policies and Procedures

We strongly encourage you to download the latest copy of *MCAT® Essentials*, available on the AAMC's website, to ensure that you have the latest information about registration and Test Day policies and procedures; this document is updated annually. A brief summary of some of the most important rules is provided here.

MCAT Registration

The only way to register for the MCAT is online. You can access AAMC's registration system at **www.aamc.org/mcat**.

The AAMC posts the schedule of testing, registration, and score release dates in the fall before the MCAT testing year, which runs from January into September. Registration for January through June is available earlier than registration for later dates, but see the AAMC's website for the exact dates each year. There is one standard registration fee, but the fee for changing your test date or test center increases the closer you get to your MCAT.

Fees and the Fee Assistance Program (FAP)

Payment for test registration must be made by MasterCard or VISA. As described earlier, the fee for rescheduling your exam or changing your testing center increases as one approaches Test Day. In addition, it is not uncommon for test centers to fill up well in advance of the registration deadline. For these reasons, we recommend identifying your preferred Test Day as soon as possible and registering. There are ancillary benefits to having a set Test Day, as well: when you know the date you're working toward, you'll study harder and are less likely to keep pushing back the exam. The AAMC offers a Fee Assistance Program (FAP) for students with financial hardship to help reduce the cost of taking the MCAT, as well as for the American Medical College Application Service (AMCAS®) application. Further information on the FAP can be found at **www.aamc.org/students/applying/fap**.

Testing Security

On Test Day, you will be required to present a qualifying form of ID. Generally, a current driver's license or United States passport will be sufficient (consult the AAMC website for the full list of qualifying criteria). When registering, take care to spell your first and last names (middle names, suffixes, and prefixes are not required and will not be verified on Test Day) precisely the same as they appear on this ID; failure to provide this ID at the test center or differences in spelling between your registration and ID will be considered a "no-show," and you will not receive a refund for the exam.

During Test Day registration, other identity data collected may include: a digital palm vein scan, a Test Day photo, a digitization of your valid ID, and signatures. Some testing centers may use a metal detection wand to ensure that no prohibited items are brought into the testing room. Prohibited items include all electronic devices, including watches and timers, calculators, cell phones, and any and all forms of recording equipment; food, drinks (including water), and cigarettes or other smoking paraphernalia; hats and scarves (except for religious purposes); and books, notes, or other study materials. If you require a medical device, such as an insulin pump or pacemaker, you must apply for accommodated testing. During breaks, you are allowed access to food and drink, but not to electronic devices, including cell phones.

Testing centers are under video surveillance and the AAMC does not take potential violations of testing security lightly. The bottom line: *know the rules and don't break them.*

Accommodations

Students with disabilities or medical conditions can apply for accommodated testing. Documentation of the disability or condition is required, and requests may take two months—or more—to be approved. For this reason, it is recommended that you begin the process of applying for accommodated testing as early as possible. More information on applying for accommodated testing can be found at **www.aamc.org/students/applying/mcat/accommodations**.

After Your Test

When your MCAT is all over, no matter how you feel you did, be good to yourself when you leave the test center. Celebrate! Take a nap. Watch a movie. Get some exercise. Plan a trip or outing. Call up all of your neglected friends or message them on social media. Go out for snacks or drinks with people you like. Whatever you do, make sure that it has absolutely nothing to do with thinking too hard—you deserve some rest and relaxation.

Perhaps most importantly, do not discuss specific details about the test with anyone. For one, it is important to let go of the stress of Test Day, and reliving your exam only inhibits you from being able to do so. But more significantly, the Examinee Agreement you sign at the beginning of your exam specifically prohibits you from discussing or disclosing exam content. The AAMC is known to seek out individuals who violate this agreement and retains the right to prosecute these individuals at their discretion. This means that you should not, under any circumstances, discuss the exam in person or over the phone with other individuals—including us at Kaplan—or post information or questions about exam content to Facebook, Student Doctor Network, or other online social media. You are permitted to comment on your "general exam experience," including how you felt about the exam overall or an individual section, but this is a fine line. In summary: *if you're not certain whether you can discuss an aspect of the test or not, just don't do it!* Do not let a silly Facebook post stop you from becoming the doctor you deserve to be.

Scores are typically released approximately one month after Test Day. The release is staggered during the afternoon and evening, ending at 5 p.m. Eastern Standard Time. This means that not all examinees receive their scores at exactly the same time. Your score report will include a scaled score for each section between 118 and 132, as well as your total combined score between 472 and 528. These scores are given as confidence intervals. For each section, the confidence interval is approximately the given score ± 1; for the total score, it is approximately the given score ± 2. You will also be given the corresponding percentile rank for each of these section scores and the total score.

AAMC Contact Information

For further questions, contact the MCAT team at the Association of American Medical Colleges:

MCAT Resource Center
Association of American Medical Colleges
www.aamc.org/mcat
(202) 828-0600
www.aamc.org/contactmcat

HOW THIS BOOK WAS CREATED

The *Kaplan MCAT Review* project began shortly after the release of the *Preview Guide for the MCAT 2015 Exam*, 2nd edition. Through thorough analysis by our staff psychometricians, we were able to analyze the relative yield of the different topics on the MCAT, and we began constructing tables of contents for the books of the *Kaplan MCAT Review* series. A dedicated staff of 30 writers, 7 editors, and 32 proofreaders worked over 5,000 combined hours to produce these books. The format of the books was heavily influenced by weekly meetings with Kaplan's learning science team.

In the years since this book was created, a number of opportunities for expansion and improvement have occurred. The current edition represents the culmination of the wisdom accumulated during that time frame, and it also includes several new features designed to improve the reading and learning experience in these texts.

These books were submitted for publication in April 2024. For any updates after this date, please visit www.kaptest.com/retail-book-corrections-and-updates.

If you have any questions about the content presented here, email KaplanMCATfeedback@kaplan.com. For other questions not related to content, email booksupport@kaplan.com.

Each book has been vetted through at least ten rounds of review. To that end, the information presented in these books is true and accurate to the best of our knowledge. Still, your feedback helps us improve our prep materials. Please notify us of any inaccuracies or errors in the books by sending an email to KaplanMCATfeedback@kaplan.com.

USING THIS BOOK

Kaplan MCAT Behavioral Sciences Review, and the other six books in the *Kaplan MCAT Review* series, bring the Kaplan classroom experience to you—right in your home, at your convenience. This book offers the same Kaplan content review, strategies, and practice that make Kaplan the #1 choice for MCAT prep.

This book is designed to help you review the behavioral sciences topics covered on the MCAT. Please understand that content review—no matter how thorough—is not sufficient preparation for the MCAT! The MCAT tests not only your science knowledge but also your critical reading, reasoning, and problem-solving skills. Do not assume that simply memorizing the contents of this book will earn you high scores on Test Day; to maximize your scores, you must also improve your reading and test-taking skills through MCAT-style questions and practice tests.

Learning Objectives

At the beginning of each section, you'll find a short list of objectives describing the skills covered within that section. Learning objectives for these texts were developed in conjunction with Kaplan's learning science team, and have been designed specifically to focus your attention on tasks and concepts that are likely to show up on your MCAT. These learning objectives will function as a means to guide your study, and indicate what information and relationships you should be focused on within each section. Before starting each section, read these learning objectives carefully. They will not only allow you to assess your existing familiarity with the content, but also provide a goal-oriented focus for your studying experience of the section.

MCAT Concept Checks

At the end of each section, you'll find a few open-ended questions that you can use to assess your mastery of the material. These MCAT Concept Checks were introduced after numerous conversations with Kaplan's learning science team. Research has demonstrated repeatedly that introspection and self-analysis improve mastery, retention, and recall of material. Complete these MCAT Concept Checks to ensure that you've got the key points from each section before moving on!

Science Mastery Assessments

At the beginning of each chapter, you'll find 15 MCAT-style practice questions. These are designed to help you assess your understanding of the chapter before you begin reading the chapter. Using the guidance provided with the assessment, you can determine the best way to review each chapter based on your personal strengths and weaknesses. Most of the questions in the Science Mastery Assessments focus on the first of the *Scientific Inquiry and Reasoning Skills* (*Knowledge of Scientific Concepts and Principles*), although there are occasional questions that fall into the second or fourth SIRS (*Scientific Reasoning and Problem-Solving* and *Data-Based and Statistical Reasoning*, respectively). You can complete each chapter's assessment in a testing interface in your online resources, where you'll also find a test-like passage set covering the same content you just studied to ensure you can also apply your knowledge the way the MCAT will expect you to!

Guided Examples with Expert Thinking

Embedded in each chapter of this book is a Guided Example with Expert Thinking. Each of these guided examples will be located in the same section as the content used in that example. Each example will feature an MCAT-level scientific article, that simulates an MCAT experiment passage. Read through the passage as you would on the real MCAT, referring to the Expert Thinking material to the right of the passage to clarify the key information you should be gathering from each paragraph. Read and attempt to answer the associated question once you have worked through the passage. There is a full explanation, including the correct answer, following the given question. These passages and questions are designed to help build your critical thinking, experimental reasoning, and data interpretation skills as preparation for the challenges you will face on the MCAT.

Sidebars

The following is a guide to the five types of sidebars you'll find in *Kaplan MCAT Behavioral Sciences Review:*

- **Bridge:** These sidebars create connections between science topics that appear in multiple chapters throughout the *Kaplan MCAT Review* series.
- **Key Concept:** These sidebars draw attention to the most important takeaways in a given topic, and they sometimes offer synopses or overviews of complex information. If you understand nothing else, make sure you grasp the Key Concepts for any given subject.
- **MCAT Expertise:** These sidebars point out how information may be tested on the MCAT or offer key strategy points and test-taking tips that you should apply on Test Day.
- **Mnemonic:** These sidebars present memory devices to help recall certain facts.
- **Real World:** These sidebars illustrate how a concept in the text relates to the practice of medicine or the world at large. While this is not information you need to know for Test Day, many of the topics in Real World sidebars are excellent examples of how a concept may appear in a passage or discrete (stand-alone) question on the MCAT.

What This Book Covers

The information presented in the *Kaplan MCAT Review* series covers everything listed on the official MCAT content lists. Every topic in these lists is covered in the same level of detail as is common to the undergraduate and postbaccalaureate classes that are considered prerequisites for the MCAT. Note that your premedical classes may include topics not discussed in these books, or they may go into more depth than these books do. Additional exposure to science content is never a bad thing, but all of the content knowledge you are expected to have walking in on Test Day is covered in these books.

Chapter profiles, on the first page of each chapter, represent a holistic look at the content within the chapter, and will include a pie chart as well as text information. The pie chart analysis is based directly on data released by the AAMC, and will give a rough estimate of the importance of the chapter in relation to the book as a whole. Further, the text portion of the Chapter Profiles includes which AAMC content categories are covered within the chapter. These are referenced directly from the AAMC MCAT exam content listing, available on the testmaker's website.

You'll also see new High-Yield badges scattered throughout the sections of this book:

In This Chapter

1.1 Amino Acids Found in Proteins

High-Yield

LEARNING OBJECTIVES

After Chapter 1.1, you will be able to:

These badges represent the top 100 topics most tested by the AAMC. In other words, according to the testmaker and all our experience with their resources, a High-Yield badge means more questions on Test Day.

This book also contains a thorough glossary and index for easy navigation of the text.

In the end, this is your book, so write in the margins, draw diagrams, highlight the key points—do whatever is necessary to help you get that higher score. We look forward to working with you as you achieve your dreams and become the doctor you deserve to be!

Studying with This Book

In addition to providing you with the best practice questions and test strategies, Kaplan's team of learning scientists are dedicated to researching and testing the best methods for getting the most out of your study time. Here are their top four tips for improving retention:

Review multiple topics in one study session. This may seem counterintuitive—we're used to practicing one skill at a time in order to improve each skill. But research shows that weaving topics together leads to increased learning. Beyond that consideration, the MCAT often includes more than one topic in a single question. Studying in an integrated manner is the most effective way to prepare for this test.

Customize the content. Drawing attention to difficult or critical content can ensure you don't overlook it as you read and re-read sections. The best way to do this is to make it more visual—highlight, make tabs, use stickies, whatever works. We recommend highlighting only the most important or difficult sections of text. Selective highlighting of up to about 10 percent of text in a given chapter is great for emphasizing parts of the text, but over-highlighting can have the opposite effect.

Repeat topics over time. Many people try to memorize concepts by repeating them over and over again in succession. Our research shows that retention is improved by spacing out the repeats over time and mixing up the order in which you study content. For example, try reading chapters in a different order the second (or third!) time around. Revisit practice questions that you answered incorrectly in a new sequence. Perhaps information you reviewed more recently will help you better understand those questions and solutions you struggled with in the past.

Take a moment to reflect. When you finish reading a section for the first time, stop and think about what you just read. Jot down a few thoughts in the margins or in your notes about why the content is important or what topics came to mind when you read it. Associating learning with a memory is a fantastic way to retain information! This also works when answering questions. After answering a question, take a moment to think through each step you took to arrive at a solution. What led you to the answer you chose? Understanding the steps you took will help you make good decisions when answering future questions.

Online Resources

In addition to the resources located within this text, you also have additional online resources awaiting you at **www.kaptest.com/booksonline**. Make sure to log on and take advantage of free practice and other resources!

Please note that access to the online resources is limited to the original owner of this book.

STUDYING FOR THE MCAT

The first year of medical school is a frenzied experience for most students. To meet the requirements of a rigorous work schedule, students either learn to prioritize their time or else fall hopelessly behind. It's no surprise, then, that the MCAT, the test specifically designed to predict success in medical school, is a high-speed, time-intensive test. The MCAT demands excellent time-management skills, endurance, and grace under pressure both during the test as well as while preparing for it. Having a solid plan of attack and sticking with it are key to giving you the confidence and structure you need to succeed.

Creating a Study Plan

The best time to create a study plan is at the beginning of your MCAT preparation. If you don't already use a calendar, you will want to start. You can purchase a planner, print out a free calendar from the Internet, use a built-in calendar or app on one of your smart devices, or keep track using an interactive online calendar. Pick the option that is most practical for you and that you are most likely to use consistently.

Once you have a calendar, you'll be able to start planning your study schedule with the following steps:

1. **Fill in your obligations and choose a day off.**

   Write in all your school, extracurricular, and work obligations first: class sessions, work shifts, and meetings that you must attend. Then add in your personal obligations: appointments, lunch dates, family and social time, etc. Making an appointment in your calendar for hanging out with friends or going to the movies may seem strange at first, but planning social activities in advance will help you achieve a balance between personal and professional obligations even as life gets busy. Having a happy balance allows you to be more focused and productive when it comes time to study, so stay well-rounded and don't neglect anything that is important to you.

   In addition to scheduling your personal and professional obligations, you should also plan your time off. Taking some time off is just as important as studying. Kaplan recommends taking at least one full day off per week, ideally from all your study obligations but at minimum from studying for the MCAT.

2. **Add in study blocks around your obligations.**

   Once you have established your calendar's framework, add in study blocks around your obligations, keeping your study schedule as consistent as possible across days and across weeks. Studying at the same time of day as your official test is ideal for promoting recall, but if that's not possible, then fit in study blocks wherever you can.

   To make your studying as efficient as possible, block out short, frequent periods of study time throughout the week. From a learning perspective, studying one hour per day for six days per week is much more valuable than studying for six hours all at once one day per week. Specifically, Kaplan recommends studying for no longer than three hours in one sitting. Within those three-hour blocks, also plan to take ten-minute breaks every hour. Use these breaks to get up from your seat, do some quick stretches, get a snack and drink, and clear your mind. Although ten minutes of break for every 50 minutes of studying may sound like a lot, these breaks will allow you to deal with distractions and rest your brain so that, during the 50-minute study blocks, you can remain fully engaged and completely focused.

3. **Add in your full-length practice tests.**

   Next, you'll want to add in full-length practice tests. You'll want to take one test very early in your prep and then spread your remaining full-length practice tests evenly between now and your test date. Staggering tests in this way allows you to form a baseline for comparison and to determine which areas to focus on right away, while also providing realistic feedback throughout your prep as to how you will perform on Test Day.

When planning your calendar, aim to finish your full-length practice tests and the majority of your studying by one week before Test Day, which will allow you to spend that final week completing a final review of what you already know. In your online resources, you'll find sample study calendars for several different Test Day timelines to use as a starting point. The sample calendars may include more focus than you need in some areas, and less in others, and it may not fit your timeline to Test Day. You will need to customize your study calendar to your needs using the steps above.

The total amount of time you spend studying each week will depend on your schedule, your personal prep needs, and your time to Test Day, but it is recommended that you spend somewhere in the range of 300–350 hours preparing before taking the official MCAT. One way you could break this down is to study for three hours per day, six days per week, for four months, but this is just one approach. You might study six days per week for more than three hours per day. You might study over a longer period of time if you don't have much time to study each week. No matter what your plan is, ensure you complete enough practice to feel completely comfortable with the MCAT and its content. A good sign you're ready for Test Day is when you begin to earn your goal score consistently in practice.

How to Study

The MCAT covers a large amount of material, so studying for Test Day can initially seem daunting. To combat this, we have some tips for how to take control of your studying and make the most of your time.

Goal Setting

To take control of the amount of content and practice required to do well on the MCAT, break the content down into specific goals for each week instead of attempting to approach the test as a whole. A goal of "I want to increase my overall score by 5 points" is too big, abstract, and difficult to measure on the small scale. More reasonable goals are "I will read two chapters each day this week." Goals like this are much less overwhelming and help break studying into manageable pieces.

Active Reading

As you go through this book, much of the information will be familiar to you. After all, you have probably seen most of the content before. However, be very careful: Familiarity with a subject does not necessarily translate to knowledge or mastery of that subject. Do not assume that if you recognize a concept you actually know it and can apply it quickly at an appropriate level. Don't just passively read this book. Instead, read actively: Use the free margin space to jot down important ideas, draw diagrams, and make charts as you read. Highlighting can be an excellent tool, but use it sparingly: highlighting every sentence isn't active reading, it's coloring. Frequently stop and ask yourself questions while you read (e.g., *What is the main point? How does this fit into the overall scheme of things? Could I thoroughly explain this to someone else?*). By making connections and focusing on the grander scheme, not only will you ensure you know the essential content, but you also prepare yourself for the level of critical thinking required by the MCAT.

Focus on Areas of Greatest Opportunity

If you are limited by only having a minimal amount of time to prepare before Test Day, focus on your biggest areas of opportunity first. Areas of opportunity are topic areas that are highly tested and that you have not yet mastered. You likely won't have time to take detailed notes for every page of these books; instead, use your results from practice materials to determine which areas are your biggest opportunities and seek those out. After you've taken a full-length test, make sure you are using your performance report to best identify areas of opportunity. Skim over content matter for which you are already demonstrating proficiency, pausing to read more thoroughly when something looks unfamiliar or particularly difficult. Begin with the Science Mastery Assessment at the beginning of each chapter. If you can get all of those questions correct within a reasonable amount of time, you may be able to quickly skim through that chapter, but if the questions prove to be more difficult, then you may need to spend time reading the chapter or certain subsections of the chapter more thoroughly.

Practice, Review, and Tracking

Leave time to review your practice questions and full-length tests. You may be tempted, after practicing, to push ahead and cover new material as quickly as possible, but failing to schedule ample time for review will actually throw away your greatest opportunity to improve your performance. The brain rarely remembers anything it sees or does only once. When you carefully review the questions you've solved (and the explanations for them), the process of retrieving that information reopens and reinforces the connections you've built in your brain. This builds long-term retention and repeatable skill sets—exactly what you need to beat the MCAT!

One useful tool for making the most of your review is the How I'll Fix It (HIFI) sheet. You can create a HIFI sheet, such as the sample below, to track questions throughout your prep that you miss or have to guess on. For each such question, figure out why you missed it and supply at least one action step for how you can avoid similar mistakes in the future. As you move through your MCAT prep, adjust your study plan based on your available study time and the results of your review. Your strengths and weaknesses are likely to change over the course of your prep. Keep addressing the areas that are most important to your score, shifting your focus as those areas change. For more help with making the most of your full-length tests, including a How I'll Fix It sheet template, make sure to check out the videos and resources in your online syllabus.

Section	Q #	Type/Topic	Why I missed it	How I'll fix it
Chem/Phys	*42*	*Nuclear chem.*	*Confused electron absorption and emission*	*Reread Physics Chapter 9.2*
Chem/Phys	*47*	K_{eq}	*Didn't know right equation*	*Memorize equation for* K_{eq}
CARS	*2*	*Detail*	*Didn't read "not" in answer choice*	*Slow down when finding match*
CARS	*4*	*Inference*	*Forgot to research answer*	*Reread passage and predict first*

Where to Study

One often-overlooked aspect of studying is the environment where the learning actually occurs. Although studying at home is many students' first choice, several problems can arise in this environment, chief of which are distractions. Studying can be a mentally draining process, so as time passes, these distractions become ever more tempting as escape routes. Although you may have considerable willpower, there's no reason to make staying focused harder than it needs to be. Instead of studying at home, head to a library, quiet coffee shop, or another new location whenever possible. This will eliminate many of the usual distractions and also promote efficient studying; instead of studying off and on at home over the course of an entire day, you can stay at the library for three hours of effective studying and enjoy the rest of the day off from the MCAT.

No matter where you study, make your practice as much like Test Day as possible. Just as is required during the official test, don't have snacks or chew gum during your study blocks. Turn off your music, television, and phone. Practice on the computer with your online resources to simulate the computer-based test environment. When completing practice questions, do your work on scratch paper or noteboard sheets rather than writing directly on any printed materials since you won't have that option on Test Day. Because memory is tied to all of your senses, the more test-like you can make your studying environment, the easier it will be on Test Day to recall the information you're putting in so much work to learn.

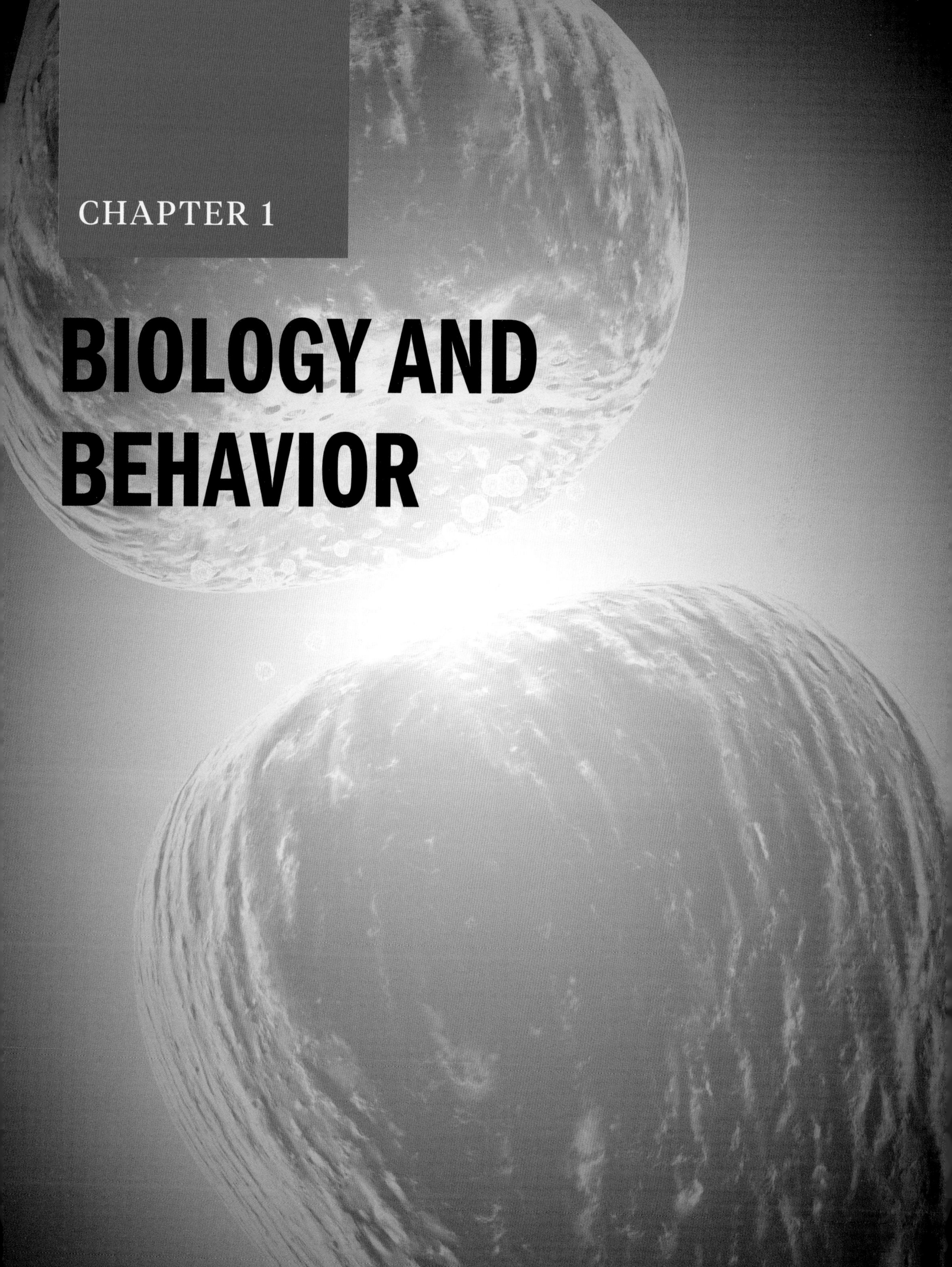

CHAPTER 1

BIOLOGY AND BEHAVIOR

SCIENCE MASTERY ASSESSMENT

Every pre-med knows this feeling: there is so much content I have to know for the MCAT! How do I know what to do first or what's important?

While the high-yield badges throughout this book will help you identify the most important topics, this Science Mastery Assessment is another tool in your MCAT prep arsenal. This quiz (which can also be taken in your online resources) and the guidance below will help ensure that you are spending the appropriate amount of time on this chapter based on your personal strengths and weaknesses. Don't worry though—skipping something now does not mean you'll never study it. Later on in your prep, as you complete full-length tests, you'll uncover specific pieces of content that you need to review and can come back to these chapters as appropriate.

How to Use This Assessment

If you answer 0–7 questions correctly:

Spend about 1 hour to read this chapter in full and take limited notes throughout. Follow up by reviewing **all** quiz questions to ensure that you now understand how to solve each one.

If you answer 8–11 questions correctly:

Spend 20–40 minutes reviewing the quiz questions. Beginning with the questions you missed, read and take notes on the corresponding subchapters. For questions you answered correctly, ensure your thinking matches that of the explanation and you understand why each choice was correct or incorrect.

If you answer 12–15 questions correctly:

Spend less than 20 minutes reviewing all questions from the quiz. If you missed any, then include a quick read-through of the corresponding subchapters, or even just the relevant content within a subchapter, as part of your question review. For questions you answered correctly, ensure your thinking matches that of the explanation and review the Concept Summary at the end of the chapter.

1. A researcher deletes a gene from an organism to determine the gene's function. This approach is most analogous to the work of which of the following scientists?
   - **A.** Paul Broca
   - **B.** Pierre Flourens
   - **C.** Franz Gall
   - **D.** Sir Charles Sherrington

2. Which component of the nervous system is NOT involved in the initial reflexive response to pain?
   - **A.** Spinal cord
   - **B.** Cerebral cortex
   - **C.** Interneuron
   - **D.** Motor neuron

3. A child has experienced nervous system damage and can no longer coordinate the movements to dribble a basketball, although the child can still walk in an uncoordinated fashion. Which region of the central nervous system was most likely affected?
   - **A.** Forebrain
   - **B.** Midbrain
   - **C.** Hindbrain
   - **D.** Spinal cord

4. The temporal lobe deals with all of the following EXCEPT:
   - **A.** language comprehension.
   - **B.** memory.
   - **C.** emotion.
   - **D.** motor skills.

5. Which part of the brain deals with both homeostasis and emotions?
   - **A.** Cerebellum
   - **B.** Pons
   - **C.** Hypothalamus
   - **D.** Thalamus

6. Which of the following activities would most likely be completed by the right hemisphere of a left-handed person?
   - **A.** Finding a car in a parking lot
   - **B.** Learning a new language
   - **C.** Reading a book for pleasure
   - **D.** Jumping rope with friends

7. Which of the following is/are true with regard to neurulation?
   - **I.** The neural tube differentiates from endoderm.
   - **II.** The neural tube becomes the peripheral nervous system.
   - **III.** Neural crest cells migrate from their original site.

   - **A.** I only
   - **B.** III only
   - **C.** II and III only
   - **D.** I, II, and III

8. Which of the following neurotransmitters is NOT classified as a catecholamine?
   - **A.** Epinephrine
   - **B.** Norepinephrine
   - **C.** Dopamine
   - **D.** Acetylcholine

9. If the amount of acetylcholinesterase, an enzyme that breaks down acetylcholine, is increased, which of the following would likely be the result?

- **A.** Weakness of muscle movements
- **B.** Excessive pain or discomfort
- **C.** Mood swings and mood instability
- **D.** Auditory and visual hallucinations

10. The adrenal glands do all of the following EXCEPT:

- **A.** promote the fight-or-flight response via estrogen.
- **B.** produce stress responses via cortisol.
- **C.** produce both hormones and neurotransmitters.
- **D.** release estrogen in males and testosterone in females.

11. A disorder of the pineal gland would most likely result in which of the following disorders?

- **A.** High blood pressure
- **B.** Diabetes
- **C.** Insomnia
- **D.** Hyperthyroidism

12. Which of the following conclusions would William James most likely support?

- **A.** Mental processes help individuals adapt to their environments.
- **B.** Psychological attributes could be measured by feeling the skull.
- **C.** Specific functional impairments can be linked to specific lesions in the brain.
- **D.** Synaptic transmission is an electrical process.

13. A scientist designs a study to determine if different regions of the brain are activated when a person speaks their native language vs. a second language. Which of the following methods would the scientist most likely choose?

- **A.** MRI
- **B.** CT scan
- **C.** fMRI
- **D.** EEG

14. During a physical examination, a physician brushes the bottom of the foot of a patient who is fifty years old with multiple sclerosis. The patient's toes are observed to curl toward the bottom of the foot, with no fanning of the toes. This response is:

- **A.** abnormal, and evidence that the patient is exhibiting a primitive reflex.
- **B.** normal, and evidence that the patient is exhibiting a primitive reflex.
- **C.** abnormal, and evidence that the patient is not exhibiting a primitive reflex.
- **D.** normal, and evidence that the patient is not exhibiting a primitive reflex.

15. Which of the following fine motor tasks would one expect to see first in an infant?

- **A.** Grasping for objects with two fingers
- **B.** Following objects with the eyes
- **C.** Scribbling with a crayon
- **D.** Moving a toy from one hand to the other

Answer Key

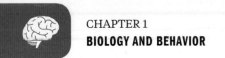

1. **B**
2. **B**
3. **C**
4. **D**
5. **C**
6. **A**
7. **B**
8. **D**
9. **A**
10. **A**
11. **C**
12. **A**
13. **C**
14. **D**
15. **B**

Detailed explanations can be found at the end of the chapter.

CHAPTER 1

BIOLOGY AND BEHAVIOR

In This Chapter

CHAPTER PROFILE

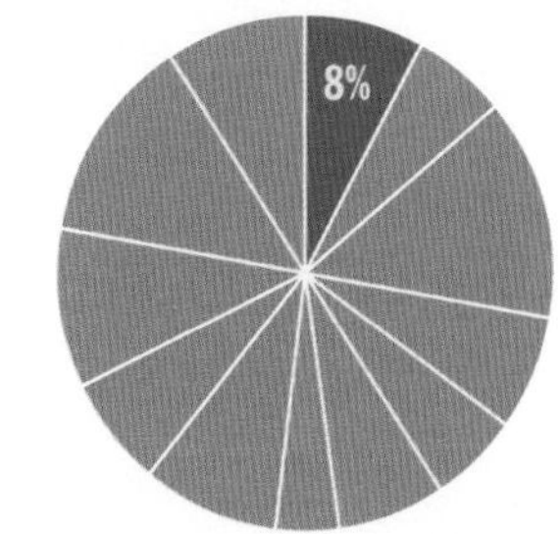

The content in this chapter should be relevant to about 8% of all questions about the behavioral sciences on the MCAT.

This chapter covers material from the following AAMC content categories:

3A: Structure and functions of the nervous and endocrine systems and ways in which these systems coordinate the organ systems

7A: Individual influences on behavior

Introduction

When you woke up this morning and got ready to start reading *MCAT Behavioral Sciences Review*, you almost certainly had specific feelings about it—perhaps you were excited to crack open the book and start learning some of the material that will get you that top score on the MCAT; perhaps you dreaded the size and rich detail of the information in the book. Either way, your body began to respond to these impulses from your mind: increasing heart rate, increasing breathing rate, dilating the eyes, and slowing down digestion. This link between the mind and the body is still a hot topic in medicine, although we've been exploring the effects of psychology on well-being for almost two centuries now.

In this chapter, we'll begin our exploration of psychology and sociology by looking at the biological side of psychology. After a quick survey of the history of neuropsychology, we'll look at the structure and organization of the human nervous system, communication between the nervous and endocrine systems, the effects of genes and environment on behavior, and some aspects of psychological development.

1.1 A Brief History of Neuropsychology

LEARNING OBJECTIVES

After Chapter 1.1, you will be able to:

- Recall the major contributions to early neuropsychology
- Connect the major contributors of early neuropsychology to their contributions

Researchers in the 19th century began to think about behavior from a physiological perspective. Many of these early thinkers formed the foundation of current knowledge about neuroanatomy, linking the functions of specific areas of the brain with thought and behavior.

Franz Gall (1758–1828) had one of the earliest theories that behavior, intellect, and even personality might be linked to brain anatomy. He developed the doctrine of phrenology. The basic idea was that if a particular trait was well-developed, then the part of the brain responsible for that trait would expand. This expansion, according to Gall, would push the area of the skull that covered that part of the brain outward and therefore cause a bulge on the head. Gall believed that one could thus measure psychological attributes by feeling or measuring the skull. Although phrenology was shown to be false, it did generate serious research on brain functions and was the impetus for the work of other psychologists through the remainder of the 19th century.

Pierre Flourens (1794–1867) was the first person to study the functions of the major sections of the brain. He did this by **extirpation**, also known as **ablation**, on rabbits and pigeons. In extirpation, various parts of the brain are surgically removed and the behavioral consequences are observed. Flourens's work led to his assertion that specific parts of the brain had specific functions, and that the removal of one part weakens the whole brain.

William James (1842–1910), known as the founder of American psychology, studied how the mind adapts to the environment. His views formed the foundation for the system of thought in psychology known as **functionalism**, which studies how mental processes help individuals adapt to their environments.

John Dewey (1859–1952) is another important name in functionalism because his 1896 article is seen as its inception. This article criticized the concept of the reflex arc, which breaks the process of reacting to a stimulus into discrete parts. Dewey believed that psychology should focus on the study of the organism as a whole as it functioned to adapt to the environment.

Around 1860, **Paul Broca** (1824–1880) added to the knowledge of physiology by examining the behavioral deficits of people with brain damage. He was the first person to demonstrate that specific functional impairments could be linked with specific brain lesions. Broca studied a person who was unable to speak and discovered that the person's disability was due to a lesion in a specific area on the left side of the person's brain. This area of the brain is now referred to as Broca's area.

Hermann von Helmholtz (1821–1894) was the first to measure the speed of a nerve impulse. He also related the measured speed of such impulses to reaction time, providing an important early link between behavior and underlying nervous system activity. Because Helmholtz provided one of the earliest measurable links between psychology and physiology, he is often credited with the transition of psychology out of the realm of philosophy and into the realm of quantifiable natural science.

Around the turn of the century, **Sir Charles Sherrington** (1857–1952) first inferred the existence of synapses. Many of his conclusions have held over time—except for one. He thought that synaptic transmission was an electrical process, but we now know that it is primarily a chemical process.

MCAT CONCEPT CHECK 1.1

Before you move on, assess your understanding of the material with this question.

1. Briefly list the main contributions of each of the following scientists to neuropsychology.
    - Franz Gall:
    - Pierre Flourens:
    - William James:
    - John Dewey:
    - Paul Broca:
    - Hermann von Helmholtz:
    - Sir Charles Sherrington:

BRIDGE

Solutions to concept checks for a given chapter in *MCAT Behavioral Sciences Review* can be found near the end of the chapter in which the concept check is located, following the Concept Summary for that chapter.

1.2 Organization of the Human Nervous System

High-Yield

MCAT EXPERTISE

The "High-Yield" badge on this section indicates that the content is frequently tested on the MCAT.

LEARNING OBJECTIVES

After Chapter 1.2, you will be able to:

- Correctly associate regions of the nervous system with the CNS or PNS
- Distinguish between afferent and efferent neurons
- Describe the functions of the somatic and autonomic nervous systems, as well as the sympathetic and parasympathetic nervous systems:

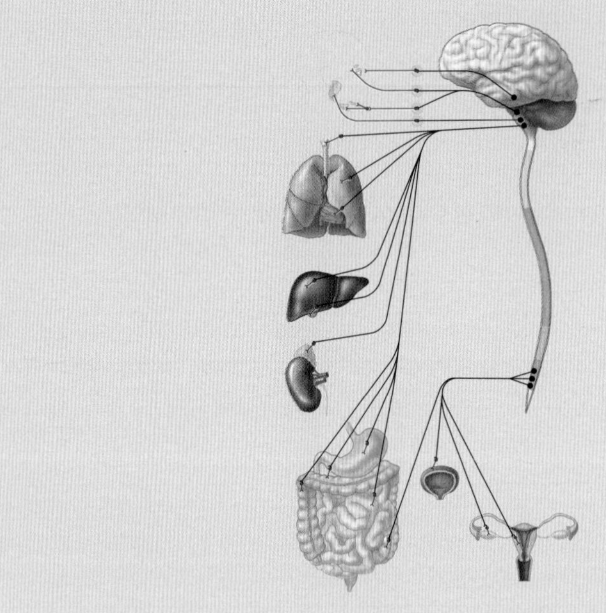

The human nervous system is a complex web of over 100 billion cells that communicate, coordinate, and regulate signals for the rest of the body. Mental and physical action occurs when the body can react to external stimuli using the nervous system. In this section, we will look at the nervous system and its basic organization.

Note: Much of the information contained in this section is also discussed in Chapter 4 of MCAT Biology Review.

Central and Peripheral Nervous Systems

MNEMONIC

Afferent neurons **a**scend in the cord toward the brain; **e**fferent neurons **e**xit the cord on their way to the rest of the body.

There are three kinds of nerve cells in the nervous system: sensory neurons, motor neurons, and interneurons. **Sensory neurons** (also known as **afferent neurons**) transmit sensory information from receptors to the spinal cord and brain. **Motor neurons** (also known as **efferent neurons**) transmit motor information from the brain and spinal cord to muscles and glands. **Interneurons** are found between other neurons and are the most numerous of the three types of neurons. Interneurons are located predominantly in the brain and spinal cord and are often linked to reflexive behavior. Neural circuits called **reflex arcs** control this type of reflexive behavior. For example, consider what occurs when someone steps on a nail. Receptors in the foot detect pain and the pain signal is transmitted by sensory neurons up to the spinal cord. At that point, the sensory neurons connect with interneurons, which then relay pain impulses up to the brain. However, rather than waiting for the brain to send out a signal, interneurons in the spinal cord send signals to the muscles of both legs directly, causing the individual to reflexively withdraw the foot in pain while simultaneously reflexively transferring weight to the other foot. The original sensory information still makes its way up to the brain; however, by that time, the muscles have already responded to the pain, thanks to the cooperation of these several reflex arcs.

Let's turn to the overall structure of the human nervous system, which is diagrammed in Figure 1.1. The nervous system can be broadly divided into two primary components: the central and peripheral nervous systems. The **central nervous system (CNS)** is composed of the brain and spinal cord. The **peripheral nervous system (PNS)**, in contrast, is made up of nerve tissue and fibers outside the brain and spinal cord. Note that the peripheral nervous system includes all 31 pairs of nerves emanating from the spinal cord, which are called **spinal nerves**, and 12 pairs of nerves emanating directly from the brain, called **cranial nerves**. The olfactory and optic nerves (cranial nerves I and II) are structurally outgrowths of the central nervous system, but are still considered components of the peripheral nervous system. The PNS thus connects the CNS to the rest of the body.

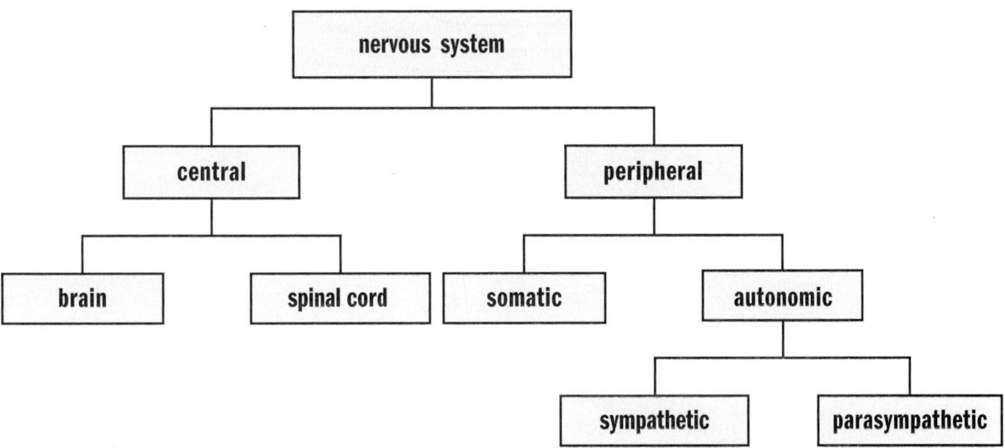

Figure 1.1 Major Divisions of the Nervous System

The peripheral nervous system is further subdivided into the somatic and autonomic nervous systems. The **somatic nervous system** consists of sensory and motor neurons distributed throughout the skin, joints, and muscles. Sensory neurons transmit information toward the CNS through afferent fibers. Motor impulses, in contrast, travel from the CNS back to the body along efferent fibers.

The **autonomic nervous system (ANS)** generally regulates heartbeat, respiration, digestion, and glandular secretions. In other words, the ANS manages the involuntary muscles associated with many internal organs and glands. The ANS also helps regulate body temperature by activating sweating or piloerection, depending on whether the body is too hot or too cold. The main thing to understand about all of these functions is that they are automatic, or independent of conscious control. Note the similarity between the words autonomic and automatic. This association makes it easy to remember that the autonomic nervous system manages automatic functions such as heartbeat, respiration, digestion, and temperature control.

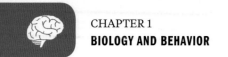

The Autonomic Nervous System

The ANS has two subdivisions: the sympathetic nervous system and the parasympathetic nervous system. These two branches often act in opposition to one another, meaning they are antagonistic. For example, the sympathetic nervous system acts to accelerate heart rate and inhibit digestion, while the parasympathetic nervous system decelerates heart rate and increases digestion.

The main role of the **parasympathetic nervous system** is to conserve energy. It is associated with resting and sleeping states, and acts to reduce heart rate and constrict the bronchi. The parasympathetic nervous system is also responsible for managing digestion by increasing peristalsis and exocrine secretions. Acetylcholine is the neurotransmitter responsible for parasympathetic responses in the body. The functions of the parasympathetic nervous system are summarized in Figure 1.2.

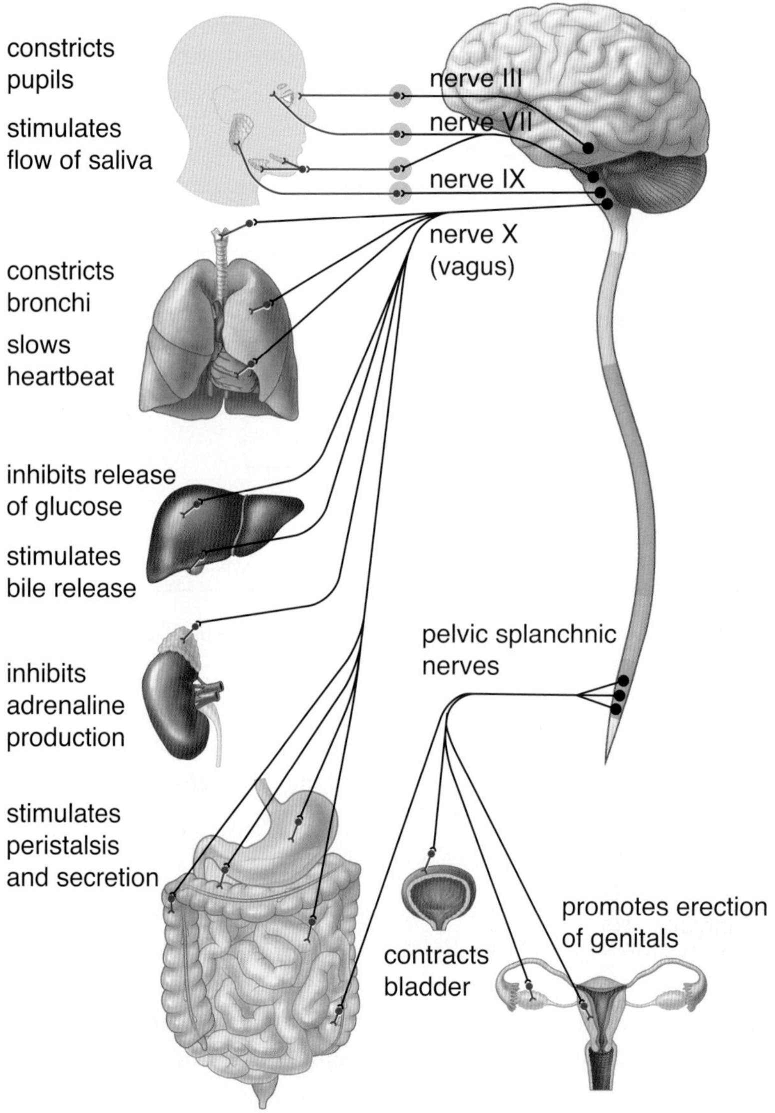

Figure 1.2 Functions of the Parasympathetic Nervous System

In contrast, the **sympathetic nervous system** is activated by stress. This can include everything from a mild stressor, such as keeping up with schoolwork, to emergencies that mean the difference between life and death. The sympathetic nervous system is closely associated with rage and fear reactions, also known as "fight-or-flight" reactions.

When activated, the sympathetic nervous system:

- Increases heart rate
- Redistributes blood to muscles of locomotion
- Increases blood glucose concentration
- Relaxes the bronchi
- Decreases digestion and peristalsis
- Dilates the eyes to maximize light intake
- Releases epinephrine into the bloodstream

The functions of the sympathetic nervous system are summarized in Figure 1.3.

MNEMONIC

Sympathetic and parasympathetic nervous systems:

- Sympathetic: "**fight**-or-**flight**"
- Parasympathetic: "**rest**-and-**digest**"

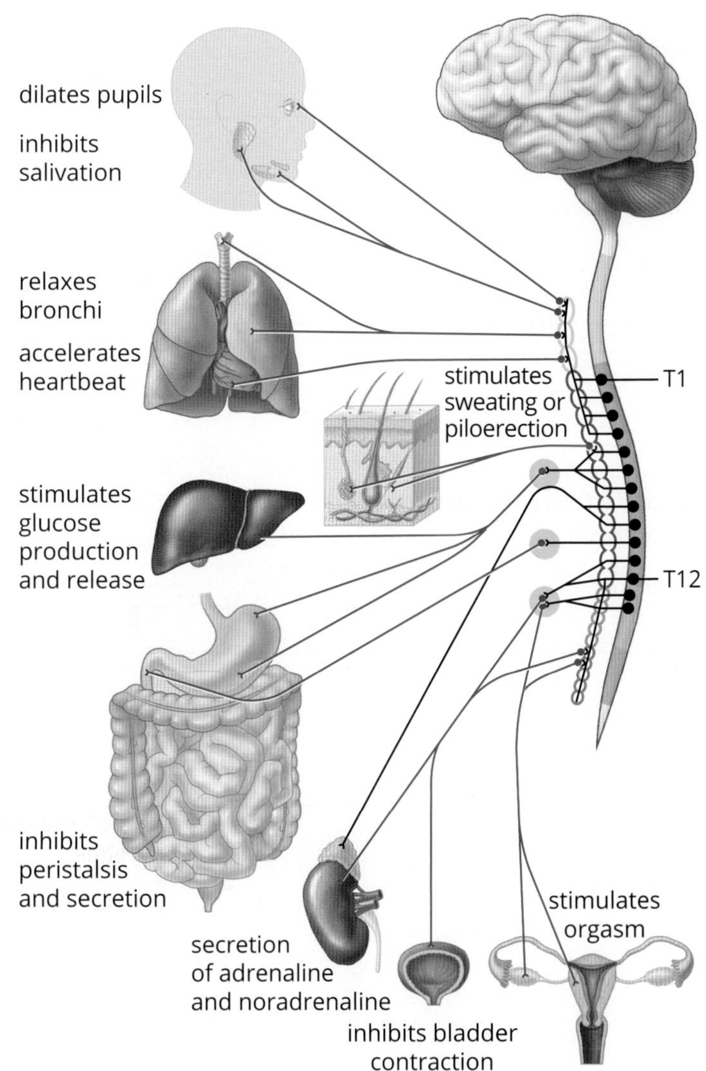

Figure 1.3 Functions of the Sympathetic Nervous System

MCAT CONCEPT CHECK 1.2

Before you move on, assess your understanding of the material with these questions.

1. What parts of the nervous system are in the central nervous system (CNS)? Peripheral nervous system (PNS)?
   - CNS:

   - PNS:

   ---

2. What do afferent neurons do? Efferent neurons?
   - Afferent:

   - Efferent:

   ---

3. What functions are accomplished by the somatic nervous system? The autonomic nervous system?
   - Somatic:

   - Autonomic:

   ---

4. What are the effects of the sympathetic nervous system? The parasympathetic nervous system?
   - Sympathetic:

   - Parasympathetic:

1.3 Organization of the Brain

High-Yield

LEARNING OBJECTIVES

After Chapter 1.3, you will be able to:

- Describe the major functions of the hindbrain, midbrain, and forebrain
- Recognize the most commonly used methods for mapping the brain
- Identify the structures protecting and surrounding the brain

Throughout this section, refer to Figure 1.4, which identifies various anatomical structures inside the human brain. As we discuss different parts of the brain, it's important to remember the functions of these brain structures. Different parts of the brain perform remarkably different functions. For instance, one part of the brain processes sensory information while an entirely different part of the brain maintains activities of the internal organs. For complex functions such as playing a musical instrument, several brain regions work together. For the MCAT, you will need to know some of the basics about how the brain integrates input from different regions.

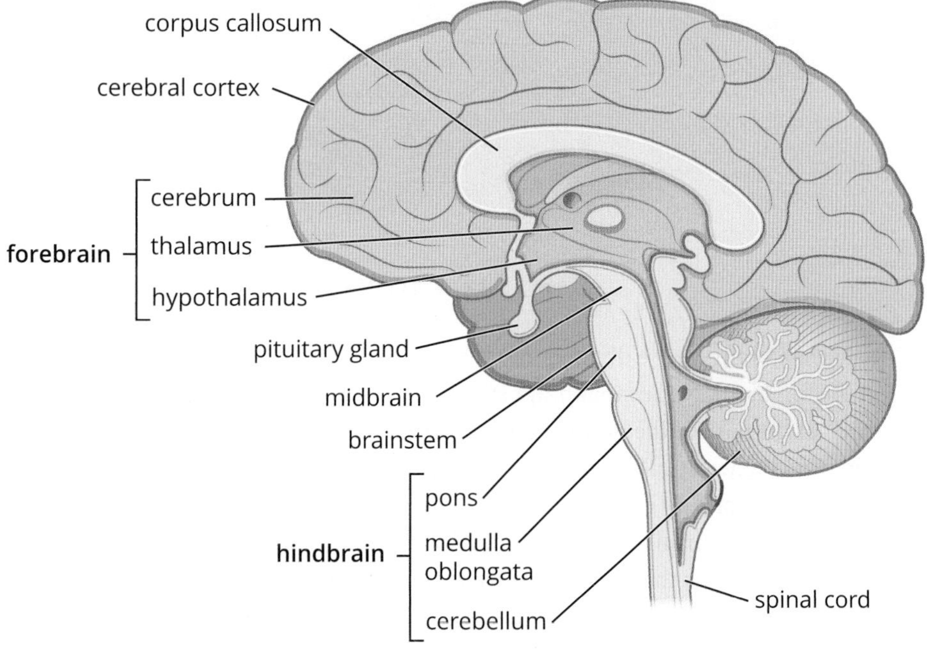

Figure 1.4 Anatomical Structures Inside the Human Brain

The brain is covered with a thick, three-layered sheath of connective tissue collectively called the **meninges**. The outer layer of connective tissue is the **dura mater**, and is connected directly to the skull. The middle layer is a fibrous, weblike structure called **arachnoid mater**. And the inner layer, connected directly to the brain, is known as the **pia mater**. These three layers of connective tissue are shown in Figure 1.5. The meninges help protect the brain by keeping it anchored within the skull, and the meninges also resorb **cerebrospinal fluid**, which is the aqueous solution that nourishes the brain and spinal cord and provides a protective cushion. Cerebrospinal fluid is produced by specialized cells that line the **ventricles** (internal cavities) of the brain.

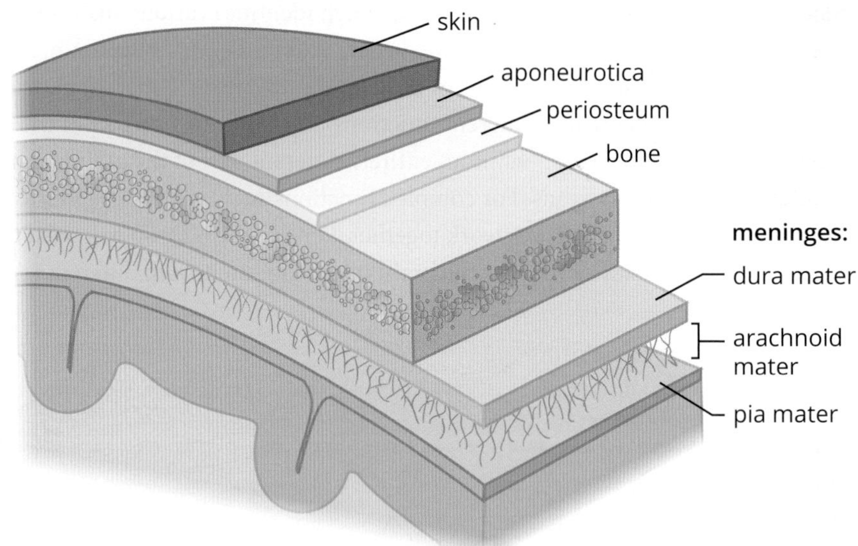

Figure 1.5 Layers of the Meninges

The human brain can be divided into three basic parts: the hindbrain, the midbrain, and the forebrain. Notice that brain structures associated with basic survival are located at the base of the brain and brain structures with more complex functions are located higher up. The meaningful connection between brain location and functional complexity is no accident. In evolutionary terms, the hindbrain and midbrain were brain structures that developed earlier. Together they form the **brainstem**, which is the most primitive region of the brain. The forebrain developed later, including the **limbic system**, a group of neural structures primarily associated with emotion and memory. Aggression, fear, pleasure, and pain are all related to the limbic system. The most recent evolutionary development of the human brain is the **cerebral cortex**, which is the outer covering of the cerebral hemispheres. In humans, the cerebral cortex is associated with everything from language processing to problem solving, and from impulse control to long-term planning. Most of the key brain regions described in the following sections are summarized in Table 1.1.

MAJOR DIVISIONS AND PRINCIPAL STRUCTURES	FUNCTIONS
Forebrain	
Cerebral cortex	Complex perceptual, cognitive, and behavioral processes
Basal ganglia	Movement
Limbic system	Emotion and memory
Thalamus	Sensory relay station
Hypothalamus	Hunger and thirst; emotion
Midbrain	Sensorimotor reflexes
Inferior and superior colliculi	
Hindbrain	
Cerebellum	Refined motor movements
Medulla oblongata	Heart, vital reflexes (vomiting, coughing)
Reticular formation	Arousal and alertness
Pons	Communication within the brain, breathing

Table 1.1 Anatomical Subdivisions of the Brain

In prenatal life, the brain develops from the neural tube. At first, the tube is composed of three swellings, which correspond to the hindbrain, midbrain, and forebrain. Both the hindbrain and forebrain later divide into two swellings, creating five total swellings in the mature neural tube. The embryonic brain is diagrammed in Figure 1.6, and its subdivisions are described further in the following sections. Understanding the relationship between the structures of the developing brain and the fully developed brain is important. So the following sections describe both the structures of the developing brain and what those structures develop into.

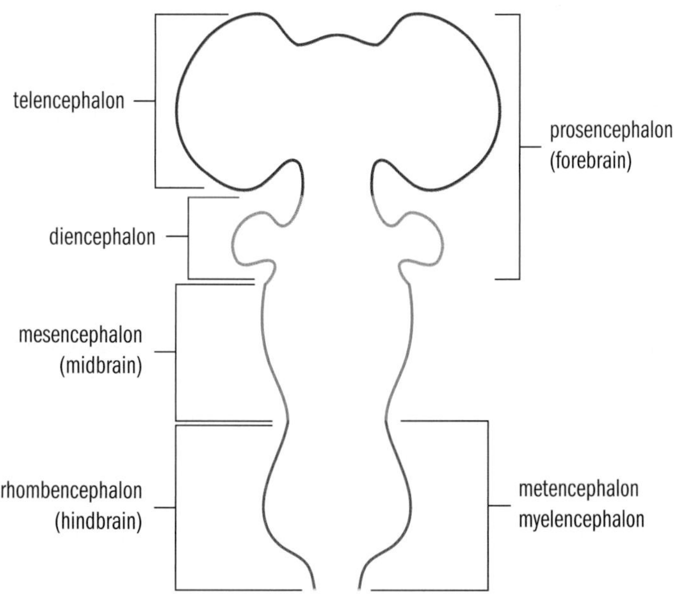

Figure 1.6 Subdivisions of the Embryonic Brain

Hindbrain

Located where the brain meets the spinal cord, the **hindbrain** (**rhombencephalon**) controls balance, motor coordination, breathing, digestion, and general arousal processes such as sleeping and waking. In short, the hindbrain manages vital functioning necessary for survival. During embryonic development, the rhombencephalon divides to form the **myelencephalon** (which becomes the medulla oblongata) and the **metencephalon** (which becomes the pons and cerebellum). In the developed brain, the **medulla oblongata** is a lower brain structure that is responsible for regulating vital functions such as breathing, heart rate, and digestion. The **pons** lies above the medulla and contains sensory and motor pathways between the cortex and the medulla. At the top of the hindbrain, mushrooming out of the back of the pons, is the **cerebellum**, a structure that helps maintain posture and balance and coordinates body movements. Damage to the cerebellum causes clumsiness, slurred speech, and loss of balance. Notably, alcohol impairs the functioning of the cerebellum, and consequently affects speech and balance.

Midbrain

Just above the hindbrain is the **midbrain** (**mesencephalon**), which receives sensory and motor information from the rest of the body. The midbrain is associated with involuntary reflex responses triggered by visual or auditory stimuli. There are several prominent nuclei in the midbrain, two of which are collectively called **colliculi**. The **superior colliculus** receives visual sensory input, and the **inferior colliculus** receives sensory information from the auditory system. The inferior colliculus has a role in reflexive reactions to sudden loud noises.

Forebrain

Above the midbrain is the **forebrain** (**prosencephalon**), which is associated with complex perceptual, cognitive, and behavioral processes. Among its other functions, the forebrain is associated with emotion and memory; it is the forebrain that has the greatest influence on human behavior. Its functions are not absolutely necessary for survival, but are associated instead with the intellectual and emotional capacities most characteristic of humans. During prenatal development, the prosencephalon divides to form the **telencephalon** (which forms the cerebral cortex, basal ganglia, and limbic system) and the **diencephalon** (which forms the thalamus, hypothalamus, posterior pituitary gland, and pineal gland).

Methods of Mapping the Brain

Neuropsychology refers to the study of functions and behaviors associated with specific regions of the brain. It is most often applied in research settings, where researchers attempt to associate very specific areas in the brain to behavior. Neuropsychology is also applied in clinical settings with evaluations of patient cognitive and behavioral functioning, as well as the diagnosis and treatment of brain disorders. Neuropsychology has its own experimental methodology and technology.

Studying human patients with brain lesions is one way that researchers have determined the functions of the brain. In order to conclude that a specific structure of the brain is responsible for a specific function, researchers look for patients that exhibit damage to that structure coupled with a loss of the function. One problem in studying human brain lesions, however, is that such lesions are rarely isolated to specific brain structures. When several brain structures are damaged, the impairment could be attributed to any of the damaged structures, and pinpointing a specific link between brain structure and function becomes difficult.

One method for studying the relationship of brain regions and behaviors is to study brain lesions in lab animals. The advantage of this approach is that precisely defined brain lesions can be created in animals by extirpation. Researchers can also produce lesions by inserting tiny electrodes inside the brain and then selectively applying intense heat, cold, or electricity to specific brain regions. Such electrodes can be placed with great precision by using stereotactic instruments, which provide high-resolution, three-coordinate images of the brain. Ethical or cruelty concerns notwithstanding, such studies have greatly increased our understanding of comparable neural structures in humans.

Another neuropsychology method involves electrically stimulating the brain and recording consequent brain activity. While operating on the brain, a surgeon can stimulate a patient's cortex with a small electrode. This stimulation causes groups of neurons to fire, thereby activating the behavioral or perceptual processes associated with those neurons. For instance, if the electrode stimulates neurons in the motor cortex, the stimulation can lead to specific muscle movements. If the electrode stimulates the visual cortex, the patient may "see" flashes of light that are not really there. By using electrical stimulation, neurosurgeons can thus create **cortical maps**. This method relies on the assistance of the patient, who is awake and alert. Because there are no pain receptors in the brain, only local anesthesia is required. Electrodes have also been used in lab animals to study deeper regions of the brain. Depending on where these electrodes are implanted, they can elicit sleep, sexual arousal, rage, or terror. Once the electrode is turned off, these behaviors cease.

Electrodes can also be used to record electrical activity produced by the brain itself. In some studies, individual neurons are recorded by inserting ultrasensitive microelectrodes into individual brain cells and recording their electrical activity. Electrical activity generated by larger groups of neurons can be studied using an **electroencephalogram (EEG)**, which involves placing several electrodes on the scalp. Broad patterns of electrical activity can thus be detected and recorded. Because this procedure is noninvasive (it does not cause any damage), electroencephalograms are commonly used with human subjects. In fact, research on sleep, seizures, and brain lesions relies heavily on EEGs, like the one shown in Figure 1.7.

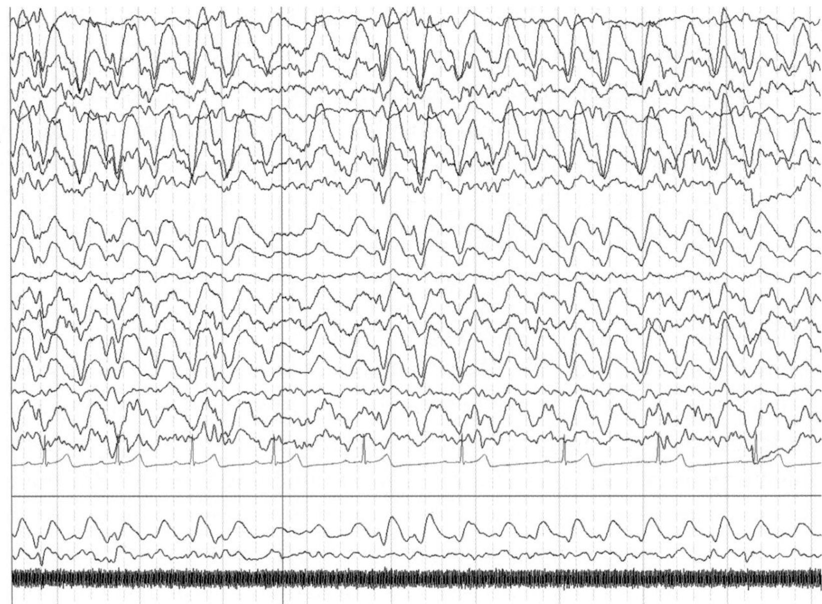

Figure 1.7 Electroencephalogram (EEG) during REM Sleep

Another noninvasive mapping procedure is **regional cerebral blood flow** (**rCBF**), which detects broad patterns of neural activity based on increased blood flow to different parts of the brain. rCBF relies on the assumption that blood flow increases to regions of the brain that are engaged in cognitive function. For example, listening to music may increase blood flow to the right auditory cortex because music is processed in that region in most individuals' brains. To measure blood flow, the patient inhales a harmless radioactive gas; a special device that can detect radioactivity in the bloodstream can then correlate radioactivity levels with regional cerebral blood flow. This research method uses noninvasive computerized scanning devices.

Some of the other common scanning devices and methods of visualization used for brain imaging include:

- **CT (computed tomography)**, also known as **CAT (computed axial tomography) scan**, in which multiple X-rays are taken at different angles and processed by a computer to produce cross-sectional images of the tissue.
- **PET (positron emission tomography) scan**, in which a radioactive sugar is injected and absorbed into the body, and its dispersion and uptake throughout the target tissue is imaged.
- **MRI (magnetic resonance imaging)**, in which a magnetic field that interacts with hydrogen atoms is used to map out hydrogen dense regions of the body.
- **fMRI (functional magnetic resonance imaging)**, which uses the same base technique as MRI, but specifically measures changes associated with blood flow. fMRI is especially useful for monitoring neural activity, since increased blood flow to a region of the brain is typically coupled with its neuronal activation.

BRIDGE

MRI techniques are dependent on the reaction of hydrogen to a magnetic field, and the scientific principles behind MRI scans are also applied in NMR techniques, which can be found in Chapter 11 of *MCAT Organic Chemistry Review.*

MCAT CONCEPT CHECK 1.3

Before you move on, assess your understanding of the material with these questions.

1. What are the main functions of the hindbrain? Midbrain? Forebrain?

Subdivision	Functions
Hindbrain	
Midbrain	
Forebrain	

2. What are some of the methods used for mapping the brain?

____

3. What structures surround and protect the brain?

____

BEHAVIORAL SCIENCES GUIDED EXAMPLE WITH EXPERT THINKING

Multiple sclerosis is a demyelinating disease that results in a host of neurological and physiological symptoms including muscle weakness, numbness, spasms, visual problems, pain, unstable mood, and fatigue. This last symptom is interesting because it is effort-independent; patients express a subjective feeling of fatigue as a result of performing physical tasks that are not typically physically or mentally taxing. To investigate a mechanism for this phenomenon, researchers used the following fMRI mask to highlight regions of interest in measuring neural resource use with respect to subjective fatigue in patients who have MS:

MS symptoms. The author finds fatigue interesting because it is subjective. Used fMRI to investigate.

Independent Variable (IV): MS (probably versus control)

Dependent Variable (DV): fMRI activity detection differences

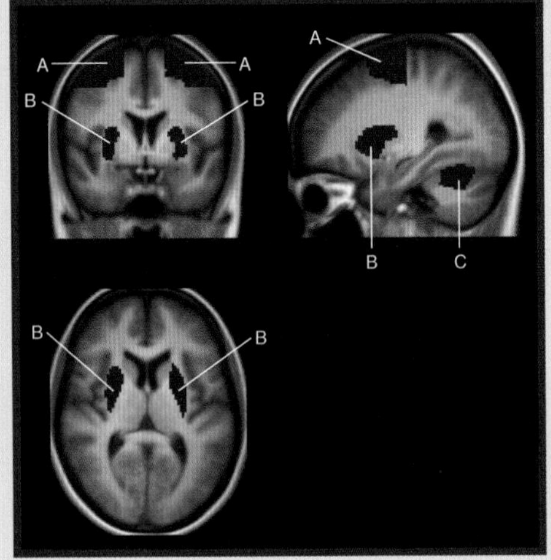

Figure 1 fMRI mask

Mask shows regions of interest for the study

Region B is a structure called the putamen, a part of the basal ganglia. It is connected to and provides pathways of communication for many structures in the brain, and generally influences and regulates motor behaviors such as planning, learning, preparation, and execution of motor sequences.

Region B = putamen: communication and motor behaviors

Researchers found no difference in activity in region B between patients with relapsing-remitting MS and controls. However, it was found that region C showed increased activation over the course of a non-fatiguing tonic grip task in patients who have MS. This increased activation correlated positively with subjective fatigue, and was not present in healthy control subjects. Furthermore, control subjects showed increased activation in region A over the course of the task, and no such activation occurred in patients who have MS.

Results: region B not implicated in MS-related fatigue, but C shows more activity and A shows less.

Adapted from: Svolgaard O, Andersen KW, Bauer C, Madsen KH, Blinkenberg M, Selleberg F, et al. (2018) Cerebellar and premotor activity during a non-fatiguing grip task reflects motor fatigue in relapsing-remitting multiple sclerosis. *PLoS ONE* 13(10): e0201162. https://doi.org/10.1371/journal.pone.0201162

Based on the functions of the regions studied, what do these results suggest about the nature of subjective fatigue in patients who have MS compared to healthy participants?

Our first step in answering this question is to identify the regions presented in the study that are referenced in the question stem. For this particular question, we don't have to worry too much about the structure of the experiment, as most of the information we need is in the results and the description of the regions. The authors give us the name and function of region B: the putamen, described in paragraph 2. This information is helpful because the actual brain isn't color coded and labeled, and due to the low structural resolution of the image, it's tough to tell exactly what region B is from the shape of the mask. On Test Day, using the image alone, we might be able to infer that region B is part of the midbrain and therefore, like other structures of the midbrain, it is probably involved in some kind of relay system. But the additional information in the passage text gives us insight that the picture alone just cannot provide. The passage also says that activity in region B is the same in patients who have MS and controls, allowing us to infer that difficulties in relaying motor signals are probably not the cause of subjective fatigue.

Based on our outside content knowledge, region C is the cerebellum, which we know is responsible for coordinating movement and for maintaining posture and balance. A differential increase in activity here implies that patients with MS may need more resources to perform motor tasks the longer these tasks are maintained.

Region A is in the forebrain. If we've studied the regions of the cerebrum, we might recognize this region as the premotor cortex, which is responsible for higher-level motor control and motivation. However, even without the specifics, we can guess that this region of the forebrain has something to do with executive motor control because of its general location. From the noted activity pattern in the final paragraph, we can guess that increased activity in region A helps to prevent subjective fatigue; thus, for patients who have MS, a lack of activation in this region may contribute to their experience of increased subjective fatigue.

We now have enough information to form a general picture of events here. In patients who have MS, the cerebellum is more active, presumably consuming more resources during maintained motor movements than the cerebellum of their healthy counterparts. This increase in resource consumption could be the patient's brain attempting to accommodate for functions from other regions that have been lost as a result of disease, or could indicate a greater overall demand on cognitive processes involving movement. This overtaxing of cerebellar resources is most likely related to the increase in subjective fatigue experienced by patients who have MS.

1.4 Parts of the Forebrain

LEARNING OBJECTIVES

After Chapter 1.4, you will be able to:

- Link the regions of the forebrain to their functions
- Describe the laterality of brain–body communication and the role of hemispheric dominance
- Describe the functions of the four lobes of the cerebral cortex:

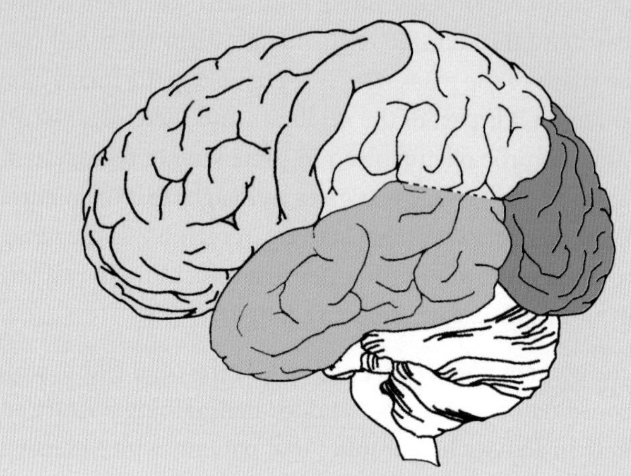

The forebrain is the most "modern" portion of the brain, and—in humans—forms the largest portion of the brain by weight and volume. The forebrain contains regions derived from the diencephalon, such as the thalamus, hypothalamus, posterior pituitary, and pineal gland; it also includes derivatives of the telencephalon, such as the cerebral cortex, basal ganglia, and limbic system.

Thalamus

The **thalamus** is a structure within the forebrain that serves as an important relay station for incoming sensory information, including all senses except for smell. After receiving incoming sensory impulses, the thalamus sorts and transmits them to the appropriate areas of the cerebral cortex. The thalamus is therefore a sensory "way station."

Hypothalamus

MNEMONIC

Functions of the Hypothalamus—The Four **F**s:

- **F**eeding
- **F**ighting
- **F**lighting
- (Sexual) **F**unctioning

The **hypothalamus**, subdivided into the lateral hypothalamus, ventromedial hypothalamus, and anterior hypothalamus, serves homeostatic functions, and is a key player in emotional experiences during high arousal states, aggressive behavior, and sexual behavior. The hypothalamus also helps control some endocrine functions, as well as the autonomic nervous system. The hypothalamus serves many homeostatic functions, which are self-regulatory processes that maintain a stable balance within the body. Receptors in the hypothalamus regulate metabolism, temperature, and water balance. When any of these functions are out of balance, the hypothalamus detects the problem and signals the body to correct the imbalance; for example, osmoreceptors in the hypothalamus may trigger the release of antidiuretic hormone to increase water reabsorption as part of fluid balance. The hypothalamus is also the primary regulator of the autonomic nervous system and is important in drive behaviors: hunger, thirst, and sexual behavior.

The **lateral hypothalamus** (LH) is referred to as the hunger center because it has special receptors thought to detect when the body needs more food or fluids. In other words, the LH triggers eating and drinking. When this part of the hypothalamus is removed in lab rats, they refuse to eat and drink and must be force-fed with tubes to survive.

MNEMONIC

When the **L**ateral **H**ypothalamus (**LH**) is removed, one **L**acks **H**unger.

The **ventromedial hypothalamus** (VMH) is identified as the "satiety center," and provides signals to stop eating. Brain lesions to this area usually lead to obesity.

MNEMONIC

When the **V**entro**M**edial **H**ypothalamus (**VMH**) is destroyed, one is **V**ery **M**uch **H**ungry.

The **anterior hypothalamus** controls sexual behavior. When the anterior hypothalamus is stimulated, lab animals will mount just about anything (including inanimate objects). In many species, damage to the anterior hypothalamus leads to permanent inhibition of sexual activity. The anterior hypothalamus also regulates sleep and body temperature.

Other Parts of the Diencephalon

The diencephalon also differentiates to form the posterior pituitary gland, pineal gland, and connecting pathways to other brain regions. The **posterior pituitary** is comprised of axonal projections from the hypothalamus and is the site of release for the hypothalamic hormones **antidiuretic hormone** (ADH, also called **vasopressin**) and **oxytocin**. The functions of these hormones are described in Chapter 5 of *MCAT Biology Review*. The **pineal gland** is the key player in several biological rhythms. Most notably, the pineal gland secretes a hormone called **melatonin**, which regulates circadian rhythms. The pineal gland receives direct signals from the retina for coordination with sunlight.

REAL WORLD

In the early 1920s, researchers first discovered the hypothalamus's role in rage and fighting through classic experiments conducted with cats. When researchers removed the cat's cerebral cortex but left the hypothalamus in place, the cat displayed a pattern of pseudoaggressive behavior that was called "sham rage"—lashing of the tail, arching of the back, clawing, and biting—except that rage was spontaneous or triggered by the mildest touch. The researchers concluded that the cortex typically inhibits this type of response. When the researchers removed the cat's cortex and hypothalamus together, the outcome was very different. The cat no longer showed any signs of sham rage, and much rougher stimulation was required before the cats showed any defensive behavior at all.

Basal Ganglia

In the middle of the brain are a group of structures known as the basal ganglia. The **basal ganglia** coordinate muscle movement as they receive information from the cortex and relay this information (via the extrapyramidal motor system) to the brain and the spinal cord. The **extrapyramidal system** gathers information about body position and carries this information to the central nervous system, but does not function directly through motor neurons. Essentially, the basal ganglia help make our movements smooth and our posture steady. **Parkinson's disease** is one chronic illness associated with destruction of portions of the basal ganglia. This disease is characterized by jerky movements and uncontrolled resting tremors. The basal ganglia may also play a role in schizophrenia and obsessive–compulsive disorder.

Limbic System

The **limbic system**, diagrammed in Figure 1.8, comprises a group of interconnected structures looping around the central portion of the brain and is primarily associated with emotion and memory. Its primary components include the septal nuclei, amygdala, hippocampus, and anterior cingulate cortex. In Chapter 5 of *MCAT Behavioral Sciences Review*, we will also explore the roles of the thalamus, hypothalamus, and cortex in the limbic system.

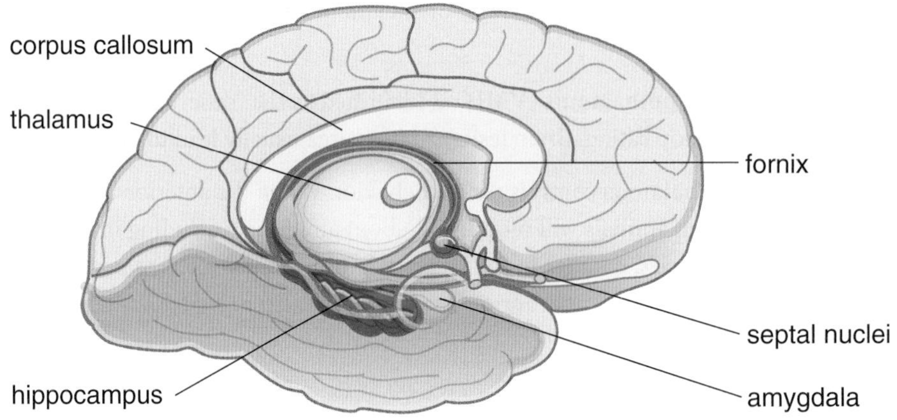

Figure 1.8 The Limbic System

Septal Nuclei

The **septal nuclei** contain one of the primary pleasure centers in the brain. Mild stimulation of the septal nuclei is reported to be intensely pleasurable; there is an association between these nuclei and addictive behavior.

Amygdala

The **amygdala** is a structure that plays an important role in defensive and aggressive behaviors, including fear and rage. Researchers base this observation on studies of animals and humans with brain lesions. When the amygdala is damaged, aggression and fear reactions are markedly reduced. Lesions to the amygdala result in docility and hypersexual states.

Hippocampus

The **hippocampus** plays a vital role in learning and memory processes; specifically, the hippocampus helps consolidate information to form long-term memories, and can redistribute remote memories to the cerebral cortex. The hippocampus communicates with other portions of the limbic system through a long projection called the **fornix**. Researchers originally discovered the connection between memory and the hippocampus through a famous patient named Henry Molaison (known as H.M. in the scientific literature until his death in 2008). Parts of H.M.'s temporal lobes—including the amygdala and hippocampus—were removed in an effort to control epileptic seizures. After surgery, H.M.'s intelligence was largely intact but he suffered

REAL WORLD

James Olds and Peter Milner discovered the association between the septal nuclei and addictive behavior in the 1950s. They demonstrated that when rats could stimulate their septal regions at will by pushing a lever, they found it so pleasurable that they preferred it to eating or any other activities, even after going 24 hours without food or sleep.

REAL WORLD

Heinrich Klüver and Paul Bucy performed studies that linked the amygdala with defensive and aggressive behavior in monkeys. When the amygdala of the Rhesus monkey was removed, they noted increased sexual behavior, decreased fear responses, and hyperorality, or the examination of inanimate or animate objects by mouth. These symptoms are now referred to as Klüver-Bucy syndrome.

a drastic and irreversible loss of memory for any new information. This kind of memory loss is called **anterograde amnesia** and is characterized by not being able to establish new long-term memories, whereas memory for events that occurred before brain injury is usually intact. The opposite kind of memory loss, **retrograde amnesia**, refers to memory loss of events that transpired before brain injury.

BRIDGE

Learning and memory are discussed thoroughly in Chapter 3 of *MCAT Behavioral Sciences Review.*

Anterior Cingulate Cortex

Due to the connection with the frontal and parietal lobes, the **anterior cingulate cortex** functions in higher order cognitive processes, including regulation of impulse control and decision-making. It also maintains connections to other parts of the limbic system, and thus plays a role in emotion and motivation.

Cerebral Cortex

The outer surface of the brain is called the **cerebral cortex**. The cortex is sometimes called the **neocortex**, a reminder that the cortex is the most recent brain region to evolve. Rather than having a smooth surface, the cortex has numerous bumps and folds called **gyri** and **sulci**, respectively. The convoluted structure of the brain provides increased surface area. The cerebrum is divided into two halves, called **cerebral hemispheres**. The surface of the cortex is divided into four lobes—the frontal lobe, parietal lobe, occipital lobe, and temporal lobe. These lobes are identified in Figure 1.9, which shows a side view of the left cerebral hemisphere.

MNEMONIC

Lobes of the brain: **F-POT**

- **F**rontal
- **P**arietal
- **O**ccipital
- **T**emporal

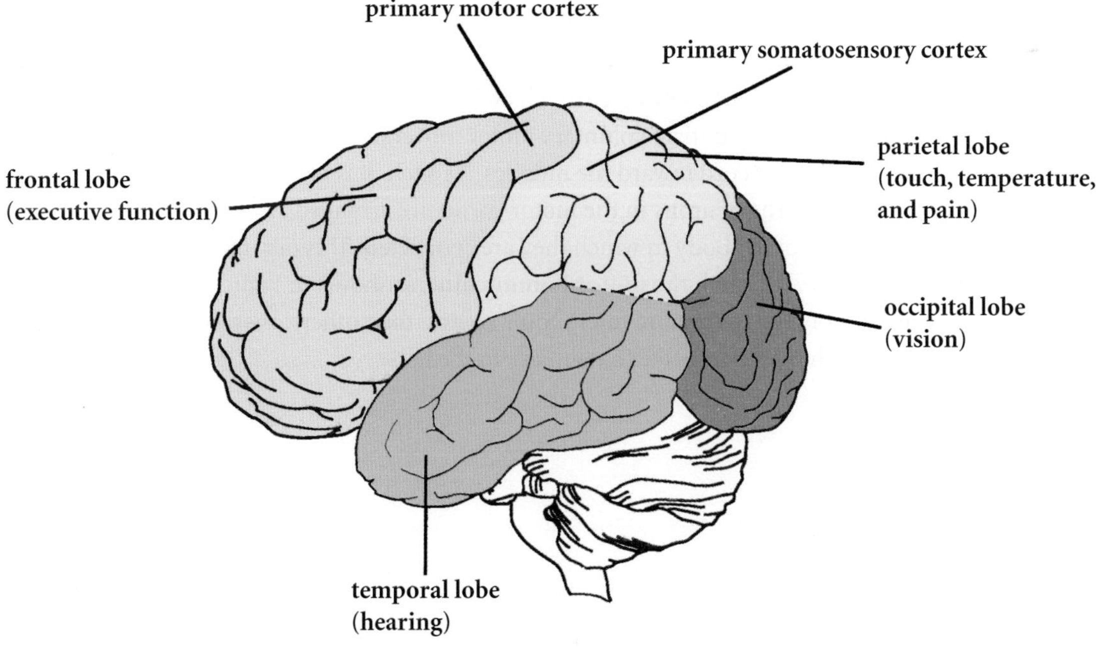

Figure 1.9 Lobes of the Brain

Frontal Lobe

The **frontal lobe** is comprised of two basic regions: the prefrontal cortex and the motor cortex. The **prefrontal cortex** manages executive function by supervising and directing the operations of other brain regions. To regulate attention and alertness, the prefrontal cortext communicates with the reticular formation in the brainstem, telling an individual either to wake up or to relax, depending on the situation. This region also supervises processes associated with perception, memory, emotion, impulse control, and long-term planning. In memory, for instance, the role of the prefrontal cortex is not to store any memory traces, but rather to remind individuals that they have something to remember at all.

Damage to the prefrontal cortex impairs its overall supervisory functions. People with prefrontal lesions may be more impulsive and generally less in control of their behavior. As a result, these individuals can have an increased tendency towards angry outbursts, as well as a higher predisposition to crying. Additionally, it is not unusual for someone with a prefrontal lesion to make vulgar and inappropriate sexual remarks, or to be apathetic to the emotional responses of others.

Because the prefrontal cortex integrates information from different cortical regions, the prefrontal cortex is a good example of an **association area**, which is an area that integrates input from diverse regions of the brain. For example, multiple inputs may be necessary to solve a complex puzzle, to plan ahead for the future, or to reach a difficult decision.

Association areas are generally contrasted with **projection areas**, which perform more rudimentary perceptual and motor tasks. An example of a projection area is the **primary motor cortex**, which is located on the **precentral gyrus**, just in front of the **central sulcus** that divides the frontal and parietal lobes. The function of the primary motor cortex is to initiate voluntary motor movements by sending neural impulses down the spinal cord toward the muscles. As such, it is considered a projection area in the brain. The neurons in the motor cortex are arranged systematically according to the parts of the body to which they are connected. This organizational pattern can be visualized through the **motor homunculus**, as shown in Figure 1.10. Because certain sets of muscles require finer motor control than others, they take up additional space in the cortex relative to their size in the body.

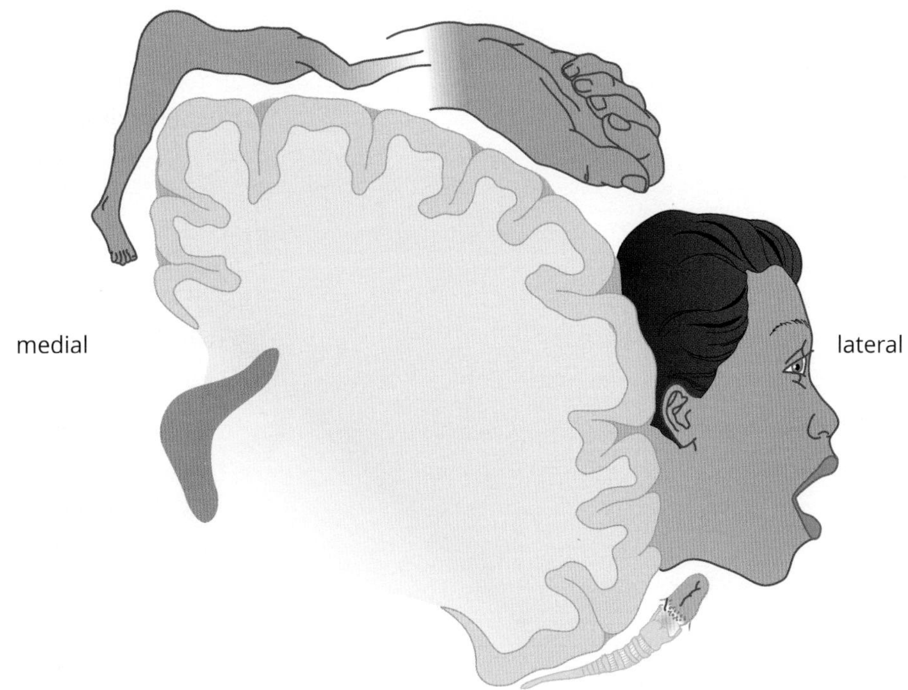

Figure 1.10 Motor Homunculus on the Precentral Gyrus of the Frontal Lobe

A third important part of the frontal lobe is **Broca's area**, which is vitally important for speech production. Broca's area is usually found in only one hemisphere, the so-called "dominant" hemisphere; for most people—both right- and left-handed—this is the left hemisphere.

Parietal Lobe

The **parietal lobe** is located to the rear of the frontal lobe. The **somatosensory cortex** is located on the **postcentral gyrus** (just behind the central sulcus) and is involved in somatosensory information processing. This projection area is the destination for all incoming sensory signals for touch, pressure, temperature, and pain. Despite certain differences, the somatosensory cortex and motor cortex are very closely related. In fact, they are so interrelated they sometimes are described as a single unit: the sensorimotor cortex. The somatosensory homunculus is shown in Figure 1.11. The central region of the parietal lobe is associated with spatial processing and manipulation. This region makes it possible to orient oneself and other objects in three-dimensional space, to do spatial manipulation of objects, and to apply spatial orientation skills such as those required for map reading.

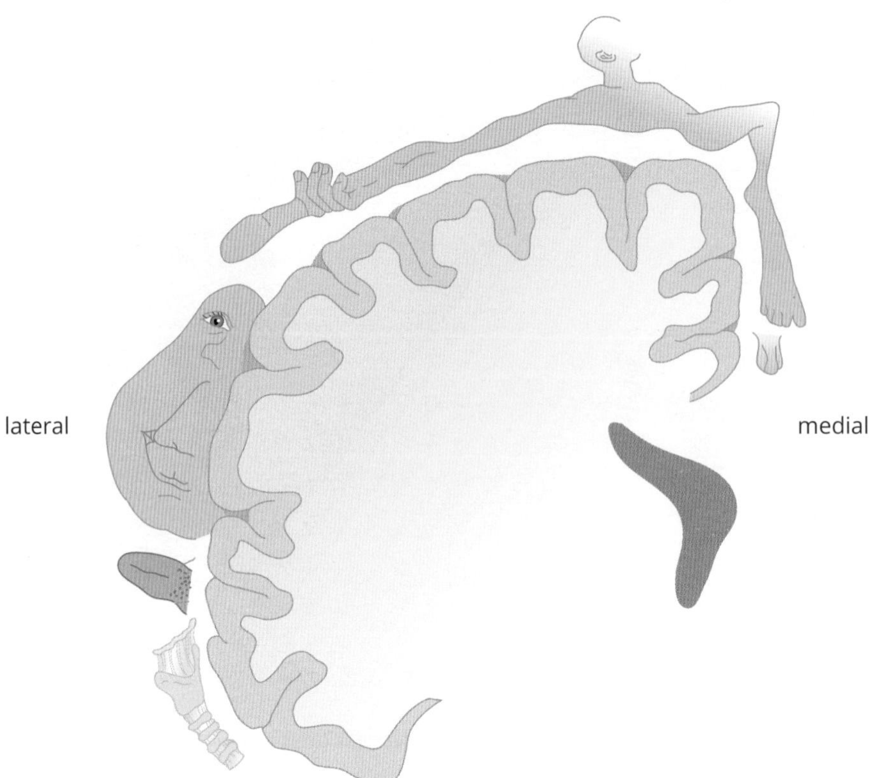

Figure 1.11 Somatosensory Homunculus on the Postcentral Gyrus of the Parietal Lobe

REAL WORLD

Several techniques have been developed to assess the function of brain regions, especially those associated with language and auditory processing. One such technique is called **speech shadowing**, which is a technique used to research both stuttering and speech perception. Speech shadowing involves participants reciting along with auditory inputs, which can be presented to one or both ears. If different messages are presented to each ear, as in a dichotic listening test, speech shadowing can ensure that the participant is paying attention to the auditory input to the correct ear. This seemingly simple experimental task requires successful functioning of the temporal lobe, parietal lobe, and frontal cortex!

Occipital Lobe

The **occipital lobes**, at the very rear of the brain, contain the **visual cortex**, which is sometimes called the **striate cortex**. *Striate* means furrowed or striped, which is how the visual cortex appears when examined under a microscope. The visual cortex is one of the best-understood brain regions, owing to the large amount of research that has been done on visual processing. Sensation and perception of visual information are discussed thoroughly in Chapter 2 of *MCAT Behavioral Sciences Review*. Areas in the occipital lobe have also been implicated in learning and motor control.

Temporal Lobe

The **temporal lobes** are associated with a number of functions. The auditory cortex and Wernicke's area are located in the temporal lobe. The **auditory cortex** is the primary site of most sound processing, including speech, music, and other sound information. **Wernicke's area** is associated with language reception and comprehension. The temporal lobe also functions in memory processing, emotion, and language. Studies have shown that electrical stimulation of the temporal lobe can evoke memories for past events. This makes sense because the hippocampus is located deep inside the temporal lobe. It is important to note that the lobes, although having seemingly independent functions, are not truly independent of one another. Often, a sensory modality may be represented in more than one area.

Cerebral Hemispheres and Laterality

In most cases, one side of the brain communicates with the opposite side of the body. In such cases, we say a cerebral hemisphere communicates **contralaterally**. For example, the motor neurons on the left side of the brain activate movements on the right side of the body. In other cases (for instance, hearing), cerebral hemispheres communicate with the same side of the body. In such cases, the hemispheres communicate **ipsilaterally**.

We distinguish between dominant and nondominant hemispheres. The dominant hemisphere is typically defined as the one that is more heavily stimulated during language reception and production. In the past, hand dominance was used as a proxy for hemispheric dominance; that is, right-handed individuals were assumed to have left-dominant brains and left-handed individuals were assumed to have right-dominant brains (because the brain communicates contralaterally with the hand). However, this correlation has not held up under scrutiny; 95 percent of right-handed individuals are indeed left brain dominant, but only 18 percent of left-handed individuals are right brain dominant.

The **dominant hemisphere** (usually the left) is primarily analytic in function, making it well-suited for managing details. For instance, language, logic, and math skills are all located in the dominant hemisphere. Again, language production (Broca's area) and language comprehension (Wernicke's area) are primarily driven by the dominant hemisphere.

The **nondominant hemisphere** (usually the right) is associated with intuition, creativity, music cognition, and spatial processing. The nondominant hemisphere simultaneously processes the pieces of a stimulus and assembles them into a holistic image. The nondominant hemisphere serves a less prominent role in language. It is more sensitive to the emotional tone of spoken language, and permits us to recognize others' moods based on visual and auditory cues, which adds to communication. The dominant hemisphere thus screens incoming language to analyze its content, and the nondominant hemisphere interprets it according to its emotional tone. The roles of the dominant and nondominant hemispheres are summarized in Table 1.2; remember that the left hemisphere is the dominant hemisphere in most individuals, regardless of handedness.

REAL WORLD

The **corpus callosum** connects and shares information between the two cerebral hemispheres; its function was discovered in patients who have epilepsy whose corpora callosa were severed in a last-ditch effort to limit their convulsive seizures. In these "split-brain" patients, in whom the corpus callosum has been severed, each hemisphere has its own function and specialization that is no longer accessible by the other. As an example of the result: an object felt only by the left hand (which projects to the right hemisphere) could not be named (because language function is usually in the left hemisphere).

FUNCTION	DOMINANT HEMISPHERE	NONDOMINANT HEMISPHERE
Visual system	Letters, words	Faces
Auditory system	Language-related sounds	Music
Language	Speech, reading, writing, arithmetic	Emotions
Movement	Complex voluntary movement	-
Spatial processes	-	Geometry, sense of direction

Table 1.2 Comparison of Dominant and Nondominant Hemispheres' Functions

MCAT CONCEPT CHECK 1.4

Before you move on, assess your understanding of the material with these questions.

1. Match the parts of the brain below to their functions:

1. Basal ganglia	A. Smooth movement
2. Cerebellum	B. Sensory relay station
3. Cerebral cortex	C. Sensorimotor reflexes
4. Hypothalamus	D. Arousal and alertness
5. Inferior and superior colliculi	E. Hunger and thirst; emotion
6. Limbic system	F. Complex perceptual, cognitive, and behavioral processes
7. Medulla oblongata	G. Vital function (breathing, digestion)
8. Reticular formation	H. Coordinated movement
9. Thalamus	I. Emotion and memory

2. What are the four lobes of the cerebral cortex, and what is the function of each?

Lobe	Function

3. What is the difference between ipsilateral and contralateral communication between the brain and the body?

____

4. How is the dominant hemisphere typically defined?

____

1.5 Influences on Behavior

High-Yield

LEARNING OBJECTIVES

After Chapter 1.5, you will be able to:

- Associate major neurotransmitters with their common functions
- Detail the links between the endocrine system and the brain
- Explain the nature *vs.* nurture debate and the different study types used to explore this question

Merely describing the functions of brain regions does not fully explain the wide variety of human behaviors that are possible. Other influences on behavior include chemical controls (neurotransmitters, hormones in the endocrine system), heredity, and the environment.

Neurotransmitters

A **neurotransmitter** is a chemical used by neurons to send signals to other neurons; more than 100 neurotransmitters have been identified. Several of the most important are described in this section and are summarized in Table 1.3. Some drugs mimic the action of neurotransmitters by binding to the same receptor to produce the same biological response. A drug that mimics the action of some neurotransmitter is called an **agonist**. Drugs can also act by blocking the action of neurotransmitters, and such drugs are called **antagonists**.

Acetylcholine

Acetylcholine is a neurotransmitter found in both the central and peripheral nervous systems. In the peripheral nervous system, acetylcholine is used to transmit nerve impulses to the muscles. It is the neurotransmitter used by the parasympathetic nervous system and a small portion of the sympathetic nervous system (in ganglia and for innervating sweat glands). In the central nervous system, acetylcholine has been linked to attention and arousal. In fact, loss of cholinergic neurons connecting with the hippocampus is associated with Alzheimer's disease, an illness resulting in progressive and incurable memory loss.

KEY CONCEPT

Acetylcholine is the neurotransmitter used by the efferent limb of the somatic nervous system and the parasympathetic nervous system. Acetylcholine can act as an excitatory or inhibitory neurotransmitter in muscle cells, dependent on the type of receptor found on the cell. For example, acetylcholine will transmit an inhibitory response in cardiac muscle cells, but it can also transmit an excitatory response if acting on skeletal muscle cells. Acetylcholine within the central nervous system largely functions as an excitatory neurotransmitter.

Epinephrine and Norepinephrine

Epinephrine, norepinephrine, and dopamine are three closely related neurotransmitters known as **catecholamines**. Due to similarities in their molecular composition, these three transmitters are also classified as **monoamines** or **biogenic amines**. The most important thing to know about the catecholamines is that they all play important roles in the experience of emotions.

Epinephrine (**adrenaline**) and **norepinephrine** (**noradrenaline**) are involved in controlling alertness and wakefulness. As the primary neurotransmitter of the sympathetic nervous system, they promote the fight-or-flight response. Whereas norepinephrine more commonly acts at a local level as a neurotransmitter, epinephrine is more often secreted from the adrenal medulla to act systemically as a hormone. Low levels of norepinephrine are associated with depression; high levels are associated with anxiety and mania.

Dopamine

Dopamine is another catecholamine that plays an important role in movement and posture. High concentrations of dopamine are normally found in the basal ganglia, which help smooth movements and maintain postural stability.

Imbalances in dopamine transmission have been found to play a role in **schizophrenia**. An important theory about the origin of this mental illness is called the **dopamine hypothesis of schizophrenia**. The dopamine hypothesis argues that delusions, hallucinations, and agitation associated with schizophrenia arise from either too much dopamine or from an oversensitivity to dopamine in the brain. Although the dopamine hypothesis of schizophrenia is an important theory, it does not account for all of the findings of the disease.

Parkinson's disease is associated with a loss of dopaminergic neurons in the basal ganglia. These disruptions of dopamine transmission lead to resting tremors and jerky movements, as well as to postural instability.

REAL WORLD

The role of dopamine in both schizophrenia and Parkinson's disease can be seen in their treatment. Antipsychotic medications used in schizophrenia are dopamine blockers, and can cause motor disturbances ("extrapyramidal symptoms") as a side effect. Parkinson's disease can be treated with L-DOPA, which increases dopamine levels in the brain; an overdose of L-DOPA can lead to psychotic symptoms similar to schizophrenia.

Serotonin

Along with the catecholamines, serotonin is classified as a monoamine or biogenic amine neurotransmitter. **Serotonin** is generally thought to play roles in regulating mood, eating, sleeping, and dreaming. Like norepinephrine, serotonin is thought to play a role in depression and mania. An oversupply of serotonin is thought to produce manic states; an undersupply is thought to produce depression.

GABA, Glycine, and Glutamate

The neurotransmitter **γ-aminobutyric acid** (**GABA**) produces inhibitory postsynaptic potentials and is thought to play an important role in stabilizing neural activity in the brain. GABA exerts its effects by causing hyperpolarization of the postsynaptic membrane.

Glycine may be better known as one of the twenty proteinogenic amino acids, but it also serves as an inhibitory neurotransmitter in the central nervous system by increasing chloride influx into the neuron. This hyperpolarizes the postsynaptic membrane, similar to the function of GABA.

Finally, **glutamate**, another of the twenty proteinogenic amino acids, also acts as a neurotransmitter in the central nervous system. In contrast to glycine, however, it is an excitatory neurotransmitter.

Peptide Neurotransmitters

Studies suggest that peptides are also involved in neurotransmission. The synaptic action of these **neuromodulators** (also called **neuropeptides**) involves a more complicated chain of events in the postsynaptic cell than that of regular neurotransmitters. Neuromodulators are therefore relatively slow and have longer effects on the postsynaptic cell than neurotransmitters. The **endorphins**, which are natural painkillers produced in the brain, are the most important peptides to know. Endorphins (and their relatives, **enkephalins**) have actions similar to morphine or other opioids in the body.

NEUROTRANSMITTER	BEHAVIOR
Acetylcholine	Voluntary muscle control, parasympathetic nervous system, attention, alertness
Epinephrine and Norepinephrine	Fight-or-flight responses, wakefulness, alertness
Dopamine	Smooth movements, postural stability
Serotonin	Mood, sleep, eating, dreaming
GABA and Glycine	Brain "stabilization"
Glutamate	Brain excitation
Endorphins	Natural painkillers

Table 1.3 Neurotransmitters and Their Functions

The Endocrine System

We've already discussed the relatively fast communication network—the nervous system—that uses chemical messages called neurotransmitters. The **endocrine system** is the other internal communication network in the body, and it uses chemical messengers called **hormones**. The endocrine system is somewhat slower than the nervous system because hormones travel to their target destinations through the bloodstream. The endocrine system is covered extensively in Chapter 5 of *MCAT Biology Review*, so our focus here will be on the role of certain endocrine organs on behavior.

The hypothalamus links the endocrine and nervous systems and, in addition to the roles described earlier, regulates the hormonal function of the pituitary gland. The hypothalamus and pituitary gland are spatially close to each other, and control is maintained through endocrine release of hormones into the **hypophyseal portal system** that directly connects the two organs, as shown in Figure 1.12.

BRIDGE

The entire endocrine system is covered in Chapter 5 of *MCAT Biology Review*.

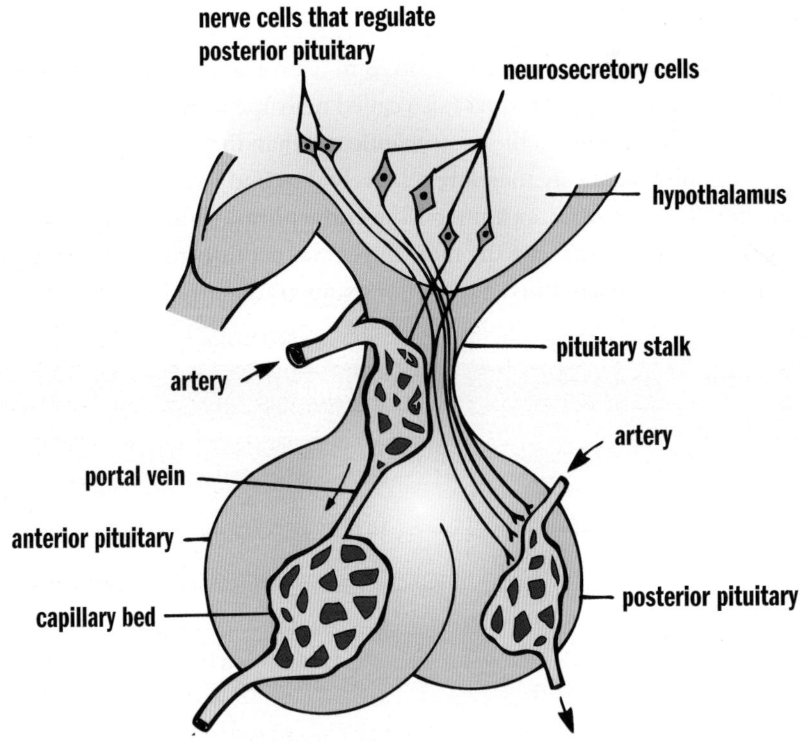

Figure 1.12 The Hypophyseal Portal System

The **pituitary gland**, sometimes referred to as the "master" gland, is located at the base of the brain and is divided into two parts: anterior and posterior. It is the **anterior pituitary** that is the "master" because it releases hormones that regulate activities of endocrine glands elsewhere in the body. However, the anterior pituitary itself is controlled by the hypothalamus. The pituitary secretes various hormones into the bloodstream that travel to other endocrine glands located elsewhere in the body to activate them. Once activated by the pituitary, a given endocrine gland manufactures and secretes its own characteristic hormone into the bloodstream.

The **adrenal glands** are located on top of the kidneys and are divided into two parts: the adrenal medulla and adrenal cortex. The **adrenal medulla** releases epinephrine and norepinephrine as part of the sympathetic nervous system. The **adrenal cortex** produces many hormones called **corticosteroids**, including the stress hormone **cortisol**. The adrenal cortex also contributes to sexual functioning by producing sex hormones, such as **testosterone** and **estrogen**.

The **gonads** are the sex glands of the body—ovaries in females and testes in males. These glands produce sex hormones in higher concentrations, leading to increased levels of testosterone in males and increased levels of estrogen in females. These sex hormones increase **libido** and contribute to mating behavior and sexual function. Higher levels of testosterone also increase aggressive behavior.

Genetics and Behavior

Just as physical traits are inherited from parents, behavioral traits can be inherited as well. Evidence for the inherited nature of behavior comes from the fact that many behaviors are species specific. For example, many animals exhibit mating behaviors only seen within their species. Behaviors can also be bred into a species; many breeds of dog have been bred for certain traits and behaviors. Behaviors are also seen to run in families. Often times, violence and aggression are observed passing along a family line, as are mental illnesses.

Innate behavior is genetically programmed as a result of evolution and is seen in all individuals regardless of environment or experience. In contrast, other behaviors are considered learned. **Learned behaviors** are not based on heredity but instead are based on experience and environment. **Adaptive value** is the extent to which a trait or behavior positively benefits a species by influencing the evolutionary fitness of the species, thus leading to **adaptation** through **natural selection**.

How much of an individual's behavior is based on genetic makeup and how much is based on environment and experiences? This controversial question is often referred to as the **nature *vs.* nurture** question. Here, **nature** is the influence of inherited characteristics on behavior. **Nurture** refers to the influence of environment and physical surroundings on behavior. There is no easy answer to this long-debated question. An individual's behavior is not only influenced by both genetics and environment, but also by how these two factors may influence each other. For example, hereditary traits may make a certain person more likely to have an addictive personality. But, the individual would still have to be exposed to drugs, alcohol, or gambling to develop an addiction.

To determine the degree of genetic influence on behavior, researchers often use one of three methods: family studies, twin studies, and adoption studies. **Family studies** rely on the fact that genetically related individuals are more similar genotypically than unrelated individuals. Researchers may compare rates of a given trait among family members to rates of that trait among unrelated individuals. For example, family studies have determined that the risk of developing schizophrenia for children of a patient who has schizophrenia is 13 times higher than in the general population. For siblings of a patient who has schizophrenia, the rate is 9 times higher. Observations such as these have led psychologists to conclude that schizophrenia has a hereditary component. Family studies are limited, however, because families share both genetics and environment. Family studies cannot distinguish shared environmental factors from shared genetic factors. For example, what if the increased rates of schizophrenia in families are a result of experiencing the same emotional climate in the home rather than genetically shared characteristics?

Twin studies, comparing concordance rates for a trait between **monozygotic** (**MZ**; identical) and **dizygotic** (**DZ**; fraternal) twins, are better able to distinguish the relative effects of shared environment and genetics. **Concordance rates** refer to the likelihood that both twins exhibit the same trait. MZ twins are genetically identical, sharing 100 percent of their genes, whereas DZ twins share approximately 50 percent of their genes. The assumption made by twin studies is that the two individuals in each MZ or DZ twin pair share the same environment; thus, differences between MZ

REAL WORLD

Bipolar disorder is considered one of the most heritable disorders, including medical illnesses. In one study, having a monozygotic (identical) twin with bipolar disorder was associated with a 43% risk of being diagnosed with the same disorder.

BRIDGE

Natural selection is discussed in greater detail in Chapter 12 of *MCAT Biology Review*.

and DZ twins are thought to reflect hereditary factors. Twin studies can also be used to measure genetic effects relative to environmental effects. In this version of the twin study, researchers compare traits in twins raised together versus twins raised apart. For example, one study of personality characteristics showed that MZ (identical) twins raised in separate families were still more similar than DZ (fraternal) twins raised together. Such a result offers convincing evidence for a strong genetic component to personality.

Finally, **adoption studies** also help us understand environmental and genetic influences on behavior. These studies compare the similarities between biological relatives and the child who was adopted to similarities between adoptive relatives and the child. For example, researchers have found that the IQ of children who were adopted is more similar to their biological parents' IQ than to their adoptive parents' IQ. This research suggests that IQ is heritable. Criminal behavior among teenagers shows a similar pattern of heritability.

MCAT CONCEPT CHECK 1.5

Before you move on, assess your understanding of the material with these questions.

1. Match the neurotransmitters below to their functions:

    1. Acetylcholine
    2. Dopamine
    3. Endorphins
    4. Epinephrine/norepinephrine
    5. GABA/glycine
    6. Glutamate
    7. Serotonin

    A. Wakefulness and alertness, fight-or-flight responses
    B. Brain "stabilizer"
    C. Mood, sleep, eating, dreaming
    D. Natural painkiller
    E. Smooth movements and steady posture
    F. Voluntary muscle control
    G. Brain excitation

2. Which endocrine organs influence behavior? What hormones do they use, and what do they accomplish?

    ___

3. Briefly discuss the influence of nature *vs.* nurture on behavior.

    ___

4. In each of the study types below, what is the sample group? The control group?

Study	Sample Group	Control Group
Family study		
Twin study		
Adoption study		

1.6 Development

LEARNING OBJECTIVES

After Chapter 1.6, you will be able to:

- Describe the process of neurulation
- Link the primitive reflexes with the behaviors to which they correspond
- Identify the main themes that dictate stages of motor development in children

The developmental process begins at the moment of conception. Physiological changes are rapid from embryonic to fetal stages, and well into infancy. Children exhibit surprisingly consistent patterns of motor abilities, as well as physiological changes based on age. Understanding these changes and when they occur is important in the discussion of developmental psychology. There are specific periods in development where children are particularly susceptible to environmental factors, called **critical periods**. Absence of the appropriate environmental factors may result in failure to learn a given skill or trait during the critical period, which may also mean learning that skill later on is difficult or even impossible.

Prenatal

The development of the nervous system starts with neurulation, at three to four weeks' gestational age. **Neurulation** occurs when the ectoderm overlying the notochord begins to furrow, forming a **neural groove** surrounded by two **neural folds**, as shown in Figure 1.13. Cells at the leading edge of the neural fold are called the **neural crest**, and will migrate throughout the body to form disparate tissues, including dorsal root ganglia, melanocytes (pigment-producing cells), and calcitonin-producing cells of the thyroid. The remainder of the furrow closes to form the **neural tube**, which will ultimately form the central nervous system (CNS). The neural tube has an **alar plate**, which differentiates into sensory neurons, and a **basal plate**, which differentiates into motor neurons. Over time, the neural tube invaginates and folds on itself many times; the embryonic brain begins as three swellings (prosencephalon, mesencephalon, rhombencephalon) that become five swellings (telencephalon, diencephalon, mesencephalon, metencephalon, myelencephalon) as it becomes the mature brain, as demonstrated in Figure 1.6 earlier in this chapter.

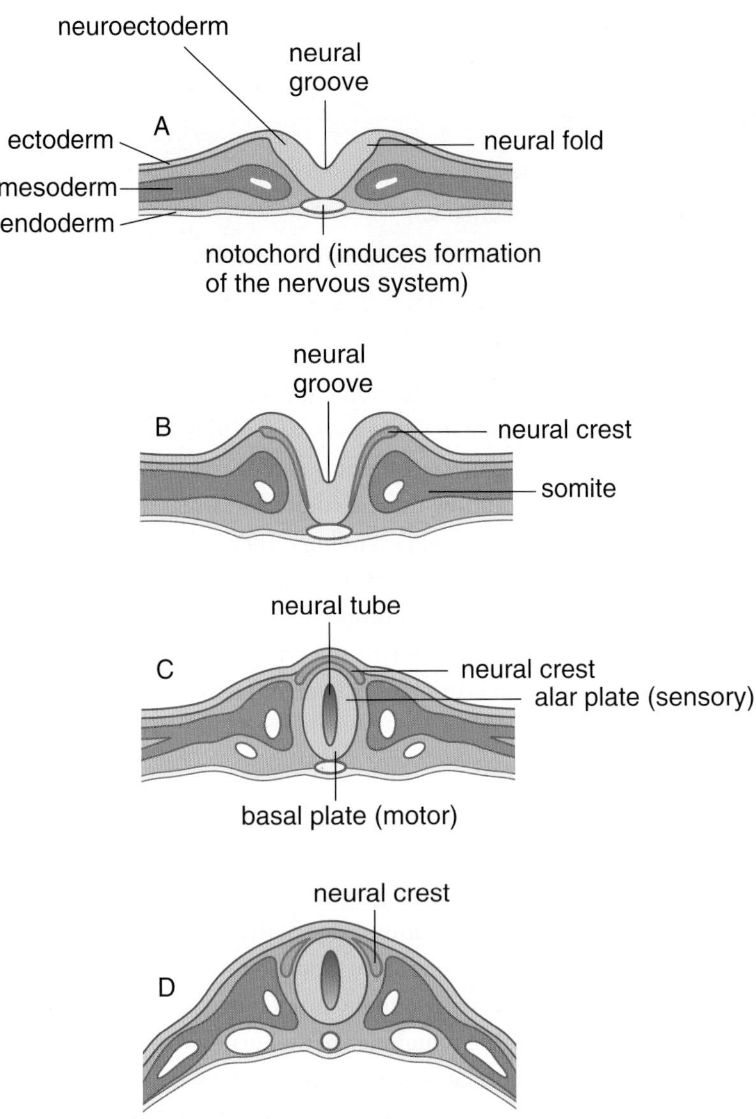

Figure 1.13 Development of the Nervous System

Prenatal development does not occur in a vacuum, of course, but in the mother's uterus. Within this environment, temperature, chemical balance, orientation of the fetus with respect to gravity, and atmospheric pressure are all carefully controlled and remain relatively constant. The fetus is attached to the uterine wall and placenta by the **umbilical cord**. The **placenta** transmits food, oxygen, and water to the fetus while returning water and waste to the mother. Maternal blood supplies many of the proteins and amino acids needed for growth, although the embryo begins to produce its own proteins and amino acids as well.

A variety of external influences can have deleterious effects on the development of the fetus. A number of viruses or bacteria can cross the placenta and cause damage to the developing fetus, including rubella (German measles), which may cause cataracts, deafness, heart defects, and intellectual disability. Other viral infections—such as measles, mumps, hepatitis, influenza, varicella (chickenpox), and herpes—have been linked to various birth defects.

An unfortunate side effect of the revolution in pharmaceutical development is that many drugs that help the mother can have damaging effects on the fetus she carries. The most notorious of these drugs is *thalidomide*, which was prescribed in the late 1950s and early 1960s to reduce morning sickness. Mothers who took this drug while pregnant often gave birth to babies with missing and malformed limbs and defects of the heart, eyes, ears, digestive tract, and kidneys. Antiepileptic medications are associated with neural tube defects, in which the neural tube fails to close completely, leading to malformations such as *spina bifida* or *anencephaly*.

A host of environmental factors and exposures may also affect maturation. Maternal malnutrition is considered to be a leading cause of abnormal development. Protein deficiency can slow growth, lead to intellectual disability, and reduce immunity to disease. Maternal narcotic addiction produces chemically dependent infants who must undergo severe withdrawal after birth. Regular cigarette smoking can lead to slowed growth, increased fetal heart rate, and a greater chance of premature birth. Daily use of alcohol also leads to slowed growth, both physically and psychologically. Finally, prenatal exposure to X-rays has been strongly linked to intellectual disabilities; defects of the skull, spinal cord, and eyes; cleft palate; and limb deformities.

Motor

Although they may seem helpless, infants are equipped with well-developed somatic structures and a broad array of reflexes that help ensure survival. A **reflex** is a behavior that occurs in response to a given stimulus without higher cognitive input. While motor and startle reflexes exist in adults, infants have a number of **primitive reflexes** that disappear with age. For example, the **rooting reflex** is the automatic turning of the head in the direction of a stimulus that touches the cheek—such as a nipple during feeding. Sucking and swallowing when an object is placed in the mouth are also examples of reflexes related to feeding.

Other primitive reflexes may have served an adaptive purpose in earlier stages of human evolution, but are currently used mainly in assessing infant neurological development. By comparing the point in time at which each of these reflexes disappears relative to the established norms, it is possible to tell whether neurological development is taking place in a normal fashion. One such reflex is the **Moro reflex**, illustrated in Figure 1.14. Infants react to abrupt movements of their heads by flinging out their arms, then slowly retracting their arms and crying. It has been speculated that this reflex may have developed during a time when our prehuman ancestors lived in trees and falling could have been prevented by instinctive clutching. The Moro reflex usually disappears after four months and its continuation at one year is a strong suggestion of developmental difficulties. Asymmetry of the Moro reflex may hint at underlying neuromuscular problems.

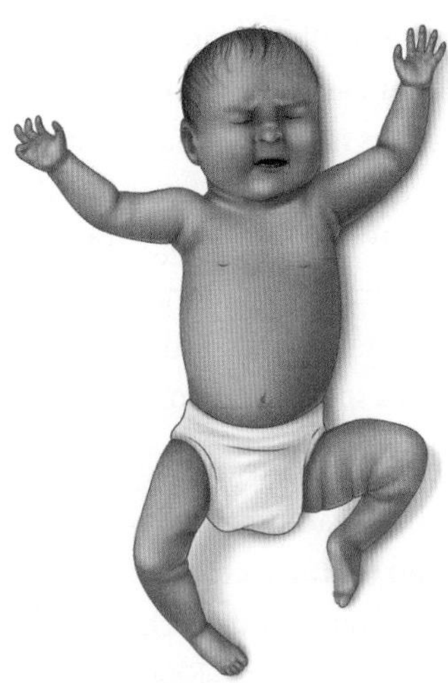

Figure 1.14 The Moro Reflex
The infant extends the arms, then slowly withdraws them and cries.

The **Babinski reflex** causes the toes to spread apart automatically when the sole of the foot is stimulated, as seen in Figure 1.15. The **grasping reflex** occurs when infants close their fingers around an object placed in the hand. Adults with neurological diseases may exhibit these primitive reflexes, especially in illnesses that cause demyelination (loss of the myelin sheath).

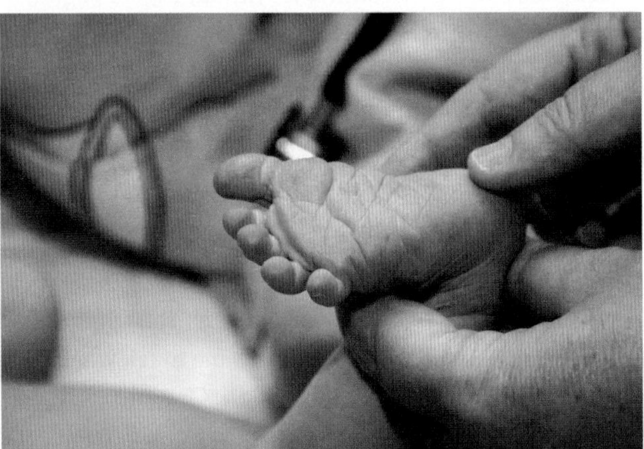

Figure 1.15 The Babinski Reflex
The big toe extends while the other toes fan outward.

Although reflexive behavior dominates the repertoire of the neonate, other behaviors occur as well. Newborn infants also kick, turn, and wave their arms. These uncoordinated, unconnected behaviors form the basis for later, more coordinated movements.

Infants typically develop motor skills at about the same age, in the same order. Due to this pattern, most psychologists and doctors agree that these are innately programmed abilities for human infants. However, the educational richness of the environment has been observed to affect the rate of learning, with more enriched environments promoting quicker development.

Motor skills are broken down into two classes: gross and fine motor skills. **Gross motor skills** incorporate movement from large muscle groups and whole body motion, such as sitting, crawling, and walking. **Fine motor skills** involve the smaller muscles of the fingers, toes, and eyes, providing more specific and delicate movement. Fine motor abilities include tracking motion, drawing, catching, and waving.

Social

In addition to motor skills, social development occurs in infancy and through adolescence. At birth, the parental figure becomes the center of the infant's world, and as the infant ages, **stranger anxiety** (a fear and apprehension of unfamiliar individuals) and **separation anxiety** (a fear of being separated from the parental figure) develop at approximately seven months and one year, respectively. During this time, play style progresses from solitary to onlooker, and at two years develops into **parallel play**, in which children will play alongside each other without influencing each other's behavior. At age three, children have an awareness of their assigned gender at birth, engage in gender-stereotyped play, and know their full name.

By age five, conformity to peers and romantic feelings for others begin to develop. From ages six through twelve, friend circles tend to be of the same gender without expression of romantic feelings. In the teenage years, children become more self-sufficient, and often express their desire for independence by rebelling against their parents. Inter-gender friendships become more common. Individuals also become more aware of their gender identity and sexual orientation and may begin sexual relationships.

In this chapter we have described several abilities and behaviors that are expected to emerge at particular times in a person's development. These skills are known as **developmental milestones**. The developmental milestones of the first three years of life are listed in Table 1.4. While this is a general timetable based on averages, most children fall within plus or minus two months of the chart. The goal is not to memorize this chart, but to recognize some themes. For example, gross motor skills progress in a head-to-toe order: children first develop the ability to lift the head, then to stabilize the trunk, and finally to walk. There is also a correlation between the development of motor skills and proximity to the center of the body, with skills being developed at the core prior to extremities. Social skills move from being parent-oriented to self-oriented to other-oriented. Language skills, discussed in Chapter 4 of *MCAT Behavioral Sciences Review*, become more complex and structured.

AGE	PHYSICAL AND MOTOR DEVELOPMENTS	SOCIAL DEVELOPMENTS	LANGUAGE DEVELOPMENTS
1st year of life	• Puts everything in mouth • Sits with support (4 mo) • Stands with help (8 mo) • Crawls, fear of falling (9 mo) • Pincer grasp (10 mo) • Follows objects to midline (4 wk) • One-handed approach/grasp of toy • Feet in mouth (5 mo) • Bang and rattle stage • Changes hands with toy (6 mo)	• Parental figure central • Issues of trust are key • Stranger anxiety (7 mo) • Play is solitary and exploratory • Pat-a-cake, peek-a-boo (10 mo)	• Laughs aloud (4 mo) • Repetitive responding (8 mo) • "mama, dada" (10 mo)
Age 1	• Walks alone (13 mo) • Climbs stairs alone (18 mo) • Emergence of hand preference (18 mo) • Kicks ball, throws ball • Pats pictures in book • Stacks three cubes (18 mo)	• Separation anxiety (12 mo) • Dependency on parental figure • Onlooker play	• Great variation in timing of language development • Uses 10 words
Age 2	• High activity level • Walks backward • Can turn doorknob, unscrew jar lid • Scribbles with crayon • Stacks six cubes (24 mo) • Stands on tiptoes (30 mo) • Able to aim thrown ball	• Selfish and self-centered • Imitates mannerisms and activities • May be aggressive • Recognizes self in mirror • "No" is favorite word • Parallel play	• Use of pronouns • Parents understand most • Two-word sentences • Uses 250 words • Identifies body parts by pointing
Age 3	• Rides tricycle • Stacks nine cubes (36 mo) • Alternates feet going up stairs • Bowel and bladder control (toilet training) • Draws recognizable figures • Catches ball with arms • Cuts paper with scissors • Unbuttons buttons	• Awareness of assigned gender • Gender-stereotyped play • Understands "taking turns" • Knows full name	• Complete sentences • Uses 900 words • Understands 3600 words • Strangers can understand • Recognizes common objects in pictures • Can answer, "Tell me what we wear on our feet?" "Which block is bigger?"

Table 1.4 Child Development Milestones

MCAT CONCEPT CHECK 1.6

Before you move on, assess your understanding of the material with these questions.

1. Describe the process of neurulation.

____

2. For each of the primitive reflexes below, briefly describe the observed behavior.

Primitive Reflex	Behavior
Rooting	
Moro	
Babinski	
Grasping	

3. What are the two main themes that dictate the stages of motor development in early childhood?

   1. ____
   2. ____

Conclusion

Behavioral psychology is the study of all physical and mental actions based on the response of the body to external stimuli, specifically the activity of the nervous and endocrine systems. The nervous system is a complex organization of structures and neurons that communicate and coordinate information. The endocrine system, in conjunction with the nervous system, controls human behavior. Aside from neurotransmitter and hormonal control of behavior, certain behaviors are genetically passed from generation to generation, as are many other physical traits. The genetic aspects of behavior are thought to interact with the learned components of behavior. Human behavior is also studied as it correlates to the development from embryo to fetus to infant and well into adolescence and adulthood. The development of motor skills and social behavior is seen to progress at a consistent rate across the species.

In the next chapter, our focus will be on the neurological systems used to interact with the world—most notably, those systems that exist for sensation and perception of the environment. These include vision, hearing, smell and taste, somatosensation, and others.

GO ONLINE

You've reviewed the content, now test your knowledge and critical thinking skills by completing a test-like passage set in your online resources!

CONCEPT SUMMARY

A Brief History of Neuropsychology

- **Neuropsychology** is the study of the connection between the nervous system and behavior. It most often focuses on the functions of various brain regions.

Organization of the Human Nervous System

- There are three types of neurons in the nervous system: **sensory** (**afferent**) neurons, **motor** (**efferent**) neurons, and **interneurons**.
- **Reflex arcs** use the ability of interneurons in the spinal cord to relay information to the source of stimuli while simultaneously routing it to the brain.
- The nervous system is made up of the **central nervous system** (**CNS**; brain and spinal cord) and **peripheral nervous system** (**PNS**; most cranial and spinal nerves).
  - The PNS is divided into the **somatic** (voluntary) and **autonomic** (automatic) divisions.
  - The autonomic system is further divided into the **parasympathetic** (rest-and-digest) and **sympathetic** (fight-or-flight) branches.

Organization of the Brain

- The brain has three subdivisions: hindbrain, midbrain, and forebrain.
  - The **hindbrain** contains the cerebellum, medulla oblongata, and reticular formation.
  - The **midbrain** contains the inferior and superior colliculi.
  - The **forebrain** contains the thalamus, hypothalamus, basal ganglia, limbic system, and cerebral cortex.
- Methods of studying the brain include studying humans and animals with lesions, electrical stimulation and activity recording (including **electroencephalography** [EEG]), and **regional cerebral blood flow**.

Parts of the Forebrain

- The **thalamus** is a relay station for sensory information.
- The **hypothalamus** maintains homeostasis and integrates with the endocrine system through the **hypophyseal portal system** that connects it to the **anterior pituitary**.
- The **basal ganglia** smoothen movements and help maintain postural stability.
- The **limbic system**, which contains the septal nuclei, amygdala, and hippocampus, controls emotion and memory.
  - The **septal nuclei** are involved with feelings of pleasure, pleasure-seeking behavior, and addiction.

    - The **amygdala** controls fear and aggression.
    - The **hippocampus** consolidates memories and communicates with other parts of the limbic system through an extension called the **fornix**.
- The **cerebral cortex** is divided into four lobes: frontal, parietal, occipital, and temporal.
    - The **frontal lobe** controls executive function, impulse control, long-term planning, motor function, and speech production.
    - The **parietal lobe** controls sensations of touch, pressure, temperature, and pain; spatial processing; orientation; and manipulation.
    - The **occipital lobe** controls visual processing.
    - The **temporal lobe** controls sound processing, speech perception, memory, and emotion.
- The brain is divided into two **cerebral hemispheres**, left and right. In most individuals, the left hemisphere is the dominant hemisphere for language.

Influences on Behavior

- **Neurotransmitters** are released by neurons and carry a signal to another neuron or effector (a muscle fiber or a gland).
    - **Acetylcholine** is used by the somatic nervous system (to move muscles), the parasympathetic nervous system, and the central nervous system (for alertness).
    - **Dopamine** maintains smooth movements and steady posture.
    - **Endorphins** and **enkephalins** act as natural painkillers.
    - **Epinephrine** and **norepinephrine** maintain wakefulness and alertness and mediate fight-or-flight responses. Epinephrine tends to act as a hormone, and norepinephrine tends to act more classically as a neurotransmitter.
    - **γ-Aminobutyric acid** (**GABA**) and **glycine** act as brain "stabilizers."
    - **Glutamate** acts as an excitatory neurotransmitter in the brain.
    - **Serotonin** modulates mood, sleep patterns, eating patterns, and dreaming.
- The endocrine system is tied to the nervous system through the hypothalamus and the anterior pituitary, as well as a few other hormones.
    - **Cortisol** is a stress hormone released by the adrenal cortex.
    - **Testosterone** and **estrogen** mediate libido; testosterone also increases aggressive behavior. Both are released by the adrenal cortex. In males, the testes also produce testosterone. In females, the ovaries also produce estrogen.
    - **Epinephrine** and **norepinephrine** are released by the adrenal medulla and cause physiological changes associated with the sympathetic nervous system.

- **Nature *vs.* nurture** is a classic debate regarding the relative contributions of genetics (nature) and environment (nurture) to an individual's traits. For most traits, both nature and nurture play a role. The relative effects of each can be studied.
  - **Family studies** look at the relative frequency of a trait within a family compared to the general population.
  - **Twin studies** compare concordance rates between monozygotic (identical) and dizygotic (fraternal) twins.
  - **Adoption studies** compare similarities between children who were adopted and their adoptive parents, relative to similarities with their biological parents.

Development

- The nervous system develops through neurulation, in which the notochord stimulates overlying ectoderm to fold over, creating a **neural tube** topped with **neural crest** cells.
  - The **neural tube** becomes the central nervous system (CNS).
  - The **neural crest** cells spread out throughout the body, differentiating into many different tissues.
- **Primitive reflexes** exist in infants and should disappear with age. Most primitive reflexes serve (or served, in earlier times) a protective role. They can reappear in certain nervous system disorders.
  - In the **rooting reflex**, infants turn their heads toward anything that brushes the cheek.
  - In the **Moro reflex**, the infant extends the arms, then slowly retracts them and cries in response to a sensation of falling.
  - In the **Babinski reflex**, the big toe is extended and the other toes fan in response to the brushing of the sole of the foot.
  - In the **grasping reflex**, infants grab anything put into their hands.
- Developmental milestones give an indication of what skills and abilities a child should have at a given age. Most children adhere closely to these milestones, deviating by only one or two months.
  - Gross and fine motor abilities progress head to toe and core to periphery.
  - Social skills shift from parent-oriented to self-oriented to other-oriented.
  - Language skills become increasingly complex.

ANSWERS TO CONCEPT CHECKS

1.1

1. - Franz Gall: phrenology; associated development of a trait with growth of its relevant part of the brain.
   - Pierre Flourens: extirpation/ablation; concluded that different brain regions have specific functions.
   - William James: "founder of American psychology"; pushed for importance of studying adaptations of the individual to the environment.
   - John Dewey: credited with the landmark article on functionalism; argued for studying the entire organism as a whole.
   - Paul Broca: correlated pathology with specific brain regions, such as speech production from Broca's area.
   - Hermann von Helmholtz: measured speed of a nerve impulse.
   - Sir Charles Sherrington: inferred the existence of synapses.

1.2

1. The central nervous system includes the brain and spinal cord. The peripheral nervous system includes most of the cranial and spinal nerves and sensors.
2. Afferent (sensory) neurons bring signals from a sensor to the central nervous system. Efferent (motor) neurons bring signals from the central nervous system to an effector.
3. The somatic nervous system is responsible for voluntary actions; most notably, moving muscles. The autonomic nervous system is responsible for involuntary actions, like heart rate, bronchial dilation, dilation of the eyes, exocrine gland function, and peristalsis.
4. The sympathetic nervous system promotes a fight-or-flight response, with increased heart rate and bronchial dilation, redistribution of blood to locomotor muscles, dilation of the eyes, and slowing of digestive and urinary functions. The parasympathetic nervous system promotes rest-and-digest functions, slowing heart rate and constricting the bronchi, redistributing blood to the gut, promoting exocrine secretions, constricting the pupils, and promoting peristalsis and urinary function.

1.3

1.

Subdivision	Functions
Hindbrain	Balance, motor coordination, breathing, digestion, general arousal processes (sleeping and waking); "vital functioning"
Midbrain	Receives sensory and motor information from the rest of the body; reflexes to auditory and visual stimuli
Forebrain	Complex perceptual, cognitive, and behavioral processes; emotion and memory

2. Methods used for mapping the brain include studying humans with brain lesions, extirpation, stimulation or recording with electrodes (cortical mapping, single-cell electrode recordings, electroencephalogram [EEG]), and regional cerebral blood flow (rCBF).
3. From most deep to most superficial, the structures surrounding the brain are the meninges, bone, periosteum, and skin.

1.4

1. 1–A, 2–H, 3–F, 4–E, 5–C, 6–I, 7–G, 8–D, 9–B
2.

Lobe	Function
Frontal	Executive function, impulse control, long-term planning (prefrontal cortex), motor function (primary motor cortex), speech production (Broca's area)
Parietal	Sensation of touch, pressure, temperature, and pain (somatosensory cortex); spatial processing, orientation, and manipulation
Occipital	Visual processing
Temporal	Sound processing (auditory cortex), speech perception (Wernicke's area), memory, and emotion (limbic system)

3. Ipsilateral communication occurs when cerebral hemispheres communicate with the same side of the body. Contralateral communication occurs when cerebral hemispheres communicate with the opposite side of the body.
4. The dominant hemisphere is typically defined as the one that is more heavily stimulated during language reception and production.

1.5

1. 1–F, 2–E, 3–D, 4–A, 5–B, 6–G, 7–C
2. The hypothalamus controls release of pituitary hormones; the pituitary is the "master gland" that triggers hormone secretion in many other endocrine glands. The adrenal medulla produces adrenaline (epinephrine), which causes sympathetic nervous system effects throughout the body. The adrenal cortex produces cortisol, a stress hormone. The adrenal cortex and testes produce testosterone, which is associated with libido.
3. Nature is defined as heredity, or the influence of inherited characteristics on behavior. Nurture refers to the influence of environment and physical surroundings on behavior. It has long been debated whether nature or nurture has the larger influence; it is a complicated situation, but for most traits, both exert some influence.
4.

Study	Sample Group	Control Group
Family study	Family of genetically related individuals	Unrelated individuals (general population)
Twin study	Monozygotic (MZ, identical) twins	Dizygotic (DZ, fraternal) twins
Adoption study	Adoptive family (relative to child who was adopted)	Biological family (relative to child who was adopted)

1.6

1. Neurulation occurs when a furrow is produced from ectoderm overlying the notochord and consists of the neural groove and two neural folds. As the neural folds grow, the cells at their leading edge are called neural crest cells. When the neural folds fuse, this creates the neural tube, which will form the CNS.
2.

Primitive Reflex	Behavior
Rooting	Turns head toward direction of any object touching the cheek
Moro	In response to sudden head movement, arms extend and slowly retract; baby usually cries
Babinski	Extension of big toe and fanning of other toes in response to brushing the sole of the foot
Grasping	Holding onto any object placed in the hand

3. Gross motor development proceeds from head to toe, and from the core to the periphery.

SCIENCE MASTERY ASSESSMENT EXPLANATIONS

1. **B**

Creating knockouts to study gene function is similar to ablation, or extirpation, in which specific parts of an organism's anatomy are removed. This approach was used by Pierre Flourens to show that specific parts of the brain served particular functions, supporting **(B)** as the correct answer. Unlike Pierre, Paul Broca, choice **(A)**, did not ablate regions of the brain; rather he studied the functional impairments in individuals with different brain lesions.

2. **B**

The cerebral cortex is not involved in the initial reflexive response to pain. Instead, the sensory receptors send information to the interneurons in the spinal cord, which stimulate a motor neuron to allow quick withdrawal. While the brain does ultimately get the signal, the reflexive withdrawal has already occurred by that time.

3. **C**

The hindbrain is responsible for balance and motor coordination, which would be necessary for dribbling a basketball. The midbrain, **(B)**, manages sensorimotor reflexes that also promote survival. The forebrain, **(A)**, is associated with emotion, memory, and higher-order cognition. The spinal cord, **(D)**, is likely not damaged as the child can still walk.

4. **D**

The temporal lobes have many functions, but motor skills are not associated with this area. The temporal lobes contain Wernicke's area, which is responsible for language comprehension, **(A)**. The temporal lobes also function in memory and emotion, **(B)** and **(C)**, because they contain the hippocampus and amygdala. Motor skills are associated with the frontal lobe (primary motor cortex), basal ganglia (smooth movements), and cerebellum (coordination).

5. **C**

The hypothalamus is responsible for homeostatic and emotional functions. The cerebellum, **(A)**, is responsible for maintaining posture and balance while the pons, **(B)**, is above the medulla and contains sensory and motor tracts between the cortex and the medulla. The thalamus, **(D)**, acts as a relay station for sensory information.

6. **A**

The right hemisphere is usually the nondominant hemisphere, even in left-handed individuals. Sense of direction is an ability of the nondominant hemisphere. The other answer choices are all abilities attributed to the dominant hemisphere.

7. **B**

Neurulation occurs when the notochord causes differentiation of overlying ectoderm into the neural tube and neural crest cells. The neural tube ultimately becomes the central nervous system (brain and spinal cord), and neural crest cells migrate to other sites in the body to differentiate into a number of different tissues. Thus, only statement III is true.

8. **D**

Catecholamines are the hormones produced by the adrenal glands during the fight-or-flight response, and include epinephrine, norepinephrine, and dopamine. Acetylcholine is produced by cholinergic neurons and is, thus, not a catecholamine.

9. **A**

If there were increased amounts of acetylcholinesterase, more acetylcholine would be degraded, lowering acetylcholine levels in the body. Low levels of acetylcholine would result in weakness or paralysis of muscles. Pain, **(B)**, could result if one were injured and endorphins were found in low levels. Mood swings, **(C)**, could be a result of varying levels of serotonin. Hallucinations, **(D)**, have been seen to result from high levels of dopamine.

10. **A**

The adrenal glands do promote the fight-or-flight response, but through epinephrine and norepinephrine, not estrogen. The adrenal cortex produces both estrogen and testosterone in both sexes, as mentioned in **(D)**, thus serving as a source of estrogen in males and testosterone in females.

11. **C**

The pineal gland is responsible for producing melatonin, which controls the body's circadian rhythm. Insomnia would be a disturbance of this circadian rhythm, and may be attributable to a pineal gland disorder in some cases. Hypertension, diabetes, and hyperthyroidism would be unrelated to issues with the pineal gland.

12. **A**

William James studied how the mind adapts to the environment and formed the foundation of functionalism. Thus, **(A)** is correct. The remaining wrong answer choices can be ascribed to other scientists. Answer **(B)** describes phrenology, the theory of Franz Gall. Answer **(C)** describes the conclusions of Paul Broca. Answer **(D)** describes one of the conclusions of Sir Charles Sherrington (although we now know that his conclusion was false).

13. **C**

Activity in different regions of the brain is assumed to positively correlate with blood flow. Thus, regions that are more active should have increased blood flow. Changes in blood flow can be measured via fMRI, supporting **(C)** as the preferred methodology. On the other hand, MRI and CT scans are used to image tissues, ruling out **(A)** and **(B)**, and EEGs are useful in measuring broad brain activity and not specific brain regions, eliminating **(D)**.

14. **D**

The Babinski reflex is a primitive reflex that refers to an extension of the big toe accompanied by fanning of the other toes. It is normal in infants, but should disappear with time—certainly by the time a child begins to walk. In a person who is fifty years old, the Babinski reflex would be abnormal. However, despite having a neurological illness, this patient is exhibiting a normal response to the brushing of the foot; that is, the patient is not showing the Babinski reflex.

15. **B**

Motor skills tend to develop from the core toward the periphery. Following objects with the eyes occurs around four weeks of age. The other actions all require movements of the hand, which do not occur in an organized fashion until later.

Consult your online resources for additional practice.

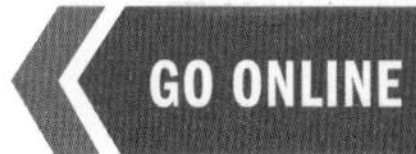

SHARED CONCEPTS

Behavioral Sciences Chapter 2
Sensation and Perception

Behavioral Sciences Chapter 3
Learning and Memory

Behavioral Sciences Chapter 4
Cognition, Consciousness, and Language

Biology Review Chapter 3
Embryogenesis and Development

Biology Review Chapter 4
The Nervous System

Biology Review Chapter 5
The Endocrine System

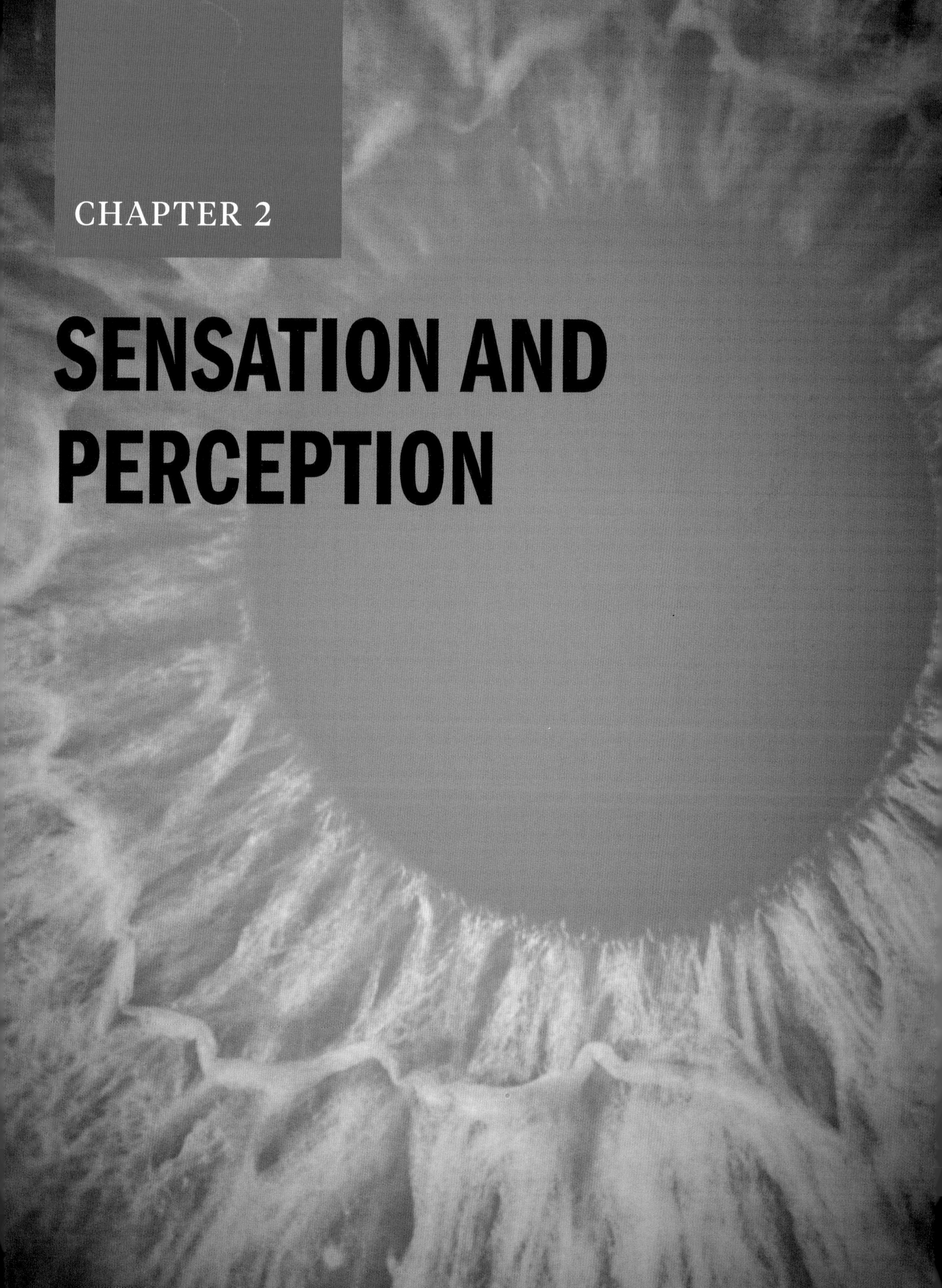

CHAPTER 2

SENSATION AND PERCEPTION

SCIENCE MASTERY ASSESSMENT

Every pre-med knows this feeling: there is so much content I have to know for the MCAT! How do I know what to do first or what's important?

While the high-yield badges throughout this book will help you identify the most important topics, this Science Mastery Assessment is another tool in your MCAT prep arsenal. This quiz (which can also be taken in your online resources) and the guidance below will help ensure that you are spending the appropriate amount of time on this chapter based on your personal strengths and weaknesses. Don't worry though—skipping something now does not mean you'll never study it. Later on in your prep, as you complete full-length tests, you'll uncover specific pieces of content that you need to review and can come back to these chapters as appropriate.

How to Use This Assessment

If you answer 0–7 questions correctly:

Spend about 1 hour to read this chapter in full and take limited notes throughout. Follow up by reviewing **all** quiz questions to ensure that you now understand how to solve each one.

If you answer 8–11 questions correctly:

Spend 20–40 minutes reviewing the quiz questions. Beginning with the questions you missed, read and take notes on the corresponding subchapters. For questions you answered correctly, ensure your thinking matches that of the explanation and you understand why each choice was correct or incorrect.

If you answer 12–15 questions correctly:

Spend less than 20 minutes reviewing all questions from the quiz. If you missed any, then include a quick read-through of the corresponding subchapters, or even just the relevant content within a subchapter, as part of your question review. For questions you answered correctly, ensure your thinking matches that of the explanation and review the Concept Summary at the end of the chapter.

1. A weight lifter is just able to tell the difference between 100 and 125 pounds. According to Weber's law, the lifter would notice a difference between:
   A. 125 and 150 pounds.
   B. 5 and 6 pounds.
   C. 25 and 35 pounds.
   D. 225 and 275 pounds.

2. A person is at a party with a friend. There is loud music in the background and the location is crowded. While listening to the music, the person hears what seems like the friend's laughter and turns around to investigate. The person is exhibiting:
   A. feature detection.
   B. bottom-up processing.
   C. vestibular sense.
   D. signal detection.

3. A person is at a restaurant and orders a spicy entrée. After the first bite, the person experiences burning in the mouth and becomes concerned that the food is too hot. The next few bites are similarly uncomfortable, but after a while the spiciness seems to subside somewhat, and by the end of the meal, the person doesn't notice the spice level. The end of the meal experience is best described as:
   A. adaptation.
   B. signal detection.
   C. a difference threshold.
   D. pain perception.

4. Which sensory receptors send signals in response to tissue damage?
   A. Chemoreceptors
   B. Nociceptors
   C. Osmoreceptors
   D. Photoreceptors

5. Which part of the eye is responsible for gathering and focusing light?
   A. Cornea
   B. Pupil
   C. Iris
   D. Retina

6. A person is looking for change to do laundry and decides to look under the seats of the household car. Even using a flashlight allows no more than an obscured look at the space below. There are various items such as wrappers and papers, but the person sees the glint of silver from an object laying flat and determines it to be a coin. To make this determination, this person used:
   A. signal detection.
   B. sensory adaptation.
   C. feature detection.
   D. kinesthetic sense.

7. Upon which part of the eye are images projected and transduced into electrical signals?
   A. Cornea
   B. Pupil
   C. Retina
   D. Lens

8. The ability to sense stimuli against one's own skin is known as:
   A. somatosensation.
   B. kinesthetic sense.
   C. vestibular sense.
   D. chemoreception.

9. Which of the following is NOT a taste modality?

A. Sweet
B. Floral
C. Savory
D. Bitter

10. Which of the following best describes the difference between endolymph and perilymph?

A. Endolymph is found in the vestibule, while perilymph is found in the cochlea.
B. Endolymph is found in the cochlea, while perilymph is found in the vestibule.
C. Endolymph is found in the membranous labyrinth, while perilymph is found in the bony labyrinth.
D. Endolymph is found in the bony labyrinth, while perilymph is found in the membranous labyrinth.

11. Chemicals that compel behavior after binding with chemoreceptors are known as:

A. pheromones.
B. olfactory receptors.
C. somatostimuli.
D. papillae.

12. Prolonged vitamin B_{12} deficiency can be associated with subacute combined degeneration of the spinal cord. Patients with this disease have difficulty walking because they lose the ability to feel where their feet are in space. This represents a loss of:

A. vestibular sense.
B. kinesthetic sense.
C. parallel processing.
D. feature detection.

13. A person proofreading a paper reads over a long, misspelled word in which an "e" is replaced with an "o." The person does not recognize the error and reads the word as correct. Which of the following could explain why the proofreader read the word as correct?

A. Parallel processing
B. Feature detection
C. Top-down processing
D. Bottom-up processing

14. A corporate logo uses five unconnected angles equally spaced in a circular fashion. When viewed, it appears to be a star. Which of the following is the logo artist using to create a complete pattern to viewers?

A. Bottom-up processing
B. Top-down processing
C. Gate theory
D. Gestalt principles

15. A patient comes in with a tumor of the pituitary gland which has grown upward into the optic chiasm and caused a visual field defect. The most likely defect from compression of the optic chiasm is:

A. complete blindness in one eye.
B. loss of the upper visual fields in both eyes.
C. loss of the nasal visual fields in both eyes.
D. loss of the temporal visual fields in both eyes.

Answer Key

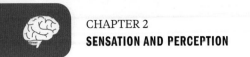

1. **C**
2. **D**
3. **A**
4. **B**
5. **A**
6. **C**
7. **C**
8. **A**
9. **B**
10. **C**
11. **A**
12. **B**
13. **C**
14. **D**
15. **D**

Detailed explanations can be found at the end of the chapter.

CHAPTER 2

SENSATION AND PERCEPTION

In This Chapter

CHAPTER PROFILE

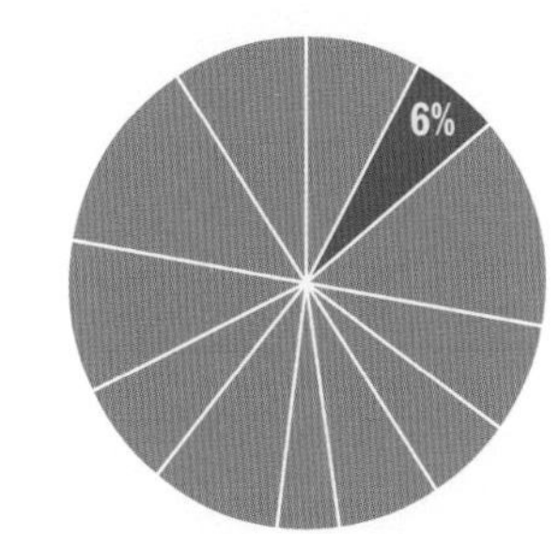

The content in this chapter should be relevant to about 6% of all questions about the behavioral sciences on the MCAT.

This chapter covers material from the following AAMC content category:

6A: Sensing the environment

Introduction

It's your first time visiting a new city. You can't wait to explore the area, try out a new café or restaurant, and see famous landmarks there. You want to take in the sights and sounds of this new place—you want to have a sensory experience. To truly experience any location, your sensory receptors—for vision, hearing, taste, smell, and somatosensation—gather all of the information from the world around you, and your brain filters and processes that information to focus on the most salient details. This activity involves a complex interplay between sensory processes, neural tracts, and the brain itself.

You finally arrive in the new city and begin to explore. You turn the corner on one street and are suddenly overwhelmed with an odd feeling of familiarity. *But . . . I've never been here before!* you think as the strange sensation of *déjà vu* sets in. Everything just seems "right": the signs are in the proper place, the cars look familiar, and everything is bizarrely where you expect it to be. *Déjà vu* (French for "already seen") comes from many sources, including processing information faster than expected. When you process an image (or other sensory input) for the first time, it actually takes longer than the next time you are exposed to that same stimulus. Thus, an exposure to the same scenery at an earlier time through a movie or television show may have primed you for *déjà vu*.

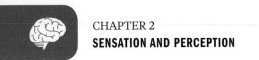

But we don't feel *déjà vu* every time we see an image again; that's where memory comes in—a topic we'll discuss in Chapter 3 of *MCAT Behavioral Sciences Review*. Indeed, this phenomenon of *déjà vu* comes from the brain's sensory receptors saying, *Yes, you have seen this before!* in tandem with the memory system saying, *But I don't know when or where!*

In this chapter, we will focus on the concept of sensation and its associated receptors, including the eyes and hair cells in the ear, as well as perception and the complex brain functions associated with processing sensory information. We'll briefly touch on the other sensory modalities, including vestibular sense, taste, smell, somatosensation, and kinesthetic sense, and consider the roles these senses play in helping us interact with the world.

2.1 Sensation *vs.* Perception

LEARNING OBJECTIVES

After Chapter 2.1, you will be able to:

- Explain the pathway for a stimulus to reach conscious perception
- Connect the common sensory receptors to their functions
- Describe absolute threshold, threshold of conscious perception, and difference threshold
- Explain Weber's law and signal detection theory
- Describe how sensory adaptation affects a difference threshold

In common parlance, we often use the terms "sensation" and "perception" interchangeably, as synonyms. However, in the field of psychology, these two terms have very specific definitions and are commonly contrasted. **Sensation** more appropriately aligns with **transduction**, which means taking the physical, electromagnetic, auditory, and other information from our internal and external environment and converting this information into electrical signals in the nervous system. Sensation is performed by receptors in the peripheral nervous system, which forward the stimuli to the central nervous system in the form of action potentials and neurotransmitters. Sensation can therefore be thought of as a raw signal, which is unfiltered and unprocessed until it enters the central nervous system.

Perception, on the other hand, refers to processing this information within the central nervous system in order to make sense of the information's significance. The complex manipulations involved in perception include both the external sensory experience and the internal activities of the brain and spinal cord. Perception thus helps us make sense of the world. The difference between sensation and perception is key to the challenge of creating artificial intelligence: we can easily create sensors for robots to pick up information from their environment, but teaching them how to comprehend and respond to that information is far more challenging.

Sensory processing is a common topic on the MCAT; you should not only understand the definitions of these terms, but also be able to apply the concepts herein to your own day-to-day sensory experiences.

Sensory Receptors

Sensory receptors are neurons that respond to stimuli by triggering electrical signals that carry information to the central nervous system. Physical objects outside of the body are referred to as **distal stimuli**. These objects often produce photons, sound waves, heat, pressure, or other stimuli that directly interact with sensory receptors; these sensory-stimulating byproducts are called **proximal stimuli**. For example, a campfire is a distal stimulus. The photons that are emitted by the fire, the sounds of crackling and popping, and the energetic gas particles that transfer heat energy are all proximal stimuli. So, proximal stimuli directly interact with and affect the sensory receptors, and thereby inform the observer about the presence of distal stimuli. Sensory receptors may encode multiple aspects of a stimulus. For example, photoreceptors respond to light and can encode not only the brightness of the light, but also its color and shape. The relationship between the physical nature of stimuli and the sensations and perceptions these stimuli evoke is studied in the field of **psychophysics**.

MNEMONIC

Distal = in the **dista**nce

Proximal = in close **proxim**ity

In order to inform the central nervous system, the signals from these stimuli must pass through specific sensory pathways. In each case, different types of receptors—generally nerve endings or specific sensory cells—receive the stimulus, transduce the stimulus into electrical signals, and transmit the data to the central nervous system through sensory ganglia. **Ganglia** are collections of neuron cell bodies found outside the central nervous system. Once transduction from these sensory ganglia occurs, the electrochemical energy is sent along neural pathways to various **projection areas** in the brain, which further analyze the sensory input.

Sensory receptors differ from one sense to another. There are over a dozen recognized sensory receptors, but the MCAT is unlikely to test even half of those. The most heavily tested receptors include:

- Photoreceptors: respond to electromagnetic waves in the visible spectrum (sight)
- Mechanoreceptors: respond to pressure or movement. Hair cells, for example, respond to movement of fluid in the inner ear structures (movement, vibration, hearing, rotational and linear acceleration)
- Nociceptors: respond to painful or noxious stimuli (somatosensation)
- Thermoreceptors: respond to changes in temperature (thermosensation)
- Osmoreceptors: respond to the osmolarity of the blood (water homeostasis)
- Olfactory receptors: respond to volatile compounds (smell)
- Taste receptors: respond to dissolved compounds (taste)

Thresholds

Perception, like sensation, is closely tied to the biology and physiology of interpreting the world around us. However, unlike sensation, perception is inextricably linked to experience as well as to internal and external biases. Sensations are relayed to the brain, which perceives the significance of the stimulus. To illustrate the significance of perception, keep in mind that all sensory information is sent to the central nervous system in the form of action potentials, which the central nervous system must then interpret and act upon. For example, the central nervous system must determine whether incoming action potentials from thermoreceptors are indicating whether an object is hot or is cold, and whether that temperature difference is enough to cause us harm. Moreover, the same sensation can produce radically different perceptions in different people, and because these variations must be explained by central nervous system activity, perception is considered a part of psychology.

A good example of the psychological element of perception is a **threshold**—the minimum amount of a stimulus that renders a difference in perception. For example, the temperature may noticeably change from warm to cool when the sun sets, but subtle fluctuations in temperature throughout the day are generally unnoticeable because they are below the difference threshold. If sound volume increases 10 dB (ten times the sound intensity), the change is usually very obvious; but, if volume increases only 0.1 dB, the change might be too small to detect. There are three main types of thresholds: the absolute threshold, the threshold of conscious perception, and the difference threshold.

MCAT EXPERTISE

On the MCAT, thresholds will frequently be used in conjunction with subjects in studies. Be on the lookout for experimental design questions when thresholds appear in a passage.

Absolute Threshold

The **absolute threshold** is the minimum of stimulus energy that is needed to activate a sensory system. This threshold is therefore a threshold in sensation, not in perception. While most human sensory systems are extremely sensitive, all systems also have an absolute threshold below which the stimulus will not be transduced into action potentials, and the information will therefore never be sent to the central nervous system. For example, sounds of extremely low intensity may still cause slight vibrations in the sensory receptors of the inner ear, but these vibrations might not be significant enough to open ion channels linked to these sensory receptors. The absolute threshold for sweet taste is a teaspoon of sucrose dissolved in two gallons of water. On a clear, dark night with no other lights shining, the eye can just detect the light of one candle burning thirty miles away. When we are talking about an absolute threshold, we're talking about how bright, loud, or intense a stimulus must be before it is sensed.

KEY CONCEPT

The absolute threshold is the minimum intensity at which a stimulus will be transduced (converted into action potentials).

BRIDGE

You may already know one of the absolute thresholds from the discussion of sound in Chapter 7 of *MCAT Physics and Math Review*. Remember that $I_0\ (10^{-12}\frac{\text{W}}{\text{m}^2})$ in the equation for sound level is the absolute threshold of normal human hearing.

Threshold of Conscious Perception

It is possible for sensory systems to send signals to the central nervous system without a person perceiving these signals. This lack of conscious perception may be because the stimulus is too subtle to demand our attention, or may last for too brief a duration for the brain to fully process the information. The level of intensity that a stimulus must pass in order to be consciously perceived by the brain is the **threshold of conscious perception**. By way of contrast, information that is received by the central nervous system but that does not cross this threshold is called **subliminal perception**. Note the difference between the absolute threshold and the threshold for conscious perception: a stimulus below the absolute threshold will not be transduced, and thus never reaches the central nervous system. A stimulus below the threshold of conscious perception arrives at the central nervous system, but does not reach the higher-order

REAL WORLD

The Latin word for "thresholds" is *limina*. Hence, something that is "subliminal" is literally "below threshold." The threshold referred to in the term "subliminal perception" is the threshold of conscious perception. So, signals that are "subliminal" are strong enough to pass the absolute threshold, but not strong enough to pass the threshold of conscious perception.

brain regions that control attention and consciousness. Contrary to common thinking, there is actually little practical value to using subliminal perception to sell products.

Difference Threshold

A third commonly studied threshold is the **difference threshold**, sometimes called the **just-noticeable difference (jnd)** between two stimuli. The difference threshold refers to the minimum change in magnitude required for an observer to perceive that two different stimuli are, in fact, different. If the difference between stimuli is below the difference threshold, the two stimuli will seem to the observer to be the same. For example, imagine two sound waves are played one after the other, the first having frequency 440 Hz and then the second having frequency 441 Hz. These sounds are different. But without formal ear training, most individuals cannot hear the difference. In this range of sound frequencies, the just-noticeable difference for most listeners is about 3 Hz. So, for the average person to hear a difference in pitch, the sound waves need to be 440 Hz and 443 Hz. Below this difference threshold, the two pitches will sound the same.

The previous example illustrates one common experimental technique researchers use to explore the difference threshold. The technique is called psychophysical discrimination testing, or sometimes just **discrimination testing**. In a common discrimination testing experiment, a participant is presented with a stimulus. The stimulus is then varied slightly and researchers ask the participant to report whether they perceive a change. Often, the difference continues to be increased until the participant reports they notice the change, and this interval is recorded as the just noticeable difference.

Returning to the example of two sounds: The difference between a 440 Hz sound and a 443 Hz sound is just noticeable for most people. But, by using discrimination testing, researchers have discovered that the absolute difference (3 Hz, in this case) is far less important than the percent difference. For this reason, the just noticeable difference is usually reported as a fraction or a percent. To compute this percent, divide the change in stimulus by the magnitude of the original stimulus. In our example, we would compute 3 Hz / 440 Hz = 0.0068 = 0.68%. To illustrate why percentages are used, consider a 1000 Hz sound. An increase of 0.68% results in a sound of frequency 1007 Hz. So, to the average person, the difference in pitch from 1000 Hz to 1007 Hz would be just noticeable. By contrast, the difference from 1000 Hz to 1003 Hz would not be noticeable. While a 3 Hz difference was noticeable in the lower frequency range, that same 3 Hz difference is not noticeable in the higher frequency range.

Ernst Heinrich Weber (1795–1878) is often credited with the observation that difference thresholds are proportional and must be computed as percentages. This idea is therefore often called **Weber's law**. Weber's law applies to the perception of a number of senses, including the perception of loudness and pitch of sounds, the perception of brightness of light, and the perception of weight of objects.

MCAT EXPERTISE

When the MCAT brings up Weber's law, questions will usually give a numerical relationship and then ask for it to be applied; typically, the solution simply amounts to applying a ratio.

Signal Detection Theory

Perception of stimuli can also be affected by nonsensory factors, such as experiences (memory), motives, and expectations. **Signal detection theory** studies how internal (psychological) and external (environmental) factors influence thresholds of sensation and perception. For example, how loud would someone need to yell your name in a crowd to get your attention? The answer depends on environmental factors, like the size of the crowd; social factors, like the makeup of the crowd and your comfort with the individuals around you; psychological factors, like whether or not you are expecting to have your name called; and personality factors, like your level of introversion or extroversion. In signal detection theory, these factors are treated like independent variables. For example, researchers can measure how likely a person is to hear their name called when the person is informed that at some point their name will be called, versus when the person is left uninformed.

A basic signal detection experiment consists of many trials; during each trial, a stimulus (signal) may or may not be presented. Trials in which the signal is presented are called **noise trials**, whereas those in which the signal is not presented are called **catch trials**. After each trial, the subject is asked to indicate whether or not a signal was presented. There are therefore four possible outcomes for each trial, as illustrated in Figure 2.1. A **hit** is a trial in which the signal is presented and the subject correctly perceives the signal; a **miss** is a trial in which the subject fails to perceive the presented signal. A **false alarm** is a trial in which the subject indicates perceiving the signal, even though the signal was not presented; a **correct negative** is a trial in which the subject correctly identifies that no signal was presented. By tracking the rates of these various outcomes, researchers are able to identify factors that influence perception.

REAL WORLD

On the surface, signal detection experiments would appear to be easy tasks—shouldn't individuals easily be able to tell if they perceived something or not? However, consider the thought processes that occur when you're quietly studying in the library with your phone on silent and you suddenly think you may have heard a buzz. *Is my phone ringing?* you wonder. You freeze in place and wait for another buzz; even if it doesn't come, you may still be so convinced you heard a signal that you still check your phone. Perception is not a passive matter!

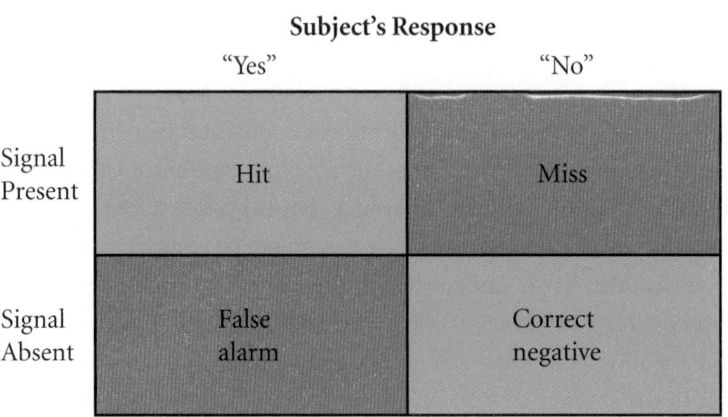

Figure 2.1 Possible Outcomes from a Signal Detection Experiment Trial

Adaptation

Our ability to detect a stimulus can change over time through **adaptation**. Adaptation can have both a physiological (sensory) component and a psychological (perceptual) component. For example, the pupils of the eyes will dilate in the dark and constrict in the light, which illustrates physiological adaptation. Similarly, in loud environments, small muscles in the middle ear will reflexively contract in order to dampen the vibration of the ossicles, reducing sound intensity. We also adapt to somatosensory stimuli; cold water no longer seems so cold once our bodies "get used to it." Once we're dressed, we stop feeling the clothes on our bodies until we have a reason to think about them. Adaptation is one way the mind and body try to focus attention on only the most relevant stimuli, which are usually changes in the environment around us.

MCAT CONCEPT CHECK 2.1

Before you move on, assess your understanding of the material with these questions.

1. What is the pathway for a stimulus to reach conscious perception?

____

2. Match each sensory receptor to its function:

1. Hair cell	A. Sense painful or bothersome physical stimuli
2. Nociceptor	B. Sense changes in temperature
3. Olfactory receptor	C. Sense electromagnetic radiation in the visible range
4. Osmoreceptor	D. Sense changes in blood concentration
5. Photoreceptor	E. Sense volatile chemicals
6. Taste receptor	F. Sense motion of fluid in the inner ear
7. Thermoreceptor	G. Sense dissolved chemicals

3. For each of the thresholds below, provide a brief description:
   - Absolute threshold:

   ____

   - Threshold of conscious perception:

   ____

   - Difference threshold:

   ____

4. What aspect of thresholds do Weber's law and signal detection theory focus on?
   - Weber's law:

   ____

   - Signal detection theory:

   ____

5. How does sensory adaptation affect a difference threshold?

____

2.2 Vision

High-Yield

LEARNING OBJECTIVES

After Chapter 2.2, you will be able to:

- List the functions of the parts of the eye, including the cornea, pupil, iris, ciliary body, canal of Schlemm, lens, retina, and sclera
- Describe parallel processing
- Identify the cell types responsible for color, shape, and motion detection
- Recall the structures in the visual pathway:

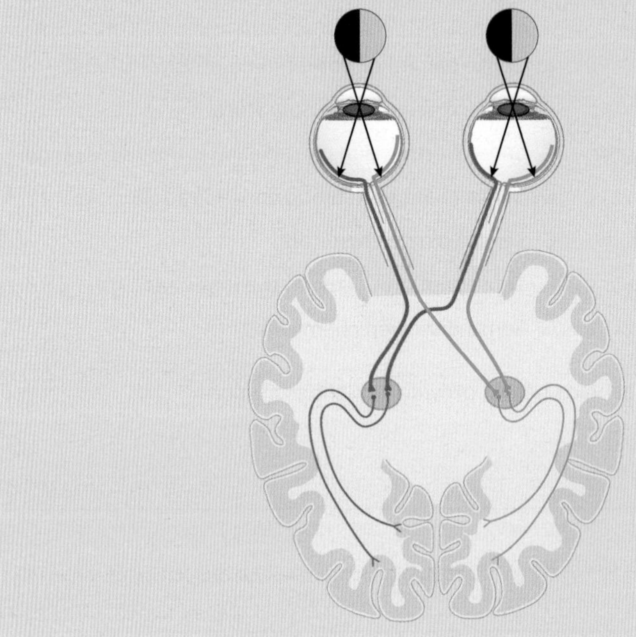

Vision is a highly adapted sense in human beings. With the ability to sense brightness, color, shape, and movement, and then to integrate this information to create a cohesive three-dimensional model of the world, the visual pathways are extremely important to everyday life. In fact, vision is the only sense to which an entire lobe of the brain is devoted: the occipital lobe. While some individuals experience vision differently, such as those who are color deficient or partially sighted, the MCAT focuses primarily on the details of unimpaired vision.

Structure and Function of the Eye

The anatomy of the eye is shown in Figure 2.2.

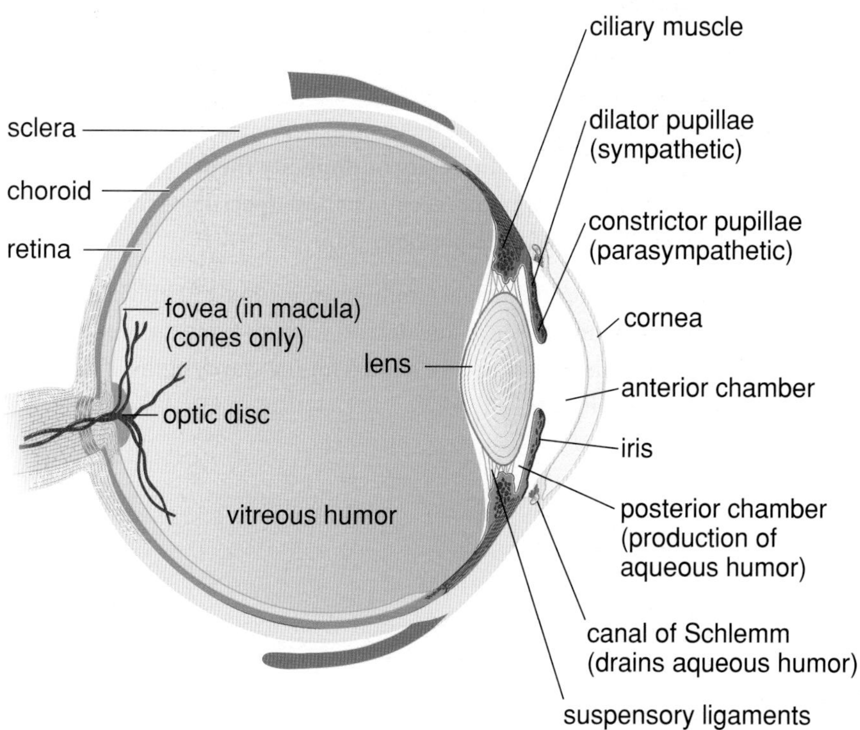

Figure 2.2 Anatomy of the Eye

The eye is a specialized organ used to detect light in the form of photons. Most of the exposed portion of the eye is covered by a thick structural layer known as the **sclera**, or the white of the eye. The sclera does not cover the frontmost portion of the eye, the cornea. The eye is supplied with nutrients by two sets of blood vessels: the **choroidal vessels**, a complex intermingling of blood vessels between the sclera and the retina, and the **retinal vessels**. The innermost layer of the eye is the **retina**, which contains the actual photoreceptors that transduce light into electrical information the brain can process.

When entering the eye, light passes first through the **cornea**, a clear, domelike window in the front of the eye, which gathers and focuses the incoming light. The front of the eye is divided into the **anterior chamber**, which lies in front of the iris, and the **posterior chamber** between the iris and the lens. The **iris**, which is the colored part of the eye, is composed of two muscles: the **dilator pupillae**, which opens the pupil under sympathetic stimulation; and the **constrictor pupillae**, which constricts the pupil under parasympathetic stimulation. The iris is continuous with the **choroid**, which is a vascular layer of connective tissue that surrounds and provides nourishment to the retina. The iris is also continuous with the the **ciliary body**, which produces the **aqueous humor** that bathes the front part of the eye before draining into the **canal of Schlemm**. The **lens** lies right behind the iris and helps control the refraction of the incoming light. Contraction of the **ciliary muscle**, a component of the ciliary body, is under parasympathetic control. As the muscle contracts, it pulls on the **suspensory ligaments** and changes the shape of the lens to focus on an image as the distance varies, a phenomenon known as **accommodation**. Behind the lens lies the **vitreous humor**, a transparent gel that supports the retina.

The **retina** is in the back of the eye and is like a screen consisting of neural elements and blood vessels. Its function is to convert incoming photons of light to electrical signals. It is actually considered part of the central nervous system and develops as an outgrowth of brain tissue. The **duplexity** or **duplicity theory of vision** states that the retina contains two kinds of photoreceptors: those specialized for light-and-dark detection and those specialized for color detection.

The retina is made up of approximately 6 million cones and 120 million rods. **Cones** are used for color vision and to sense fine details. Cones are most effective in bright light and come in three forms, which are named for the wavelengths of light they best absorb, as shown in Figure 2.3.

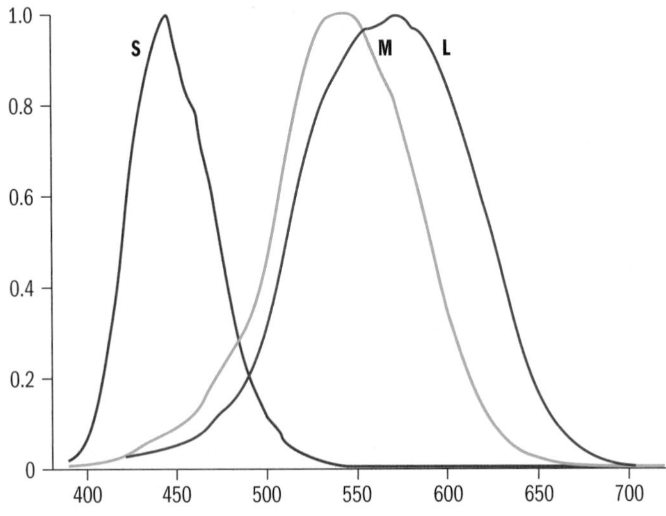

Figure 2.3 Relative Absorption of the Three Types of Cones at Different Wavelengths

The cones are named for the wavelengths at which they have highest light absorption: short (S, also called blue), medium (M, green), and long (L, red).

MNEMONIC

Cones are for **c**olor vision. **Rods** function best in "**rod**uced" light.

In reduced illumination, **rods** are more functional than cones because each rod cell is highly sensitive to photons and is somewhat easier to stimulate than a cone cell. In part, the sensitivity of rods has to do with the fact that all rods contain only a single pigment type called **rhodopsin**. In general, color vision requires far more light because each cone responds only to certain wavelengths of light. By contrast, a rod can be stimulated by light of any color. However, while rods permit vision in reduced light, the tradeoff is that rods only allow sensation of light and dark. Also, even though individual rods are highly sensitive to light, as a whole they are less useful for detecting fine details because rods are spread over a much larger area of the retina.

While there are many more rods than cones in the human eye, the central section of the retina, called the **macula**, has a high concentration of cones; in fact, the center-most region of the macula, called the **fovea**, contains only cones. As one moves further away from the fovea, the concentration of rods increases while the concentration of cones decreases. Therefore, visual acuity is best at the fovea, and the fovea is most sensitive in normal daylight vision. Some distance away from the center of the retina, the optic nerve leaves the eye. This region of the retina, which is devoid of photoreceptors, is called the optic disk, and gives rise to a **blind spot**, as shown in Figure 2.4.

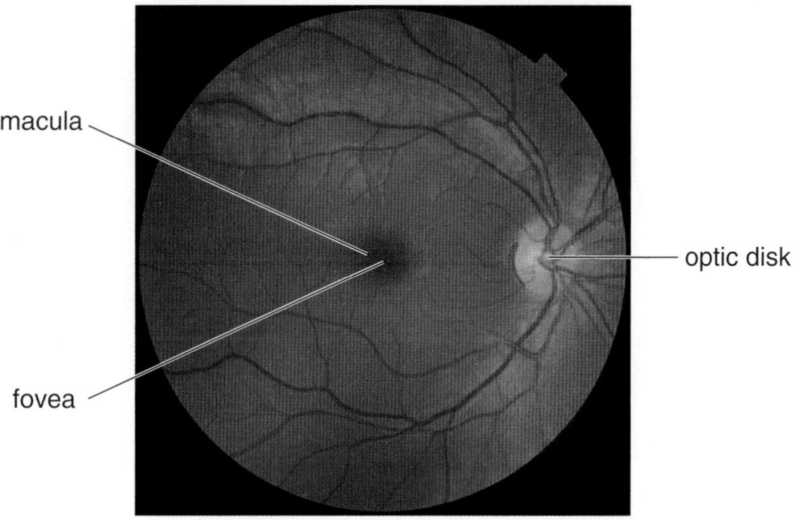

Figure 2.4 Specialized Regions of the Retina

Rods and cones are specialized neurons and, like most neurons, connect with other neurons through synapses. However, rods and cones do not connect directly to the optic nerve. Rather, there are several layers of neurons in between, as shown in Figure 2.5. Rods and cones synapse directly with **bipolar cells**, which highlight gradients between adjacent rods or cones. Bipolar cells then synapse with **ganglion cells**, the axons of which group together to form the **optic nerve**. These bipolar and ganglion cells not only fall "in between" the rods and cones and the optic nerve, but also the bipolar and ganglion cells are actually located in front of the rods and cones, closer to the front of the eye. This arrangement means that a photon must actually navigate past several layers of cells to reach the rods and cones at the "back" of the retina; the information is then transmitted "forward" (in the form of action potentials) from the rod and cone cells until the signal reaches the ganglion cells. Observe in Figure 2.5 that there are significantly more photoreceptor cells than ganglion cells, so the output from each ganglion cell represents the combined activity of many rods and cones. The result is a pruning of details as information from the photoreceptors is combined. As the number of receptors that converge through the bipolar neurons onto one ganglion cell increases, the resolution decreases. On average, the number of cones converging onto an individual ganglion cell is smaller than for rods. This arrangement helps explain why color vision has a greater sensitivity to fine detail than black-and-white vision does.

Also shown in Figure 2.5 are **amacrine** and **horizontal cells**, which receive input from multiple retinal cells in the same area before the information is passed on to ganglion cells. Amacrine and horizontal cells can thereby accentuate slight differences between the visual information in each bipolar cell. For example, these cells are important for edge detection, as they increase our perception of contrasts.

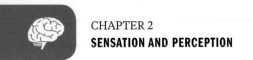

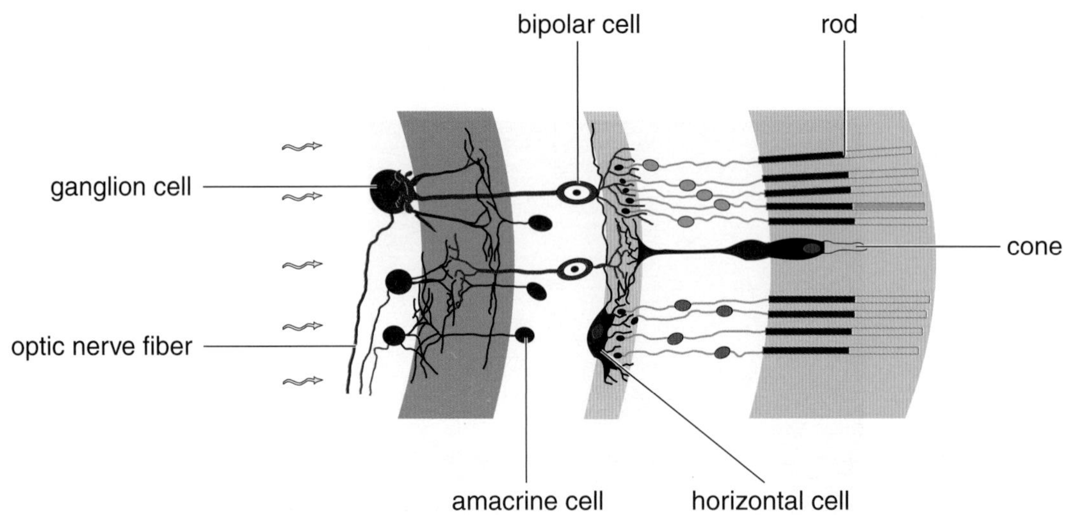

Figure 2.5 Cells of the Retina

Visual Pathways

Visual pathways refer to both the anatomical connections between the eyes and brain and to the flow of visual information along these connections. These pathways are illustrated in Figure 2.6. The terminology for these pathways can seem confusing. So here is a simple rule of thumb to start: If an object is to your left, then photons from that object stimulate the right side of the retina in each eye. To help visualize this result, look at Figure 2.6. In this figure, an object to your left is represented by the black patches. Photons from that object enter each eye and stimulate the retinal fibers that are illustrated using the color orange. This result is due to simple geometry! If an object is to your left, then photons from that object must travel to the right to reach your eyes, and in so doing, these photons enter your eyes and continue travelling to the right, thereby stimulating the right side of each of your eyes' retinas. From there, retinal fibers from each eye project to the right side of your brain. All of this leads to a famous result: Visual information from objects to your left is processed by the right side of your brain. Similarly, if an object is to your right (the blue patches in the figure), then photons from this object stimulate the left side of each retina (the green nerves), and this visual data is sent to the left side of your brain.

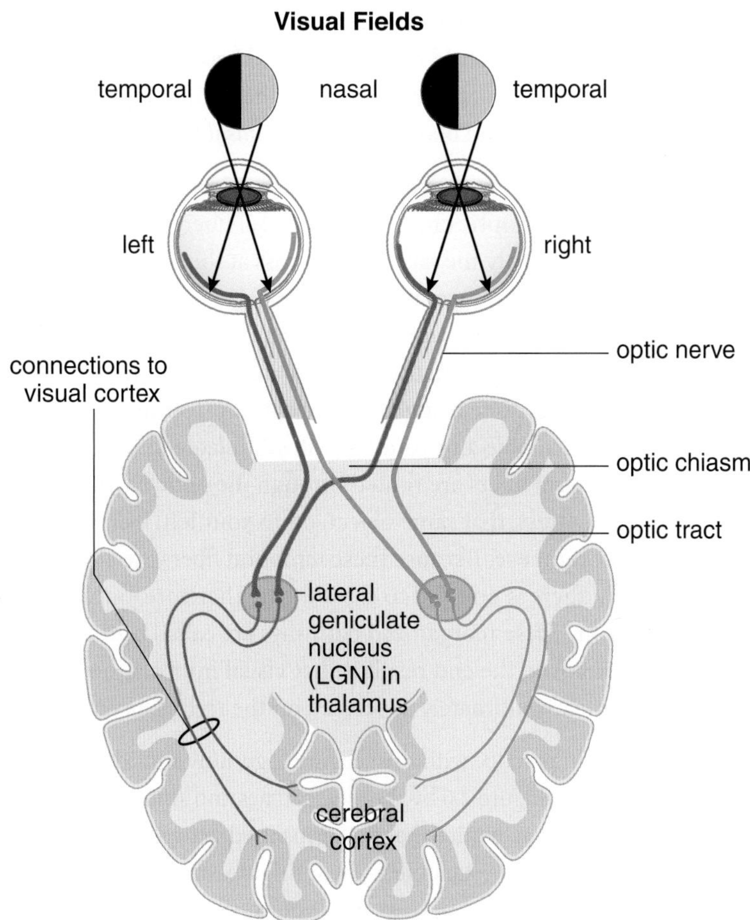

Figure 2.6 Visual Pathways

To help describe these phenomena with more specificity, scientists use the terms **temporal** and **nasal retinal fibers**. Here's an example. Consider an object to your left, which, again, is represented in Figure 2.6 by the black patches. In each eye, photons from this object strike the retinal fibers that are illustrated in Figure 2.6 using the color orange. But notice: In the right eye, those orange fibers are on the lateral side of the retina, closer to your temple, and so these fibers are called the temporal fibers of the right eye. In the left eye, the orange fibers are medial, closer to your nose, and so are called the nasal fibers of the left eye.

Scientists also describe the placement of the object in space, relative to your eyes, using the terms "temporal" and "nasal," but now referring to the **temporal** and **nasal visual fields**. Again, this terminology is best explained by example. Imagine an object is a little to your left. Then photons from that object must cross in front of your nose in order to reach your right eye. So, this object is described as being in the nasal field of your right eye. Think carefully about what happens to these photons next. They enter your right eye, continue heading to the right, and ultimately strike the temporal fibers on the right side of your right eye. Pay close attention to the terminology: We're saying that an object in the *nasal field* of the right eye stimulates the *temporal fibers* of that eye! This mixing of terminology—nasal field stimulates temporal fibers—illustrates the crossing over of visual information. By contrast, for that same object, which, remember, is

MNEMONIC

If photons from an object would need to cross your nose to reach your eye (or if the object is right in front of your nose) then the object is said to be in the nasal visual field for that eye.

spatially to your left, photons can enter your left eye directly. Therefore, this object is described as being in the temporal field of your left eye. Then, these photons strike the nasal fibers of the left eye. So again, there is mixing of the terminology—the object is in the *temporal field* of the left eye, and photons strike the *nasal fibers* of the left eye.

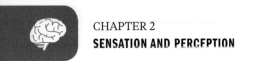

As the retinal fibers from each eye travel through the optic nerves toward the brain, a significant event occurs at the **optic chiasm**. Here, nasal fibers from the left and right eyes cross paths. Notice that only the nasal fibers cross at the optic chiasm. Again, an example helps illustrate: Visual information from an object to your left is ultimately processed by the right side of your brain. So if photons from that object are captured by the left eye, then this visual information needs to be routed to the opposite side of your brain. That routing is done using the optic chiasm. To use the "nasal" and "temporal" terminology: An object to your left will stimulate the nasal fibers of your left eye, and then these nasal fibers are routed through the optic chiasm to the right side of your brain. By contrast, that same object, off to your left, will stimulate the temporal fibers of your right eye. Because these temporal fibers are already on the right side of your body, there is no need to cross them. Therefore, only nasal fibers cross at the optic chiasm. These reorganized pathways are called **optic tracts** after leaving the optic chiasm. And the end result is that visual information emanating from an object to your left is ultimately processed on the right side of your brain.

MNEMONIC

For each eye, the temporal visual field stimulates the nasal retinal fibers. The nasal visual field stimulates the temporal retinal fibers.

To summarize, and to gain practice with this terminology, now consider an object off to your right, represented in Figure 2.6 by the blue patches and stimulating the fibers illustrated using the color green. Notice that the terms used depend on which eye is considered. Let's start with the left eye. An object to your right would be in the nasal field of your left eye, because photons would have to cross your nose to reach your left eye. These photons would stimulate the temporal fibers of the left eye—nasal field stimulates temporal fibers. Then, because these temporal fibers are already on the left side of your body, they would be routed directly back to the left side of your brain. Now consider the right eye. An object spatially to your right is in the temporal field of your right eye, and stimulates the nasal fibers of that eye. Then, in order to bring this visual information to the left side of your brain, these nasal fibers are routed through the optic chiasm. Through all of this, keep these two trends in mind: First, the temporal field of each eye stimulates the nasal fibers of each eye, and vice versa. Second, the nasal fibers cross at the optic chiasm.

REAL WORLD

When there is a sudden, bright flash of light, the superior colliculus aligns the eyes with the likely stimulus. In other words, it's the superior colliculus (as well as the sympathetic nervous system) that gives us the "deer in the headlights" appearance during the startle response.

From the optic chiasm, the information goes to several different places in the brain: some nerve fibers pass to the **lateral geniculate nucleus (LGN)** of the **thalamus** where they synapse with nerves that then pass through radiations in the temporal and parietal lobes to the **visual cortex** in the **occipital lobe**. This connectivity makes sense because the thalamus is a well-known connecting and routing center of the forebrain. Other nerve fibers branch off from the optic tracts, skip the thalamus, and head directly to the **superior colliculi** in the midbrain, which control some reflexive responses to visual stimuli and reflexive eye movements.

Processing

The ability to sense light information in the environment around us is useful in its own right. But, to effectively interact with the environment, we must also be able to make sense of visual stimuli. The connections between optic tract, LGN, and visual cortex help create a cohesive image of the world through a phenomenon known as parallel processing. Visual **parallel processing** is the brain's ability to analyze information regarding color, form, motion, and depth simultaneously, i.e. "in parallel," using independent pathways in the brain. For example, most people can quickly and easily recognize a moving car from a distance. The speed of recognition is facilitated, in part, by the fact that the form of the car (i.e. its shape) and the motion of the car are processed simultaneously in separate, parallel pathways in the brain.

Now let's explore where each of these four aspects of vision is processed and the specialized cells that contribute to their detection. As described previously, cones are responsible for our perception of color. **Form** refers not only to the shape of an object, but also our ability to discriminate an object of interest from the background by detecting its boundaries. Neurons carrying information from the fovea and surrounding central portion of the retina synapse with **parvocellular cells** in the lateral geniculate nucleus. These cells have very high color **spatial resolution**; that is, these cells permit us to detect very fine detail when thoroughly examining an object. However, parvocellular cells can only work with stationary or slow-moving objects because these cells have very low **temporal resolution**.

Conversely, **magnocellular cells** are well-suited for detecting motion because these cells have very high temporal resolution. Reflecting the fact that form and motion are processed in parallel, magnocellular cells and parvocellular cells are located in distinct layers of the lateral geniculate nucleus. Also, magnocellular cells predominantly receive inputs from the periphery of our vision, allowing more rapid detection of objects approaching us from the sides. However, magnocellular cells have low spatial resolution, so much of the rich detail of an object can no longer be seen once the object is motion. Magnocellular cells therefore provide a blurry but moving image of an object.

MNEMONIC

Magnocellular cells specialize in **m**otion detection.

Depth perception, our ability to discriminate the three-dimensional shape of our environment and judge the distance of objects within it, is largely based on discrepancies between the inputs the brain receives from our two eyes (more on this to follow in *MCAT Behavioral Sciences Review*, Section 2.5, Object Recognition). Specialized cells in the visual cortex known as **binocular neurons** are responsible for comparing the inputs to each hemisphere and detecting these differences.

Finally, our brains wouldn't be very good at processing visual information if they didn't learn to associate certain patterns of stimuli with expected behaviors or outcomes. To assist in this, a whole slew of even more specialized cells called **feature detectors** exist in the visual cortex. Each feature detector cell type detects a very particular, individual feature of an object in the visual field. For example, if we were to look at a stop sign we would activate: a feature detector for the color red, while another feature detector would respond to the white border and letters. Yet another type of feature detector would recognize the horizontal lines, while still others would be activated by the angled lines of the octagon. Rather than needing to individually process each of these features every time, the overall combination of feature detectors become activated in parallel. Finally, our response to the stop sign, i.e. to STOP, also is stored for future retrieval.

MCAT CONCEPT CHECK 2.2

Before you move on, assess your understanding of the material with these questions.

1. List the functions of the various parts of the eye:
   - Cornea:
   - Pupil:
   - Iris:
   - Ciliary body:
   - Canal of Schlemm:
   - Lens:
   - Retina:
   - Sclera:
2. List the structures in the visual pathway, from where light enters the cornea to the visual projection areas in the brain.
3. What is parallel processing?

4. In feature detection, what type of cells are responsible for color? Form? Motion? Depth?
    - Color:

    - Form:

    - Motion:

    - Depth:

2.3 Hearing and Vestibular Sense

LEARNING OBJECTIVES

After Chapter 2.3, you will be able to:

- Identify the structures used to detect linear acceleration and rotational acceleration
- Explain how the structural features of the cochlea and the hair cells are able to transmit information about pitch of an incoming sound to the brain
- List the structures in the auditory pathway

The ear is a complex organ responsible not only for our sense of hearing, but also for our **vestibular sense**, which is our ability to both detect rotational and linear acceleration and to use this information to inform our sense of balance and spatial orientation. These senses are critically important to our ability to get around in the world, and their associated structures are encased in some of the densest bone of the body to protect these structures from damage. While recognizing that individuals can differ in their hearing and vestibular senses, as with vision, the MCAT primarily focuses on the standard healthy operations of these senses.

Structure and Function of the Ear

The ear is divided into three parts, as shown in Figure 2.7: the outer, middle, and inner ear. A sound wave first reaches the cartilaginous outside part of the ear, called the **pinna** or **auricle**. The main function of the pinna is to channel sound waves into the **external auditory canal**, which directs the sound waves to the **tympanic membrane (eardrum)**. The membrane vibrates in phase with the incoming sound waves. The frequency of the sound wave determines the rate at which the tympanic membrane vibrates: it moves back and forth at a high rate for high-frequency sounds and more slowly for low-frequency sounds. Louder sounds have greater **intensity**, which corresponds to an increased amplitude of vibration.

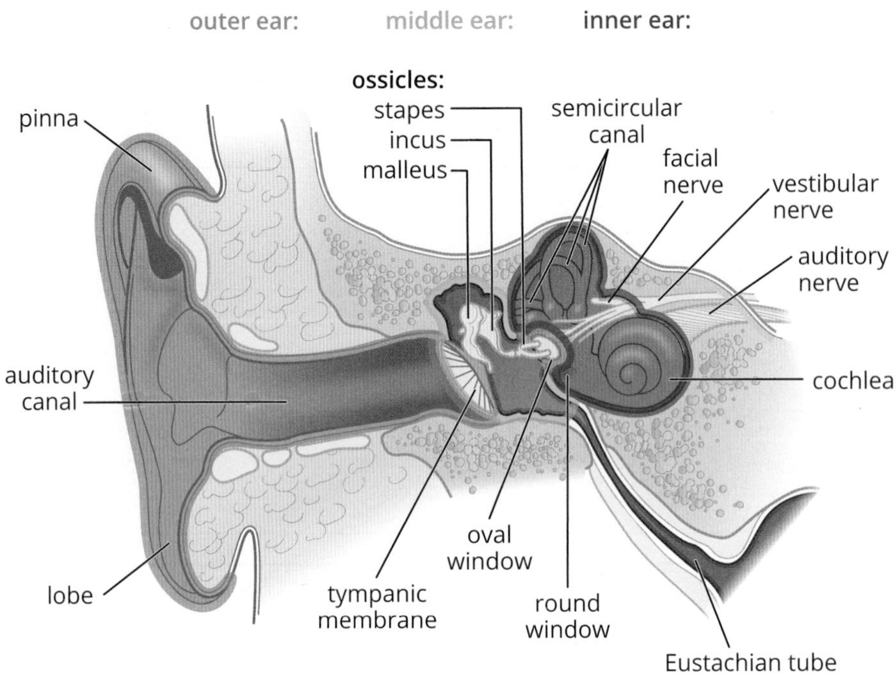

Figure 2.7 Anatomy of the Ear

BRIDGE

Remember that sound is a longitudinal wave carried through air (or another medium), which causes displacement of particles parallel to the axis of sound propagation. In other words, when a sound wave hits your eardrum, it literally causes it to oscillate back and forth because of moving air particles. Sound is discussed in Chapter 7 of *MCAT Physics and Math Review*.

The tympanic membrane divides the outer ear from the middle ear. The middle ear houses the three smallest bones in the body, called **ossicles**. The ossicles help transmit and amplify the vibrations from the tympanic membrane to the inner ear. The **malleus** (**hammer**) is affixed to the tympanic membrane; it acts on the **incus** (**anvil**), which acts on the **stapes** (**stirrup**). The baseplate of the stapes rests on the oval window of the cochlea, which is the entrance to the inner ear. The middle ear is connected to the nasal cavity via the **Eustachian tube**, which helps equalize pressure between the middle ear and the environment.

The inner ear sits within a **bony labyrinth**, which is a hollow region of the temporal bone containing the cochlea, vestibule, and semicircular canals, as shown in Figure 2.8. Inside the bony labyrinth rests a continuous collection of tubes and chambers called the **membranous labyrinth**. This collection of structures contains receptors for the sense of equilibrium and hearing. The membranous labyrinth is filled by a potassium-rich fluid called **endolymph**, and is suspended within the bony labyrinth by a thin layer of another fluid called perilymph. **Perilymph** simultaneously transmits vibrations from the outside world and cushions the inner ear structures.

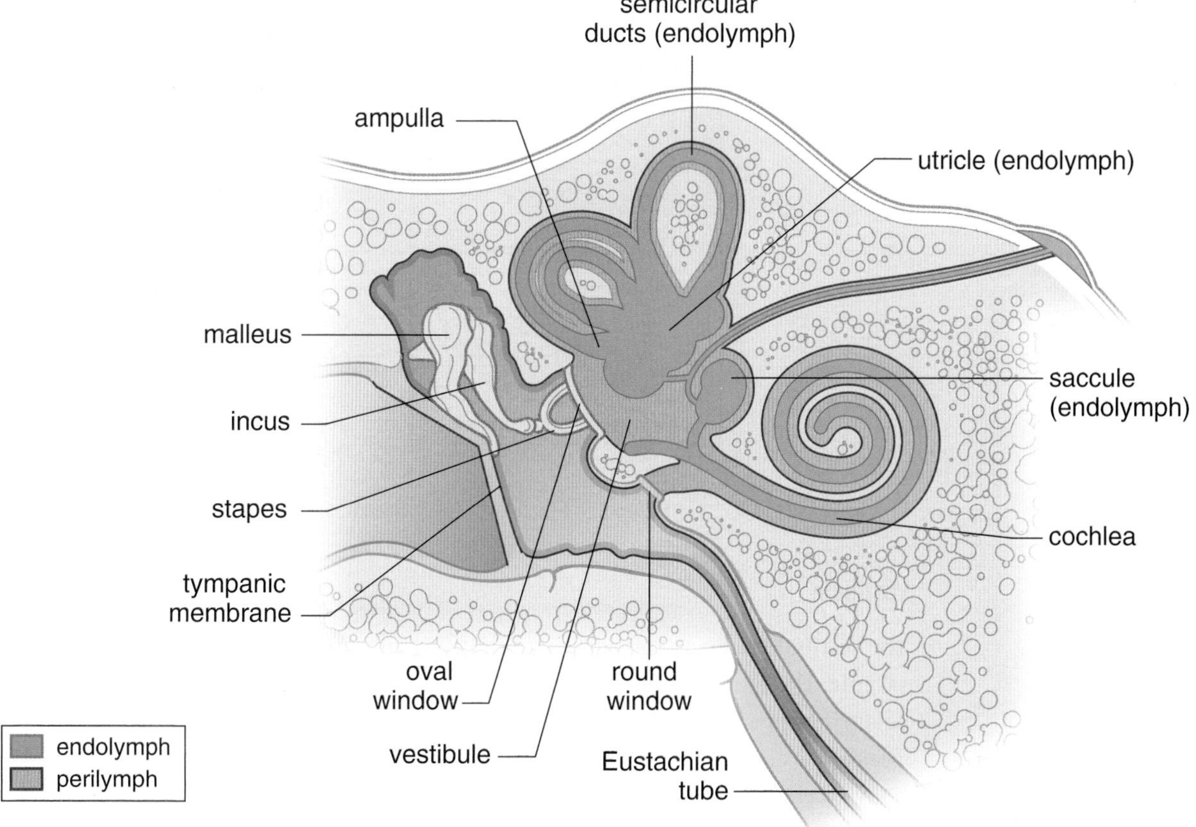

Figure 2.8 The Membranous and Bony Labyrinth
The membranous labyrinth is filled with endolymph (blue); it is suspended within the bony labyrinth, which is filled with perilymph (purple).

Cochlea

The **cochlea** is a spiral-shaped organ that contains the receptors for hearing; it is divided into three parts called **scalae**, as shown in Figure 2.9. All three scalae run the entire length of the cochlea. The middle scala houses the actual hearing apparatus, called the **organ of Corti**, which rests on a thin, flexible membrane called the **basilar membrane**. The organ of Corti is composed of thousands of hair cells, which are bathed in endolymph. On top of the organ of Corti is a relatively immobile membrane called the **tectorial membrane**. The other two scalae, filled with perilymph, surround the hearing apparatus and are continuous with the oval and round windows of the cochlea. Thus, sound entering the cochlea through the oval window causes vibrations in perilymph, which are transmitted to the basilar membrane. Because fluids are essentially incompressible, the **round window**, a membrane-covered hole in the cochlea, permits the perilymph to actually move within the cochlea. Like the rods and cones of the eye, the hair cells in the organ of Corti transduce the physical stimulus into an electrical signal, which is carried to the central nervous system by the **auditory (vestibulocochlear) nerve**.

BRIDGE

The junction between the stapes and the oval window is extremely similar to a thermodynamic gas-piston system, as described in Chapter 3 of *MCAT Physics and Math Review*. However, fluids are not as compressible as gases; therefore, the round window must be present to allow the perilymph in the cochlea to actually move back and forth with the stapedial footplate.

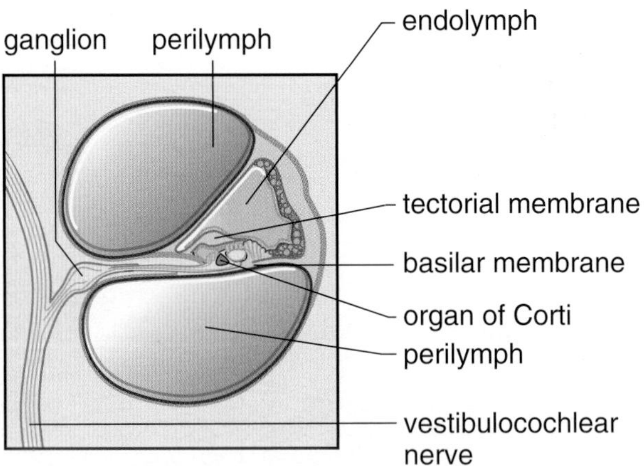

Figure 2.9 Structure of the Cochlea (Cross-Section)

Vestibule

The **vestibule** refers to the portion of the bony labyrinth that contains the **utricle** and **saccule**. These structures are sensitive to linear acceleration, so are used as part of the balancing apparatus and to determine one's orientation in three-dimensional space. The utricle and saccule contain modified hair cells covered with **otoliths**. As the body accelerates, these otoliths will resist that motion. This bends and stimulates the underlying hair cells, which send a signal to the brain.

Semicircular Canals

While the utricle and saccule are sensitive to linear acceleration, the three **semicircular canals** are sensitive to rotational acceleration. The semicircular canals are arranged perpendicularly to each other, and each ends in a swelling called an **ampulla**, where hair cells are located. When the head rotates, endolymph in the semicircular canal resists this motion, bending the underlying hair cells, which send a signal to the brain.

MNEMONIC

The **l**ateral geniculate nucleus (**L**GN) is for **l**ight; the **m**edial geniculate nucleus (**M**GN) is for **m**usic.

Auditory Pathways

The **auditory pathways** in the brain are a bit more complex than the visual pathways. Most sound information passes through the vestibulocochlear nerve to the brainstem, where it ascends to the **medial geniculate nucleus** (**MGN**) of the thalamus. From there, nerve fibers project to the **auditory cortex** in the temporal lobe for sound processing. Some information is also sent to the **superior olive**, which localizes the sound, and the **inferior colliculus**, which is involved in the startle reflex and helps keep the eyes fixed on a point while the head is turned (vestibulo–ocular reflex).

Hair Cells

Hair cells are named for the long tufts of **stereocilia** on their top surface, shown in Figure 2.10. As vibrations reach the basilar membrane underlying the organ of Corti, the stereocilia adorning the hair cells begin to sway back and forth within the endolymph. The swaying causes the opening of ion channels, which cause a receptor potential. Certain hair cells are also directly connected to the immobile tectorial membrane; these hair cells are involved in amplifying the incoming sound.

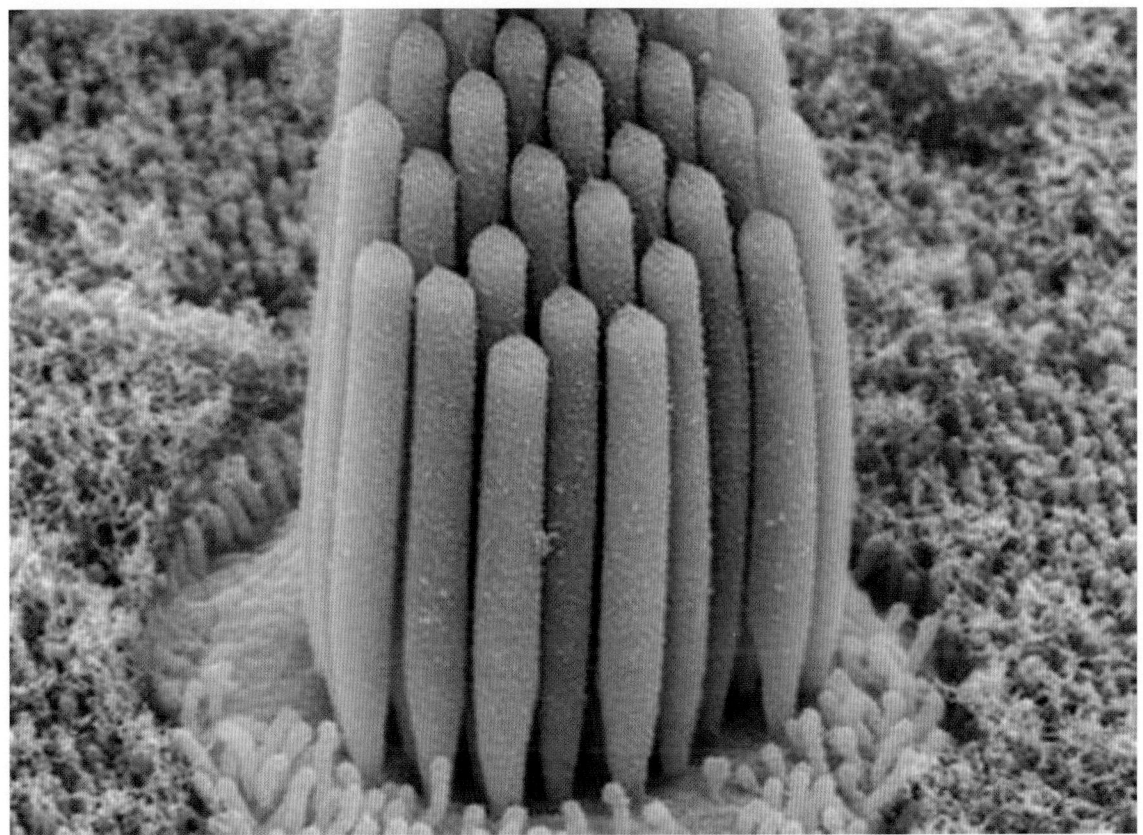

Figure 2.10 Stereocilia of a Hair Cell
Movement of fluid inside the cochlea leads to depolarization of the neuron associated with the hair cell.

The basilar membrane changes thickness depending on its location in the cochlea. The accepted theory on sound perception is **place theory**, which states that the location of a hair cell on the basilar membrane determines the perception of pitch when that hair cell is vibrated. The highest-frequency pitches cause vibrations of the basilar membrane very close to the oval window, whereas low-frequency pitches cause vibrations at the apex, away from the oval window. Thus, the cochlea is **tonotopically** organized: which hair cells are vibrating gives the brain an indication of the pitch of the sound.

BEHAVIORAL SCIENCES GUIDED EXAMPLE WITH EXPERT THINKING

Researchers sought to explain a puzzling phenomenon: that differences in frequency of auditory stimuli sometimes correspond with greater than proportional increases in perceived pitch, even when taking Weber's law into account.

Translated: even if a noise "should" sound high (high frequency), it isn't necessarily perceived as high (high pitch).

This is a reference to my outside knowledge: I'll need to recall what Weber's law is!

The researchers hypothesized that differences in pitch specifically for sounds that are familiar, such as human speech, would be amplified perceptually when compared to those of unfamiliar sounds. Participants listened to recordings of both human vocalizations and synthesized tones at 165 Hz and 175 Hz.

There are a couple of independent variables here: type of sound and frequency. The dependent variable must be what the participants report hearing.

Adapted from: Monson BB, Han S, Purves D (2013) Are Auditory Percepts Determined by Experience? *PLoS ONE* 8(5): e63728. https://doi.org/10.1371/journal.pone.0063728

What results would help to support the researchers' hypothesis? What effect would these results have on the place theory of sound perception?

When we encounter experimental passages on the MCAT, there are a few things we should always be on the lookout for: what is the hypothesis, and what are the independent and dependent variables? What results are presented, and what effect do they have on the hypothesis? What conclusion did the researchers come to, and is this conclusion reasonable given the results? If any of these pieces are missing, we should anticipate questions asking us to identify possibilities that would fit with the passage and our outside content knowledge. Here, we are given a hypothesis without results, and the question asks us for results that would support the hypothesis. We'll need to dive into the scenario to identify what is being tested and why, and then use that information in conjunction with our content knowledge to answer this question.

The researchers hypothesized that differences in perceived pitch are amplified for more commonly encountered sounds. Take a moment to consider what the researchers should expect if this hypothesis is correct. It would stand to reason that the difference between perceived pitch for the 165 Hz and 175 Hz samples would be larger for voices (which we hear all the time) than for synthesized tones (which we don't hear too often). In other words, a result that would help support the researchers' hypothesis would be experimental evidence that human vocalizations sounded more different in pitch than the synthesized tones, despite the absolute difference in frequency remaining the same.

For the second part of the question, we need to recall that place theory predicts that perceived pitch results directly from the location of the hair cells that are vibrated when exposed to that frequency; higher frequencies vibrate hair cells closer to the oval window, and lower frequencies vibrate hair cells that are farther away. What is important here is that, according to place theory, the type of sound is irrelevant; sounds of the same frequency should be perceived in the same way. However, the hypothesis of this study (and the hypothetical results we've just imagined in support of that hypothesis), call place theory into question as, according to the passage, similar changes in vibrations are not perceived as the same change in pitch. According to this study, the perceived pitch instead depends at least in part on the nature of the sound perceived by the listener. Note that this new finding wouldn't affect a place theory of sound sensation/detection, which could still be accurate: this experiment is specifically about how sounds are *perceived*, not about how they are *sensed*.

In sum, to support the researchers, the results must find a greater perceived difference in pitch between the vocal samples as compared to the synthesized audio samples. This finding would contradict the place theory of sound perception.

MCAT CONCEPT CHECK 2.3

Before you move on, assess your understanding of the material with these questions.

1. What structures are used to detect linear acceleration? Rotational acceleration?
   - Linear acceleration:

   ____

   - Rotational acceleration:

   ____

2. List the structures in the auditory pathway, from where sound enters the pinna to the auditory projection areas in the brain.

   ____

3. How does the organization of the cochlea indicate the pitch of an incoming sound?

   ____

2.4 Other Senses

High-Yield

LEARNING OBJECTIVES

After Chapter 2.4, you will be able to:

- List the structures in the olfactory pathway
- Distinguish between the chemicals detected by the nose and mouth
- Recall the four main modalities of somatosensation

While vision and hearing are, by far, the most heavily tested senses on the MCAT, the other senses are still considered fair game on Test Day. These include the chemical senses of smell and taste; somatosensation, which includes all of the modalities of "touch"; and kinesthetic sense.

REAL WORLD

Smell is an impressive motivator for behavior. Food aromas may make a person hungry, a familiar fragrance may remind a person of a significant other from years ago, and an unpleasant smell may signify that an unknown bottle contains a dangerous chemical rather than water. Smell is the only sense that does not pass through the thalamus, but rather travels—unfiltered—into higher-order brain centers.

Smell

Smell is considered one of the chemical senses, which means that it responds to incoming chemicals from the outside world. Specifically, smell responds to volatile or aerosolized compounds. **Olfactory chemoreceptors** (**olfactory nerves**) are located in olfactory epithelium in the upper part of the nasal cavity. Chemical stimuli must bind to their respective chemoreceptors to cause a signal. There are a tremendous number of specific chemoreceptors, which allows us to recognize subtle differences in similar scents, such as lavender and jasmine.

Smell can also carry interpersonal information through the medium of **pheromones**, which are chemicals secreted by one animal, and which, once bonded with chemoreceptors, compel or urge another animal to behave in a specific way. Pheromones have debatable effects on humans, but play an enormous role in many animals' social, foraging, and sexual behaviors.

As is true with all senses, there is a defined **olfactory pathway** to the brain. Odor molecules are inhaled into the nasal passages and then contact the olfactory nerves in the olfactory epithelium. These receptor cells are activated, sending signals to the **olfactory bulb**. These signals are then relayed via the **olfactory tract** to higher regions of the brain, including the limbic system.

Taste

As a sense, taste is often simpler than we imagine. There are five basic tastes: sweet, sour, salty, bitter, and umami (savory). Flavor is not synonymous with taste, but rather refers to the complex interplay between smell and taste, which can be affected by nonchemical stimuli like texture and the individual's mood.

Tastes are detected by **chemoreceptors**, which are sensitive to dissolved compounds. Saltiness, for example, is a reaction to alkali metals, and is generally triggered by the sodium found in table salt. Sourness, on the other hand, is a reaction to acid, such as lemon juice or vinegar. Sweet, bitter, and savory flavors are also triggered by specific molecules binding to receptors. The receptors for taste are groups of cells called **taste buds**, which are found in little bumps on the tongue called **papillae**. Taste information travels from taste buds to the brainstem, and then ascends to the **taste center** in the thalamus before traveling to higher-order brain regions.

Somatosensation

Somatosensation is often reduced to "touch" when listed as a sense, but is actually quite complex. Somatosensation is usually described as having four modalities: pressure, vibration, pain, and temperature. At least five different types of receptor receive tactile information, including:

- **Pacinian corpuscles:** respond to deep pressure and vibration
- **Meissner corpuscles:** respond to light touch
- **Merkel cells (discs):** respond to deep pressure and texture
- **Ruffini endings:** respond to stretch
- **Free nerve endings:** respond to pain and temperature

Transduction occurs in the receptors, which send the signal to the central nervous system where it eventually travels to the **somatosensory cortex** in the parietal lobe.

There are three additional concepts related to touch perception that are important to know: two-point thresholds, physiological zero, and gate theory of pain. A **two-point threshold** refers to the minimum distance necessary between two points of stimulation on the skin such that the points will be felt as two distinct stimuli. Below the two-point threshold, the two stimuli will be felt as one. The size of the two-point threshold depends on the density of nerves in the particular area of skin being tested.

REAL WORLD

Pain and temperature actually use a different pathway than pressure and vibration through the spinal cord. This can be seen in Brown-Séquard syndrome, in which half of the spinal cord is severed. Patients lose pressure and vibration sense on the same side as the lesion, but lose pain and temperature sensation on the opposite side.

Temperature is judged relative to **physiological zero**, or the normal temperature of the skin (between 86° and 97°F). Thus, an object feels "cold" because it is under physiological zero; an object feels "warm" because it is above physiological zero.

REAL WORLD

The gate theory of pain explains why rubbing an injury (like bumping your knee on a table) seems to reduce the pain of the injury.

Pain perception is part of the somatosensory system and can result from signals sent from a variety of sensory receptors, most commonly **nociceptors**. Pain also relies on thresholds, which may vary greatly from person to person. For example, the temperature of water that is perceived to be "so hot it hurts" may vary by several degrees between individuals. The **gate theory of pain** proposes that a special "gating" mechanism can turn pain signals on or off, affecting whether or not we perceive pain. In this theory, the spinal cord is able to preferentially forward the signals from other touch modalities (pressure, temperature) to the brain, thus reducing the sensation of pain. Gate theory has been superseded by other theories, but is still a useful model for understanding how touch is processed at the spinal cord.

Kinesthetic Sense

Kinesthetic sense is also called **proprioception** and refers to the ability to tell where one's body is in space. For example, even with your eyes closed, you could still describe the location and position of your hand. The receptors for proprioception, called **proprioceptors**, are found mostly in muscle and joints, and play critical roles in hand–eye coordination, balance, and mobility.

MCAT CONCEPT CHECK 2.4

Before you move on, assess your understanding of the material with these questions.

1. List the structures in the olfactory pathway, from where odor molecules enter the nose to where olfactory signals project in the brain.

____

____

2. Both smell and taste are sensitive to chemicals. What is different about the types of chemicals each one can sense?

____

____

3. What are the four main modalities of somatosensation?

    1. ______________________________
    2. ______________________________
    3. ______________________________
    4. ______________________________

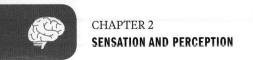

2.5 Object Recognition

High-Yield

LEARNING OBJECTIVES

After Chapter 2.5, you will be able to:

- Compare and contrast bottom-up processing and top-down processing
- Describe each of the Gestalt principles: proximity, similarity, good continuation, subjective contours, closure, and prägnanz

Modern theories of object recognition assume at least two major types of psychological processing: bottom-up processing and top-down processing. **Bottom-up (data-driven) processing** refers to object recognition by parallel processing and feature detection, as described earlier. Essentially, the brain takes the individual sensory stimuli and combines them together to create a cohesive image before determining what the object is. **Top-down (conceptually driven) processing** is driven by memories and expectations that allow the brain to recognize the whole object and then recognize the components based on these expectations. In other words, top-down processing allows us to quickly recognize objects without needing to analyze their specific parts. Neither system is sufficient by itself: if we only performed bottom-up processing, we would be extremely inefficient at recognizing objects; every time we looked at an object, it would be like looking at the object for the first time. On the other hand, if we only performed top-down processing, we would have difficulty discriminating slight differences between similar objects. This distinction is also partially responsible for the feeling of *déjà vu* described in the introduction to this chapter: when we believe we are experiencing something for the first time, we expect to rely on bottom-up processing; however, when the mind is able to recognize an experience more quickly than expected (through top-down processing), the mind searches for a reason for this recognition. In other words, *déjà vu* is often evoked when we have recognition without an obvious reason: *I know that person from somewhere . . . but where?* The distinction between top-down and bottom-up processing is relevant for all senses, but is most commonly applied in the context of vision.

Perceptual organization refers to the ability to create a complete picture or idea by combining top-down and bottom-up processing with all of the other sensory clues gathered from an object. Most of the images we see in everyday life are incomplete; often, we may only be able to see a part of an object and we must infer what the rest of the object looks like. By using what information is available in terms of depth, form, motion, constancy, and other clues, we can often "fill in the gaps" using Gestalt principles (described below).

Depth perception relies on a number of visual cues that are interpreted by the brain to deduce an object's distance. These visual cues are separated into monocular and binocular cues. **Monocular cues** only require one eye and include relative size, interposition, linear perspective, motion parallax, and other minor cues. **Relative size** refers to the idea that objects appear larger the closer they are. **Interposition** means that when two objects overlap, the one in front is closer. **Linear perspective** refers

REAL WORLD

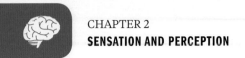

If you've ever looked out the side window of a car, bus, or train on a clear night, you've experienced motion parallax. Parallax is the reason why the objects on the side of the road are a blur as you go past, why objects further away move more slowly as you pass them, and why the moon seems to follow you as you ride along.

to the convergence of parallel lines at a distance: the greater the convergence, the further the distance. **Motion parallax** is the perception that objects closer to us seem to move faster when we change our field of vision (look at something else).

Binocular cues primarily involve **retinal disparity** which refers to the slight difference in images projected on the two retinas. This feature of depth perception is exploited in virtual reality (VR) devices: the images supplied to each eye are slightly different, giving the perception of depth even though the VR device displays 2D images. A secondary binocular cue is **convergence**, in which the brain detects the angle between the two eyes required to bring an object into focus. If a person was looking at a distant object, both of their eyes would stare straight ahead. However if they were looking at something nearby (perhaps their own nose!) the left and right eyes would be held at an extreme angle. This difference in the degree of convergence is used to perceive distance.

The form of an object is usually determined through parvocellular cells and feature detection, and the motion of an object is perceived through magnocellular cells, as described earlier. **Constancy** refers to our ability to perceive that certain characteristics of objects remain the same, despite changes in the environment. For example, we perceive a white piece of paper as essentially the same color whether the paper is illuminated by fluorescent lights, incandescent bulbs, or sunlight—this type of constancy is called color constancy. We also have constancy for brightness, size, and shape, depending on context.

Gestalt Principles

The brain constantly uses incomplete information to try to create a complete picture of the environment. **Gestalt principles** are a set of general rules that account for the fact that the brain tends to view incomplete stimuli in organized, patterned ways. There are dozens of Gestalt principles, but the highest-yield are summarized below and can be visualized in Figure 2.11.

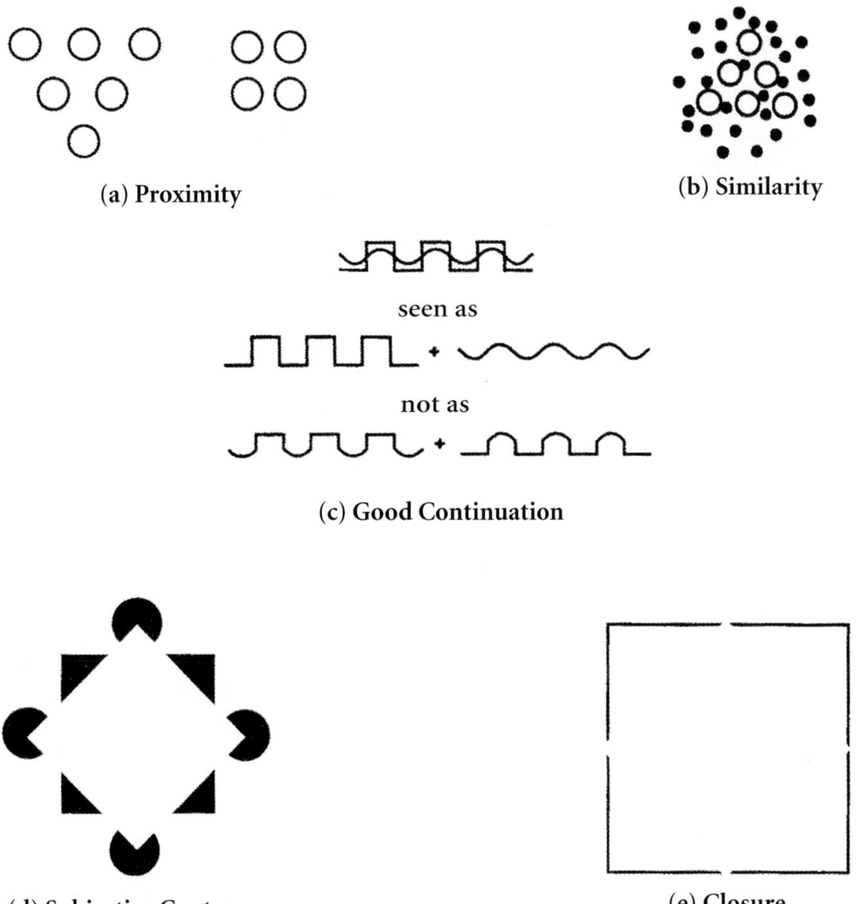

Figure 2.11 Gestalt Principles

The **law of proximity** says that elements close to one another tend to be perceived as a unit. In Figure 2.11a, we do not see ten unrelated dots; rather, we see a triangle and a square, each composed of a certain number of dots. The **law of similarity** says that objects that are similar tend to be grouped together. In Figure 2.11b, we see the big hollow dots as being distinct from the others, forming a triangle against a background of small filled-in dots. The **law of good continuation** says that elements that appear to follow in the same pathway tend to be grouped together. That is, there is a tendency to perceive continuous patterns in stimuli rather than abrupt changes. As seen in Figure 2.11c, our mind tends to break down this complex figure into a sawtooth line and a wavy line, rather than two lines that contain both sawtooth and wavy elements. Some researchers have argued that the phenomena of subjective contours may arise from this law. **Subjective contours** have to do with perceiving contours and, therefore, shapes that are not actually present in the stimulus. In Figure 2.11d, subjective contours lead to the perception of a white diamond on a black square with its corners lying on the four circles. Finally, the **law of closure** says that when a space is enclosed by a contour, the space tends to be perceived as a complete figure. Closure also refers to the fact that certain figures tend to be perceived as more complete (or closed) than they really are. In Figure 2.11e, we don't see four right angles; instead, we see a square, even though the four sides aren't complete. All these laws operate to create the most stable, consistent, and simplest figures possible within a given visual field. Taken altogether, the Gestalt principles are governed by the **law of prägnanz**, which says that perceptual organization will always be as regular, simple, and symmetric as possible.

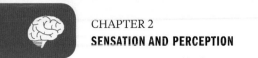

MCAT CONCEPT CHECK 2.5

Before you move on, assess your understanding of the material with these questions.

1. How is sensory information integrated in bottom-up processing? Top-down processing?
   - Bottom-up processing:

   ____

   - Top-down processing:

   ____

2. Briefly describe each of the Gestalt principles below:

Gestalt Principle	Description
Proximity	
Similarity	
Good continuation	
Subjective contours	
Closure	
Prägnanz	

Conclusion

The sensory systems described in this chapter are key to your success on Test Day. Not only are the eye, ear, and other senses high-yield in their own right, but connections to topics in physics, biology, research design, and other concepts in the behavioral sciences make these key topics for passages. But sensation is only one part of the system; we must then take this raw information and process it in the brain to truly perceive the world around us. We use complex neurological pathways to integrate and sort sensory information. We then process sensory information through multiple systems, analyzing individual features and components of the environment while building expectations based on our memories and past experiences. We fill in gaps in our sensorium using Gestalt principles. And what reaches our conscious awareness is the final product: a cohesive concept of the world around us.

You've completed your visit to the new city. You used your rods and cones to see the sites, your chemoreceptors to taste and smell the local food, your hair cells to listen to the local sounds, and your kinesthetic and vestibular senses to help navigate through physical space. As you get ready to head home, all you're left with are your memories—a topic we'll turn to in the next chapter.

GO ONLINE

You've reviewed the content, now test your knowledge and critical thinking skills by completing a test-like passage set in your online resources!

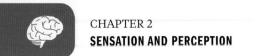

CONCEPT SUMMARY

Sensation *vs.* Perception

- **Sensation** is the conversion, or transduction, of physical, electromagnetic, auditory, and other information from the internal and external environment into electrical signals in the nervous system.
- **Perception** is the processing of sensory information to make sense of its significance.
- **Sensory receptors** are nerves that respond to stimuli and trigger electrical signals.
  - Sensory neurons are associated with **sensory ganglia**: collections of cell bodies outside the central nervous system.
  - Sensory stimuli are transmitted to **projection areas** in the brain, which further analyze the sensory input.
  - Common sensory receptors include photoreceptors, hair cells, nociceptors, thermoreceptors, osmoreceptors, olfactory receptors, and taste receptors.
- A **threshold** is the minimum stimulus that causes a change in signal transduction.
  - The **absolute threshold** is the minimum of stimulus energy that is needed to activate a sensory system.
  - The **threshold of conscious perception** is the minimum of stimulus energy that will create a signal large enough in size and long enough in duration to be brought into awareness.
  - The **difference threshold** or **just-noticeable difference** (**jnd**) is the minimum difference in magnitude between two stimuli before one can perceive this difference.
  - **Weber's law** states that the jnd for a stimulus is proportional to the magnitude of the stimulus, and that this proportion is constant over most of the range of possible stimuli.
  - **Signal detection theory** refers to the effects of nonsensory factors, such as experiences, motives, and expectations, on perception of stimuli.
  - Signal detection experiments allow us to look at **response bias**. In a signal detection experiment, a stimulus may or may not be given, and the subject is asked to state whether or not the stimulus was given. There are four possible outcomes: hits, misses, false alarms, or correct negatives.
- **Adaptation** refers to a decrease in response to a stimulus over time.

Vision

- The eye is an organ specialized to detect light in the form of photons.
  - The **cornea** gathers and filters incoming light.
  - The **iris** divides the front of the eye into the **anterior** and **posterior chambers**. It contains two muscles, the **dilator** and **constrictor pupillae**, which open and close the **pupil**.
  - The **lens** refracts incoming light to focus it on the retina and is held in place by **suspensory ligaments** connected to the **ciliary muscle**.
  - The ciliary body produces **aqueous humor**, which drains through the **canal of Schlemm**.
  - The retina contains rods and cones. **Rods** detect light and dark; **cones** come in three forms (short-, medium-, and long-wavelength) to detect colors.
  - The retina contains mostly cones in the **macula**, which corresponds to the central visual field. The center of the macula is the **fovea**, which contains only cones.
  - Rods and cones synapse on **bipolar cells**, which synapse on **ganglion cells**. Integration of the signals from ganglion cells and edge-sharpening is performed by **horizontal** and **amacrine cells**.
  - The bulk of the eye is supported by the **vitreous** on the inside and the **sclera** and **choroid** on the outside.
- The visual pathway starts from the eye, and travels through the **optic nerves**, **optic chiasm**, **optic tracts**, **lateral geniculate nucleus** (**LGN**) of the thalamus, and **visual radiations** to get to the **visual cortex**.
  - The optic chiasm contains fibers crossing from the nasal side of the retina (temporal visual fields) of both eyes.
  - The visual radiations run through the temporal and parietal lobes.
  - The visual cortex is in the occipital lobe.
- Vision, like all senses, is processed through **parallel processing**: the ability to simultaneously analyze and combine information regarding color, form, motion, and depth.
  - Color is detected by **cones**.
  - Form is detected by **parvocellular cells**, with high spatial resolution and low temporal resolution.
  - Motion is detected by **magnocellular cells**, with low spatial resolution and high temporal resolution.
  - Depth is detected by **binocular neurons**.

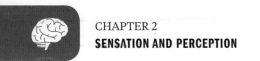

Hearing and Vestibular Sense

- The ear is divided into the outer, middle, and inner ear.
  - The **outer ear** consists of the **pinna** (**auricle**), **external auditory canal**, and **tympanic membrane**.
  - The **middle ear** consists of the **ossicles**: **malleus** (hammer), **incus** (anvil), and **stapes** (stirrup). The footplate of the stapes rests on the **oval window** of the cochlea. The middle ear is connected to the nasal cavity by the **Eustachian tube**.
  - The **inner ear** contains the **bony labyrinth**, within which is the **membranous labyrinth**. The bony labyrinth is filled with **perilymph**; the membranous labyrinth is filled with **endolymph**. The membranous labyrinth consists of the **cochlea**, which detects sound; **utricle** and **saccule**, which detect linear acceleration; and **semicircular canals**, which detect rotational acceleration.
- The auditory pathway starts from the cochlea and travels through the **vestibulocochlear nerve** and **medial geniculate nucleus** (**MGN**) of the thalamus to get to the **auditory cortex** in the temporal lobe.
- Sound information also projects to the **superior olive**, which localizes the sound, and the **inferior colliculus**, which is involved in the startle reflex.

Other Senses

- Smell is the detection of volatile or aerosolized chemicals by the **olfactory chemoreceptors** (**olfactory nerves**) in the olfactory epithelium.
  - The olfactory pathway starts from the olfactory nerves and travels through the **olfactory bulb** and **olfactory tract** to get to higher-order brain areas, such as the limbic system.
  - Pheromones are chemicals given off by animals that have an effect on social, foraging, and sexual behavior in other members of that species.
- Taste is the detection of dissolved compounds by **taste buds** in **papillae**. It comes in five modalities: sweet, sour, salty, bitter, and umami (savory).
- **Somatosensation** refers to the four touch modalities: pressure, vibration, pain, and temperature.
  - A **two-point threshold** is the minimum distance necessary between two points of stimulation on the skin such that the points will be felt as two distinct stimuli.
  - **Physiological zero** is the normal temperature of the skin to which objects are compared to determine if they feel "warm" or "cold."
  - **Nociceptors** are responsible for pain perception. The **gate theory of pain** states that pain sensation is reduced when other somatosensory signals are present.
- **Kinesthetic sense** (**proprioception**) refers to the ability to tell where one's body is in three-dimensional space.

Object Recognition

- **Bottom-up (data-driven) processing** refers to recognition of objects by parallel processing and feature detection. It is slower, but less prone to mistakes.
- **Top-down (conceptually driven) processing** refers to recognition of an object by memories and expectations, with little attention to detail. It is faster, but more prone to mistakes.
- Perceptual organization refers to our synthesis of stimuli to make sense of the world, including integration of depth, form, motion, and constancy.
- **Gestalt principles** are ways that the brain can infer missing parts of a picture when a picture is incomplete.
  - The **law of proximity** says that elements close to one another tend to be perceived as a unit.
  - The **law of similarity** says that objects that are similar appear to be grouped together.
  - The **law of good continuation** says that elements that appear to follow the same pathway tend to be grouped together.
  - **Subjective contours** refer to the perception of nonexistent edges in figures, based on surrounding visual cues.
  - The **law of closure** says that when a space is enclosed by a group of lines, it is perceived as a complete or closed line.
  - The **law of prägnanz** says that **perceptual organization** will always be as regular, simple, and symmetric as possible.

ANSWERS TO CONCEPT CHECKS

2.1

1. Sensory receptor → afferent neuron → sensory ganglion → spinal cord → brain (projection areas)
2. 1–F, 2–A, 3–E, 4–D, 5–C, 6–G, 7–B
3. Absolute threshold is the minimum stimulus that can evoke an action potential in a sensory receptor. Threshold of conscious perception is the minimum stimulus that can evoke enough action potentials for a long enough time that the brain perceives the stimulus. The difference threshold (just-noticeable difference) is the minimum difference between two stimuli that can be detected by the brain.
4. Weber's law explains that just-noticeable differences are best expressed as a ratio, which is constant over most of the range of sensory stimuli. Signal detection theory concerns the threshold to sense a stimulus, given obscuring internal and external stimuli.
5. Adaptation generally raises the difference threshold for a sensory response; as one becomes used to small fluctuations in the stimulus, the difference in stimulus required to evoke a response must be larger.

2.2

1. Cornea: gathers and focuses the incoming light
   Pupil: allows passage of light from the anterior to posterior chamber
   Iris: controls the size of the pupil
   Ciliary body: produces aqueous humor; accommodation of the lens
   Canal of Schlemm: drains aqueous humor
   Lens: refracts the incoming light to focus it on the retina
   Retina: detects images
   Sclera: provides structural support
2. Cornea → pupil → lens → vitreous → retina (rods and cones → bipolar cells → ganglion cells) → optic nerve → optic chiasm → optic tract → lateral geniculate nucleus (LGN) of thalamus → radiations through parietal and temporal lobes → visual cortex (occipital lobe)
3. Parallel processing is the ability to simultaneously analyze color, shape, and motion of an object and to integrate this information to create a cohesive image of the world. Parallel processing also calls on memory systems to compare a visual stimulus to past experiences to help determine the object's identity.
4. Cones are responsible for color. Parvocellular cells are responsible for form. Magnocellular cells are responsible for motion. Binocular neurons are responsible for depth.

2.3

1. Linear acceleration is detected by the utricle and saccule. Rotational acceleration is detected by the semicircular canals.
2. Pinna → external auditory canal → tympanic membrane → malleus → incus → stapes → oval window → perilymph in cochlea → basilar membrane → hair cells → vestibulocochlear nerve → brainstem → medial geniculate nucleus (MGN) of thalamus → auditory cortex (temporal lobe)
3. The basilar membrane is tonotopically organized: high-pitched sounds cause vibrations at the base of the cochlea, whereas low-pitched sounds cause vibrations at the apex of the cochlea.

2.4

1. Nostril → nasal cavity → olfactory chemoreceptors (olfactory nerves) on olfactory epithelium → olfactory bulb → olfactory tract → higher-order brain regions, including limbic system
2. Smell is sensitive to volatile or aerosolized compounds; taste is sensitive to dissolved compounds.
3. The four main modalities of somatosensation are pressure, vibration, pain, and temperature.

2.5

1. Bottom-up processing requires each component of an object to be interpreted through parallel processing and then integrated into one cohesive whole. Top-down processing starts with the whole object and, through memory, creates expectations for the components of the object.
2.

Gestalt Principle	Description
Proximity	Components close to one another tend to be perceived as a unit.
Similarity	Components that are similar (in color, shape, size) tend to be grouped together.
Good continuation	Components that appear to follow in the same pathway tend to be grouped together; abrupt changes in form are less likely than continuation of the same pattern.
Subjective contours	Edges or shapes that are not actually present can be implied by the surrounding objects (especially if good continuation is present).
Closure	A space enclosed by a contour tends to be perceived as a complete figure; such figures tend to be perceived as more complete (or closed) than they really are.
Prägnanz	Perceptual organization will always be as regular, simple, and symmetric as possible.

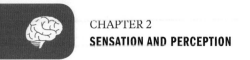

SCIENCE MASTERY ASSESSMENT EXPLANATIONS

1. **C**

Weber's law posits that thresholds are proportional. Going from 100 to 125 pounds is a 25 percent increase. **(C)** is a 40 percent increase while all the rest are all under 25 percent.

2. **D**

The person is discerning a specific noise within a field of many noises. This is the definition of signal detection. In an experimental setup, the person's response would be considered a hit if the friend was indeed laughing; it would be considered a false alarm if the friend was not laughing.

3. **A**

The spicy food can be considered an extreme stimulus because it eclipses what the person believes can be handled in terms of heat. However, after experiencing the stimulus over and over, the experience of spice drops to barely perceptible. This is sensory adaptation: a reduction in response to a stimulus over time.

4. **B**

Nociceptors are important for pain sensation, which would be expected during tissue damage. Chemoreceptors, **(A)**, respond to chemicals, whether volatile or aerosolized (olfaction) or dissolved (taste). Osmoreceptors, **(C)**, respond to changes in blood osmolarity, and photoreceptors, **(D)**, respond to light.

5. **A**

The cornea is responsible for gathering and focusing light. The pupil and iris, **(B)** and **(C)**, are both involved in regulating the amount of light coming into the eye but not in focusing it. The retina, **(D)**, transduces the light into electrical signals that are sent to the brain. The lens serves a similar function to the cornea and would also be a valid answer to this question.

6. **C**

This person was able to distinguish the coin from other items by recognizing specific features of the coin; in this case, it was the glint of the metal surface and its position in the car. This phenomenon is called feature detection.

7. **C**

The retina is the part of the eye upon which images are projected. Rods and cones in the retina then convert the electromagnetic radiation into electrical signals.

8. **A**

Somatosensation refers to the various modalities of touch: pressure, vibration, temperature, and pain. Kinesthetic sense, **(B)**, refers to the ability to tell where one's body is in space. Vestibular sense, **(C)**, refers to the detection of linear and rotational acceleration in the middle ear. Finally, chemoreception, **(D)**, refers to sensing chemicals in the environment.

9. **B**

The five tastes are sweet, sour, salty, bitter, and umami. Floral would be related to smell rather than taste.

10. **C**

Endolymph is the potassium-rich fluid that bathes the hair cells of the inner ear, all of which are found within the membranous labyrinth. Perilymph is found in the space between the membranous labyrinth and the bony labyrinth. Both the membranous labyrinth and bony labyrinth contribute to the cochlea and the vestibule, eliminating **(A)** and **(B)**.

11. **A**

Pheromones are the volatile chemicals given off by organisms that bind with olfactory chemoreceptors and influence behavior. It is debatable if pheromones serve a role in humans, but they are known to affect foraging and sexual behavior in some animals.

12. **B**

Kinesthetic sense, or proprioception, refers to the ability to tell where body parts are in three-dimensional space. The sensors for proprioception are found predominantly in the muscles and joints. Loss of vestibular sense, **(A)**, would also cause difficulty walking, but this would be due to a sense of dizziness or vertigo, not an inability to feel one's feet.

13. **C**

The proofreader used a larger pattern to identify the word and expected to see an "e," thus missing the error. This is related to top-down processing; the proofreader used recognition and expectations, which led to missing a detail. Bottom-up processing, **(D)**, would be the analysis of each detail individually before creating a cohesive image.

14. **D**

Gestalt principles are the basis for many optical illusions and include the tendency of people to see continuity even when lines are unconnected. Specifically, this logo appears to rely on the law of closure to create one complete star from five nontouching angles.

15. **D**

The optic chiasm houses the crossing fibers from each optic nerve. Specifically, the fibers coming from the nasal half of the retina in each eye cross in the chiasm to join the optic tract on the opposite side. Remember that the lens of the eye causes inversion, so images on the nasal half of the retina actually originate in the temporal visual field. This condition is called bitemporal hemianopsia.

Consult your online resources for additional practice. GO ONLINE

SHARED CONCEPTS

Behavioral Sciences Chapter 1
Biology and Behavior

Biology Chapter 4
The Nervous System

Biology Chapter 10
Homeostasis

Physics and Math Chapter 3
Thermodynamics

Physics and Math Chapter 7
Waves and Sound

Physics and Math Chapter 8
Light and Optics

CHAPTER 3

LEARNING AND MEMORY

SCIENCE MASTERY ASSESSMENT

Every pre-med knows this feeling: there is so much content I have to know for the MCAT! How do I know what to do first or what's important?

While the high-yield badges throughout this book will help you identify the most important topics, this Science Mastery Assessment is another tool in your MCAT prep arsenal. This quiz (which can also be taken in your online resources) and the guidance below will help ensure that you are spending the appropriate amount of time on this chapter based on your personal strengths and weaknesses. Don't worry though—skipping something now does not mean you'll never study it. Later on in your prep, as you complete full-length tests, you'll uncover specific pieces of content that you need to review and can come back to these chapters as appropriate.

How to Use This Assessment

If you answer 0–7 questions correctly:

Spend about 1 hour to read this chapter in full and take limited notes throughout. Follow up by reviewing **all** quiz questions to ensure that you now understand how to solve each one.

If you answer 8–11 questions correctly:

Spend 20–40 minutes reviewing the quiz questions. Beginning with the questions you missed, read and take notes on the corresponding subchapters. For questions you answered correctly, ensure your thinking matches that of the explanation and you understand why each choice was correct or incorrect.

If you answer 12–15 questions correctly:

Spend less than 20 minutes reviewing all questions from the quiz. If you missed any, then include a quick read-through of the corresponding subchapters, or even just the relevant content within a subchapter, as part of your question review. For questions you answered correctly, ensure your thinking matches that of the explanation and review the Concept Summary at the end of the chapter.

1. Researchers repeatedly startle a participant with a loud buzzer. After some time, the participant stops being startled by the buzzer. If the researchers interrupt the study with the sound of pans banging together, which of the following would likely be observed?
   - **A.** Increased startle response to the buzzer
   - **B.** Decreased startle response to the buzzer
   - **C.** No change in the response to the buzzer
   - **D.** Generalization to previously nonaversive stimuli

2. Many pets will run toward the kitchen when they hear the sound of a can opener opening a can of pet food. The sound of the can opener is a(n):
   - **A.** conditioned response.
   - **B.** unconditioned response.
   - **C.** conditioned stimulus.
   - **D.** unconditioned stimulus.

3. A person suffers from food poisoning after eating a spoiled orange and later finds that the smell of lemon and other citrus fruits causes a feeling of nausea. This is an example of:
   - **A.** acquisition.
   - **B.** generalization.
   - **C.** discrimination.
   - **D.** negative reinforcement.

4. Which of the following processes would increase the likelihood of a behavior?
   - **A.** Extinction
   - **B.** Negative punishment
   - **C.** Positive punishment
   - **D.** Avoidance learning

5. The presynaptic neuron becomes more efficient at releasing neurotransmitters while receptor density increases on the postsynaptic neuron. These changes are consistent with:
   - **A.** long term potentiation.
   - **B.** synaptic pruning.
   - **C.** an increase in neuroplasticity.
   - **D.** amnesia.

6. A rat is trained to press a lever to obtain food under a fixed-interval schedule. Which of the following behaviors would the rat most likely exhibit?
   - **A.** Pressing the lever continuously whenever it is hungry.
   - **B.** Pressing the lever exactly once and waiting for the food pellet before pressing it again.
   - **C.** Pressing the lever slowly at first, but with increasing frequency as the end of the interval approaches.
   - **D.** None of the above; the association formed by fixed-interval schedules is too weak to increase behavior.

7. Which of the following is true of teaching an animal a complicated, multistage behavior?
   - **I.** The individual parts of the behavior should not run counter to the animal's natural instincts.
   - **II.** The behaviors must be tied to a food reward of some kind.
   - **III.** Rewarding individual parts of the behavior on their own interferes with reinforcement of the entire behavior.

   - **A.** I only
   - **B.** I and III only
   - **C.** II and III only
   - **D.** I, II, and III

8. Which of the following is true of controlled processing?
   - **A.** It is the means through which information enters short-term memory.
   - **B.** Information that requires controlled processing cannot become automatic.
   - **C.** It always requires active attention to the information being encoded.
   - **D.** Most information we can later recall is encoded using controlled processing.

9. Which of the following methods of encoding is most conducive to later recall?
   A. Semantic
   B. Visual
   C. Iconic
   D. Acoustic

10. A hemisphere of the brain can be removed to prevent the recurrence of severe seizures. How would the ability of the other hemisphere to adopt functionalities of the removed section differ between a two year old and a twenty year old?
    A. The brain of the twenty year old would be more adaptable due to higher neuroplasticity.
    B. The brain of the twenty year old would be more adaptable due to lower neuroplasticity.
    C. The brain of the two year old would be more adaptable due to higher neuroplasticity.
    D. The brain of the two year old would be more adaptable due to lower neuroplasticity.

11. An individual memorizes a shopping list by associating each item with an image that corresponds with a number. This individual is using which of the following mnemonics?
    A. Clustering
    B. Method of loci
    C. Elaborative rehearsal
    D. Peg-words

12. A researcher uses a partial-report procedure after presenting participants with an array of nine numbers for a fraction of a second. Which of the following is the most likely result of this procedure?
    A. The participants will be able to recall any of the rows or columns in great detail but only immediately after presentation.
    B. The participants will only be able to recall the first few numbers in the array due to the serial position effect.
    C. The participants will be able to recall approximately seven of the numbers for a few seconds following presentation of the stimulus.
    D. The participants will not be able to recall any of the numbers verbally, but will be able to draw the full array under hypnosis.

13. Which of the following is an example of a semantic memory?
    A. Having the ability to drive a car
    B. Knowing the parts of a car engine
    C. Remembering the experience of learning to drive
    D. Associating a car with other vehicles in a semantic network

14. Which of the following is an example of a circumstance that could cause a state-dependent recall effect?
    I. The individual is outside on a rainy day.
    II. The individual is intoxicated on cannabis.
    III. The individual is experiencing a manic episode.

    A. I only
    B. III only
    C. II and III only
    D. I, II, and III

15. Which of the following would an older adult be most likely to have trouble recalling?
    A. The circumstances of meeting a significant other in college
    B. A doctor's appointment scheduled for 1:00 p.m.
    C. The names of the characters in a favorite television show
    D. That a library book needs to be returned when passing by the library on a morning walk

Answer Key

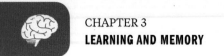

1. **A**
2. **C**
3. **B**
4. **D**
5. **A**
6. **C**
7. **A**
8. **C**
9. **A**
10. **C**
11. **D**
12. **A**
13. **B**
14. **C**
15. **B**

Detailed explanations can be found at the end of the chapter.

CHAPTER 3

LEARNING AND MEMORY

In This Chapter

CHAPTER PROFILE

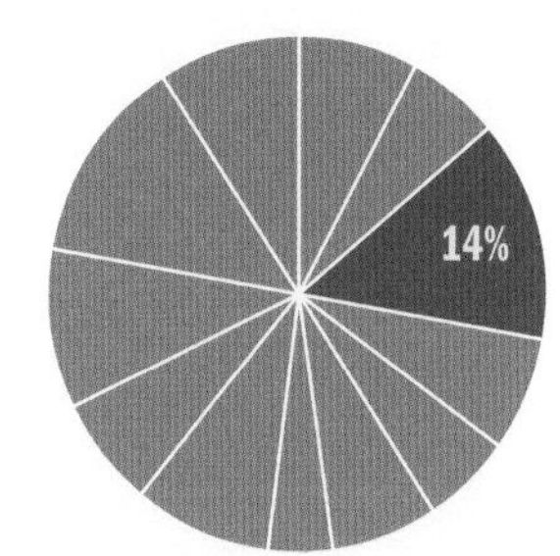

The content in this chapter should be relevant to about 14% of all questions about the behavioral sciences on the MCAT.

This chapter covers material from the following AAMC content categories:

6B: Making sense of the environment

7C: Attitude and behavior change

Introduction

A college student sits hunched over a desk in a quiet library, poring over a small stack of textbooks. It's 11 p.m. the night before the organic chemistry midterm, and there seems to be a near endless list of reactions to commit to memory before tomorrow. The situation seems bleak, but the student has been here before and has taken every precaution to make sure that this study session will be successful: drinking coffee will increase alertness, while the quiet of the library will reduce distractions to aid concentration. It's stressful, to be sure, but the student has been able to study this way before and do quite well, reinforcing this current set of behaviors. The student works through a set of flashcards—reactants on one side, products on the other—and is able to identify most of them but misses a few. Those cards are placed in a separate pile to be reviewed later. Sure, this rehearsal will most likely help for the exam tomorrow, but will the student be able to recall this information again for the final in two months? The student takes another sip of coffee and tries to put everything else out of mind, focusing intently on the book sitting open on the desk.

Sound familiar? If you're like most students, you've found yourself in a similar scenario at least once. This chapter will discuss the ways in which you both memorize new information and learn new behaviors. This will not only help you to directly prepare to answer MCAT questions about this content, but also to learn a few new tricks about how to effectively commit all of the MCAT content to memory. This is a skill that will be helpful both now and later in your career as a doctor.

3.1 Learning

LEARNING OBJECTIVES

After Chapter 3.1, you will be able to:

- Apply principles of habituation, dishabituation, and sensitization to real-life scenarios
- Identify the conditioned stimulus, unconditioned stimulus, conditioned response, and unconditioned response in a Pavlovian learning paradigm
- Distinguish between negative reinforcement, positive reinforcement, negative punishment, and positive punishment
- Predict how reinforcement schedule will affect relative frequency of behavioral response in an operant conditioning scenario:

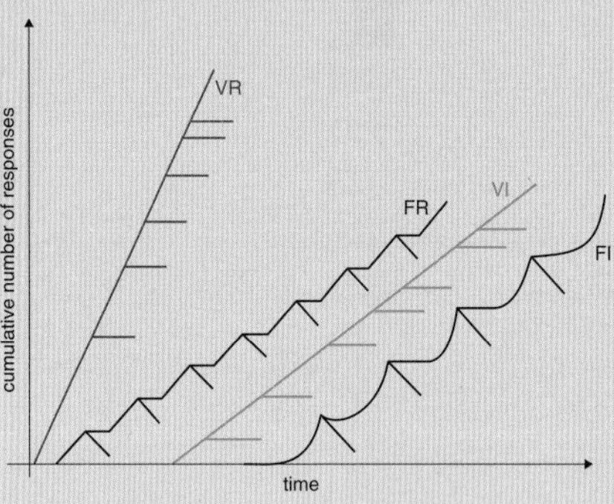

To a psychologist, **learning** refers specifically to the way in which we acquire new behaviors. To understand learning, we must start with the concept of a stimulus. A **stimulus** can be defined as anything to which an organism can respond, including all of the sensory inputs we discussed in Chapter 2 of *MCAT Behavioral Sciences Review*. The combination of stimuli and responses serves as the basis for all behavioral learning.

Responses to stimuli can change over time depending on the frequency and intensity of the stimulus. For instance, repeated exposure to the same stimulus can cause a decrease in response called **habituation**. This is seen in many first-year medical students: students often have an intense physical reaction the first time they see a cadaver or treat a severe laceration, but as they get used to these stimuli, the reaction lessens until they are unbothered by these sights. Note that a stimulus too weak to elicit a response is called **subthreshold** stimulus.

KEY CONCEPT

Dishabituation is the recovery of a response to a stimulus, usually after a different stimulus has been presented. Note that the term refers to changes in response to the original stimulus, not the new one.

The opposite process can also occur. **Dishabituation** is defined as the recovery of a response to a stimulus after habituation has occurred. Dishabituation is often noted when, late in the habituation of a stimulus, a second stimulus is presented. The second stimulus interrupts the habituation process and thereby causes an increase in response to the original stimulus. Imagine, for example, that you're taking a long car trip and driving for many miles on a highway. After a while, your brain will get used to the sights, sounds, and sensations of highway driving: the dashed lines dividing the lanes, the sound of the engine and the tires on the road, and so on. Habituation has occurred. At some point you use an exit ramp, and these sensations change. As you merge onto the new highway, you pay more attention to the sensory stimuli coming in. Even if the stimuli are more or less the same as on the previous highway, the presentation of a different stimulus (using the exit ramp) causes dishabituation and a new awareness of—and response to—these stimuli. Dishabituation is temporary and always refers to changes in response to the original stimulus, not the new one.

Learning, then, is a change in behavior that occurs in response to a stimulus. While there are many types of learning, the MCAT focuses on two types: associative learning and observational learning.

Associative Learning

Associative learning is the creation of a pairing, or association, either between two stimuli or between a behavior and a response. On the MCAT, you'll be tested on two kinds of associative learning: classical and operant conditioning.

Classical Conditioning

Classical conditioning is a type of associative learning that takes advantage of biological, instinctual responses to create associations between two unrelated stimuli. For many people, the first name that comes to mind for research in classical conditioning is Ivan Pavlov. His experiments on dogs were not only revolutionary, but also provide a template for the way the MCAT will test classical conditioning.

Classical conditioning works, first and foremost, because some stimuli cause an innate or reflexive physiological response. For example, we reflexively salivate when we smell bread baking in an oven, or we may jump or recoil when we hear a loud noise. Any stimulus that brings about such a reflexive response is called an **unconditioned stimulus**, and the innate or reflexive response is called an **unconditioned response**. Many stimuli do not produce a reflexive response and are known as **neutral stimuli**.

In Pavlov's experiment, the unconditioned stimulus was meat, which would cause the dogs to salivate reflexively, and the neutral stimulus was a ringing bell. Through the course of the experiment, Pavlov repeatedly rang the bell before placing meat in the dogs' mouths. Initially, the dogs did not react much when they only heard the bell ring without receiving meat. However, after this procedure was repeated several times, the dogs began to salivate when they heard the bell ring. In fact, the dogs would salivate even if Pavlov only rang the bell and did not deliver any meat. Pavlov thereby turned a neutral stimulus into a **conditioned stimulus**: a normally neutral stimulus that, through association, now causes a reflexive response called a **conditioned response**. The process of using a reflexive, unconditioned stimulus to turn a neutral stimulus into a conditioned stimulus is termed **acquisition**, as shown in Figure 3.1.

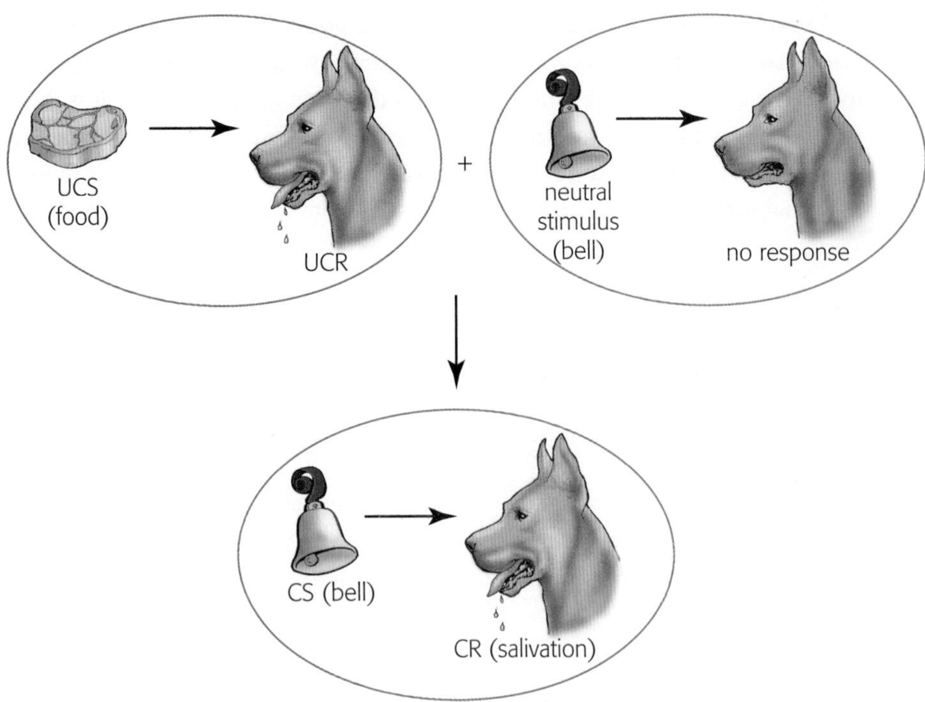

Figure 3.1 Acquisition in Classical Conditioning
UCS = unconditioned stimulus, UCR = unconditioned response, CS = conditioned stimulus, CR = conditioned response

MCAT EXPERTISE

On the MCAT, the key to telling conditioned and unconditioned responses apart will be to look at which stimulus is causing them: unconditioned stimuli cause an unconditioned response, while conditioned stimuli cause a conditioned response.

Notice that the stimuli change in this experiment, but the response is the same throughout. Because salivation in response to food is natural and requires no conditioning, it is an unconditioned response in this context. On the other hand, when paired with the conditioned stimulus of the bell, salivation is considered a conditioned response.

However, it is important to recognize that just because a conditioned response has been acquired, that does not mean that the conditioned response is permanent. **Extinction** refers to the loss of a conditioned response, and can occur if the conditioned stimulus is repeatedly presented without the unconditioned stimulus. Applying this concept to the Pavlov example, if the bell rings often enough without the dog getting meat, the dog may stop salivating when the bell sounds. Interestingly, this extinction of a response is not always permanent; after some time, presenting subjects again with an extinct conditioned stimulus will sometimes produce a weak conditioned response, a phenomenon called **spontaneous recovery**.

There are a few processes that can modify the response to a conditioned stimulus after acquisition has occurred. **Generalization** is a broadening effect by which a stimulus similar enough to the conditioned stimulus can also produce the conditioned response. In one famous experiment, researchers conditioned a child called Little Albert to be afraid of a white rat by pairing the presentation of the rat with a loud noise. Subsequent tests showed that Little Albert's conditioning had generalized such that he also exhibited a fear response to a white stuffed rabbit, a white sealskin coat, and even a man with a white beard.

Finally, in **stimuli discrimination** (sometimes referred to as just **discrimination**), an organism learns to distinguish between similar stimuli. Discrimination is the opposite of generalization. Pavlov's dogs could have been conditioned to discriminate between bells of different tones by having one tone paired with meat, and another tone presented without meat. In this case, association could have occurred with one tone but not the other.

MCAT EXPERTISE

Classical conditioning is a favorite topic on the MCAT. Expect at least one question to describe a Pavlovian experiment and ask you to identify the role of one of the stimuli or responses described.

Operant Conditioning

Whereas classical conditioning is concerned with instincts and biological responses, the study of **operant conditioning** examines the ways in which consequences of voluntary behaviors change the frequency of those behaviors. Just as the MCAT will test you on the difference between conditioned and unconditioned responses and stimuli, it will ask you to distinguish between reinforcement and punishment too. Operant conditioning is associated with B. F. Skinner, who is considered one of the founders of **behaviorism**, the theory that all behaviors are conditioned. The four possible relationships between stimulus and behavior are summarized in Figure 3.2.

Behavior	Stimulus: Added	Stimulus: Removed
Continues	Positive reinforcement	Negative reinforcement
Stops	Positive punishment	Negative punishment

Figure 3.2 Terminology of Operant Conditioning

Reinforcement

Almost all animals will innately search for resources in their environment. These **reward-seeking behaviors**, such as foraging and approach behaviors, are modified over time as the animal interacts with various stimuli and adjusts its behaviors accordingly. **Reinforcement** is the process of increasing the likelihood that an animal will perform a behavior. Reinforcers are divided into two categories. **Positive reinforcers** increase the frequency of a behavior by adding a positive consequence or incentive following the desired behavior. Money is an example of a common and

REAL WORLD

This concept of learning by consequence forms the foundation for behavioral therapies for many disorders including phobias, anxiety disorders, and obsessive-compulsive disorder.

strong positive reinforcer: employees will continue to work if they are paid. **Negative reinforcers** act similarly in that they increase the frequency of a behavior, but they do so by removing something unpleasant. For example, taking an aspirin reduces a headache, so the next time you have a headache, you are more likely to take one. Negative reinforcement is often confused with punishment, which will be discussed in the next section, but remember that the frequency of the behavior is the distinguishing factor: any reinforcement—positive or negative—increases the likelihood that a behavior will be performed.

Negative reinforcement can be subdivided into escape learning and avoidance learning, which differ in whether the unpleasant stimulus occurs or not. **Escape learning** describes a situation where the animal experiences the unpleasant stimulus and, in response, displays the desired behavior in order to trigger the removal of the stimulus. So, in this type of learning, the desired behavior is used to escape the stimulus. In contrast, **avoidance learning** occurs when the animal displays the desired behavior in anticipation of the unpleasant stimulus, thereby avoiding the unpleasant stimulus.

Avoidance learning often develops from multiple experiences of escape learning. An example of this progression from escape learning to avoidance learning is the seat belt warning in a car. If a driver begins driving without buckling the seat belt, then the car will produce an annoying beeping noise, which only ends when the seat belt is buckled. In this example, the desired behavior is to buckle the seat belt. This behavior is reinforced by the removal of an unpleasant stimulus (the audible beeping), so this type of learning is negative reinforcement. More specifically, this example illustrates escape learning, since the driver first experiences the unpleasant stimulus, then exhibits the desired behavior in order to escape the unpleasant stimulus. However, after forgetting to buckle the seat belt several times, the driver will eventually learn to preemptively buckle up before driving the car in order to avoid the beeping sound. At that point, the escape learning has progressed to avoidance learning. Finally, this example illustrates an important misconception about the term *negative reinforcement:* Buckling one's seat belt is generally considered a "positive" behavior, in that it protects one's health. Nevertheless, the terms "positive" and "negative" in operant conditioning only refer to the addition or removal of a stimulus. So even though buckling up is a "good" thing, this example illustrates several types of negative reinforcement!

Classical and operant conditioning can be used hand-in-hand. For example, some dog trainers take advantage of reinforcers when training dogs to perform tricks. Sometimes, the trainers will feed the dog a bit of meat after it performs a trick. The meat can be said to be a **primary reinforcer** because the meat is a treat that the dog responds to naturally. Dog trainers also use tiny handheld devices that emit a clicking sound. This clicker would not normally be a reinforcer on its own, but the trainers use classical conditioning to pair the clicker with meat to elicit the same response. The clicker is thus a **conditioned reinforcer**, which is sometimes called a **secondary reinforcer**. Eventually, the dog may even associate the presence of the trainer with the possibility of reward, making the presence of the trainer a **discriminative stimulus**. A discriminative stimulus indicates that reward is potentially available in an operant conditioning paradigm.

Punishment

In contrast to reinforcement, **punishment** uses conditioning to reduce the occurrence of a behavior. **Positive punishment** adds an unpleasant consequence in response to a behavior to reduce that behavior; for example, receiving a ticket and having to pay a fine for parking illegally. Because positive punishment involves using something unpleasant to discourage a behavior, it is sometimes referred to as **aversive conditioning**. By contrast, **negative punishment** is removing a stimulus in order to cause reduction of a behavior. For example, a parent or guardian may forbid a child from watching television as a consequence for bad behavior, with the goal of preventing the behavior from happening again.

KEY CONCEPT

Negative reinforcement is often confused with positive punishment. Negative reinforcement is the *removal* of a bothersome stimulus to *encourage* a behavior; positive punishment is the *addition* of a bothersome stimulus to *reduce* a behavior.

BRIDGE

Sociological institutions often rely on punishments and rewards to adjust behavior. Within a society, **formal sanctions**, or rules and laws, can be used to reinforce or punish behavior. Likewise, **informal sanctions**, such as ostracization, praise, and shunning, can be used to reinforce or punish social behavior without depending on rules established by social institutions. Socialization and social institutions are discussed in Chapters 8 and 11 of *MCAT Behavioral Sciences Review*, respectively.

Reinforcement Schedules

The presence or absence of reinforcing or punishing stimuli is just a part of the story. The rate at which desired behaviors are acquired is also affected by the **reinforcement schedule** being used to deliver the stimuli. There are two key factors to reinforcement schedules: whether the schedule is fixed or variable, and whether the schedule is based on a ratio or an interval.

- **Fixed-ratio (FR) schedules** reinforce a behavior after a specific number of performances of that behavior. For example, in a typical operant conditioning experiment, researchers might reward a rat with a food pellet every third time it presses a bar in its cage. **Continuous reinforcement** is a fixed-ratio schedule in which the behavior is rewarded every time it is performed.
- **Variable-ratio (VR) schedules** reinforce a behavior after a varying number of performances of the behavior, but such that the average number of performances to receive a reward is relatively constant. With this type of reinforcement schedule, researchers might reward a rat first after two button presses, then eight, then four, then finally six.
- **Fixed-interval (FI) schedules** reinforce the first instance of a behavior after a specified time period has elapsed. For example, once our rat gets a pellet, it has to wait 60 seconds before it can get another pellet. The first lever press after 60 seconds gets a pellet, but subsequent presses during those 60 seconds accomplish nothing.
- **Variable-interval (VI) schedules** reinforce a behavior the first time that behavior is performed after a varying interval of time. Instead of waiting exactly 60 seconds, for example, our rat might have to wait 90 seconds, then 30 seconds, then three minutes. In each case, once the interval elapses, the next press gets the rat a pellet.

Of these schedules, variable-ratio works the fastest for learning a new behavior, and is also the most resistant to extinction. The effectiveness of the various reinforcement schedules is demonstrated in Figure 3.3.

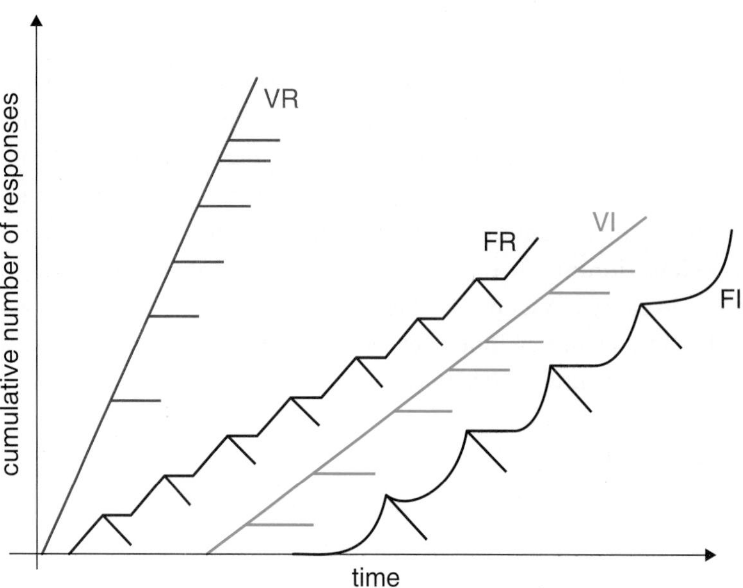

Figure 3.3 Reinforcement Schedules

Hatches correspond to instances of reinforcement. The start of each line corresponds to time zero for that schedule.

MNEMONIC

VR stands for **V**ariable-**R**atio, but it can also stand for **V**ery **R**apid and **V**ery **R**esistant to extinction.

REAL WORLD

Gambling (and gambling addiction) is so difficult to extinguish because most gambling games are based on variable-ratio schedules. While the probability of winning the jackpot on any individual pull of a slot machine is the same, we get caught in the idea that the next pull will be the "right one."

There are a few things to note in this graph. First, variable-ratio schedules have the fastest response rate: the rat will continue pressing the bar quickly with the hope that the next press will be the "right one." Also note that fixed schedules (fixed-ratio and fixed-interval) often have a brief moment of no responses after the behavior is reinforced: the rat will stop hitting the lever until it wants another pellet, once it has figured out what behavior is necessary to receive the pellet.

One final idea associated with operant conditioning is the concept of **shaping**, which is the process of rewarding increasingly specific behaviors that become closer to a desired response. For example, if you wanted to train a bird to spin around in place and then peck a key on a keyboard, you might first give the bird a treat for turning slightly to the left, then only for turning a full 90 degrees, then 180, and so on, until the bird has learned to spin around completely. Then you might only reward this behavior if done near the keyboard until eventually the bird is only rewarded once the full set of behaviors is performed. While it may take some time, the use of shaping in operant conditioning can allow for the training of extremely complicated behaviors.

Cognitive and Biological Factors in Associative Learning

It would be incorrect to say that classical and operant conditioning are the only factors that affect behavior, nor would it be correct to say that we are all mindless and robotic, unable to resist the rewards and punishments that occur in our lives. Since Skinner's initial perspectives, it has been found that many cognitive and biological factors are at work that can change the effects of associative learning or allow us to resist them altogether.

Many organisms undergo **latent learning**, which is learning that occurs without a reward but that is spontaneously demonstrated once a reward is introduced. The classic experiment associated with latent learning involves rats running a maze. Rats that were simply carried through the maze and then incentivized with a food reward for completing the maze on their own performed just as well—and in some cases better—than those rats that had been trained to run the maze using more standard operant conditioning techniques by which they were rewarded along the way.

Problem solving is another method of learning that steps outside the standard behaviorist approach. Think of the way young children put together a jigsaw puzzle: often, they will take pieces one-by-one and try to make them fit together until they find the correct match. Many animals will also use this kind of trial-and-error approach, testing behaviors until they yield a reward. As we get older, we gain the ability to analyze the situation and respond correctly the first time, as when we seek out the correct puzzle piece and orientation based on the picture we are forming. Humans and chimpanzees alike will often avoid trial-and-error learning and instead take a step back, observe the situation, and take decisive action to solve the challenges they face.

Not all behaviors can be taught using operant conditioning techniques. Many animals are predisposed to learn (or not learn) behaviors based on their own natural abilities and instincts. Animals are most able to learn behaviors that coincide with their natural behaviors: birds naturally peck when searching for food, so rewarding them with food in response to a pecking-based behavior works well. This predisposition is known as **preparedness**. Similarly, it can be very difficult to teach animals behaviors that work against their natural instincts. When animals revert to an instinctive behavior after learning a new behavior that is similar, the animal has undergone **instinctive** (or **instinctual**) **drift**. For example, researchers used behavioral techniques to train raccoons to place coins in a piggy bank. Their efforts were ultimately unsuccessful as the learned behaviors were only temporary. Eventually, rather than placing the coins in the bank, the raccoons would pick up the coins, rub them together, and dip them into the bank before pulling them back out. The researchers concluded that the task they were trying to train the raccoons to perform was conflicting with their natural food-gathering instinct, which was to rub seeds together and wash them in a stream to clean them before eating. The researchers had far better luck training the raccoons to place a ball in a basketball net, as the ball was too large to trigger the food-washing instinct.

Observational Learning

Observational learning is the process of learning a new behavior or gaining information by watching others. The most famous and perhaps most controversial study into observational learning is Albert Bandura's Bobo doll experiment, in which children watched an adult in a room full of toys punching and kicking an inflatable clown toy. When the children were later allowed to play in the room, many of them ignored the other toys in the room and inflicted similar violence on the Bobo doll

REAL WORLD

The connection between violent video games and aggressive behavior is still under active debate. While there are many interest groups on both sides of the controversy, the American Academy of Pediatrics (a major medical society) published one report in which they attributed a 13 to 22% increase in aggressive behavior to observational learning from video games.

just as they had seen the adult do. It's important to note that observational learning is not simply imitation because observational learning can be used to teach individuals to avoid behavior as well. In later iterations of the Bobo doll experiment, children who watched the adult get scolded after attacking the Bobo doll were less likely to be aggressive toward the Bobo doll themselves.

Like associative learning, there are a few neurological factors that affect observational learning. The most important of these are **mirror neurons**. These neurons are located in the frontal and parietal lobes of the cerebral cortex and fire both when an individual performs an action and when that individual observes someone else performing that action. Mirror neurons are largely involved in motor processes, but additionally are thought to be related to empathy and vicarious emotions; some mirror neurons fire both when we experience an emotion and also when we observe another experiencing the same emotion. Mirror neurons also play a role in imitative learning by a number of primates, as shown in Figure 3.4.

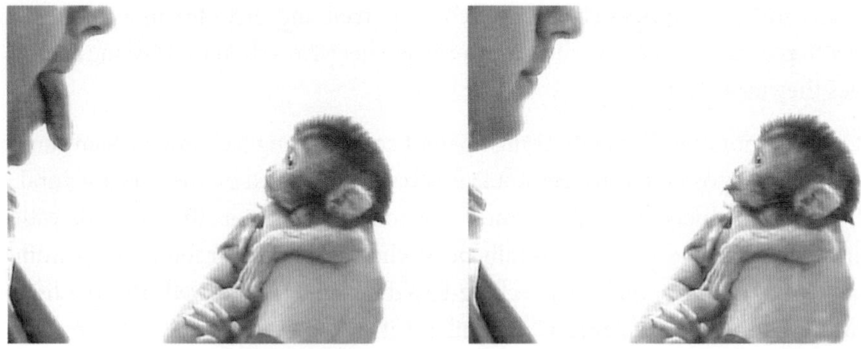

Figure 3.4 Use of Mirror Neurons in a Macaque
Many neonatal primates imitate facial expressions using mirror neurons.

Research suggests that observational learning through **modeling** is an important factor in determining people's behavior throughout their lifetime. People learn what behaviors are acceptable by watching others perform them. Much attention is focused on violent media as a model for antisocial behavior, but prosocial modeling can be just as powerful. Of course, observational learning is strongest when a model's words are consistent with actions. Many parents and guardians adopt a *Do as I say, not as I do* approach when teaching their children, but research suggests that children will disproportionately imitate what the model *did*, rather than what the model *said*.

MCAT CONCEPT CHECK 3.1

Before you move on, assess your understanding of the material with these questions.

1. Which of the following might cause a person to eat more food during a meal: eating each course separately and moving to the next only when finished with the current course, or interrupting the main course several times by eating side dishes?

___

2. A college student plays a prank on a roommate by popping a balloon behind the roommate's head after every time making popcorn. Before long, the smell of popcorn makes the roommate nervous. Which part of the story corresponds to each of the classical conditioning concepts below?
   - Conditioned stimulus:

   ___

   - Unconditioned stimulus:

   ___

   - Conditioned response:

   ___

   - Unconditioned response:

   ___

3. What is the difference between negative reinforcement and positive punishment? Provide an example of each.
   - Negative reinforcement:

   ___

   ___

   - Positive punishment:

   ___

   ___

3.2 Memory

LEARNING OBJECTIVES

After Chapter 3.2, you will be able to:

- Order the three modes of encoding from strongest to weakest
- Distinguish between maintenance rehearsal and elaborative rehearsal
- Predict how learning environments may impact recall
- Describe factors and phenomena that can lead to flaws in memory
- Define each type of human memory:

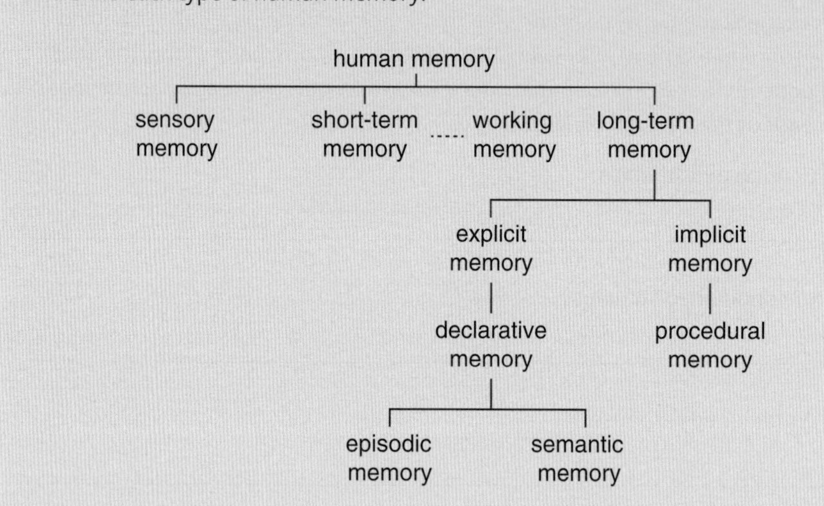

While learning is mostly concerned with behavior, the study of memory focuses on how we gain the knowledge that we accumulate over our lifetimes. The formation of memories can be divided into three major processes: encoding, storage, and retrieval.

MCAT EXPERTISE

Do not allow yourself to study for the MCAT using automatic processing! Just reading the text "to get through it" won't cut it for the MCAT. Engage with the text: fill out the MCAT Concept Checks, write notes in the margins, ask yourself questions. Scientific studies of learning have demonstrated, time and time again, that controlled processing improves comprehension, retention, and speed and accuracy on Test Day.

Encoding

Encoding refers to the process of putting new information into memory. Information gained without any effort is the result of **automatic processing**. This type of cognitive processing is unintentional, and information is passively absorbed from the environment. As you walk down the street, you are constantly bombarded with information that seeps into your brain: you notice the temperature; you keep track of the route that you're taking; you might stop for coffee and realize that the same cashier has been working each day this week.

There are, however, times when we must actively work to gain information. In studying for the MCAT, for example, you may create flashcards to memorize the enzymes of digestion or the functions of endocrine hormones. This active memorization is known as **controlled (effortful) processing**.

With practice, controlled processing can become automatic. Think back to a time when you were learning a foreign language. At first, each word required a great deal of processing to decipher: you had to hear the word and consciously translate it into your native language in order to understand what was being said. This process took an amount of time and effort that was probably difficult to maintain for prolonged periods. However, as you gained more experience with the language, this process became easier until you may have been able to understand those same words intuitively, without having to think very hard about them at all. At that point, this skill that once required controlled processing became automatic.

There are a few different ways that we encode the meaning of information when controlled processing is required. We can visualize information (**visual encoding**), store the way it sounds (**acoustic encoding**), link it to knowledge that is already in memory (**elaborative encoding**), or put it into a meaningful context (**semantic encoding**). Of these, semantic encoding is the strongest and visual encoding is the weakest. When using semantic encoding, the more vivid the context, the better. In fact, we tend to recall information best when we can put it into the context of our own lives, a phenomenon called the **self-reference effect**.

REAL WORLD

The purpose of the Real World sidebars in your *MCAT Review* books is semantic encoding: by putting content into a meaningful context, retention of the information is improved. Most of our Real World sidebars are related to medicine because of the self-reference effect.

Of course, grouping information into a meaningful context is only one trick that we can use to aid in encoding. Another such aid is **maintenance rehearsal**, which is the repetition of a piece of information to either keep it within working memory (to prevent forgetting) or to store it in short-term and eventually long-term memory—topics discussed in the next section.

Mnemonics are another common way to memorize information, particularly lists. As you've seen in your Kaplan study materials, mnemonics are often acronyms or rhyming phrases that provide a vivid organization of the information we are trying to remember. Two other mnemonic techniques are commonly employed by memory experts. The **method of loci** involves associating each item in a list with a location along a route through a building that has already been memorized. For example, in memorizing a grocery list, someone might picture a carton of eggs sitting on their doorstep, a person spilling milk in the front hallway, a giant stick of butter in the living room, and so on. Later, when the person wishes to recall the list, they simply take a mental walk through the locations and recall the images they formed earlier. Similarly, the **peg-word system** associates numbers with items that rhyme with or resemble the numbers. For example, one might be associated with the sun, two with a shoe, three with a tree, and so on. As groundwork, the individual memorizes their personal peg-list. When another list needs to be memorized, the individual can simply pair each item in the list with their peg-list. In this example, the individual may visualize eggs being fried by the sun (1), a pair of shoes (2) filled with milk, and a tree (3) with leaves made of butter. Because of the serial nature of both the method-of-loci and peg-word systems, they are very useful for memorizing large lists of objects in order.

REAL WORLD

Many feats of memory are accomplished via mnemonic techniques. In fact, the method of loci is a favorite among participants in the World Memory Championships.

Finally, **chunking** (sometimes referred to as **clustering**) is a memory trick that involves taking individual elements of a large list and grouping them together into groups of elements with related meaning. For example, consider the following list of 16 letters: E-N-A-L-P-K-C-U-R-T-R-A-C-S-U-B. Memorizing the list in order by rote might prove difficult until we realize that we can reverse the items and group them into meaningful chunks: BUS, CAR, TRUCK, PLANE.

Storage

Following encoding, information must be stored if it is to be remembered. There are several types of memory storage.

Sensory Memory

MCAT EXPERTISE

Sensory memory theoretically encompasses all five major senses, but studies have been mostly limited to sight, hearing, and touch (haptic memory).

The first and most fleeting kind of memory storage is **sensory memory**, which preserves information in its original sensory form (auditory, visual, etc.) with high accuracy and lasts only a very short time, generally less than one second. Sensory memory consists of both **iconic memory** (fast-decaying memory of visual stimuli) and **echoic memory** (fast-decaying memory of auditory stimuli). Sensory memories are maintained by the major projection areas of each sensory system, such as the occipital lobe for vision and the temporal lobe for hearing. Of course, sensory memory fades very quickly, so unless the information is attended to, it will be lost.

MCAT EXPERTISE

Eidetic memory refers to the ability to recall, with high precision, an image after only a brief exposure. It is hypothesized that eidetic memory represents an extreme example of iconic memory that endures for a few minutes. Although generally not observed in adults, it is reported to occur in a small percentage of children.

The nature of sensory memory can be demonstrated experimentally. Consider the following procedure: a research participant is presented with a three-by-three array of letters, such as that presented in Figure 3.5, that is flashed onto a screen for a mere fraction of a second. When asked to list all of the letters in the array, the participant is able to correctly identify three or four (a procedure known as **whole-report**). However, when asked to list the letters of a particular row immediately after the presentation of the stimulus (known as **partial-report**), the participant can do so with 100 percent accuracy, no matter which row is chosen. This is iconic memory in action: in the time it takes to list out a few of the items, the entire list fades; yet it is clear that all of the letters do make their way into iconic memory because any small subset can be recalled at will.

B	X	O
R	T	P
W	Q	L

Figure 3.5 A Sample 3-by-3 Array for Studying Sensory Memory

Short-Term Memory

Of course, we do pay attention to some of the information that we are exposed to, and that information enters our **short-term memory**. Similar to sensory memory, short-term memory fades quickly, over the course of approximately 30 seconds without rehearsal. In addition to having a limited duration, the number of items we can hold in our short-term memory at any given time, our **memory capacity**, is limited to approximately seven items, usually stated as the 7 ± 2 rule. As discussed in the previous section, the capacity of short-term memory can be increased by clustering information, and the duration can be extended using maintenance rehearsal. Short-term memory is housed primarily in the **hippocampus**, which is also responsible for the consolidation of short-term memory into long-term memory.

Working Memory

Working memory is closely related to short-term memory and is similarly supported by the hippocampus. It enables us to keep a few pieces of information in our con sciousness simultaneously and to manipulate that information. To do this, one must integrate short-term memory, attention, and executive function; accordingly, the frontal and parietal lobes are also involved. This is the form of memory that allows us to do simple math in our heads.

Long-Term Memory

With enough rehearsal, information moves from short-term to **long-term memory**, an essentially limitless warehouse for knowledge that we are then able to recall on demand, sometimes for the rest of our lives. One of the ways that information is consolidated into long-term memory is **elaborative rehearsal**. Unlike maintenance rehearsal, which is simply a way of keeping the information at the forefront of consciousness, elaborative rehearsal is the association of the information to knowledge already stored in long-term memory. Elaborative rehearsal is closely tied to the self-reference effect noted earlier; those ideas that we are able to relate to our own lives are more likely to find their way into our long-term memory. While long-term memory is primarily controlled by the hippocampus, it should be noted that memories are moved, over time, back to the cerebral cortex. Thus, very long-term memories—our names and birthdates, the faces of our parents—are generally not affected by damage to the hippocampus.

REAL WORLD

Have you ever looked at a picture of a simple unfinished puzzle, and been able to fit the pieces together mentally? This skill is explained by one of the major theories that underlies working memory, which includes the concept of a visuospatial sketchpad. The visuospatial sketchpad was proposed by Baddeley and Hitch as part of their three-part working memory model along with the other two components they proposed: the central executive and the phonological loop. The visuospatial sketchpad explains our ability to not only store visual and spatial information, but to manipulate it as well.

BEHAVIORAL SCIENCES GUIDED EXAMPLE WITH EXPERT THINKING

A musical melody is a series of tones played in succession that form a coherent musical idea. Melodies can be played in many different keys: absolute pitch of each of the notes changes as one changes keys, but as long as the relative pitches remain the same, the melody should still be recognizable. For example, the familiar melody of the song "Happy Birthday" should sound the same regardless of whether the first note is a C, an F, or a B-flat. While most people cannot name the key of a piece of music just by listening to it, they might still implicitly remember the way it "is supposed to sound" such that they could distinguish between the same melody played in different keys.

The passage provides the definition of melody, and the difference between absolute and relative pitch. Helpful if I'm not a musician, but really just setting the stage for the experiment.

Researchers interested in musical memory conducted an experiment with three groups of participants: 7- to 8-year-olds, 9- to 11-year-olds, and adults. Twenty-four British and Irish folk melodies were selected to be used as stimuli. The researchers expected that because these melodies share the sorts of patterns that are common in much North American and European music, they should thereby be recognizable as melodies rather than simply a series of tones. It was also expected that these melodies were uncommon enough to be unfamiliar to the participants prior to the beginning of the study.

Setup: Three groups, two children, one adult, presented with melodies. The melodies were intended to be unfamiliar.

IV: age

In the exposure phase, participants were presented with twelve of the twenty-four melodies to listen to. In the recognition phase, all twenty-four melodies were presented. Of the melodies that were presented in the exposure phase, six were then presented in the recognition phase transposed to a higher or lower key. Participants were asked to rate whether each melody they heard in the recognition phase had been presented in the exposure phase, and rated this recognition on a scale from 1 ("definitely new") to 6 ("definitely heard before").

Finally, the description of the procedure. I'll want to parse this carefully. In the second phase, some melodies were heard earlier and some weren't. Some of those that have been heard before were presented in a different key.

Second IV: transposition

DV: recognition

To eliminate bias, researchers used an Area Under the Curve (AUC) score, which measures the relative difference in recognition rather than the absolute difference. If all of the recognition scores for old melodies are higher than the recognition scores for new melodies for a given participant, the AUC score is 1.0, which represents perfect discrimination. A score of 0.5 represents chance performance, such that the old and new melodies are indistinguishable.

I've never heard of an AUC score before, but the passage explains it. I'll probably need this to interpret the results.

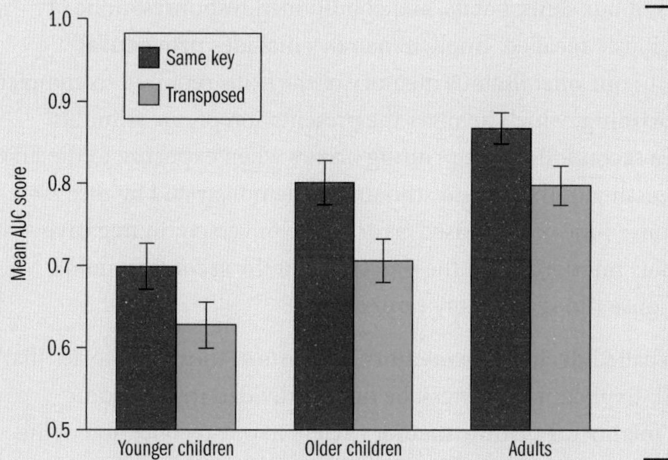

IV: age and transposition

DV: recognition (AUC)

Higher AUC means better recognition, so it looks like people get better as they get older. It also seems people were slightly worse at recognizing the transposed melodies.

Figure 1 Mean AUC score for each group

Adapted from: Schellenberg EG, Poon J, Weiss MW (2017) Memory for melody and key in childhood. *PLoS ONE* 12(10): e0187115. https://doi.org/10.1371/journal.pone.0187115

What do these results suggest about between-group and within-group trends with respect to explicit and implicit memory for music?

To answer this question, we're going to have to make use of the results of this study, but we'll also have to recall and apply content knowledge from outside this passage. The prompt asks about differences in explicit and implicit memory, so we'll want to start by figuring out how these terms apply in this general context, then apply these ideas specifically to the experiment. For the MCAT, we should know that explicit and implicit memory are both subdivisions of long-term memory. Explicit memory is the encoding of facts, and particularly relevant to the present study is episodic memory, the kind of explicit memory that involves experiences. The question "have I heard this melody before?" is answered by accessing an explicit memory. The relevance of implicit memory is more difficult here, since we typically think of implicit memories as procedural, involving skills and conditioned responses. Whenever we're not sure of how a concept in a question is related to the passage, we should go back to the passage and search for clues. The passage does provide a clue: in the first paragraph, the author describes the memory of the key of a melody as implicit rather than explicit. Now that we know what we're looking for, we can examine the results of the study with these concepts, and the way in which they relate to the passage, already in mind.

The AUC score system used by the researchers might be unfamiliar, but the concept isn't that much different from what we might normally see in a study like this one. A score of 0.5 represents random chance, and a score higher than that means that the participants were able to distinguish the melodies they'd heard from the ones they hadn't. The higher the score, the better the recognition. From the figure, we can see two trends: as participants get older, recognition gets better, but when the melody is transposed, recognition gets worse for all participants by approximately the same amount.

We must apply the memory vocabulary words to these trends. We can conclude from these results that explicit memory for music improved by age across groups. Further, within each group, implicit memory did play a significant role in recognition, because melodies that matched those heard earlier explicitly but not implicitly (i.e., they were the same melody but transposed to a different key) were less readily recognized. The role of implicit memory on recognition seems to be consistent between groups.

There are two types of long-term memory. **Implicit memory** (also called **nondeclarative memory**) consists of our skills, habits, and conditioned responses, none of which need to be consciously recalled. Implicit memory includes **procedural memory**, which relates to our unconscious memory of the skills required to complete procedural tasks, and **priming**, which involves the presentation of one stimulus affecting perception of a second. **Positive priming** occurs when exposure to the first stimulus improves processing of the second stimulus, as demonstrated by measures such as decreased response time or decreased error rate. Conversely, in **negative priming** the first stimulus interferes with the processing of the second stimulus, resulting in slower response times and more errors.

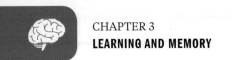

MCAT EXPERTISE

Although semantic and episodic memory are differentiated and can be separate, they can also co-occur. One type of explicit memory with components of both episodic and semantic memory is **flashbulb memory**, which is the detailed recollection of stimuli immediately surrounding an important (or emotionally arousing) event. Flashbulb memory helps you answer the question "Do you remember where you were when...?"

Explicit memory (also called **declarative memory**) consists of those memories that require conscious recall. Explicit memory can be further divided into episodic memory and semantic memory. **Episodic memory** refers to our recollection of life experiences. By contrast, **semantic memory** refers to ideas, concepts, or facts that we know, but are not tied to specific life experiences. **Autobiographical memory** is the name given to our explicit memories about our lives and ourselves, and includes all of our episodic memories of our own life experiences, but also includes semantic memories that relate to our personal traits and characteristics. Interestingly, memory disorders can affect one type of memory but leave others alone. For example, a patient who has amnesia might not remember learning to ride a bicycle (episodic memory) or the names of the parts of a bicycle (semantic memory), but may, surprisingly, retain the skill of riding a bicycle when given one (procedural memory). The various major categories of memory are summarized in Figure 3.6.

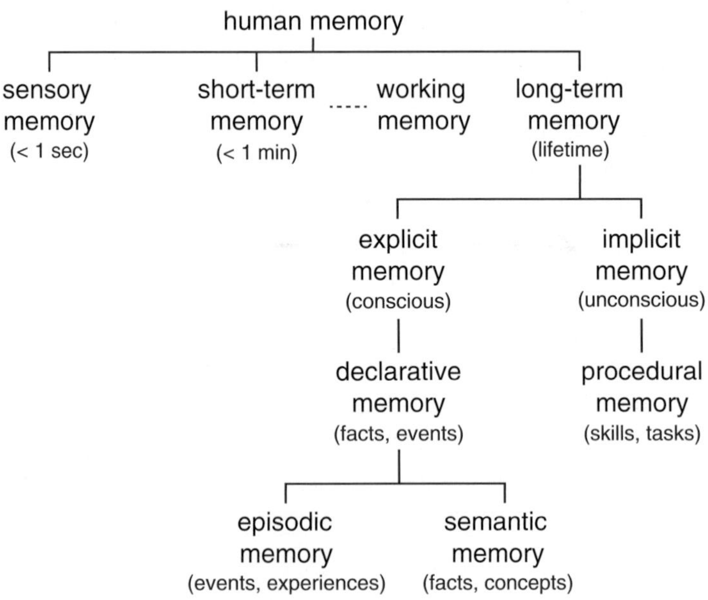

Figure 3.6 Types of Memory

Retrieval

Of course, memories that are stored are of no use unless we can pull them back out to use them. **Retrieval** is the name given to the process of demonstrating that something that has been learned has been retained. Most people think about retrieval in terms of **recall**, or the retrieval and statement of previously learned information, but learning can be additionally demonstrated by recognizing or quickly relearning information.

Recognition, the process of merely identifying a piece of information that was previously learned, is far easier than recall. This difference is something you can take advantage of because the MCAT, as a multiple-choice test, is largely based on recognizing information. If the MCAT were a fill-in-the-blank style exam, your approach to studying would have to be vastly different and far more in-depth.

REAL WORLD

Think back to elementary school. How many of your classmates do you think you could list? Chances are, not many. On the other hand, glancing through your class photo, you would probably recognize the vast majority of your former classmates. This gap is the difference between recall and recognition.

Relearning is another way of demonstrating that information has been stored in long-term memory. In studying the memorization of lists, Hermann Ebbinghaus found that his recall of a list of short words he had learned the previous day was often quite poor. However, he was able to rememorize the list much more quickly the second time through. Ebbinghaus interpreted this to mean that the information had been stored, even though it wasn't readily available for recall. Through additional research, he discovered that the longer the amount of time between sessions of relearning, the greater the retention of the information later on. Ebbinghaus dubbed this phenomenon the **spacing effect**, and it helps to explain why cramming is not nearly as effective as spacing out studying over an extended period of time.

Recalling a fact at a moment's notice can be difficult. Fortunately, the brain has ways of organizing information so that it can take advantage of environmental cues to tell it where to find a given memory. Psychologists think of memory not as simply a stockpile of unrelated facts, but rather as a network of interconnected ideas. The brain organizes ideas into a **semantic network**, as shown in Figure 3.7, in which concepts are linked together based on similar meaning, not unlike an Internet encyclopedia wherein each page includes links for similar topics. For example, the concept of *red* might be closely linked to other colors, like *orange* and *green*, as well as objects, like *fire engine* and *roses*. When one node of our semantic network is activated, such as seeing the word *red* on a sign, the other linked concepts around it are also unconsciously activated, a process known as **spreading activation**. Spreading activation is at the heart of the previously mentioned positive priming, as recall is aided by first being presented with a word or phrase, a **recall cue**, that is close to the desired semantic memory.

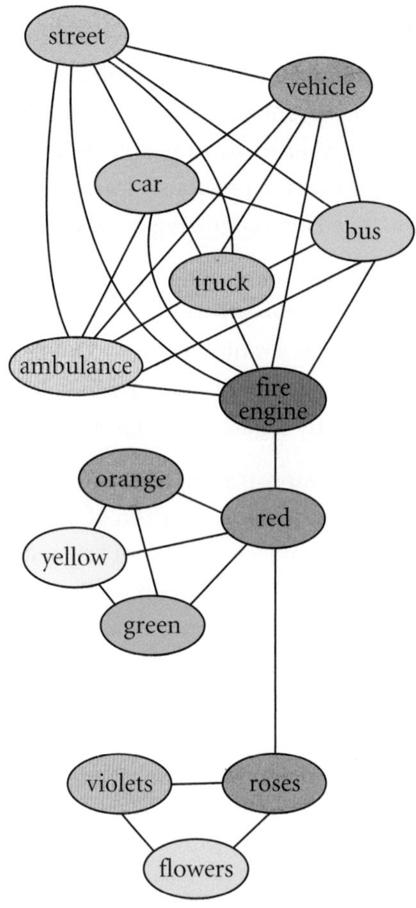

Figure 3.7 An Example Semantic Network

In spreading activation, the concept of red will also unconsciously activate other linked concepts.

Another common retrieval cue is **context effect**, where memory is aided by being in the physical location where encoding took place. Psychologists have shown a person will score better when they take an exam in the same room in which they learned the information. Context effects can go even further than this; facts learned underwater are better recalled when underwater than when on land. Similarly, **source monitoring** is a part of the retrieval process that involves determining the origin of memories, and whether they are factual (real and accurate) or fictional (from a dream, novel, or movie).

A person's mental state can also affect their ability to recall. **State-dependent memory,** alternately referred to as a **state-dependent effect**, is a retrieval cue based on performing better when in the same mental state as when the information was learned. People who learn facts or skills while intoxicated, for example, will show better recall or proficiency when performing those same tasks while intoxicated as compared to performing them while sober. Emotions work in a similar way: being in a foul mood primes negative memories, which in turn work to sustain the foul mood. So not only will memory be better for information learned when in a similar mood, but recall of negative or positive memories will lead to the persistence of the mood.

Finally, studies on list memorization have indicated that an item's position in the list affects participants' ability to recall, which Ebbinghaus termed the **serial-position effect**. When researchers give participants a list of items to memorize, the participants have much higher recall for both the first few and last few items on the list. The tendency to remember early and late items in the list is known as the **primacy** and **recency effect**, respectively. However, when asked to remember the list later, people show strong recall for the first few items while recall of the last few items fades. Psychologists interpret this to mean that the recency effect is a result of the last items still being in short-term memory on initial recall.

Forgetting

Unfortunately, even long-term memory is not always permanent. Several phenomena can result in **amnesia**, a significant loss of memorized information. The inability to remember where, when, or how one has obtained knowledge is called **source amnesia**.

Brain Disorders

There are several disorders that can lead to decline in memory. The most common is **Alzheimer's disease**, which is a degenerative brain disorder thought to be linked to a loss of acetylcholine in neurons that link to the hippocampus, although its exact causes are not well understood. Alzheimer's is marked by progressive **dementia** (a loss of cognitive function) and memory loss, with atrophy of the brain, as shown in Figure 3.8. While not perfectly linear, memory loss in Alzheimer's disease tends to proceed in a retrograde fashion, with loss of recent memories before distant memories. Microscopic findings of Alzheimer's include **neurofibrillary tangles** and **β-amyloid plaques**. One common phenomenon that occurs in individuals with middle- to late-stage Alzheimer's is **sundowning**, an increase in dysfunction in the late afternoon and evening.

BRIDGE

The β-amyloid plaques of Alzheimer's disease are incorrectly folded copies of the amyloid precursor protein, in which insoluble β-pleated sheets form and then deposit in the brain. Protein folding is discussed in detail in Chapter 1 of *MCAT Biochemistry Review*.

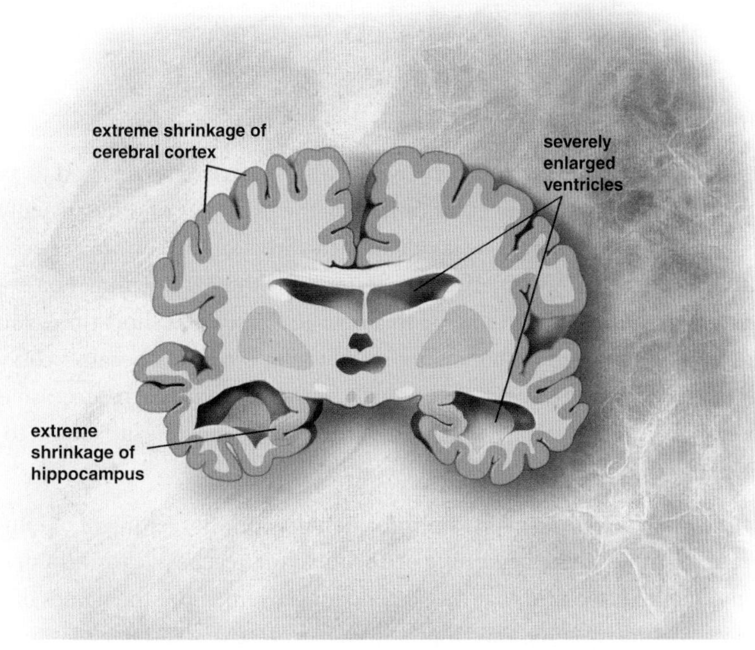

Figure 3.8 Findings of Alzheimer's Disease

Korsakoff's syndrome is another form of memory loss caused by thiamine deficiency in the brain. The disorder is marked by both **retrograde amnesia** (the loss of previously formed memories) and **anterograde amnesia** (the inability to form new memories). Another common symptom is **confabulation**, or the process of creating vivid but fabricated memories, typically thought to be an attempt made by the brain to fill in the gaps of missing memories.

Agnosia is the loss of the ability to recognize objects, people, or sounds, though usually only one of the three. Agnosia is usually caused by physical damage to the brain, such as that caused by a stroke or a neurological disorder such as multiple sclerosis.

Decay

Of course, not all memory loss is due to a disorder. Through a process known as **decay**, memories are simply lost naturally over time as the neurochemical trace of a short-term memory fades. In his word memorization experiment, Ebbinghaus noted what he called a "curve of forgetting," formally called the **retention function**, as shown in Figure 3.9. For a day or two after learning the list, recall fell sharply but then leveled off.

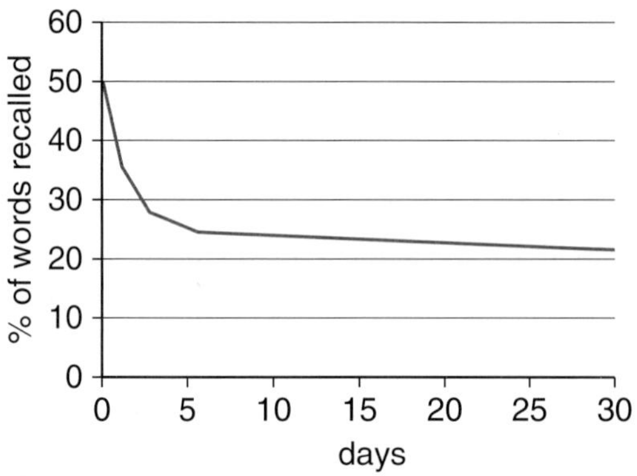

Figure 3.9 Ebbinghaus's Curve of Forgetting

Interference

Another common reason for memory loss is **interference** (also referred to as an **interference effect**), a retrieval error caused by the existence of other, usually similar, information. Interference can be classified by its direction. When we experience **proactive interference**, old information is interfering with new learning. For example, think back to a time when you moved to a new address. For a short time, you may have had trouble recalling individual pieces of the new address because you were so used to the old one. Similarly, Ebbinghaus found that with each successive list he learned, his recall for new lists decreased over time, an effect he attributed to interference caused by older lists.

Retroactive interference is when new information causes forgetting of old information. For example, at the beginning of a school year, teachers learning a new set of students' names often find that they can no longer remember the names of the previous year's students. One way of preventing retroactive interference is to reduce the number of interfering events, which is why it is often best to study in the evening about an hour before falling asleep (although this also depends on your personal style!).

Aging and Memory

Contrary to popular belief, aging does not necessarily lead to significant memory loss; while there are many individuals whose memory fades in old age, this is not always the case. In fact, studies show that there is a larger range of memory ability for 70-year-olds than there is for 20-year-olds. There are, however, some trends that can be demonstrated when evaluating the memories of older individuals. When asked about the most pivotal events in their lives, people in their 70s and 80s tend to say that their most vivid memories are of events that occurred in their teens and 20s, a fact that psychologists interpret to mean that this time is a peak period for encoding in a person's life.

Even for older adults, certain types of memory remain quite strong. People tend not to demonstrate much degeneration in recognition or skill-based memory as they age. Even certain types of recall will remain strong for most people; semantically meaningful material can be easily learned and recalled, most likely due to older individuals having a larger semantic network than their younger counterparts. **Prospective memory** (remembering to perform a task at some point in the future) remains mostly intact when it is event-based—that is, primed by a trigger event, such as remembering to buy milk when walking past the grocery store. On the other hand, time-based prospective memory, such as remembering to take a medication every day at 7:00 a.m., does tend to decline with age.

Memory Reconstruction

We often think of memory as a record of our experiences or a kind of video recording that is stored to be accessed later; this accurate recall of past events is defined as **reproductive memory**. Nothing could be further from the truth. **Reconstructive memory** is a theory of memory recall in which cognitive processes such as imagination, semantic memory, and perception affect the act of remembering. This theory explains how two people can recall the same event as occurring in completely different ways. A memory that incorrectly recalls actual events or recalls events that never occurred is known as a **false memory**. Despite their unsettling nature, false memories are common and are to be expected when we consider the many factors that can affect memory. Most memories are encoded with little detail, only focusing on the details deemed important in the moment. Also, as previously discussed, if a person repeatedly rehearses the memory in their mind, then that person may fill in missing details with unreliable information. Repressed memories, memories stored in the unconscious mind and blocked from recall, have also been a topic of controversy. Some psychologists believe repressed memories can be brought back into our conscious mind either spontaneously or through psychotherapy. Such memories are called **recovered memories**. However, it is not possible to distinguish between false memories and recovered memories without evidence and some research psychologists believe psychotherapy is more likely to lead to the creation of false memories. So, the act of recalling a memory can result in the production of a false memory.

False memory production is not only limited to internal factors. Memories can also be affected by outside sources as well. In a famous experiment, participants were shown several pictures including one picture of a car stopped at a yield sign. Later, these participants were presented with written descriptions of the pictures, and some of these descriptions contained misinformation, such as a description of a car stopped at a stop sign. When asked to recall the details of the pictures, many participants insisted they had seen a stop sign in the picture. This example illustrates the **misinformation effect**, where a person's recall of an event becomes less accurate due to the injection of outside information into the memory.

The misinformation effect can also be seen at the point of recall. In another experiment, participants were shown a video of an automobile accident. Some participants were then asked, *How fast were the cars moving when they collided?*, while others were asked about the accident using more descriptive language such as *How fast were the cars moving when they crashed?* Those participants who were asked the question with leading language were much more likely to overstate the severity of the accident than those who had been asked the question with less descriptive language.

Intrusion errors refers to false memories that have included a false detail into a particular memory. This is similar to the misinformation effect but distinct in that the intrusion error is not from an outside source. Instead, the intruding memory is injected into original memory due to both memories being related or sharing a theme. Upon memory recall, the brain incorrectly associates the intruding memory with the source memory, leading to a false memory. For example, if over the years you've attended multiple New Year's Eve celebrations in two different cities, then your memories of the two cities are linked. A possible intrusion error could be recalling that a particular restaurant is located in Vancouver, because you recall eating at the restaurant on New Year's Eve and celebrating New Year's Eve in Vancouver. However, the restaurant is really in Toronto, where you have also celebrated on New Year's.

REAL WORLD

During a Congressional Medal of Honor ceremony in 1983, Ronald Reagan relayed a vivid story about a heroic World War II pilot who received a posthumous medal. Skeptical reporters, unaware of any incident matching the details of Reagan's story, checked into the story and found that the pilot had existed—in the 1944 movie *A Wing and a Prayer*. Reagan had remembered the details of the pilot's heroic actions but had forgotten their source.

Source-monitoring error involves confusion between semantic and episodic memory: a person remembers the details of an event, but confuses the context under which those details were gained. Source-monitoring error often manifests when a person hears a story of something that happened to someone else, and later recalls the story as a personal memory.

MCAT CONCEPT CHECK 3.2

Before you move on, assess your understanding of the material with these questions.

1. List the three modes in which information can be encoded, from strongest to weakest.

    1. ______________________________
    2. ______________________________
    3. ______________________________

2. In what ways is maintenance rehearsal different from elaborative rehearsal?

    ____

3. In terms of recall, why might it be a bad idea to study for the MCAT while listening to music?

    ____

4. What are some factors that might cause eyewitness courtroom testimony to be unreliable?

    ____

3.3 Neurobiology of Learning and Memory

LEARNING OBJECTIVES

After Chapter 3.3, you will be able to:

- Describe neuroplasticity and changes that occur throughout the human life span
- Recall the terms involved with removing and strengthening memory connections

Even as you read this text, your brain is changing. Memory, and therefore learning, involves changes in brain physiology, such that with each new concept you learn your brain is altering its synaptic connections in response. You may have heard that it is far easier for children to learn a new language than it is for adults. Indeed, the saying *you can't teach an old dog new tricks*, while not strictly true, does have its roots in neurobiology.

As infants, we are born with many more neurons than we actually need. As our brains develop, neural connections form rapidly in response to stimuli via a phenomenon called **neuroplasticity** (also known as **neural plasticity**). In fact, the brains of young children are so plastic that they can reorganize drastically in response to injury, as evidenced by studies of children who have had entire hemispheres of their brains removed to prevent severe seizures. The remaining hemisphere will change to take over functions of the missing parts of the brain, allowing these children to grow up to lead unimpaired lives. While our brains do maintain a degree of plasticity throughout our lives, adult brains display nowhere near the degree of plasticity as those of a child. Another way our brains change is through a process called **synaptic pruning**. As we grow older, weak neural connections are broken while strong ones are bolstered, increasing the efficiency of our brains' ability to process information.

MCAT EXPERTISE

Recent research has begun to elucidate the mechanism of long-term potentiation. It has been observed that a specific type of glutamate receptor, the **NMDA receptor**, is required for the strengthening of synaptic connections.

MNEMONIC

The word *potentiate* means to increase the potency or strength of something. Long-term potentiation can be thought of as the strengthening of a "long-term" synaptic connection.

This concept of plasticity is important because it is closely linked to learning and memory. As you learned in Chapter 4 of *MCAT Biology Review*, stimuli cause activation of neurons, which release their neurotransmitters into the **synaptic cleft**, the gap between a neuron and a target cell. These neurotransmitters continue to stimulate activity until degradation, reuptake, or diffusion out of the synaptic cleft. In the interim, this neural activity forms a memory trace that is thought to be the cause of short-term memory. As discussed earlier, if the stimulus isn't repeated or rehearsed, the memory trace disappears, and the consequence is the loss of the short-term memory. However, as the stimulus is repeated, the stimulated neurons become more efficient at releasing their neurotransmitters and at the same time receptor sites on the other side of the synapse increase, increasing receptor density. The strengthening of neural connections through repeated use is known as **long-term potentiation**, and is believed to be the neurophysiological basis of long-term memory.

As described in the previous section, a memory begins its life as a sensory memory in the projection area of a given sensory modality. This sensory memory is brief, unless maintained in consciousness and moved, as a short-term memory, into the hippocampus in the temporal lobe. The memory can then be manipulated through working memory while in the hippocampus (in tandem with the frontal and parietal lobes), and even stored for later recall. Over very long periods of time, memories are gradually moved from the hippocampus back to the cerebral cortex. Note that this general pathway is a drastic oversimplification of the complex interplay of brain regions involved in memory, but is a useful paradigm for Test Day.

MCAT CONCEPT CHECK 3.3

Before you move on, assess your understanding of the material with these questions.

1. What is neuroplasticity? How does neuroplasticity change during life?

____

____

2. What is the term for removing weak neural connections? What is the term for strengthening memory connections through increased neurotransmitter release and receptor density?
   - Removing weak connections:

   ____

   - Strengthening connections:

   ____

Conclusion

In this chapter, we discussed two very important ways that we react to our environments. We are constantly receiving input from the world around us, and the way we memorize that information depends greatly on both the nature of the information and its importance to us individually. That information can also have a profound effect on us, causing us to increase or decrease the frequency of certain behaviors, sometimes without our conscious knowledge. Because the concepts of learning and memory are both used heavily in research, we can expect the MCAT to place many of its passages testing these topics within an experimental context.

GO ONLINE

You've reviewed the content, now test your knowledge and critical thinking skills by completing a test-like passage set in your online resources!

CONCEPT SUMMARY

Learning

- **Habituation** is the process of becoming used to a stimulus. **Dishabituation** can occur when a second stimulus intervenes, causing a resensitization to the original stimulus.
- **Associative learning** is a way of pairing together stimuli and responses, or behaviors and consequences.
- In **classical conditioning**, an unconditioned stimulus that produces an instinctive, unconditioned response is paired with a neutral stimulus. With repetition, the neutral stimulus becomes a conditioned stimulus that produces a conditioned response.
- In **operant conditioning**, behavior is changed through the use of consequences.
    - **Reinforcement** increases the likelihood of a behavior.
    - **Punishment** decreases the likelihood of a behavior.
    - The schedule of reinforcement affects the rate at which the behavior is performed. Schedules can be based either on a ratio of behavior to reward or on an amount of time, and can be either fixed or variable. Behaviors learned through variable-ratio schedules are the hardest to extinguish.
- **Observational learning**, or **modeling**, is the acquisition of behavior by watching others.

Memory

- **Encoding** is the process of putting new information into memory. It can be **automatic** or **effortful**. Semantic encoding is stronger than both acoustic and visual encoding.
- **Sensory** and **short-term memory** are transient and are based on neurotransmitter activity. **Working memory** requires short-term memory, attention, and executive function to manipulate information.
- **Long-term memory** requires elaborative rehearsal and is the result of increased neuronal connectivity.
    - **Explicit** (**declarative**) **memory** stores facts and stories.
    - **Implicit** (**nondeclarative**) **memory** stores skills and conditioning effects.
- Facts are stored via **semantic networks**.
- **Recognition** of information is stronger than **recall**.
- **Retrieval** of information is often based on **priming** interconnected nodes of the semantic network.

- Memories can be lost through disorders such as Alzheimer's disease, Korsakoff's syndrome, or agnosia; decay; or interference.
- Memories are highly subject to influence by outside information and mood both at the time of encoding and at recall.

Neurobiology of Learning and Memory

- Both learning and memory rely on changes in brain chemistry and physiology, the extent of which depends on **neuroplasticity**, which decreases as we age.
- **Long-term potentiation**, responsible for the conversion of short-term to long-term memory, is the strengthening of neuronal connections resulting from increased neurotransmitter release and adding of receptor sites.

ANSWERS TO CONCEPT CHECKS

3.1

1. Eating each course of a meal before moving on to the next causes habituation; each bite causes less pleasurable stimulation, so people feel less desire to keep eating. On the other hand, mixing up the courses of a meal causes dishabituation for taste, which would cause people to eat more overall.
2. The conditioned stimulus is the smell of popcorn. The unconditioned stimulus is the popping of the balloon. The conditioned response is nervousness (fear) in response to the presence of popcorn. The unconditioned response is fear in response to the popping of the balloon.
3. Negative reinforcement causes an increase of a given behavior by removing something unpleasant, while positive punishment reduces behavior by adding something unpleasant. Examples will vary, but common negative reinforcers include medicines that reduce pain or avoiding uncomfortable situations to reduce anxiety. Common examples of positive punishments include having to pay a fine for speeding while driving or getting detention in school for bad behavior.

3.2

1. Of the three modes in which information can be encoded, semantic is the strongest, followed by acoustic. Visual is the weakest.
2. Maintenance rehearsal is the repetition of information to keep it within short-term memory for near-immediate use. Elaborative rehearsal is the association of information to other stored knowledge and is a more effective way to move information from short-term to long-term memory.
3. Because you will be taking the MCAT in a quiet room, studying under similar circumstances will aid recall due to context effects. Music may also compete for attention, reducing your ability to focus on the relevant study material.
4. Several factors can affect the accuracy of eyewitness testimony, including the manner in which questions are asked; the nature of information shared with the witness by police, lawyers, and other witnesses following the event; the misinformation effect; source-monitoring error; and the amount of time elapsed between the event and the trial. Even watching crime dramas, watching the news, or witnessing similar events can cause source-monitoring error.

3.3

1. Neuroplasticity is the ability of the brain to form new connections rapidly. The brain is most plastic in young children, and plasticity quickly drops off after childhood.
2. Pruning is the term for removing weak neural connections. Long-term potentiation is the strengthening of memory connections through increased neurotransmitter release and receptor density.

SCIENCE MASTERY ASSESSMENT EXPLANATIONS

1. **A**

After a while, the participant became habituated to the sound of the buzzer. Introducing a new stimulus, such as the banging pans, should dishabituate (resensitize) the original stimulus, causing a temporary increase in response to the sound of the buzzer.

2. **C**

The sound of a can opener would not normally produce a response on its own, making it a stimulus that must have been conditioned by association with food.

3. **B**

Generalization is the process by which similar stimuli can produce the same conditioned response. Here, the response to the taste and smell of oranges has generalized to that of all citrus.

4. **D**

Avoidance learning is a type of negative reinforcement in which a behavior is increased to prevent an unpleasant future consequence. Extinction, **(A)**, is a decreased response to a conditioned stimulus when it is no longer paired with an unconditioned stimulus. Punishment, **(B)** and **(C)**, leads to decreased behaviors in operant conditioning.

5. **A**

Long term potentiation is believed to be the neurophysiological basis of long-term memory, making **(A)** the correct answer. As synapses are reinforced, the neuroplasticity of the brain decreases, eliminating **(C)**. Also, synaptic pruning, **(B)**, refers to the removal of infrequently used synapses, which is not consistent with the description in the question stem.

6. **C**

In a fixed-interval schedule, the desired behavior is rewarded the first time it is exhibited after the fixed interval has elapsed. Both fixed-interval and fixed-ratio schedules tend to show this phenomenon: almost no response immediately after the reward is given, but the behavior increases as the rat gets close to receiving the reward.

7. **A**

Complicated, multistage behaviors are typically taught through shaping, so statement III must not be part of the correct answer. Reinforcers do not necessarily need to be food-based, and instinctive drift can interfere with learning of complicated behaviors; therefore, only statement I is accurate.

8. **C**

This is the definition of controlled processing and is the only answer choice that is necessarily true of controlled processing. Effortful processing is used to create long-term memories, and—with practice—can become automatic, invalidating **(A)** and **(B)**. Most of our day-to-day activities are processed automatically, making **(D)** incorrect.

9. **A**

Semantic encoding, or encoding based on the meaning of the information, is the strongest of the methods of encoding. Visual encoding, **(B)**, is the weakest, and acoustic encoding, **(D)**, is intermediate between the two. Iconic memory, **(C)**, is a type of sensory memory.

10. **C**

We are born with an overabundance of neurons and quickly form many new synapses in the first few years of life. As we age, the plasticity of our brains decreases, although we do retain some plasticity throughout adulthood. Thus, the brain of a two year old, due to its higher neuroplasticity, would better adapt, supporting **(C)** as the correct answer.

11. **D**

The association of words on a list to a preconstructed set of ideas is common to both the method-of-loci and peg-word mnemonics. Method-of-loci systems, **(B)**, associate items with locations, while peg-word systems use images associated with numbers.

12. **A**

Partial-report procedures, in which the individual is asked to recall a specific portion of the stimulus, are incredibly accurate, but only for a very brief time. This is a method of studying sensory (specifically, iconic) memory. Both the serial position effect, **(B)**, and the 7 ± 2 rule, **(C)**, are characteristics of short-term memory.

13. **B**

Semantic memory is the category of long-term memory that refers to recall of facts, rather than experiences or skills. Be careful not to confuse semantic memory with semantic networks, **(D)**, which are the associations of similar concepts in the mind to aid in their retrieval.

14. **C**

State-dependent recall is concerned with the internal rather than external states of the individual. As such, both statements II and III are examples of state-dependent circumstances, while statement I might cause a context effect instead.

15. **B**

Older adults may have trouble with time-based prospective memory, which is remembering to do an activity at a particular time. Other forms of memory are generally preserved, or may decline slightly but less significantly than time-based prospective memory.

Consult your online resources for additional practice. GO ONLINE

SHARED CONCEPTS

Behavioral Sciences Chapter 1
Biology and Behavior

Behavioral Sciences Chapter 2
Sensation and Perception

Behavioral Sciences Chapter 4
Cognition, Consciousness, and Language

Behavioral Sciences Chapter 5
Motivation, Emotion, and Stress

Behavioral Sciences Chapter 7
Psychological Disorders

Biology Chapter 4
The Nervous System

CHAPTER 4

COGNITION, CONSCIOUSNESS, AND LANGUAGE

SCIENCE MASTERY ASSESSMENT

Every pre-med knows this feeling: there is so much content I have to know for the MCAT! How do I know what to do first or what's important?

While the high-yield badges throughout this book will help you identify the most important topics, this Science Mastery Assessment is another tool in your MCAT prep arsenal. This quiz (which can also be taken in your online resources) and the guidance below will help ensure that you are spending the appropriate amount of time on this chapter based on your personal strengths and weaknesses. Don't worry though—skipping something now does not mean you'll never study it. Later on in your prep, as you complete full-length tests, you'll uncover specific pieces of content that you need to review and can come back to these chapters as appropriate.

How to Use This Assessment

If you answer 0–7 questions correctly:

Spend about 1 hour to read this chapter in full and take limited notes throughout. Follow up by reviewing **all** quiz questions to ensure that you now understand how to solve each one.

If you answer 8–11 questions correctly:

Spend 20–40 minutes reviewing the quiz questions. Beginning with the questions you missed, read and take notes on the corresponding subchapters. For questions you answered correctly, ensure your thinking matches that of the explanation and you understand why each choice was correct or incorrect.

If you answer 12–15 questions correctly:

Spend less than 20 minutes reviewing all questions from the quiz. If you missed any, then include a quick read-through of the corresponding subchapters, or even just the relevant content within a subchapter, as part of your question review. For questions you answered correctly, ensure your thinking matches that of the explanation and review the Concept Summary at the end of the chapter.

1. Which of the following terms describes how existing schemata are modified to incorporate new information?
   - **A.** Assimilation
   - **B.** Adaptation
   - **C.** Affirmation
   - **D.** Accommodation

2. After completing final exams, a student reports to student health complaining of paranoia, lack of appetite, and elevated heart rate. After some questioning, the student admits to having been awake for over 48 hours after having bought pills from another student in the library. Which of the following drugs has this student likely taken?
   - **A.** Phenobarbitol
   - **B.** Dextroamphetamine
   - **C.** Oxycodone
   - **D.** Lysergic acid diethylamide

3. A student is volunteering in a hospital with a stroke center. When asked what the prevalence of stroke is among those greater than 65 years old, the student states that it is probably about 40% even though data analysis indicates that it is significantly lower. What accounts for this error?
   - **A.** Deductive reasoning
   - **B.** Representativeness heuristic
   - **C.** Base rate fallacy
   - **D.** Confirmation bias

4. Which of the following types of intelligence is NOT described by Gardner's theory of multiple intelligences?
   - **A.** Fluid intelligence
   - **B.** Bodily–kinesthetic intelligence
   - **C.** Visual–spatial intelligence
   - **D.** Linguistic intelligence

5. EEG waveforms during REM sleep most resemble which of the following states of consciousness?
   - **A.** Alertness
   - **B.** Slow-wave sleep
   - **C.** Stage 1 sleep
   - **D.** Meditation

6. Which of the following indicates the pattern of sleep stages during a complete sleep cycle early in the night?
   - **A.** 1–2–3–4–1–2–REM
   - **B.** 1–2–3–4–3–2–REM
   - **C.** 4–3–2–1–2–3–REM
   - **D.** 4–3–2–4–3–1–REM

7. Which of the following best explains a student's ability to sit on the couch and watch reruns of a favorite TV show while studying for a chemistry exam?
   - **A.** Selective attention
   - **B.** Divided attention
   - **C.** Shadowing
   - **D.** Parallel processing

8. Which theory of dreaming states that dreams and thoughts during wakeful periods use the same stream-of-consciousness system?
   - **A.** Activation–synthesis theory
   - **B.** Problem solving theory
   - **C.** Cognitive process theory
   - **D.** Neurocognitive theory

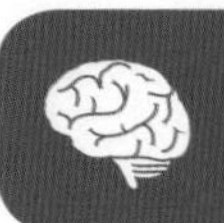

9. A 19-year-old college student with bloodshot eyes is picked up by campus police after shoplifting a large bag of corn chips and a dozen ice cream sandwiches. During questioning, the student cannot stop giggling and repeatedly asks for water, complaining of dryness in the mouth. What is the psychoactive substance in the drug this student has most likely recently taken?
   - **A.** Alprazolam
   - **B.** 3,4-Methylenedioxy-*N*-methylamphetamine
   - **C.** Diacetylmorphine
   - **D.** Tetrahydrocannabinol

10. Language consists of multiple components. Which of the following involves the order in which words are put together?
   - **A.** Phonology
   - **B.** Semantics
   - **C.** Syntax
   - **D.** Pragmatics

11. A child speaks in sentences of at least 3 words, but makes grammatical errors including misuse of the past tense. How old is this child likely to be?
   - **A.** 14 months
   - **B.** 22 months
   - **C.** 30 months
   - **D.** 5 years

12. Which language theory states that language development occurs due to preferential reinforcement of certain phonemes by parents and caregivers?
   - **A.** Nativist theory
   - **B.** Learning theory
   - **C.** Social interactionist theory
   - **D.** Neurocognitive theory

13. A stroke patient comprehends speech but cannot properly move the mouth to form words. Which of the following brain areas is likely affected?
   - **A.** Broca's area
   - **B.** Wernicke's area
   - **C.** Arcuate fasciculus
   - **D.** Superior temporal gyrus

14. A person sits at a terminal of the airport and works on a challenging sudoku puzzle. The individual ignores most of the intercom announcements but, after an announcement indicates that boarding for the flight has begun, quickly gets in line. Which of the following best explains the person's rapid response for boarding?
   - **A.** Cocktail party phenomenon
   - **B.** Divided attention
   - **C.** Automatic processing
   - **D.** Effortful processing

15. During which of the following stages does dreaming occur?
   - **I.** Stage 3
   - **II.** Stage 4
   - **III.** REM

   - **A.** I only
   - **B.** II only
   - **C.** III only
   - **D.** I, II, and III

Answer Key

1. **D**
2. **B**
3. **C**
4. **A**
5. **A**
6. **B**
7. **B**
8. **C**
9. **D**
10. **C**
11. **C**
12. **B**
13. **A**
14. **A**
15. **D**

Detailed explanations can be found at the end of the chapter.

CHAPTER 4

COGNITION, CONSCIOUSNESS, AND LANGUAGE

In This Chapter

CHAPTER PROFILE

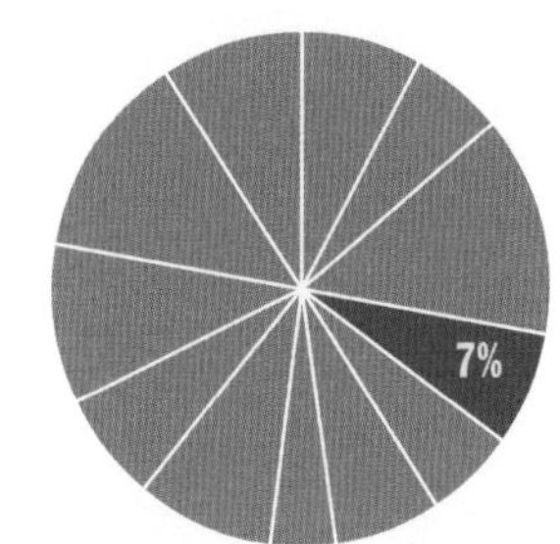

The content in this chapter should be relevant to about 7% of all questions about the behavioral sciences on the MCAT.

This chapter covers material from the following AAMC content category:

6B: Making sense of the environment

Introduction

As you think and move through the world, you often take your brain for granted. As you read, speak, ponder, make decisions, and perform complex motor functions, your brain is rapidly using electrical and chemical impulses to encode, store, and retrieve information. Most of these processes occur without your awareness or conscious thought. Imagine going to the grocery store. You fill your cart while comparing prices, assessing the produce, and planning what meals to make in the near future. After the cashier totals your purchases, you pull out a debit card, punch in a PIN, and leave with your groceries. While you were shopping in that grocery store, your brain was busy taking in all of the information around it and deciding which stimuli required attention. At the same time, you were making conscious decisions about your purchases, likely daydreaming, and maybe even singing along to music playing in the background. But, to your awareness, this was still just a simple trip to the store because most of the time, you don't even notice the tremendous processing power of your brain as you navigate the world.

But in some ways, this capacity for simultaneous conscious thinking, daydreaming, and decision making is what makes us human. Many of these functions are under the province of the frontal lobe, which—in comparison to other species on this planet—is disproportionately large in *H. sapiens sapiens*. Your frontal lobe enables you to eschew instantaneous reward and to seek out delayed gratification, like studying for the MCAT to get that high score you deserve. The frontal lobe also controls your production of language, which permits you to transmit ideas between individuals, cultures, and time. Finally, the frontal lobe helps you coordinate your thinking by deciding which stimuli deserve your attention. These are functions that are indispensable to your daily functioning and will be the focus of this chapter.

4.1 Cognition

High-Yield

LEARNING OBJECTIVES

After Chapter 4.1, you will be able to:

- List the steps in the information processing model of cognition
- Describe the effects of aging, heredity, and environment on cognitive function
- Recall Piaget's four stages of cognitive development and their key features

The study of **cognition** looks at how our brains process and react to the incredible information overload presented to us by the world. Cognition, overall, is not a uniquely human trait, but we are certainly the most advanced species on the planet in terms of complex thought. As described in the introduction, the frontal lobe is disproportionately large in our subspecies; a comparison to our recent anthropological ancestors demonstrates that our skull is shaped to accommodate this enlarged lobe, as shown in Figure 4.1.

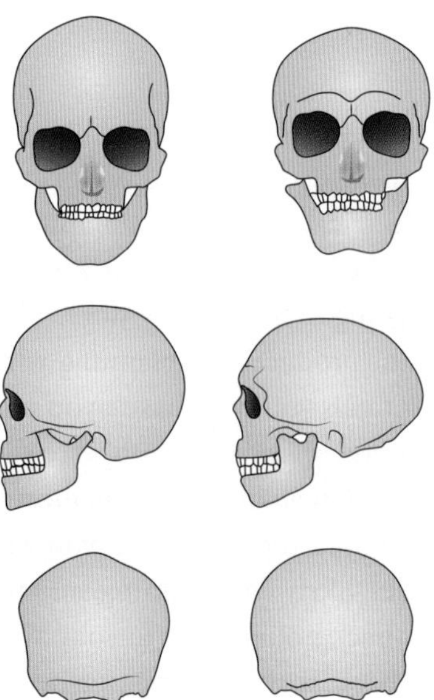

Figure 4.1 Skulls of *H. sapiens* (left) and *H. neanderthalensis* (right)

Information Processing Model

In the 1950s, much of science and engineering turned toward the production of computers and artificial intelligence. It was noted that certain steps were required in order to use a computer to store and process information. First, the information must be encoded in a language that the computer understands. Then, the information must be stored in such a way that it can be found later. Finally, the computer must be able to retrieve that information when required.

Psychologists took this model of information processing and applied it to the human brain. They theorized that the brain is somewhat like a computer. It must encode information into a series of chemical and electrical signals. Then, the brain must be able to store this information such that it can be retrieved when needed. Then, there must be a process by which the brain is able to retrieve information.

The manner in which information is encoded, stored, and retrieved has been a consistent source of debate. One prominent theory, Paivio's **dual-coding theory**, states that both verbal association and visual images are used to process and store information. For example, the word "tree" can recall some information, and a picture of a tree can recall that same information. The fact that we can code this information in two different ways builds redundancy and increases the chance that the information can be retrieved and used effectively when cued, much like search engine optimization within a computer program.

However, the human brain is not a computer. While the computer analogy creates a simple paradigm by which information is processed by the brain, it does not tell the whole story. The human brain doesn't just handle information in the form of facts: it also handles emotions, sensations such as smell and taste, as well as memories. As discussed in Chapter 3 of *MCAT Behavioral Sciences Review*, encoding, storage, and retrieval are often flavored by context and emotion.

BRIDGE

The key memory processes of encoding, storage, and retrieval are covered in Chapter 3 of *MCAT Behavioral Sciences Review.*

The **information processing model** has four key components, or pillars:

- Thinking requires sensation, encoding, and storage of stimuli.
- Stimuli must be analyzed by the brain (rather than responded to automatically) to be useful in decision making.
- Decisions made in one situation can be extrapolated and adjusted to help solve new problems (also called situational modification).
- Problem solving is dependent not only on the person's cognitive level, but also on the context and complexity of the problem.

Cognitive Development

Cognitive development is the development of one's ability to think and solve problems across the life span. Interestingly, during childhood, cognitive development is limited by the pace of brain maturation. Early cognitive development includes learning control of one's own body as well as learning how to interact with and manipulate the environment. Early cognitive development is characterized by mastering the

REAL WORLD

Abstract thought, the ability to think about things that are not physically present, can be lost in some mental disorders. For example, a common cognitive test with patients who have schizophrenia is to ask them to interpret a cliché, such as *Don't count your chickens before they hatch.* These patients have concrete thinking and will give an answer focused on the chickens themselves—not the underlying concept.

KEY CONCEPT

Piaget's stages of cognitive development:

- Sensorimotor
- Preoperational
- Concrete operational
- Formal operational

physical environment. As physical tasks are mastered, a new challenge looms for a developing child: abstract thinking. As discussed in Chapter 1 of *MCAT Behavioral Sciences Review*, social skills also develop during the lifetime.

As you will see during our review of Piaget's stages of cognitive development, the development of the ability to think abstractly is developed throughout childhood. The development of abstract thinking is also dependent upon increases in working memory and mental capacities. As the brain develops, the ability to process information in an abstract manner also develops.

Piaget's Stages of Cognitive Development

Jean Piaget was one of the most influential figures in developmental psychology. Piaget's model of cognitive development proposes that there are qualitative differences between the way that children and adults think, and that these differences can be explained by dividing the life span into four stages of cognitive development: sensorimotor, preoperational, concrete operational, and formal operational. Piaget believed that passage through each of these stages was a continuous and sequential process in which completion of each stage prepares the individual for the stage that follows.

Before delving into the actual stages, we have to look at how Piaget explained learning. According to Piaget, infants learn mainly through instinctual interaction with the environment. For example, infants possess a grasping reflex. Through experience with this reflex, the infant learns that it is possible to grasp objects. Piaget referred to these organized patterns of behavior and thought as schemata. A **schema** can include a concept (What is a *dog*?), a behavior (What do you do when someone asks you your name?), or a sequence of events (What do you normally do in a sit-down restaurant?). As a child proceeds through the stages, new information has to be placed into the different schemata. Piaget theorized that new information is processed via **adaptation**. According to Piaget, adaptation to information comes about by two complementary processes: assimilation and accommodation. **Assimilation** is the process of classifying new information into existing schemata. If the new information does not fit neatly into existing schemata, then accommodation occurs. **Accommodation** is the process by which existing schemata are modified to encompass this new information.

The first stage in Piaget's model is the **sensorimotor stage**, starting at birth and lasting until about two years of age. In this stage, a child learns to manipulate the environment in order to meet physical needs and learns to coordinate sensory input with motor actions (hence the name *sensorimotor*). To explore their surroundings, infants in the sensorimotor stage begin to exhibit two types of behavior patterns called **circular reactions**, named for their repetitive natures. **Primary circular reactions** are repetitions of body movements that originally occurred by chance, such as sucking the thumb. Usually such behaviors are repeated because the child finds these behaviors soothing. **Secondary circular reactions** occur when manipulation is focused on something outside the body, such as repeatedly throwing toys from a high chair. These behaviors are often repeated because the child gets a response from the environment, such as a parent picking up the dropped toy. The key milestone that ends the sensorimotor stage is the development of **object permanence**, which is the understanding that objects continue to exist even when out of view. Object permanence is the idea behind "peek-a-boo," shown in Figure 4.2. This game is so entertaining to young infants because they lack object permanence. Each time an adult reveals the face from behind hands, the child interprets it as though the adult has just come into existence. Object permanence marks the beginning of **representational thought**, in which the child has begun to create mental representations of external objects and events.

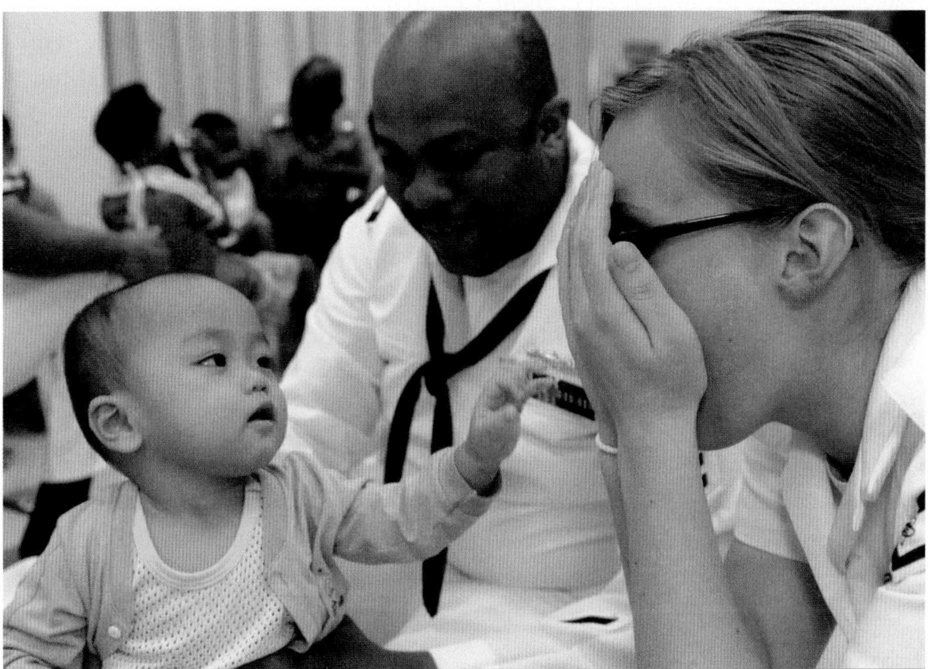

Figure 4.2 Peek-a-Boo
This game depends on the child being in the sensorimotor stage, prior to the development of object permanence.

The **preoperational stage** lasts from about two to seven years of age and is characterized by symbolic thinking and egocentrism. **Symbolic thinking** refers to the ability to pretend, play make-believe, and have an imagination. **Egocentrism** refers to the inability to imagine what another person may think or feel. The preoperational stage also includes the inability to grasp the concept of **conservation**, which is the understanding that a physical amount remains the same, even if there is a change in shape or appearance. For example, imagine a child presented with two equal quantities of pizza. On one plate is a single large slice, while the other plate has the exact same quantity in two slices. A child in the preoperational stage will be unable to tell that the quantities are equal and will focus mainly on the number of slices on the plate rather than the actual quantity. Piaget believed that this flaw in cognition was due to **centration**, which is the tendency to focus on only one aspect of a phenomenon, for example the number of slices, while ignoring other important elements.

The **concrete operational stage** lasts from about 7 to 11 years of age. In this stage, children can understand conservation and consider the perspectives of others. This consideration results in the loss of egocentrism. Additionally, they are able to engage in logical thought as long as they are working with concrete objects or information that is directly available. These children have not yet developed the ability to think abstractly.

The **formal operational stage** starts around 11 years of age, and is marked by the ability to think logically about abstract ideas. Generally coinciding with adolescence, this stage is marked by the ability to reason about abstract concepts and problem solve. The difference between this type of thought and concrete operations is illustrated by Piaget's pendulum experiment. Children were given a pendulum in which they could vary the length of the string, the weight of the pendulum, the force of the push, and the initial angle of the swing. They were asked to find out what determined the frequency of the swing. Children in the concrete operational stage manipulated the variables at random and even distorted the data to fit preconceived hypotheses. Adolescents, on the other hand, were able to hold all variables but one constant at a given time, proceeding methodically to discover that only the length of the string affects the frequency. The ability to mentally manipulate variables in a number of ways, generally within the scope of scientific experiments, is an important component of the formal operational stage, and is termed **hypothetical reasoning**.

BRIDGE

Culture has profound effects on cognitive development, as well as social structure, rules, and mores. Culture is discussed in detail in Chapter 11 of *MCAT Behavioral Sciences Review.*

BRIDGE

Lev Vygotsky is also a key figure in the psychology of identity. Along with Kohlberg, Freud, and Erikson, he proposed a staged system of identity formation. These theorists are discussed in Chapter 6 of *MCAT Behavioral Sciences Review.*

Role of Culture in Cognitive Development

Cognitive development is very much related to culture, as one's culture will determine what one is expected to learn. Some cultures will place a higher value on social learning, including cultural traditions and roles, while other cultures will value knowledge. In addition, one's culture will also influence the rate of cognitive development as children can be treated very differently from culture to culture.

Lev Vygotsky, a prominent educational psychologist, proposed that the engine driving cognitive development is children's internalization of their culture, including interpersonal and societal rules, symbols, and language. As children develop, their skills and abilities are still in formative stages. With help from adults or other children, those skills can develop further. That help may come in the form of instruction from a teacher or even watching another child perform the skill.

Cognitive Changes in Late Adulthood

Aging brings about many changes in cognition. Reaction time increases steadily in early adulthood, while time-based prospective memory—the ability to remember to perform a task at a specific time in the future—declines with age. Intellectual changes also occur; however, IQ changes have been found to be misleading. Early research into the field of intelligence and aging indicated that a substantial decline in IQ occurs between the ages of 30 and 40. In order to further elucidate what specific changes were occurring, intelligence itself was separated into two subtypes: fluid intelligence and crystallized intelligence. **Fluid intelligence** consists of solving new or novel problems, possibly using creative methods. Figuring out how to navigate through a new video game world involves the usage of fluid intelligence. **Crystallized intelligence** is more related to solving problems using acquired knowledge, and often

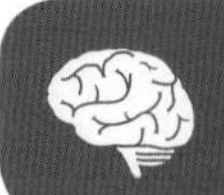

can be procedural. For instance, working through a General Chemistry stoichiometry problem requires crystallized intelligence since it involves recall of the proper equations and the steps taken to work through the calculations. Fluid intelligence was shown to peak in early adulthood but decline with age, while crystallized intelligence peaked in middle adulthood and remains stable with age.

Decline in intellectual abilities in adulthood has been linked with how long an older adult retains the ability to function in what are known as **activities of daily living** (eating, bathing, toileting, dressing, and ambulation). It appears, however, that this decline is not uniform. Certain characteristics, such as higher level of education, more frequent performance of intellectual activities, socializing, and a stimulating environment have been found to be protective against intellectual decline.

Intellectual decline is not always benign. Some types of intellectual decline are very common and indicate a progressive loss of function beyond that of old age. Disorders and conditions that are characterized by a general loss of cognitive function are collectively known as **dementia**. Dementia often begins with impaired memory, but later progresses to impaired judgment and confusion. Personality changes are also very common as dementia progresses. The most common cause of dementia is Alzheimer's disease. Vascular (multi-infarct) dementia, caused by high blood pressure and repeated microscopic clots in the brain, is also a very common cause. It is also important to note that people with dementia often require full-time supportive care in order to carry out activities of daily living. This causes tremendous stress on families, including children and spouses of those with dementia, as the care for the person with dementia often falls on family members.

REAL WORLD

Alzheimer's disease accounts for approximately 60 to 80% of all dementia cases.

Heredity, Environment, and Biologic Factors

Cognition can be affected by a wide variety of conditions. These may include actual problems with the brain itself (organic brain disorders), genetic and chromosomal conditions, metabolic derangements, and long-term drug use. The environment can also affect both cognitive development and day-to-day cognition.

Parenting styles may influence cognitive development by reward, punishment, or indifference for an emerging skill. In addition, genetics can predispose to a state that may make cognitive development difficult. For example, many genetic and chromosomal diseases such as Down syndrome and Fragile X syndrome are associated with delayed cognitive development. Antisocial personality disorder has also been shown to have a strong genetic component. The presence of genes for this disorder may make it difficult for a child to appreciate the rights of others.

Intellectual disabilities in children can also be caused by chemical exposures, illness, injury, or trauma during birth. Alcohol use during pregnancy can cause fetal alcohol syndrome, which results in slowed cognitive development and distinct craniofacial features, shown in Figure 4.3. Infections in the brain may result in electrical abnormalities and slowed development. Complications during birth—especially those causing reduced oxygen delivery to the brain—may also affect cognition. Finally, reduced cognition can also occur following trauma to the brain, as occurs with shaken baby syndrome.

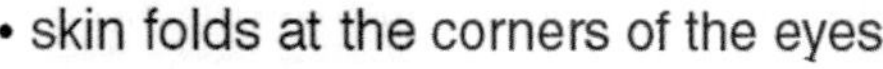
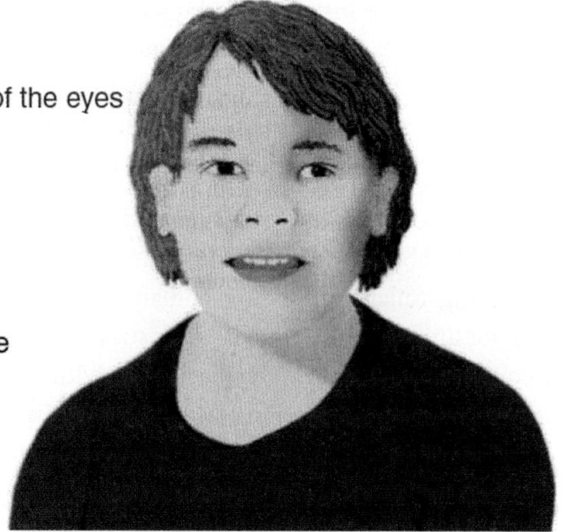

Figure 4.3 Craniofacial Features of Fetal Alcohol Syndrome

REAL WORLD

The delirium associated with alcohol withdrawal, called *delirium tremens*, can be life threatening. As a depressant, alcohol is the only major drug of abuse in which both overdose and withdrawal can be lethal.

However, not all cognitive decline in adulthood is slow. If there has been a rapid decline in cognition, this may be the result of delirium. **Delirium** is rapid fluctuation in cognitive function that is reversible and caused by medical (nonpsychological) causes. It can be caused by a variety of issues, including electrolyte and pH disturbances, malnutrition, low blood sugar, infection, a drug reaction, alcohol withdrawal, and pain.

MCAT CONCEPT CHECK 4.1

Before you move on, assess your understanding of the material with these questions.

1. The three steps in the information processing model are:

   1. ____________________
   2. ____________________
   3. ____________________

2. A person brings a parent, who is an older adult, to the doctor. During the past two days, the parent has been overheard speaking to a spouse who has been deceased for four years. Prior to that, the parent was completely normal. The parent most likely has:

   ____

3. List Piaget's four stages of cognitive development and the key features of each.

Stage	Key Features

4.2 Problem Solving and Decision Making

LEARNING OBJECTIVES

After Chapter 4.2, you will be able to:

- Identify examples of functional fixedness, mental sets, trial-and-error problem solving, algorithms, and deductive reasoning
- Recall key fallacies and biases, including base rate fallacy, disconfirmation principle, confirmation bias, overconfidence, and belief perseverance
- Describe models of intellectual functioning and tests of intellectual ability
- Explain the availability and representativeness heuristics

Every day you are faced with problems. Many of these problems you solve without any real conscious thought about what is happening. However, much like the scientific method, problem solving itself has a process. First, we must frame the problem; that is, we create a mental image or schematic of the issue. Then, we generate potential solutions and begin to test them. These potential solutions may be derived from a **mental set**, which is the tendency to approach similar problems in the same way. Once solutions have been tested, we evaluate the results, considering other potential solutions that may have been easier or more effective in some way.

KEY CONCEPT

The first step in problem solving (framing the problem) may seem obvious; however, when we get "stuck" on a problem, it is most often because the manner in which we have framed the problem is inefficient or not useful.

Problem solving can be impeded by an inappropriate mental set, as well as by functional fixedness, which is demonstrated by Duncker's candle problem. Consider the following scenario: You walk into a room and see a box of matches, some tacks, and a candle. Your task is to mount the candle on the wall so that it can be used without the wax dropping on the floor. Before reading on, try to solve the problem.

Most people find the task challenging. You might have thought of tacking the candle to the wall, but that solution doesn't work because the wax would still drop to the floor. The key is to realize that the matchbox can serve not just as a container for the matches, but as a holder for the candle. The solution, therefore, is to tack the box to the wall and put the candle in the box. **Functional fixedness** can thus be defined as the inability to consider how to use an object in a nontraditional manner.

Types of Problem Solving

In psychology, different approaches to problem solving include trial-and-error, algorithms, deductive reasoning, and inductive reasoning.

Trial-and-Error

Trial-and-error is a less sophisticated type of problem solving in which various solutions are tried until one is found that seems to work. While an educated approach may be used, this type of problem solving is usually only effective when there are relatively few possible solutions.

Algorithms

An **algorithm** is a formula or procedure for solving a certain type of problem. Algorithms can be mathematical or a set of instructions, designed to automatically produce the desired solution.

Deductive Reasoning

Deductive (top-down) reasoning starts from a set of general rules and draws conclusions from the information given. An example of deductive reasoning is a logic puzzle, as shown in Figure 4.4. In these puzzles, one has to synthesize a list of logical rules to come up with the single possible solution to the problem.

	Black	Turnip	Spade	Yuka	Patch	Liam	1	2	3
Jim	●	X	X	●	X	X	X	●	X
Tylor	X	●	X	X			X	X	●
Meera	X	X	●	X			●	X	X
1				X	X	●			
2						X			
3	X		X			X			
Yuka			X						
Patch									
Liam									

Figure 4.4 A Logic Puzzle Grid
Logic puzzles are applications of deductive reasoning in which only one possible solution can be deduced based on the information given.

MCAT EXPERTISE

Remember that a deduction is a solution that *must* be true based on the information given. This is why answers on the MCAT that merely *might* be true (but don't *have* to be) are never the correct answer.

Inductive Reasoning

Inductive (bottom-up) reasoning seeks to create a theory via generalizations. This type of reasoning starts with specific instances, and then draws a conclusion from them.

Heuristics, Biases, Intuition, and Emotion

We make decisions every day. Some are related to our daily routines: *What should I wear today?* Others concern our broader life goals: *Where am I going to apply to medical school?* Decision making is a complicated process, but we use a number of tools, such as heuristics, biases, intuition, and emotions, to speed up or simplify the process. While useful from a time and complexity standpoint, these tools can also lead us to short-sighted or problematic solutions.

Heuristics

Heuristics are simplified principles used to make decisions; they are colloquially called "rules of thumb." The **availability heuristic** is a heuristic used when we base the likelihood of an event on how easily examples of that event come to mind. Often, the use of this heuristic leads us to a correct decision, but not always. As an example, answer the following question: *Are there more words in the English language that start with the letter "K" or that have "K" as their third letter?*

Most people respond that there are more words that begin with the letter "K" than have "K" as their third letter. In fact, there are actually at least twice as many words in English that have "K" as the third letter than begin with "K." Most people approach this question by trying to think of words that fit into each category. Because we so often classify words by their first letter, most people can easily think of words beginning with "K." However, most people have a harder time thinking of words with "K" as their third letter. Thus, in this case, the availability heuristic tends to lead to an incorrect answer.

The **representativeness heuristic** involves categorizing items on the basis of whether they fit the prototypical, stereotypical, or representative image of the category. For example, consider a standard coin that is flipped ten times in a row and lands on heads every time. What is the probability of the coin landing on heads the next time? Mathematically, the probability must still be 50 percent, but most individuals will either overestimate the probability based on the pattern that has been established, or underestimate the probability with the logic that the number of heads and tails must "even out." Hence, like the availability heuristic, the use of the representativeness heuristic can sometimes lead us astray. Using prototypical or stereotypical factors while ignoring actual numerical information is called the **base rate fallacy**.

While heuristics can lead us astray, they are essential to speedy and effective decision making. Heuristics are often used by experts in a given field. For instance, to win at chess, one must be able to think several moves ahead. On any particular turn, there may be 15 or 20 possible moves, each one of which may have multiple consequences; analyzing every possibility would take far too long. There are heuristics, however, that can quickly rule out some of the possible moves: the king must be protected, it is generally good to control the center squares, and pieces should not be put in danger when possible. In this way, heuristics provide a more efficient—although sometimes inaccurate—method for problem solving.

MCAT EXPERTISE

Detail questions on the MCAT often have wrong answer choices that are stated in the passage, but that fail to answer the question posed. According to the availability heuristic, students who do not truly problem solve on MCAT questions will be tempted by these familiar-sounding answers merely because they can recall that statement being mentioned in the passage. Don't forget to use your Distillation effectively, as described in Chapter 6 of *MCAT Critical Analysis and Reasoning Skills Review*!

Bias and Overconfidence

When a potential solution to a problem fails during testing, this solution should be discarded. This is known as the **disconfirmation principle**: the evidence obtained from testing demonstrated that the solution does not work. However, the presence of a confirmation bias may prevent an individual from eliminating this solution. **Confirmation bias** is the tendency to focus on information that fits an individual's beliefs, while rejecting information that goes against them. Confirmation bias also contributes to **overconfidence**, or a tendency to erroneously interpret one's decisions, knowledge, and beliefs as infallible. An additional type of bias is **hindsight bias**, which is the tendency for people to overestimate their ability to predict the outcome of events that already happened. The similar phenomenon of **belief perseverance** refers to the inability to reject a particular belief despite clear evidence to the contrary. Together, confirmation bias, overconfidence, hindsight bias, and belief perseverance can seriously impede a person's analysis of available evidence.

Intuition

Intuition can be defined as the ability to act on perceptions that may not be supported by available evidence. Often, people may have beliefs that are not necessarily supported by evidence, but that a person "feels" to be correct. Intuition is often developed by experience. For example, an emergency room physician, over the course of seeing thousands of patients with chest pain, may develop a keen sense of which patients are actually having a heart attack without even looking at an electrocardiogram (EKG) or a patient's vital signs. This intuition can be more accurately described by the **recognition-primed decision model**: the doctor's brain is actually sorting through a wide variety of information to match a pattern. Over time, the doctor has gained an extensive level of experience that can be accessed without awareness.

Emotion

Emotion is the subjective experience of a person in a certain situation. How a person feels often influences how a person thinks and makes decisions. For example, a person who is angry is often more likely to engage in more risky decision making. In addition, emotions in decision making are not limited to the emotion experienced while the decision is being made; emotions that a person *expects* to feel from a particular decision are also involved. For example, if people believe a car will make them feel more powerful, they may be more likely to purchase that car.

Intellectual Functioning

Intellectual functioning is a highly studied area of psychology. How is intelligence defined? What makes someone more intelligent than someone else? These are multifaceted questions that are difficult to answer; however, theorists have proposed models for some aspects of intelligence.

Theories of Intelligence

There has been much debate concerning the definition of intelligence. Howard Gardner's theory of **multiple intelligences** is one of the most all-encompassing definitions, with at least eight defined types of intelligence: linguistic, logical–mathematical, musical, visual–spatial, bodily–kinesthetic, interpersonal, intrapersonal, and naturalist. Gardner argues that some cultures value the first two abilities over the others. After all, linguistic ability and logical–mathematical ability are the two abilities tested on traditional intelligence quotient (IQ) tests. Despite not being the central focus of the cultures Gardner discusses, people's interpersonal and intrapersonal intelligence can heavily impact their quality of life. **Interpersonal intelligence** is the ability to detect and navigate the moods and motivations of others. Gardner believed that people with high interpersonal intelligence would make great sales representatives and therapists. While intrapersonal intelligence centers around being mindful of one's own emotions, strengths, and weaknesses, which can provide clear guidance what role one should take in a group or society.

KEY CONCEPT

Gardner's multiple intelligences include linguistic, logical-mathematical, musical, visual-spatial, bodily-kinesthetic, interpersonal, intrapersonal, and naturalist.

Robert Sternberg pioneered a cognitive perspective that focused on how people use their intelligence, rather than taking the traditional approach of trying to measure an individual's level of intelligence. More specifically, Sternberg's triarchic theory of human intelligence defines three subtypes: **analytical intelligence**, which involves the ability to evaluate and reason; **creative intelligence**, which is the ability to solve problems using novel methods; and **practical intelligence**, which involves dealing with everyday problems at home or at work.

Successful navigation of our social world also requires us to have a good understanding of both our own emotions and the emotions of those around us. The theory of **emotional intelligence** addresses our emotional awareness in four components: the ability to express and perceive emotions in ourself and others, the ability to comprehend and analyze our emotions, the ability to regulate our emotions, and awareness of how emotions shape our thoughts and decisions. Empathy is often given as an example of emotional intelligence because empathy requires individuals to understand their own emotions well enough to recognize those emotions in other people.

Variations in Intellectual Ability

There are a number of tests and studies that have historically attempted to quantify intelligence. A founding concept behind these tests is Charles Spearman's "g factor," or general intelligence factor. The theory behind the existence of a g factor is based on the observation that performance on different cognitive tasks is in many cases positively correlated, indicating an underlying factor or variable is playing a role. This underlying variable of intelligence is often measured with standardized tests that generate an **intelligence quotient (IQ)** for the test taker. IQ tests were largely pioneered by Alfred Binet in the early twentieth century. A professor at Stanford University took Binet's work and created what is known as the **Stanford–Binet IQ test**.

BEHAVIORAL SCIENCES GUIDED EXAMPLE WITH EXPERT THINKING

Many decisions we make use the affect heuristic; that is, we make quick judgments that depend on our emotional state as well as the emotional content of the decision. Specifically, options that provoke positive emotions are more likely to be chosen than those that evoke negative emotions when the choices are quantitatively identical, and even in cases in which the choice associated with negative feelings is quantitatively better. The following three studies were conducted:

Definition of new term: affect heuristic

Study 1: Researchers provided participants with quantitative information about one of two issues: violent crime (steadily declining) and deer overpopulation (steadily increasing). The participants were asked to rate the severity of each problem on a scale from "very bad" to "very good."

Two problems which differ in severity and emotional content.

IV: issue

DV: rating

Study 2: Participants were asked to set a price for a listing of retail electronics. One listing included ten unused items and another included 18 items, 14 of which were unused and 4 of which were described as being used and in "poor working order."

Comparison between two sets of items, varying in content.

IV: list contents (broken items)

DV: price

Study 3: Participants were asked to determine punishments for a number of hypothetical college students who had committed some form of academic dishonesty. The participants were provided a photograph and description for each student. The same student was used, but one group of participants was provided a picture of the student smiling, and the other group was provided a picture of the student with a neutral expression.

Meting out punishment for people, some of whom are smiling in pictures.

IV: photograph

DV: punishment

Adapted from: Wilson, Robyn S.; Arvai, Joseph L. (March 2006). "When Less Is More: How Affect Influences Preferences When Comparing Low and High-risk Options". *Journal of Risk Research.* **9** (2): 165–178.

LaFrance, M.; Hecht, M.A. (March 1995). "Why Smiles Generate Leniency". *Personality and Social Psychology Bulletin.* **21** (3): 207–214.

If the affect heuristic is shown to be a factor in each of the studies described, what can be predicted about the results of each?

When the MCAT asks us to predict the outcome of an experiment, we want to first consider any content, either from the passage or from our outside knowledge, that might be relevant. Fortunately, this passage provides us with what we need: a definition of the affect heuristic in paragraph one. For each study, then, you'll want to select the result that prioritizes emotional content. For Study 1, the notion of "violent crime" likely provokes a strong negative feeling, while for most people, the idea of "deer overpopulation" would not be expected to be associated with the same emotional response. It is likely, then, that participants would rate violent crime as a worse problem than deer overpopulation regardless of the actual outcome of analyzing the quantitative impact of each issue. The same heuristic explains why people are more afraid of being attacked by a shark at the beach than of other, more likely causes of death, even though statistically injury or death by shark attack is extremely unlikely.

In Study 2, we would predict that the broken items would cause the second listing to be valued the same as or lower than the first listing, even though the second listing actually contains more items that are unused. This is because the additional broken items would be expected to provoke a negative feeling about the second listing. In fact, further research supports the idea that participants do not add up the value of each item separately, but rather evaluate heuristically what they perceive as an average value of the set of items as a whole.

Finally, in Study 3, we should expect that the smiling students were judged less harshly than those presented with a neutral expression. This is based on the same decision-making heuristic we've been using all along: the smiling student is more likely to be associated with positive emotions, and the decision is likely to be altered based on this emotional input.

To summarize: we should predict based on the affect heuristic that in Study 1 the violent crime will be rated as the more problematic of the two issues, that in Study 2 the batch of items containing broken goods will be rated as less valuable, and that in Study 3 the smiling student will be judged less harshly than the neutral student.

While later iterations of the test use different methodologies to arrive at a score, it is useful to know the original formula for calculating IQ:

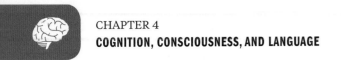

$$IQ = \frac{\text{mental age}}{\text{chronological age}} \times 100$$

Equation 4.1

Using this equation, a four-year-old with intelligence abilities at the level of the average six-year-old would have an IQ of 150. The distribution of IQ scores from the original study of the Stanford–Binet IQ test is shown in Figure 4.5.

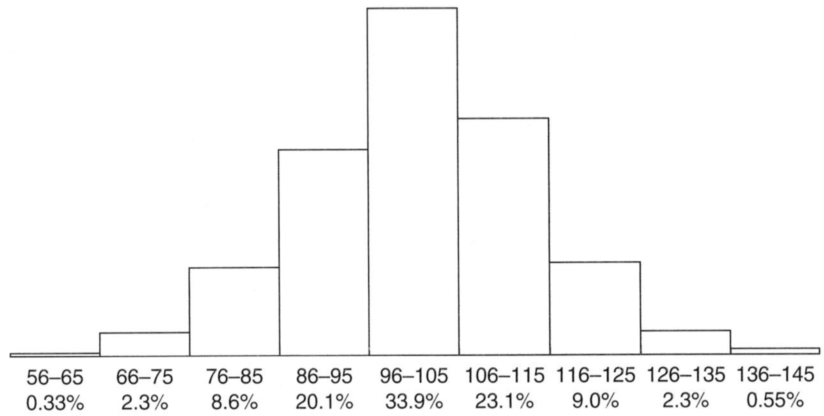

Figure 4.5 Distribution of IQ Scores for Children 5 to 14 Years of Age
Mean = 100; SD = 15

REAL WORLD

The Stanford-Binet IQ test, while still popular, has been found to have variable levels of success in assessing intelligence with different ages and cultural groups. In practice, a variety of intelligence assessments can be found in use around the world. Some of these tests still use measures similar to IQ but are specialized for a subpopulation, such as the Wechsler Intelligence Scale for Children (WISC) exam. Other assessments eschew the concept of IQ entirely and instead follow alternative theories, including Gardner's theory of multiple intelligences.

Some theorists have argued heavily for intelligence as a hereditary trait, most notably Galton in his novel *Hereditary Genius.* In reality, variations in intellectual ability can be attributed to many determinants, including genes, environment, and educational experiences. Intellectual ability does appear to run in families, which may be due to both genetics and the environment; some environments are simply more enriching than others. Parental expectations, socioeconomic status, and nutrition have all been shown to correlate with intelligence.

The educational system plays a significant role in the development of intelligence. Children who attend school tend to have greater increases in IQ, and IQ actually decreases slightly during summer vacations. Early intervention in childhood also improves IQ, especially for children in low-enrichment environments. Finally, both children with high IQs and those with cognitive disabilities benefit from specialized educational environments. For students with cognitive disabilities, this is often defined as the least restrictive environment, in which they are encouraged to participate as much as possible in the regular mainstream classroom, with individualized help as needed.

MCAT CONCEPT CHECK 4.2

Before you move on, assess your understanding of the material with these questions.

1. A child plays with a tool set, noting that a nail can only be hit with a hammer. When a friend suggests that the handle of a screwdriver can be used to hit a nail, the child passionately objects. This is an example of:

____

2. A doctor uses a flowchart to treat a patient with sepsis. Given its use in problem solving, a flowchart is an example of a(n):

____

3. A patient in a mental health facility believes that the sky is pink. Despite several trips outside, the patient still declares the sky is pink. Which psychological principle does this represent?

____

4. Provide a brief definition of the availability and representativeness heuristics.
   - Availability heuristic:

____

   - Representativeness heuristic:

____

4.3 Consciousness

High-Yield

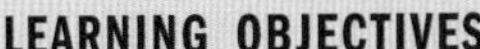

LEARNING OBJECTIVES

After Chapter 4.3, you will be able to:

- Identify the two hormones most associated with circadian rhythms
- Distinguish between dyssomnia and parasomnia
- Associate the stages of sleep with their EEG waveforms and other main features:

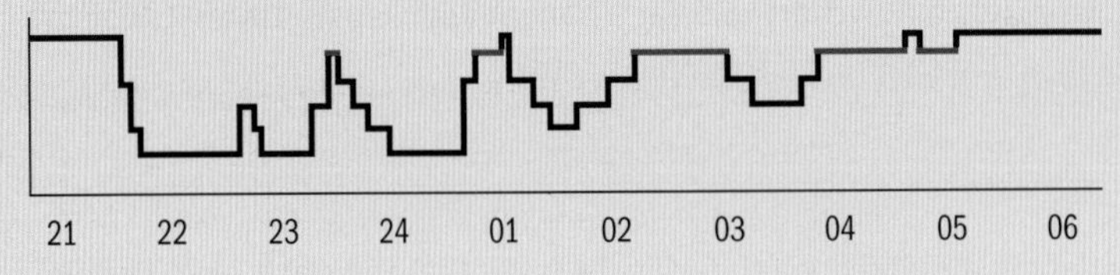

Consciousness is one's level of awareness of both the world and one's own existence within that world.

States of Consciousness

The accepted states of consciousness are alertness, sleep, dreaming, and altered states of consciousness. Technically, sleep and dreaming are also considered altered states, but we will consider these states separately from hypnosis, meditation, and drug-induced altered states of consciousness. Altered states of consciousness may also result from sickness, dementia, delirium, and coma.

Alertness

Alertness is a state of consciousness in which we are awake and able to think. In this state, we are able to perceive, process, access, and verbalize information. In the alert state, we also experience a certain level of **physiological arousal**, which is characterized by physiological reactions such as increased heart rate, breathing rate, blood pressure, and so on. Cortisol levels tend to be higher, and electroencephalogram (EEG) waves indicate a brain in the waking state.

Alertness is maintained by neurological circuits in the prefrontal cortex at the very front of the brain. Fibers from the prefrontal cortex communicate with the **reticular formation** (reticular activation system), a neural structure located in the brainstem, to keep the cortex awake and alert. A brain injury that results in disruption of these connections results in coma.

MCAT EXPERTISE

One of the best ways to enhance your recall and test performance is to maintain a regular schedule of sleep. Regular sleep, exercise, and a healthy diet help to make Test Day successful.

Sleep

Sleep is important to consider while studying for the MCAT or any other major exam. While it may be tempting to pull all-nighters in an attempt to maximize your test score, this may not be the best strategy for success. In fact, long-term sleep deprivation has been linked with diminished cognitive performance as well as the development of chronic diseases such as diabetes and obesity.

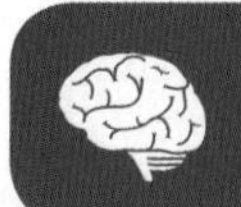

Stages of Sleep

Sleep is studied by recording brain wave activity occurring during the course of a night's sleep. This is done with **electroencephalography**, or **EEG**, which records an average of the electrical patterns within different portions of the brain. There are four characteristic EEG patterns correlated with different stages of waking and sleeping: beta, alpha, theta, and delta waves. There is a fifth wave that corresponds to REM sleep, which is the time during the night when we have most of our dreams. These sleep stages form a complete cycle lasting about 90 minutes.

REAL WORLD

An electroencephalogram (EEG) is a test used to monitor electrical activity in the brain. It consists of 19 recording electrodes placed on the scalp for 20 to 40 minutes, or continuously in the inpatient setting. This technique is used to study sleep and to identify areas of unusual brain activity, as seen during seizures.

Beta and alpha waves characterize brain wave activity when we are awake and are shown in Figure 4.6. **Beta waves** have a high frequency and occur when the person is alert or attending to a mental task that requires concentration. Beta waves occur when neurons are randomly firing. **Alpha waves** occur when we are awake but relaxing with our eyes closed, and are somewhat slower than beta waves. Alpha waves are also more synchronized than beta waves.

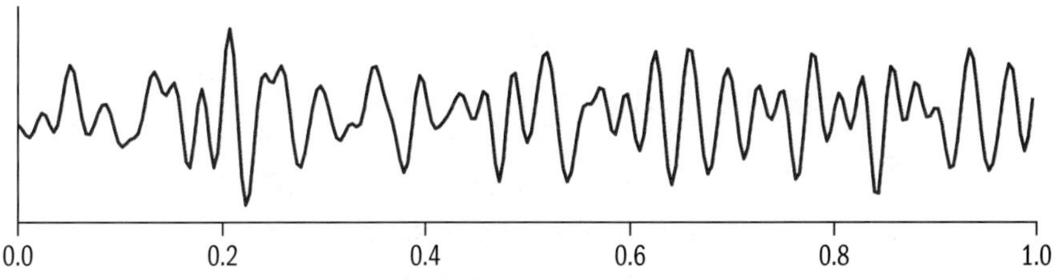

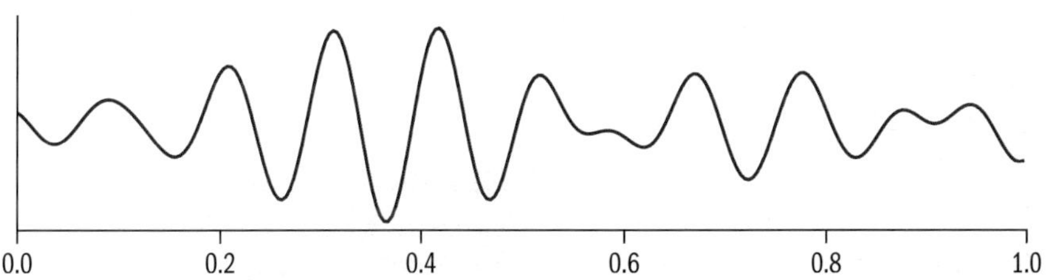

Figure 4.6 Beta (top) and Alpha (bottom) Waves on EEG
Beta and alpha waves are seen during alertness.

As soon as you doze off, you enter **Stage 1** (also known as NREM1), which is detected on the EEG by the appearance of **theta waves**, shown in Figure 4.7. At this point, EEG activity is characterized by irregular waveforms with slower frequencies and higher voltages.

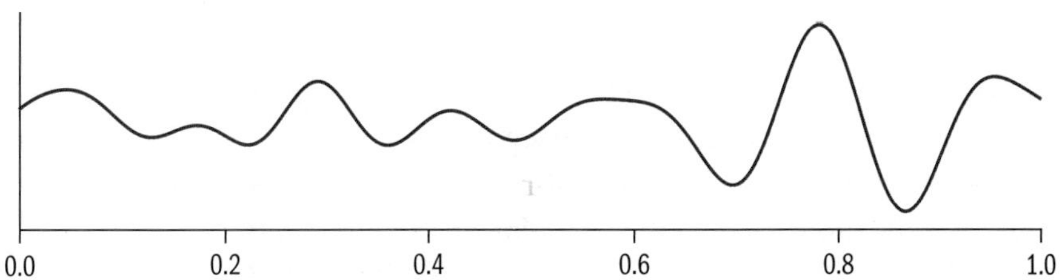

Figure 4.7 Theta Waves
Theta waves are seen during Stage 1 and 2 sleep.

As you fall more deeply asleep, you enter **Stage 2** (NREM2). The EEG shows theta waves along with **sleep spindles**, which are bursts of high-frequency waves, and **K complexes**, which are singular high-amplitude waves, shown in Figure 4.8.

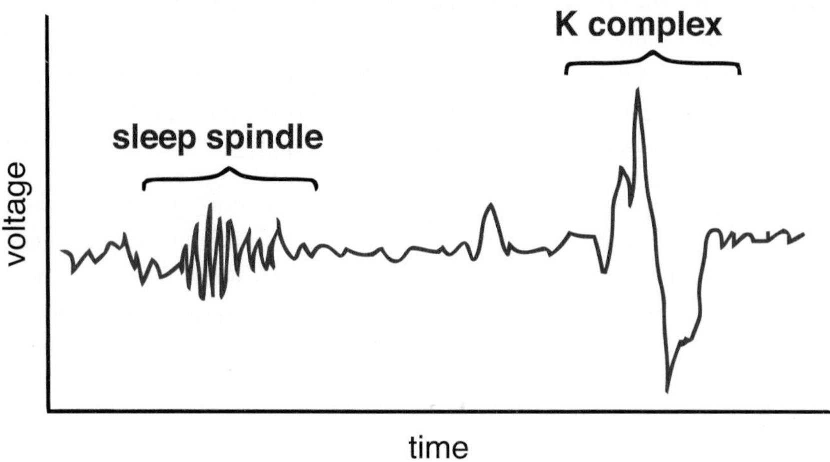

Figure 4.8 Sleep Spindle and K Complex in Stage 2 Sleep

As you fall even more deeply asleep, you enter **Stage 3** (NREM3), also known as **slow-wave sleep** (**SWS**). EEG activity grows progressively slower until only a few sleep waves per second are seen. These low-frequency, high-voltage sleep waves are called **delta waves**, shown in Figure 4.9. During this stage, rousing someone from sleep becomes exceptionally difficult. SWS has been associated with cognitive recovery and memory consolidation, as well as increased growth hormone release.

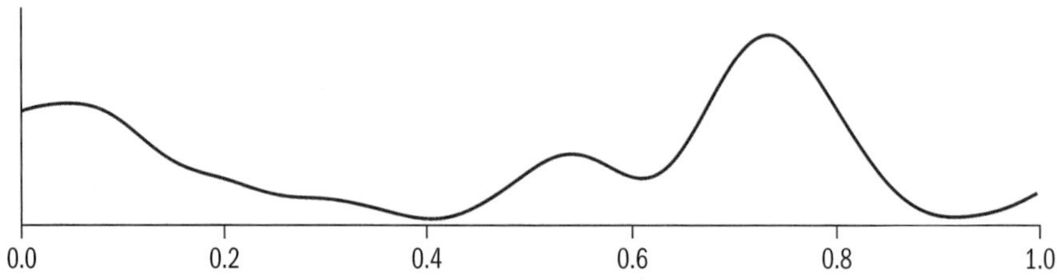

Figure 4.9 Delta Waves of Slow-Wave Sleep

MNEMONIC

Remember the sequential order of these brain waves—**b**eta, **al**pha, **t**heta, **d**elta—by combining their first letters to form **BAT-D** and remember that a **bat** sleeps during the **d**ay.

The stages above are collectively called **non-rapid eye movement sleep**, which is where the acronym *NREM* comes from. Interspersed between cycles of the NREM stages is **rapid eye movement** (**REM**) **sleep**. In REM sleep, arousal levels reach that of wakefulness, but the muscles are paralyzed. REM sleep is also called **paradoxical sleep** because one's heart rate, breathing patterns, and EEG mimic wakefulness, but the individual is still asleep. This is the stage in which dreaming is most likely to occur and is also associated with memory consolidation. Recent studies have associated REM more with procedural memory consolidation and SWS with declarative memory consolidation.

Sleep Cycles and Changes to Sleep Cycles

A **sleep cycle** refers to a single complete progression through the sleep stages. The makeup of a sleep cycle changes during the course of the night, as shown in Figure 4.10. Early in the night, SWS predominates as the brain falls into deep sleep and then into more wakeful states. Later in the night, REM sleep predominates.

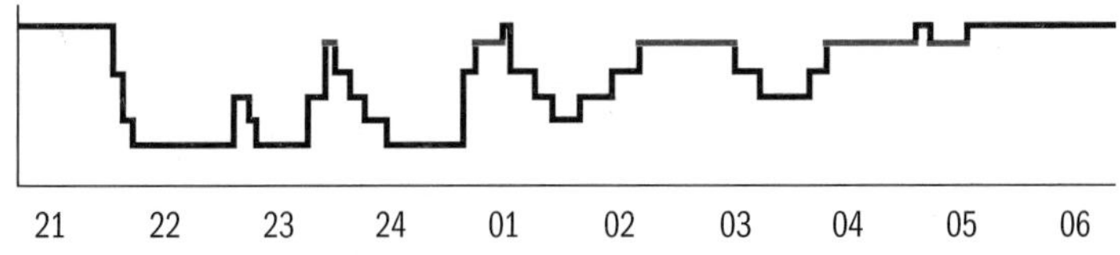

Figure 4.10 Hypnogram of Sleep Cycles

Over the life span, the length of the sleep cycle increases from approximately 50 minutes in children to 90 minutes in adults. Children also spend more time in SWS than adults. Changes to sleep cycles from disrupted sleep or disordered work schedules can cause many health problems. Disruption of SWS and REM can result in diminished memory. Sleep deprivation also causes diminished cognitive performance, although the person who is sleep-deprived is unlikely to recognize that performance has been subpar. Sleep deprivation also negatively affects mood, problem solving, and motor skills.

Sleep and Circadian Rhythms

Our daily cycle of waking and sleeping is regulated by internally generated rhythms or **circadian rhythms**. In humans and other animals, the circadian rhythm approximates a 24-hour cycle that is somewhat affected by external cues such as light. Biochemical signals underlie circadian rhythms. Sleepiness can partially be attributed to blood levels of **melatonin**, a serotonin-derived hormone from the **pineal gland**. The retina has direct connections to the hypothalamus, which controls the pineal gland; thus, decreasing light can cause the release of melatonin.

Cortisol, a steroid hormone produced in the **adrenal cortex**, is also related to the sleep–wake cycle. Its levels slowly increase during early morning because increasing light causes the release of **corticotropin-releasing factor** (**CRF**) from the hypothalamus. CRF causes release of **adrenocorticotropic hormone** (**ACTH**) from the anterior pituitary, which stimulates cortisol release. Cortisol contributes to wakefulness.

MNEMONIC

Melatonin **mell**ows you out. Corti**sol** helps you get up with the sun (*sol* is Latin for sun).

BRIDGE

The hypothalamic-pituitary-adrenal axis is an example of how the endocrine system can regulate behavior. The endocrine system is discussed in Chapter 5 of *MCAT Biology Review.*

Dreaming

Philosophers and those interested in the human experience have hypothesized about the purpose, meaning, and function of dreaming since antiquity. The ancient Egyptians believed that dreams were messages sent from the supernatural world to tell of future events. In ancient Greece, people believed dreams to carry messages from the gods, but the dream required the help of a priest to interpret. Dreams have long been a subject of wonder.

Most dreaming occurs during REM; however, soon after we enter Stage 2 sleep, our mental experience starts to shift to a dreamlike state. Throughout the night, approximately 75% of dreaming occurs during REM. REM dreams tend to be longer and more vivid than those experienced during NREM sleep.

BRIDGE

Other psychologists have proposed explanations for dreams. One of the most notable is Freud, who separated dreams into their *manifest content* (what one actually sees and hears) and their *latent content* (the underlying significance of these dream elements). Freud's model of personality is discussed in Chapter 6 of *MCAT Behavioral Sciences Review.*

While the purpose and meaning of dreams is not fully understood, a few theories have been proposed. In the **activation–synthesis theory**, dreams are caused by widespread, random activation of neural circuitry. This activation can mimic incoming sensory information, and may also consist of pieces of stored memories, current and previous desires, met and unmet needs, and other experiences. The cortex then tries to stitch this unrelated information together, resulting in a dream that is both bizarre and somewhat familiar. In the **problem solving dream theory**, dreams are a way to solve problems while you are sleeping. Dreams are untethered by the rules of the real world, and thus allow interpretation of obstacles differently than during waking hours. Finally, in the **cognitive process dream theory**, dreams are merely the sleeping counterpart of stream-of-consciousness. Just as you may be thinking about an upcoming weekend trip when your consciousness quickly shifts to your upcoming MCAT Test Day, so too does the content of a dream rapidly shift and change. Ultimately, the question is less *Which group is right?* and more *How can we unify these theories?* The study of dreaming is limited by the difference between the brain and the mind: dreaming must have a neurological component, but is still highly subjective. **Neurocognitive models of dreaming** seek to unify biological and psychological perspectives on dreaming by correlating the subjective, cognitive experience of dreaming with measurable physiological changes.

Sleep-Wake Disorders

Sleep-wake disorders are divided into two categories: dyssomnias and parasomnias. **Dyssomnias** refer to disorders that make it difficult to fall asleep, stay asleep, or avoid sleep, and include insomnia, narcolepsy, and sleep apnea. **Parasomnias** are abnormal movements or behaviors during sleep, and include night terrors and sleepwalking. Most sleep-wake disorders occur during NREM sleep.

MNEMONIC

Hypna**go**gic hallucinations occur when one is **go**ing to bed. Hypno**pom**pic hallucinations occur when one is **pop**ping up out of bed.

Insomnia is difficulty falling asleep or staying asleep. It is the most common sleep-wake disorder and may be related to anxiety, depression, medications, or disruption of sleep cycles and circadian rhythms. **Narcolepsy**, in contrast, is a condition characterized by lack of voluntary control over the onset of sleep. The symptoms of narcolepsy are unique and include **cataplexy**, a loss of muscle control and sudden intrusion of REM sleep during waking hours, usually caused by an emotional trigger; **sleep paralysis**, a sensation of being unable to move despite being awake; and **hypnagogic** and **hypnopompic hallucinations**, which are hallucinations when going to sleep or awakening. Another dyssomnia is **sleep apnea**, which is an inability to breathe during sleep. People with this disorder awaken often during the night in order to breathe. Sleep apnea can be either obstructive or central. Obstructive sleep apnea occurs when a physical blockage in the pharynx or trachea prevents airflow; central sleep apnea occurs when the brain fails to send signals to the diaphragm to breathe.

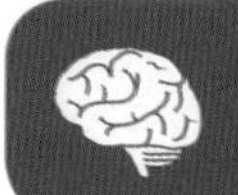

Night terrors, which are most common in children, are periods of intense anxiety that occur during slow-wave sleep. Children will often thrash and scream during these terrors, and will show signs of sympathetic overdrive, with a high heart rate and rapid breathing. Because these usually occur during SWS, the child experiencing the episode is very difficult to wake, and usually does not remember the dream the next morning. **Sleepwalking**, or **somnambulism**, also usually occurs during SWS. Some people who sleepwalk may eat, talk, have sexual intercourse, or even drive great distances while sleeping with absolutely no recollection of the event. Most return to their beds and awake in the morning, with no knowledge of their nighttime activities. Contrary to popular belief, awakening a person who is sleepwalking will not harm the person; however, it is generally suggested to quietly guide the individual back to bed to avoid disturbing SWS.

Sleep deprivation can result from as little as one night without sleep, or from multiple nights with poor-quality, short-duration sleep. Sleep deprivation results in irritability, mood disturbances, decreased performance, and slowed reaction time. Extreme deprivation can cause psychosis. While one cannot make up for lost sleep, people who are permitted to sleep normally after sleep deprivation often exhibit **REM rebound**, an earlier onset and greater duration of REM sleep compared to normal.

Hypnosis

Hypnosis, named after the ancient Greek god of sleep, *Hypnos*, was first documented in the eighteenth century. **Hypnosis** can be defined as a state in which a person appears to be in control of normal functions, but is in a highly suggestible state. In other words, a hypnotized person easily succumbs to the suggestions of others. Hypnosis starts with **hypnotic induction**, in which the hypnotist seeks to relax the subject and increase the subject's level of concentration. Then, the hypnotist can suggest perceptions or actions to the hypnotized person. In practice, hypnosis is not the same as its sensationalized version in the media, in which a hypnotist will snap his fingers and cause an individual to exhibit bizarre behavior. Rather, hypnosis has been used successfully for pain control, psychological therapy, memory enhancement, weight loss, and smoking cessation. Brain imaging has indicated that hypnotic states are indeed real; however, effective hypnosis requires a willing personality and lack of skepticism on the part of the patient.

BRIDGE

Hypnosis has been used to recover repressed memories of trauma; however, these memories are not admissible in a court of law. This is because the suggestible state of hypnotism makes an individual vulnerable to creating false memories, which can be perceived as completely real. False memories are discussed in Chapter 3 of *MCAT Behavioral Sciences Review*.

Meditation

Defining **meditation** can be tricky and is highly dependent on the practitioners of meditation and their beliefs. Meditation has been a central practice in many reli gions. Meditation usually involves quieting of the mind for some purpose, whether spiritual, religious, or related to stress reduction. In the secular Western tradition, meditation is often used for counseling and psychotherapy because it produces a sense of relaxation and relief from anxiety and worrying. To that end, meditation causes physiological changes such as decreased heart rate and blood pressure. On EEG, meditation resembles Stage 1 sleep with theta and slow alpha waves.

MCAT EXPERTISE

Recent studies have demonstrated that mindful meditation not only improves psychological well-being, but may even help improve test scores and student performance. Take time for yourself while studying for the MCAT; keep your mind calm to keep it sharp.

MCAT CONCEPT CHECK 4.3

Before you move on, assess your understanding of the material with these questions.

1. For each of the sleep stages below, list its EEG waveforms and main features.

Stage	EEG Waves	Features
Awake		
Stage 1		
Stage 2		
Stage 3		
Stage 4		
REM		

2. Which two hormones are most associated with maintaining circadian rhythms?

-

-

3. What is the difference between a dyssomnia and a parasomnia? Provide an example of each.

- Dyssomnia:

- Parasomnia:

4.4 Consciousness-Altering Drugs

LEARNING OBJECTIVES

After Chapter 4.4, you will be able to:

- List the drugs (or drug classes) known to increase GABA activity in the brain
- Recall the drugs (or drug classes) known to upregulate dopamine, norepinephrine, or serotonin activity
- Identify the three main structures in the mesolimbic reward pathway and the primary neurotransmitter of the pathway:

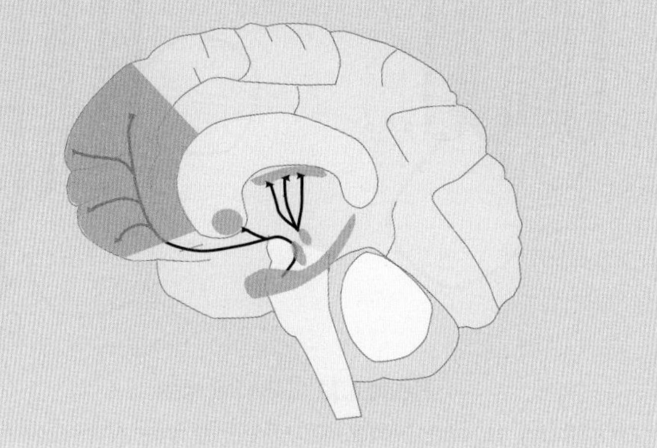

Consciousness-altering drugs, also known as psychoactive drugs, are generally described in four different groups: depressants, stimulants, opiates, and hallucinogens. Biologically speaking, marijuana has depressant, stimulant, and hallucinogenic effects, and will be considered separately.

Depressants

Depressants reduce nervous system activity, resulting in a sense of relaxation and reduced anxiety. Of the depressants, alcohol is certainly the most common. Another is sedatives, or "downers," which calm and induce sleep.

Alcohol

Alcohol has several different effects on the brain. It increases activity of the GABA receptor, a chloride channel that causes hyperpolarization of the membrane, as shown in Figure 4.11. This hyperpolarization causes generalized brain inhibition at the physiological level, resulting in diminished arousal at moderate doses. The changes in brain activity also cause changes in outward behavior. For example, excessive consumption of alcohol may be associated with a notable a lack of self-control known as **disinhibition**, which occurs because the centers of the brain that prevent inappropriate behavior are also depressed. Alcohol also increases dopamine levels, causing a sense of mild euphoria. At higher doses, brain activity becomes more disrupted. Logical reasoning and motor skills are affected, and fatigue may result. One of the main effects on logical reasoning is the inability to recognize consequences of actions, creating a short-sighted view of the world called **alcohol myopia**.

Alcohol use is implicated in many automobile accidents, homicides (for both perpetrator and victim), and hospital admissions. Intoxication with alcohol is often measured using blood alcohol level.

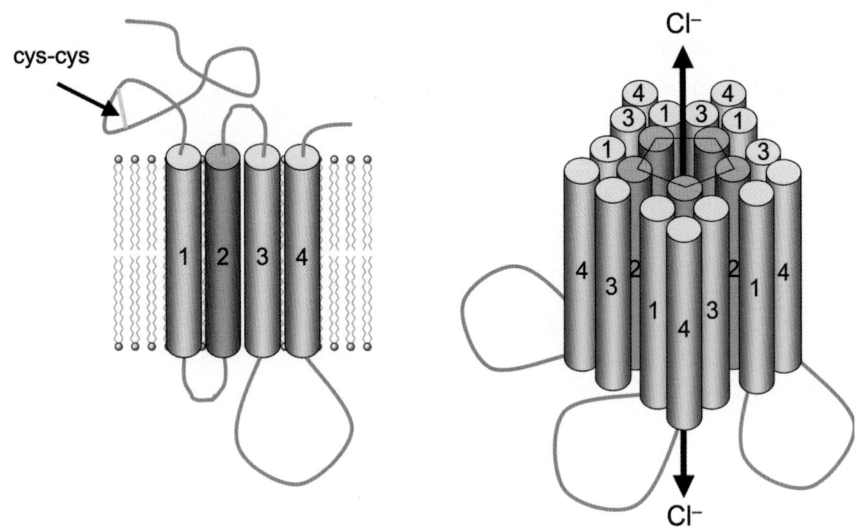

Figure 4.11 GABA Receptor

GABA is the primary inhibitory neurotransmitter in the brain; its receptor is a chloride channel that causes hyperpolarization of the membrane.

Alcohol is one of the most widely abused drugs. Alcoholism rates tend to be higher for those of lower socioeconomic status (SES), but people with a lower SES who have alcoholism tend to enter recovery sooner and at higher rates. Alcoholism tends to run in families, and children of people with alcoholism are also likely to suffer from major depressive disorder. Long-term consequences of alcoholism include cirrhosis and liver failure, pancreatic damage, gastric or duodenal ulcers, gastrointestinal cancer, and brain disorders including **Wernicke–Korsakoff syndrome**, caused by a deficiency of thiamine (vitamin B_1) and characterized by severe memory impairment with changes in mental status and loss of motor skills.

Sedatives

Sedatives tend to depress central nervous system activity, resulting in feelings of calm, relaxation, and drowsiness. Two types of sedatives are barbiturates and benzodiazepines. **Barbiturates** were historically used as anxiety-reducing (anxiolytic) and sleep medications, but have mostly been replaced by **benzodiazepines**, which are less prone to overdose. Barbiturates include *amobarbital* and *phenobarbital*; benzodiazepines include *alprazolam*, *lorazepam*, *diazepam*, and *clonazepam*. These drugs also increase GABA activity, causing a sense of relaxation. However, both of these drug types can be highly addictive. If taken with alcohol, overdoses of barbiturates or benzodiazepines may result in coma or death.

Stimulants

Stimulants cause an increase in arousal in the nervous system. Each drug increases the frequency of action potentials, but does so by different mechanisms.

Amphetamines

Amphetamines cause increased arousal by increasing release of dopamine, norepinephrine, and serotonin at the synapse and decreasing their reuptake. This increases arousal and causes a reduction in appetite and decreased need for sleep. Physiological effects include an increase in heart rate and blood pressure. Psychological effects include euphoria, hypervigilance (being "on edge"), anxiety, delusions of grandeur, and paranoia. Prolonged use of high doses of amphetamines can result in stroke or brain damage. People who take amphetamines often suffer from withdrawal after discontinuation, leading to depression, fatigue, and irritability.

Cocaine

Cocaine originates from the coca plant, grown in the high-altitude regions of South America. Cocaine can be purified from these leaves or created synthetically. Similar to amphetamines, cocaine also acts on dopamine, norepinephrine, and serotonin synapses, but cocaine decreases reuptake of the neurotransmitters instead. Hence, the effects of cocaine intoxication and withdrawal are therefore similar to amphetamines, as listed above. Cocaine also has anesthetic and vasoconstrictive properties, and is therefore sometimes used in surgeries in highly vascularized areas, such as the nose and throat. These vasoconstrictive properties can also lead to heart attacks and strokes when used recreationally. **Crack** is a form of cocaine that can be smoked. With quick and potent effects, this drug is highly addictive.

Ecstasy (3,4-methylenedioxy-*N*-methylamphetamine, MDMA)

Ecstasy, commonly called "E," acts as a hallucinogen combined with an amphetamine. As a designer amphetamine, its mechanism and effects are similar to other amphetamines. Physiologically, ecstasy causes increased heart rate, increased blood pressure, blurry vision, sweating, nausea, and hyperthermia. Psychologically, ecstasy causes feelings of euphoria, increased alertness, and an overwhelming sense of well-being and connectedness. Ecstasy is an example of a club or rave drug, and is often packaged in colorful pills, as shown in Figure 4.12.

Figure 4.12 Pills of Ecstasy (MDMA)

Opiates and Opioids

Opiates and opioids are types of narcotics, also known as painkillers. Derived from the poppy plant, **opium** has been used and abused for centuries. Today, we have numerous drugs, used both recreationally and therapeutically, derived from opium. Naturally occurring forms, called **opiates**, include *morphine* and *codeine*. Semisynthetic derivatives, called **opioids**, include *oxycodone, hydrocodone*, and *heroin*. These compounds bind to opioid receptors in the peripheral and central nervous system. They act as endorphin agonists and cause a decreased reaction to pain and a sense of euphoria. Overdose, however, can cause death by respiratory suppression, in which the brain stops sending signals to breathe.

Heroin, or *diacetylmorphine*, was originally created as a substitute for morphine. Once injected, the body rapidly metabolizes heroin to morphine. Usually smoked or injected, heroin was once the most widely abused opioid; however, this designation has shifted to prescription opioids like oxycodone and hydrocodone. Treatment for opioid addiction may include use of *methadone*, a long-acting opioid with lower risk of overdose.

Hallucinogens

Hallucinogens are drugs which typically cause introspection, distortions of reality and fantasy, and enhancement of sensory experiences. Physiologic effects include increased heart rate and blood pressure, dilation of pupils, sweating, and increased body temperature. Examples of hallucinogens include **lysergic acid diethylamide (LSD)**, shown in Figure 4.13, *peyote, mescaline, ketamine*, and *psilocybin*-containing mushrooms. The exact mechanism of most hallucinogens is unknown, but is thought to be a complex interaction between various neurotransmitters, especially serotonin.

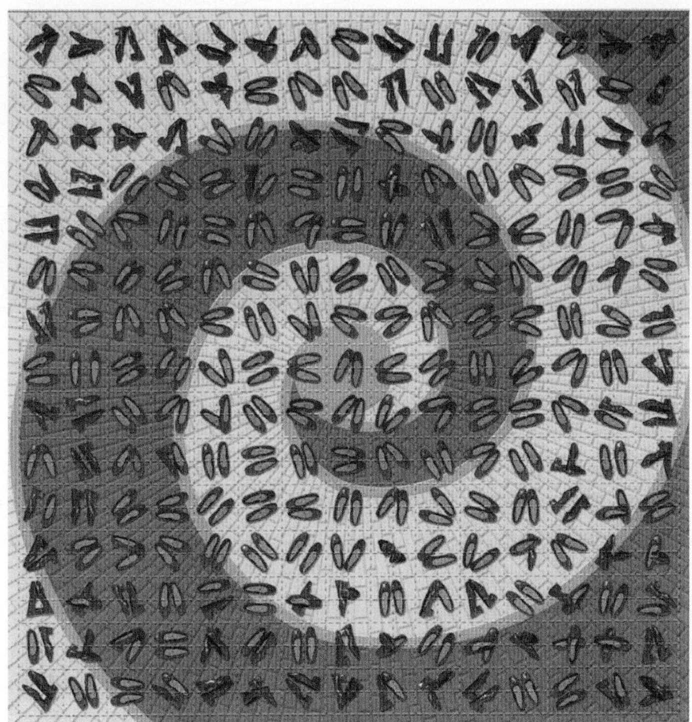

Figure 4.13 Sheet of LSD Blotter Paper
LSD is often sold on colorful paper, reflecting the fact that, like ecstasy, LSD is considered a club drug.

Marijuana

Marijuana, shown in Figure 4.14, primarily refers to the leaves and flowers of two plant species: *Cannabis sativa* and *Cannabis indica*. It has been the subject of many news reports in the last few years as many states move toward the legalization of marijuana for medical or recreational use. While talks about the legal status of marijuana in the United States are fairly recent, marijuana has been used for centuries, with the earliest known accounts originating from approximately 3 B.C.E.

Figure 4.14 Cannabis

The active chemical in marijuana is known as **tetrahydrocannabinol** (**THC**). THC exerts its effects by acting at cannabinoid receptors, glycine receptors, and opioid receptors. How these receptors interact to create the "high" achieved from marijuana use is unknown. It is known, however, that THC inhibits GABA activity and indirectly increases dopamine activity (causing pleasure). Physiological effects are mixed, including eye redness, dry mouth, fatigue, impairment of short-term memory, increased heart rate, increased appetite, and lowered blood pressure. Psychologically, effects seem to fall into the categories of stimulant, depressant, and hallucinogen.

Drug Addiction

Drug addiction is highly related to the **mesolimbic reward pathway**, one of four dopaminergic pathways in the brain, as shown in Figure 4.15. This pathway includes the **nucleus accumbens** (NAc), the **ventral tegmental area** (**VTA**), and the connection between them called the **medial forebrain bundle** (**MFB**). This pathway is normally involved in motivation and emotional response, and its activation accounts for the positive reinforcement of substance use. This addiction pathway is activated by all substances that produce psychological dependence. Gambling and falling in love also activate this pathway.

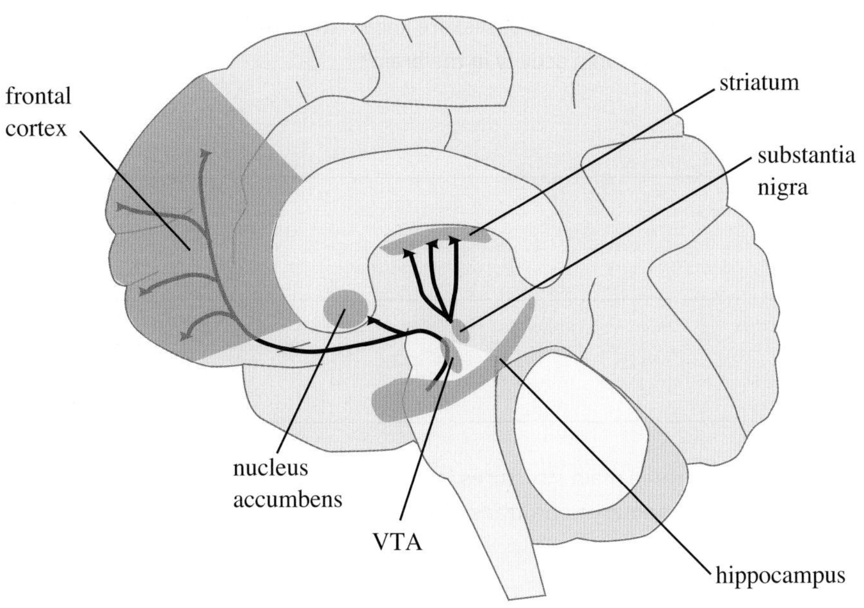

Figure 4.15 Dopaminergic Pathways in the Brain

The reward pathway is composed of the nucleus accumbens, ventral tegmental area (VTA), and the medial forebrain bundle between them (not labeled).

MCAT CONCEPT CHECK 4.4

Before you move on, assess your understanding of the material with these questions.

1. Which three drugs (or drug classes) are known to increase GABA activity in the brain?
   - ____________________
   - ____________________
   - ____________________

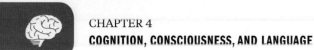

2. Which three drugs (or drug classes) are known to increase dopamine, norepinephrine, and serotonin activity in the brain?
   -
   -
   -

3. What are the three main structures in the mesolimbic reward pathway? What is this pathway's primary neurotransmitter?
   - Structure:
   - Structure:
   - Structure:
   - Neurotransmitter:

4.5 Attention

LEARNING OBJECTIVES

After Chapter 4.5, you will be able to:

- Compare and contrast controlled processing and automatic processing
- Describe the role of the "filter" used in selective attention

Attention refers to concentrating on one aspect of the sensory environment, or **sensorium**. While this definition is straightforward, an understanding of how attention works and the mechanism by which we can shift our attention from one set of stimuli to another is still somewhat unclear.

Selective Attention

Selective attention is focusing on one part of the sensorium while ignoring other stimuli. It therefore acts as a filter between sensory stimuli and our processing systems. If a stimulus is attended to, it is passed through a filter and analyzed further. If the stimulus is not attended to, it is lost. In its original conceptualization, selective attention was viewed as an all-or-nothing process: if we choose a particular stimulus to give our attention to, the other stimuli are lost. However, recent evidence indicates that this is not the case.

Imagine this: You are at a party, talking with a friend. However, your ears perk up when you hear your name spoken halfway across the room. Even though you were engaged in conversation and presumably paying attention, you were able to perceive your name being mentioned. This is sometimes called the **cocktail party phenomenon** and is evidence of a different interpretation of selective attention. Selective attention is probably more of a filter that allows us to focus on one thing while allowing other stimuli to be processed in the background. Only if the other stimuli are particularly important—one's name being mentioned, a sudden flash of light, pain—do we shift our attention to them.

Dichotic listening tests are designed to test selective attention. Participants are given headphones that have distinct auditory stimuli going to each ear. Participants are then asked to pay attention to either or both stimuli, then asked to repeat out loud what they heard in the attended ear, which is termed **shadowing**. This task tests selective attention because participants are asked to filter out information from the unattended ear. Alternatively, the task can test whether participants can subconsciously gain information from the unattended ear.

Divided Attention

Divided attention is the ability to perform multiple tasks at the same time. Most new or complex tasks require undivided attention and utilize **controlled** (**effortful**) **processing**, discussed in Chapter 3 of *MCAT Behavioral Sciences Review*. In contrast, familiar or routine actions can be performed with **automatic processing**, which permits the brain to focus on other tasks with divided attention. Consider learning to drive: at first, drivers intensely grip the steering wheel and pay undivided attention to the road ahead. But as you become more accustomed to driving, you can relegate some aspects of driving—like knowing how hard to push on the pedal—to automatic processing. This lets a driver perform secondary tasks such as changing the radio station. That being said, automatic processing is far from perfect. It does not allow for innovation or rapid response to change, which may contribute to the high incidence of car accidents that result from distracted driving.

MCAT CONCEPT CHECK 4.5

Before you move on, assess your understanding of the material with these questions.

1. Compare and contrast controlled (effortful) processing and automatic processing:
   - Controlled (effortful) processing:

   ____

   - Automatic processing:

   ____

2. Briefly describe the function of the "filter" used in selective attention:

   ____

   ____

4.6 Language

High-Yield

LEARNING OBJECTIVES

After Chapter 4.6, you will be able to:

- Recall the expected milestones of language development and the ages at which they should occur
- Identify the primary characteristics of the nativist, behaviorist, and social interactionist theories of language development
- Describe the symptoms and brain regions associated with Broca's, Wernicke's, and conduction aphasia:

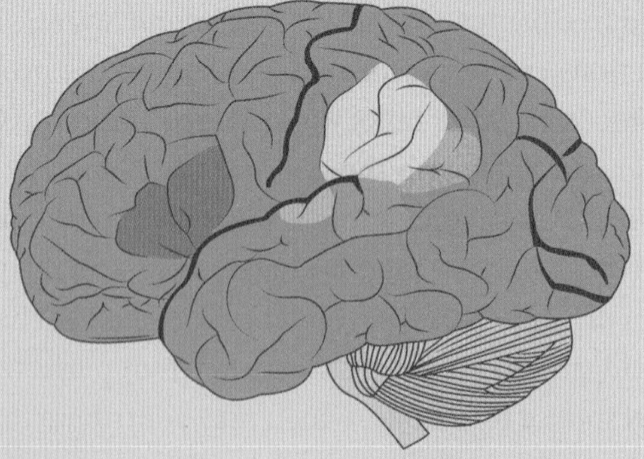

Whether it is written, spoken, or signed, **language** is fundamental to the creation of communities. As humans began to live in groups, the ability to communicate became essential. Division of labor and a sense of shared history require that the meaning of the language be the same for all speakers of the language.

Components of Language

There are five basic components of language: phonology, morphology, semantics, syntax, and pragmatics.

Phonology

Phonology refers to the actual sound of language. There are about 40 speech sounds or **phonemes** in English, although many more exist in other languages, as shown in Figure 4.16. Children must learn to produce and recognize the sounds of language, separating them from environmental noises and other human-created sounds, like coughing. Additionally, when a language has subtle differences in speech sounds that represent a change in meaning, children learn to distinguish those phonemes; this ability is called **categorical perception**. The ability to recognize a word as being the same, even if the pronunciation of the word varies between people is an auditory example of **constancy**, which is described in Chapter 2 of *MCAT Behavioral Sciences Review*.

International Phonetic Alphabet

Plosive		**Tap or Flap**		**Lateral Fricative**	
Name	**Text**	**Name**	**Text**	**Name**	**Text**
Voiceless bilabial	p	Voiced alveolar	ɾ	Voiceless alveolar	ɬ
Voiced bilabial	b	Voiced retroflex	ɽ	Voiced alveolar	ɮ
Voiceless alveolar	t	**Fricative**		**Approximant**	
Voiced alveolar	d	**Name**	**Text**	**Name**	**Text**
Voiceless retroflex	ʈ	Voiceless bilabial	ɸ	Voiced labiodental	ʋ
Voiced retroflex	ɖ	Voiced bilabial	β	Voiced alveolar	ɹ
Voiceless palatal	c	Voiceless labiodental	f	Voiced retroflex	ɻ
Voiced palatal	ɟ	Voiced labiodental	v	Voiced palatal	j
Voiceless velar	k	Voiceless dental	θ	Voiced velar	ɰ
Voiced velar	ɡ	Voiced dental	ð	**Lateral Approximant**	
Voiceless uvular	q	Voiceless alveolar	s	**Name**	**Text**
Voiced uvular	ɢ	Voiced alveolar	z	Voiced alveolar	l
Voiceless glottal	ʔ	Voiceless postalveolar	ʃ	Voiced retroflex	ɭ
Nasal		Voiced postalveolar	ʒ	Voiced palatal	ʎ
Name	**Text**	Voiceless retroflex	ʂ	Voiced velar	ʟ
Voiced bilabial	m	Voiced retroflex	ʐ	**Clicks**	
Voiced labiodental	ɱ	Voiceless palatal	ç	**Name**	**Text**
Voiced alveolar	n	Voiced palatal	ʝ	Bilabial	ʘ
Voiced retroflex	ɳ	Voiceless velar	x	Dental	ǀ
Voiced palatal	ɲ	Voiced velar	ɣ	(Post)alveolar	ǃ
Voiced velar	ŋ	Voiceless uvular	χ	Palatoalveolar	ǂ
Voiced uvular	ɴ	Voiced uvular	ʁ	Alveolar lateral	ǁ
Trill		Voiceless pharyngeal	ħ		
Name	**Text**	Voiced pharyngeal	ʕ		
Voiced bilabial	ʙ	Voiceless glottal	h		
Voiced alveolar	r	Voiced glottal	ɦ		
Voiced uvular	ʀ	Voiceless dorso-palatal	ɧ		

Figure 4.16 A Section of the International Phonetic Alphabet
The IPA is an unambiguous system of writing all of the known phonemes of all human languages.

Morphology

Morphology refers to the structure of words. Many words are composed of multiple building blocks called **morphemes**, each of which connotes a particular meaning. Consider the word *redesigned*, which can be broken into three morphemes: *re–*, indicating *to do again*; *–design–*, the verb root; and *–ed*, indicating an action in the past.

Semantics

Semantics refers to the association of meaning with a word. A child must learn that certain combinations of phonemes represent certain physical objects or events, and that words may refer to entire categories, such as *animal*, while others refer to specific members of categories, such as *doggy*. One can see this skill developing in young children as they may refer to all animals as *doggy*.

Syntax

Syntax refers to how words are put together to form sentences. A child must notice the effects of word order on meaning: *Nathan has only three pieces of candy* has a very different meaning than *Only Nathan has three pieces of candy.*

Pragmatics

Finally, **pragmatics** refers to the dependence of language on context and preexisting knowledge. In other words, the manner in which we speak may differ depending on the audience and our relationship to that audience. Imagine asking to share a seat on a bus. Depending on whom we ask, we may word this request in wildly different ways. To a stranger, we may be more formal: *Pardon me, do you mind if I share this seat?* To a close friend, we may be less so: *Hey, move over!* Pragmatics are also affected by **prosody**—the rhythm, cadence, and inflection of our voices.

Language Development

To effectively interact with society, a child must learn to communicate through language, whether oral or signed. An important precursor to language is **babbling**. Almost without exception, children—including children who are hearing impaired—spontaneously begin to babble during their first year. For children who can hear, babbling reaches its highest frequency between nine and twelve months. For children who are hearing impaired or deaf, verbal babbling often ceases soon after it begins.

KEY CONCEPT

Timeline of language acquisition:

- 9 to 12 months: babbling
- 12 to 18 months: about one word per month
- 18 to 20 months: "explosion of language" and combining words
- 2 to 3 years: longer sentences (3 words or more)
- 5 years: language rules largely mastered

The timeline of language acquisition is fairly consistent among children. From 12 to 18 months, children add about one word per month. Starting around 18 months, an explosion of language begins. During this **naming explosion**, the child quickly learns dozens of words, and uses each word with varying inflection and gestures to convey a desired meaning. For example, a child may ask, *Apple?* while pointing at an apple in a bowl of fruit, in an effort to request the apple. During this naming explosion, children may also frequently fall into **overextension**, in which they inappropriately apply a term to an object that bears cursory similarities to the term. For example, in an attempt to request a kiwi from a fruit bowl, a child might point at the kiwi and

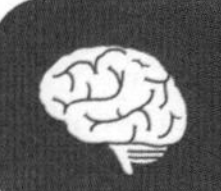

ask, *Apple?* For children at this age, gestures, inflection, and context are essential for the parent or caregiver to identify the meaning.

Between 18 and 20 months of age, children begin to combine words. For example, children may say, *Eat apple* to indicate that they would like to eat an apple. In the grocery store, the same children may ask, *That apple?* to distinguish between fruit. In this way, context and gesture becomes less important as the ability to assemble sentences develops.

By the age of two or three years, children can speak in longer sentences. Vocabulary grows exponentially. As a child creates longer sentences, grammatical errors increase as the child internalizes the complex rules of grammar. These include **errors of growth** in which a child applies a grammatical rule (often a morpheme) in a situation where it does not apply: *runned* instead of *ran*, or *funner* instead of *more fun*. Interestingly, caregivers are less likely to correct errors of grammar than errors of word choice.

For the most part, language is substantially mastered by the age of five. The acquisition of language appears easy for most children, which has led to significant speculation on exactly how this occurs.

REAL WORLD

Pediatricians often monitor language development to determine if there is a developmental delay. For example, a two-year-old child who uses fewer than 10 words has a significant developmental delay and should be referred for speech therapy. This would also prompt a search for other developmental issues.

Nativist (Biological) Theory

The **nativist (biological) theory**, largely credited to linguist Noam Chomsky, advocates for the existence of some innate capacity for language. Chomsky is known for his study of **transformational grammar**. He focused on syntactic transformations, or changes in word order that retain the same meaning; for example, *I took the MCAT* vs. *The MCAT was taken by me*. Chomsky noted that children learn to make such transformations effortlessly at an early age. He therefore concluded that this ability must be innate. In this theory, this innate ability is called the **language acquisition device (LAD)**, a theoretical pathway in the brain that allows infants to process and absorb language rules.

Nativists believe in a **critical period** for language acquisition between two years and puberty. If no language exposure occurs during this time, later training is largely ineffective. This idea came to light through an unfortunate test case: a victim of child abuse. This child had been isolated from all human contact from age two to thirteen, when she was discovered by authorities. Even with later language exposure, she was unable to master many rules of language, although she was able to learn some aspects of syntax. The fact that this child was able to learn some rules may indicate that there is a sensitive period for language development, rather than a critical period. A **sensitive period** is a time when environmental input has maximal effect on the development of an ability. Most psychologists consider the sensitive period for language development to be before the onset of puberty.

Learning (Behaviorist) Theory

The **learning (behaviorist) theory**, proposed by B. F. Skinner, explained language acquisition by operant conditioning. Very young babies are capable of distinguishing between phonemes of all human languages, but by six months of age show a strong preference for phonemes in the language spoken by their parents. Skinner explained language acquisition by **reinforcement**. That is, caregivers repeat and reinforce sounds that sound most like the language they speak. Thus, over time, the infant perceives that certain sounds have little value and are not reinforced, while other sounds have value and are reliably reinforced by caregivers. While this may account for the development of words and speech, many psycholinguists point out that this theory cannot fully explain the explosion in vocabulary that occurs during early childhood.

Social Interactionist Theory

The **social interactionist theory** of language development focuses on the interplay between biological and social processes. That is, language acquisition is driven by the child's desire to communicate and behave in a social manner, such as interacting with caretakers and other children. Interactionist theory allows for the role of brain development in the acquisition of language. As the biological foundation for language develops and children are exposed to language, the brain groups sounds and meanings together. Then, as the child interacts with others, certain brain circuits are reinforced, while others are de-emphasized, resulting in atrophy of those circuits.

Influence of Language on Cognition

Psycholinguistics has long focused on the relationship between language and thinking. Linguist Benjamin Whorf proposed the **Whorfian hypothesis**, also called the **linguistic relativity hypothesis**, which suggests that our perception of reality—the way we think about the world—is determined by the content of language. In essence, language affects the way we think rather than the other way around. For instance, some Inuit dialects have a wide variety of names for different types of snow, whereas the English language has very few. Therefore, according to the Whorfian hypothesis, people who speak one of those Inuit dialects are better at discriminating subtleties between different types of snow than English speakers are. This is a somewhat controversial notion that depends on the definition of "a word" but most linguists agree that language can influence how we think to some degree. Word choice, inflection, context, and speaker all play a role in our perception of a message. In addition, language often provides an original framework for understanding information. A more expansive framework with more specific vocabulary allows for more sophisticated processing of that information and enhanced communication of that information to others.

Brain Areas and Language

Two different areas of the brain are responsible for speech production and language comprehension, as shown in Figure 4.17. Both, however, are located in the dominant hemisphere, which is usually the left hemisphere. **Broca's area**, located in the inferior frontal gyrus of the frontal lobe, controls the motor function of speech via connections with the motor cortex. **Wernicke's area**, located in the superior temporal gyrus of the temporal lobe, is responsible for language comprehension. Broca's area and Wernicke's area are connected by the **arcuate fasciculus**, a bundle of axons that allows appropriate association between language comprehension and speech production.

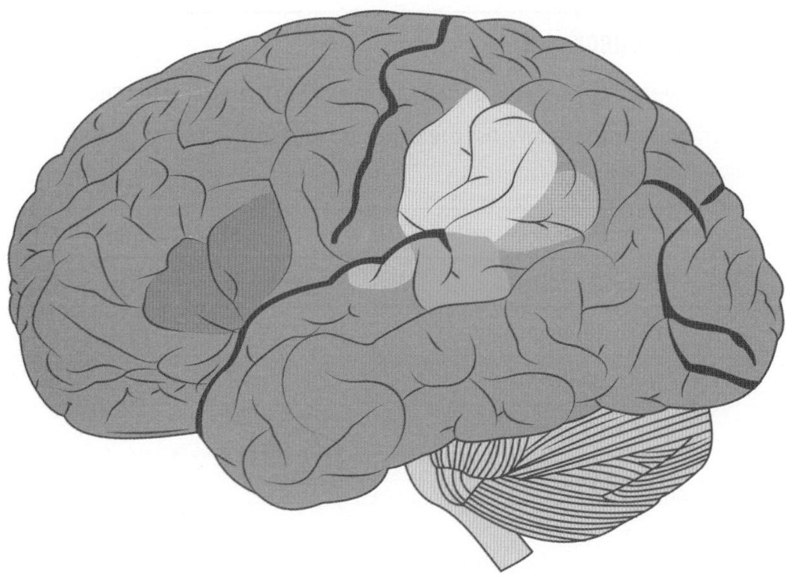

Figure 4.17 Brain Areas Associated with Language

Blue = Broca's area; Green = Wernicke's area. Other colored regions are associated with other aspects of language beyond the scope of the MCAT (yellow = supramarginal gyrus; orange = angular gyrus; light pink = primary auditory cortex).

Aphasia is a deficit of language production or comprehension. Much of what we know regarding language and aphasia is through observations of people with damage to speech-related areas. When damage occurs to Broca's area, speech comprehension is intact but the patient will have a reduced or absent ability to produce spoken language. This is known as **Broca's (expressive) aphasia**. These patients are often very frustrated because they are stuck with the sensation of having every word on the tip of their tongue.

On the other hand, when Wernicke's area is damaged, motor production and fluency of speech is retained but comprehension of speech is lost. This is known as **Wernicke's (receptive) aphasia**. Because speech comprehension is lost, these patients speak nonsensical sounds and inappropriate word combinations devoid of meaning. Patients with Wernicke's aphasia often believe that they are speaking and understanding perfectly well, even though the people around them have no comprehension of what is being said. This can also be very frustrating to patients.

Finally, if the arcuate fasciculus is affected, the resulting aphasia is known as **conduction aphasia**. Because Broca's and Wernicke's areas are unaffected, speech production and comprehension are intact. However, the patient is unable to repeat something that has been said because the connection between these two regions has been lost. This is a very rare form of aphasia.

MCAT CONCEPT CHECK 4.6

Before you move on, assess your understanding of the material with these questions.

1. For each of the ages below, list the expected milestone(s) of language development:

Age	Milestone(s)
9 to 12 months	
12 to 18 months	
18 to 20 months	
2 to 3 years	
5 years	

2. For each of the theories of language acquisition below, what is the primary motivation or trigger for language development?
   - Nativist (Biological):

   ____

   - Learning (Behaviorist):

   ____

   - Social interactionist:

   ____

3. Briefly describe the clinical features of each type of aphasia listed below:
   - Broca's aphasia:

   ____

   - Wernicke's aphasia

   ____

   - Conduction aphasia:

   ____

Conclusion

One of the biggest questions that psychology and biology seek to answer is how the brain, an organ consisting of lipids, water, and neurotransmitters, becomes the mind. Cognition and consciousness allow us to think about who we are, where we are, and what we are doing at a given moment, and this all occurs due to a complex interaction between individual neurons within the brain. Not only do we experience consciousness, but our behaviors are also intricately intertwined with physiological brain function. Language is one of the most complex cognitive processes, requiring intact comprehension and production mechanisms and an understanding of the rules of our native language.

As much as we research what the mind is and how it works, there is as much interest in *why* we do what we do and how we *feel* about it. This is the function of motivation (both internal and external) and emotion, which we will explore in detail in the next chapter.

GO ONLINE

You've reviewed the content, now test your knowledge and critical thinking skills by completing a test-like passage set in your online resources!

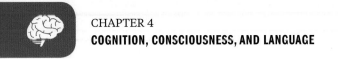

CONCEPT SUMMARY

Cognition

- Thought is more than just that of which we are conscious. The brain processes and makes decisions about the importance of various stimuli below the level of conscious awareness.
- The **information processing model** states that the brain encodes, stores, and retrieves information much like a computer.
- The ability to think abstractly develops over the life span. Early cognitive development is limited by brain maturation. Culture, genes, and environment also influence cognitive development.
- **Piaget's stages of cognitive development** are sensorimotor, preoperational, concrete operational, and formal operational.
  - The **sensorimotor stage** focuses on manipulating the environment to meet physical needs through **circular reactions. Object permanence** ends this stage.
  - The **preoperational stage** focuses on **symbolic thinking**, **egocentrism**, and **centration**.
  - The **concrete operational stage** focuses on understanding the feelings of others and manipulating physical (concrete) objects.
  - The **formal operational stage** focuses on abstract thought and problem solving.
- A mild level of cognitive decline while aging is normal; significant changes in cognition may signify an underlying disorder.
- Biological factors that affect cognition include organic brain disorders, genetic and chromosomal conditions, metabolic derangements, and drug use.

Problem Solving and Decision Making

- **Problem solving** requires identification and understanding of the problem, generation of potential solutions, testing of potential solutions, and evaluation of results.
  - A **mental set** is a pattern of approach for a given problem. An inappropriate mental set may negatively impact problem solving.
  - **Functional fixedness** is the tendency to use objects only in the way they are normally utilized, which may create barriers to problem solving.
- Types of problem solving include **trial-and-error**, **algorithms**, **deductive reasoning** (deriving conclusions from general rules), and **inductive reasoning** (deriving generalizations from evidence).
- Heuristics, biases, intuition, and emotions may assist decision making but may also lead to erroneous or problematic decisions.

    - **Heuristics** are shortcuts or rules of thumb used to make decisions.
    - **Biases** exist when an experimenter or decision maker is unable to objectively evaluate information.
    - **Intuition** is a "gut feeling" regarding a particular decision. However, intuition can often be attributed to experience with similar situations.
    - Emotional state often plays a role in decision making.
- Gardner's theory of **multiple intelligences** proposes at least eight areas of intelligence including: linguistic, logical–mathematical, musical, visual–spatial, bodily–kinesthetic, interpersonal, intrapersonal, and naturalist.
- Variations in intellectual ability can be attributed to combinations of environment, education, and genetics.

Consciousness

- States of consciousness include alertness, sleep, dreaming, and altered states of consciousness.
- **Alertness** is the state of being awake and able to think, perceive, process, and express information. **Beta** and **alpha waves** predominate on **electroencephalography** (**EEG**).
- **Sleep** is important for health of the brain and body.
    - **Stage 1** is light sleep and is dominated by **theta waves** on EEG. **Stage 2** is slightly deeper and includes theta waves, **sleep spindles**, and **K complexes**.
    - **Stages 3** and **4** are deep (**slow-wave**) sleep (**SWS**). **Delta waves** predominate on EEG. Most sleep-wake disorders occur during Stage 3 and 4 **non-rapid eye movement** (**NREM**) **sleep**. Dreaming in SWS focuses on consolidating declarative memories.
    - **Rapid eye movement** (**REM**) **sleep** is sometimes called **paradoxical sleep**: the mind appears close to awake on EEG, but the person is asleep. Eye movements and body paralysis occur in this stage. Dreaming in REM focuses on consolidating procedural memories.
    - The **sleep cycle** is approximately 90 minutes for adults; the normal cycle is Stage 1–2–3–4–3–2–REM or just 1–2–3–4–REM, although REM becomes more frequent toward the morning.
    - Changes in light in the evening trigger release of **melatonin** by the **pineal gland**, resulting in sleepiness. **Cortisol** levels increase in the early morning and help promote wakefulness. **Circadian rhythms** normally trend around a 24-hour day.
    - Most **dreaming** occurs during REM, but some dreaming occurs during other sleep stages. There are many different models that attempt to account for the content and purpose of dreaming.
    - Sleep-wake disorders include **dyssomnias**, such as insomnia, narcolepsy, sleep apnea, and sleep deprivation; and **parasomnias**, such as night terrors and sleepwalking (somnambulism).

- **Hypnosis** is a state of consciousness in which individuals appear to be in control of their normal faculties but are in a highly suggestible state. Hypnosis is often used for pain control, psychological therapy, memory enhancement, weight loss, and smoking cessation.
- **Meditation** involves a quieting of the mind and is often used for relief of anxiety. It has also played a role in many of the world's religions.

Consciousness-Altering Drugs

- Consciousness-altering drugs are grouped by effect into depressants, stimulants, opiates, and hallucinogens.
  - **Depressants** include alcohol, barbiturates, and benzodiazepines. They promote or mimic GABA activity in the brain.
  - **Stimulants** include amphetamines, cocaine, and ecstasy. They increase dopamine, norepinephrine, and serotonin concentration at the synaptic cleft.
  - **Opiates** and **opioids** include heroin, morphine, opium, and prescription pain medications such as oxycodone and hydrocodone. They can cause death by respiratory depression.
  - **Hallucinogens** include lysergic acid diethylamide (LSD), peyote, mescaline, ketamine, and psilocybin-containing mushrooms.
  - **Marijuana** has depressant, stimulant, and hallucinogenic effects. Its active ingredient is **tetrahydrocannabinol**.
- Drug addiction is mediated by the **mesolimbic pathway**, which includes the **nucleus accumbens**, **medial forebrain bundle**, and **ventral tegmental area**. Dopamine is the main neurotransmitter in this pathway.

Attention

- **Selective attention** allows one to pay attention to a particular stimulus while determining if additional stimuli in the background require attention.
- **Divided attention** uses **automatic processing** to pay attention to multiple activities at one time.

Language

- Language consists of phonology, morphology, semantics, syntax, and pragmatics.
  - **Phonology** refers to the actual sound of speech.
  - **Morphology** refers to the building blocks of words, such as rules for pluralization (*–s* in English), past tense (*–ed*), and so forth.
  - **Semantics** refers to the meaning of words.
  - **Syntax** refers to the rules dictating word order.
  - **Pragmatics** refers to the changes in language delivery depending on context.

- Theories of language development focus on different reasons or motivations for language acquisition.
  - The **nativist (biological) theory** explains language acquisition as being innate and controlled by the **language acquisition device (LAD)**.
  - The **learning (behaviorist) theory** explains language acquisition as being controlled by operant conditioning and reinforcement by caregivers.
  - The **social interactionist theory** explains language acquisition as being caused by a motivation to communicate and interact with others.
- The **Whorfian (linguistic relativity) hypothesis** states that the lens through which we view and interpret the world is created by language.
- Speech areas in the brain are found in the dominant hemisphere, which is usually the left.
  - The motor function of speech is controlled by **Broca's area**. Damage results in **Broca's aphasia**, a nonfluent aphasia in which generating each word requires great effort.
  - Language comprehension is controlled by **Wernicke's area**. Damage results in **Wernicke's aphasia**, a fluent, nonsensical aphasia with lack of comprehension.
  - The **arcuate fasciculus** connects Wernicke's area and Broca's area. Damage results in **conduction aphasia**, marked by the inability to repeat words heard despite intact speech generation and comprehension.

ANSWERS TO CONCEPT CHECKS

4.1

1. The three steps in the information processing model are encoding, storage, and retrieval.
2. The older adult most likely has delirium. The time course is incompatible with the slow decline of dementia.
3.

Stage	Key Features
Sensorimotor	Focuses on manipulating environment for physical needs; circular reactions; ends with object permanence
Preoperational	Symbolic thinking, egocentrism, and centration
Concrete operational	Understands conservation and the feelings of others; can manipulate concrete objects logically
Formal operational	Can think abstractly and problem solve

4.2

1. Functional fixedness
2. Algorithm
3. Belief perseverance
4. The availability heuristic is used for making decisions based on how easily similar instances can be imagined. The representativeness heuristic is used for making decisions based on how much a particular item or situation fits a given prototype or stereotype.

4.3

1.

Stage	EEG Waves	Features
Awake	Beta and alpha	Able to perceive, process, access information, and express that information verbally
Stage 1	Theta	Light sleep and dozing
Stage 2	Theta	Sleep spindles and K complexes
Stage 3	Delta	Slow-wave sleep; dreams; declarative memory consolidation; sleep-wake disorders occur in this stage
Stage 4	Delta	Slow-wave sleep; dreams; declarative memory consolidation; sleep-wake disorders occur in this stage
REM	Mostly beta	Appears awake physiologically; dreams; procedural memory consolidation; body is paralyzed

2. The two hormones most associated with maintaining circadian rhythms are melatonin and cortisol.
3. Dyssomnias are disorders in which the duration or timing of sleep is disturbed. Examples include insomnia, narcolepsy, and sleep apnea. Parasomnias are disorders in which abnormal behaviors occur during sleep. Examples include night terrors and sleepwalking (somnambulism).

4.4

1. Drugs known to increase GABA activity in the brain include alcohol, barbiturates, and benzodiazepines. Note that marijuana inhibits GABA activity.
2. Drugs known to increase dopamine, norepinephrine, and serotonin activity in the brain include amphetamines, cocaine, and ecstasy (MDMA). Ecstasy is a designer amphetamine; it is mentioned separately here because of its hallucinogenic properties.
3. The three main structures in the mesolimbic reward pathway are the nucleus accumbens, medial forebrain bundle, and ventral tegmental area. The neurotransmitter of this pathway is dopamine.

4.5

1. Controlled (effortful) processing is used when maintaining undivided attention on a task, and is usually used for new or complex actions. Automatic processing is used for less critical stimuli in divided attention, and is usually used for familiar or repetitive actions.
2. The filter in selective attention permits us to focus on one set of stimuli while scanning other stimuli in the background for important information (such as our name or a significant change in the environment).

4.6

1.

Age	Milestone(s)
9 to 12 months	Babbling
12 to 18 months	Increase of about one word per month
18 to 20 months	"Explosion of language" and combining words (two-word sentences)
2 to 3 years	Longer sentences of three or more words
5 years	Language rules largely mastered

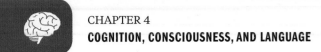

2. The primary trigger in the nativist theory is an innate ability to pick up language via the language acquisition device. In the learning theory, it is operant conditioning with reinforcement by caregivers. In the social interactionist theory, it is a desire to communicate and act socially.
3. Broca's aphasia is marked by difficulty producing language, with hesitancy and great difficulty coming up with words. Wernicke's aphasia is fluent, but includes nonsensical sounds and words devoid of meaning; language comprehension is lost. Conduction aphasia is marked by difficulty repeating speech, with intact speech production and comprehension.

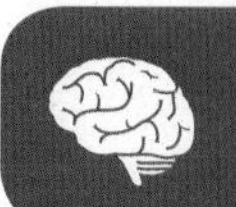

SCIENCE MASTERY ASSESSMENT EXPLANATIONS

1. **D**

Jean Piaget hypothesized that new information is processed by adaptation, **(B)**. Adaptation is too broad of an answer because it includes both assimilation, **(A)**, and accommodation, **(D)**. Assimilation is incorporation of new information into existing schemata. If the new information doesn't fit, then accommodation occurs. Accommodation is the modification of existing schemata to account for new information and is thus the correct answer.

2. **B**

While many drugs can cause paranoia, the elevation in heart rate combined with prolonged alertness and a decrease in appetite suggests amphetamines, matching choice **(B)**.

3. **C**

The base rate fallacy occurs when prototypical or stereotypical factors are used for analysis rather than actual data. Because the student is volunteering in a hospital with a stroke center, the student sees more patients who have experienced a stroke than would be expected in a hospital without a stroke center. Thus, this experience changes the student's perception and results in base rate fallacy. Deductive reasoning, **(A)**, refers to drawing conclusions by integrating different pieces of evidence. The representativeness heuristic, **(B)**, involves categorization and classification based on how well an individual example fits its category. Confirmation bias, **(D)**, occurs when people only seek information that reinforces their opinions.

4. **A**

Fluid intelligence consists of problem-solving skills and is not one of Gardner's eight multiple intelligences. Gardner's theory lists linguistic, logical–mathematical, musical, visual–spatial, bodily–kinesthetic, interpersonal, intrapersonal, and naturalist intelligences.

5. **A**

EEG during REM is composed mainly of beta waves, which are present during alertness. SWS, **(B)**, consists mainly of delta waves, which are not typically present during REM sleep. Stage 1 sleep, **(C)**, consists mainly of theta waves. Meditation, **(D)**, is quieting of the mind, and consists mainly of slow alpha and theta waves.

6. **B**

Early in the evening, sleep cycles include deepening of sleep (Stages 1–2–3–4), followed either by lightening of sleep (Stages 4–3–2) and then REM, or just directly moving from Stage 4 into REM. Later in the evening, the cycle may be shortened as slow-wave sleep becomes less common.

7. **B**

Divided attention is the ability to perform several tasks simultaneously. Routine actions, such as rewatching familiar episodes of a favorite TV show, are completed via automatic processing which allows the brain to focus on other tasks. Working on two tasks simultaneously means the student is using divided attention, supporting **(B)** as the correct answer. By contrast, parallel processing, **(D)**, refers to the brain's ability to analyze imagery by processing several aspects of the image (shape, color, motion, etc.) independently and simultaneously.

8. **C**

Cognitive theorists proposed in the cognitive process dream theory that wakeful and dreaming states use the same mental systems within the brain, particularly stream-of-consciousness. The activation–synthesis theory, **(A)**, states that dreams are caused by widespread, random activation of neural circuitry. The problem solving dream model, **(B)**, indicates that dreams are used to solve problems while sleeping due to untethering of dreams from obstacles perceived while awake. The neurocognitive theorists, **(D)**, seek to unify cognitive and biological perspectives by correlating the subjective dream experience with the physiological experience of dreaming.

9. **D**

The description of the student matches the clinical features of marijuana (cannabis) use: hunger (presumably, based on what was shoplifted), redness of the eyes, dry mouth, and euphoria. Marijuana may also cause an increased heart rate, short-term memory loss, paranoia, and—in high doses—hallucinations. Tetrahydrocannabinol is the primary active substance in marijuana.

10. **C**

Syntax refers to how words are put together to form sentences and create meaning. Phonology, **(A)**, refers to the actual sounds of a language. Semantics, **(B)**, refers to the association of meaning with a word. Pragmatics, **(D)**, refers to changes in usage, wording, and inflection based on context.

11. **C**

A child who speaks in three-word sentences but has not yet mastered most of the fundamental rules of language, including past tense, is likely to be between two and three years old.

12. **B**

Learning theory, largely based on the work of B. F. Skinner, states that parents reinforce phonemes that sound most like their language, resulting in preferential preservation of these phonemes. Nativist theory, **(A)**, posits a critical period during which language acquisition occurs. Social interactionist theory, **(C)**, indicates that language develops via interaction with parents and caregivers as well as a desire of the child to communicate. Neurocognitive theory, **(D)**, is concerned with the subjective experience of dreaming and the physiology of dreaming.

13. **A**

Broca's area governs the motor function of language. A stroke that affects Broca's area will leave receptive language intact, but word formation will be affected. A stroke affecting Wernicke's area, **(B)**, will make it so the individual is unable to comprehend speech. A stroke affecting the arcuate fasciculus, **(C)**, will result in an inability to repeat words heard but spontaneous language production is intact. The superior temporal gyrus, **(D)**, is where Wernicke's area is located.

14. **A**

With selective attention, a person focuses on a specific task while filtering out background stimuli. In this way, the individual in the airport ignored most of the background noise while working on a puzzle. However, the brain can shift attention quickly when a stimulus is particularly important, such as the specific call for boarding the flight. This shift is known as the cocktail party phenomenon, named for a person's ability to quickly reorient their focus when they hear their name at a party. Thus, **(A)** is the correct answer. By contrast, while effortful processing, **(D)**, describes this person's intense focus on the puzzle, it does not explain the ability to respond quickly to the boarding announcement. Similarly, divided attention, **(B)**, works best on simple tasks and the question stem uses the word "challenging" to describe the puzzle.

15. **D**

About 75% of dreaming occurs during REM, but dreams occur in all other stages of sleep as well. More bizarre dreams are likely to occur during REM.

Consult your online resources for additional practice.

EQUATION TO REMEMBER

(4.1) **Stanford–Binet intelligence quotient:** $\text{IQ} = \frac{\text{mental age}}{\text{chronological age}} \times 100$

SHARED CONCEPTS

Behavioral Sciences Chapter 1
Biology and Behavior

Behavioral Sciences Chapter 3
Learning and Memory

Behavioral Sciences Chapter 6
Identity and Personality

Behavioral Sciences Chapter 11
Social Structure and Demographics

Biology Chapter 4
The Nervous System

Critical Analysis and Reasoning Skills Chapter 6
Formal Logic

CHAPTER 5

MOTIVATION, EMOTION, AND STRESS

SCIENCE MASTERY ASSESSMENT

Every pre-med knows this feeling: there is so much content I have to know for the MCAT! How do I know what to do first or what's important?

While the high-yield badges throughout this book will help you identify the most important topics, this Science Mastery Assessment is another tool in your MCAT prep arsenal. This quiz (which can also be taken in your online resources) and the guidance below will help ensure that you are spending the appropriate amount of time on this chapter based on your personal strengths and weaknesses. Don't worry though—skipping something now does not mean you'll never study it. Later on in your prep, as you complete full-length tests, you'll uncover specific pieces of content that you need to review and can come back to these chapters as appropriate.

How to Use This Assessment

If you answer 0–7 questions correctly:

Spend about 1 hour to read this chapter in full and take limited notes throughout. Follow up by reviewing **all** quiz questions to ensure that you now understand how to solve each one.

If you answer 8–11 questions correctly:

Spend 20–40 minutes reviewing the quiz questions. Beginning with the questions you missed, read and take notes on the corresponding subchapters. For questions you answered correctly, ensure your thinking matches that of the explanation and you understand why each choice was correct or incorrect.

If you answer 12–15 questions correctly:

Spend less than 20 minutes reviewing all questions from the quiz. If you missed any, then include a quick read-through of the corresponding subchapters, or even just the relevant content within a subchapter, as part of your question review. For questions you answered correctly, ensure your thinking matches that of the explanation and review the Concept Summary at the end of the chapter.

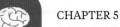

1. A college student strives for excellent grades and hopes to graduate with a better GPA than an older sibling. This type of motivation is considered:
   A. extrinsic motivation.
   B. intrinsic motivation.
   C. a primary drive.
   D. a secondary drive.

2. When practicing a recital song at home, a student sounds perfectly in pitch to family and friends. However, when performing at the recital in front of a large audience of peers, strangers, and coaches, the student's pitch and tone are off, resulting in a poor performance. This second performance is best explained by:
   A. drive reduction theory.
   B. instinct approach theory.
   C. Maslow's hierarchy of needs.
   D. the Yerkes–Dodson law.

3. Seeking homeostasis to reduce an uncomfortable internal state is associated with which motivational theory?
   A. Drive reduction theory
   B. Instinct theory
   C. Arousal theory
   D. Incentive theory

4. People from cultures around the world can identify which of the following emotions?
   A. Happiness, sadness, and surprise
   B. Happiness, anger, and apathy
   C. Sadness, anticipation, and happiness
   D. Excitement, anger, and disgust

5. Experiencing emotion involves three components, which are:
   A. behavioral, reactionary, and cognitive.
   B. emotional, physical, and mental.
   C. physiological, cognitive, and behavioral.
   D. emotional, cognitive, and behavioral.

6. The statement "I noticed my heart racing and breathing rate increasing when I saw a bear, so I am afraid" corresponds most closely with which theory of emotion?
   A. Schachter–Singer theory
   B. Yerkes–Dodson theory
   C. Cannon–Bard theory
   D. James–Lange theory

7. Which theory of motivation is most significantly informed by Darwin's theory of evolution?
   A. Arousal theory
   B. Drive reduction theory
   C. Instinct theory
   D. Incentive theory

8. Simultaneous processing of conscious emotions and physiological activation is the defining feature of which theory of emotion?
   A. Schachter–Singer theory
   B. James–Lange theory
   C. Incentive theory
   D. Cannon–Bard theory

9. All of the following brain regions are primarily responsible for the experience of emotions EXCEPT the:
   A. amygdala.
   B. prefrontal cortex.
   C. basal ganglia.
   D. thalamus.

10. A person with high left frontal lobe activity is most likely experiencing which emotion?
    A. Happiness
    B. Sadness
    C. Surprise
    D. Disgust

11. Determination of the intensity and risk of a stressor occurs during which stage(s) of stress appraisal?
    A. Primary appraisal only
    B. Secondary appraisal only
    C. Both primary and secondary appraisal
    D. Neither primary nor secondary appraisal

12. A medical student is feeling a high level of stress due to upcoming exams and pressure from family to engage in activities at home. The student chooses to go to the gym for a workout to help relax. This workout is which type of stress?
    A. Hassle
    B. Frustration
    C. Distress
    D. Eustress

13. Which type of conflict is associated with the LEAST amount of stress?
    A. Approach–approach conflict
    B. Avoidance–avoidance conflict
    C. Approach–avoidance conflict
    D. Avoidance–escape conflict

14. While cleaning your house, you notice a large spider on the wall by your head and feel your heart rate jump up and your skin temperature grow warm. Which stage of stress response are you experiencing?
    A. Alarm
    B. Resistance
    C. Exhaustion
    D. Homeostasis

15. Each of the following responses to stress is considered maladaptive EXCEPT:
    A. drug use.
    B. social withdrawal.
    C. progressive muscle relaxation.
    D. avoiding the stressor.

Answer Key

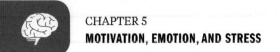

1. **A**
2. **D**
3. **A**
4. **A**
5. **C**
6. **D**
7. **C**
8. **D**
9. **C**
10. **A**
11. **B**
12. **D**
13. **A**
14. **A**
15. **C**

Detailed explanations can be found at the end of the chapter.

CHAPTER 5

MOTIVATION, EMOTION, AND STRESS

In This Chapter

CHAPTER PROFILE

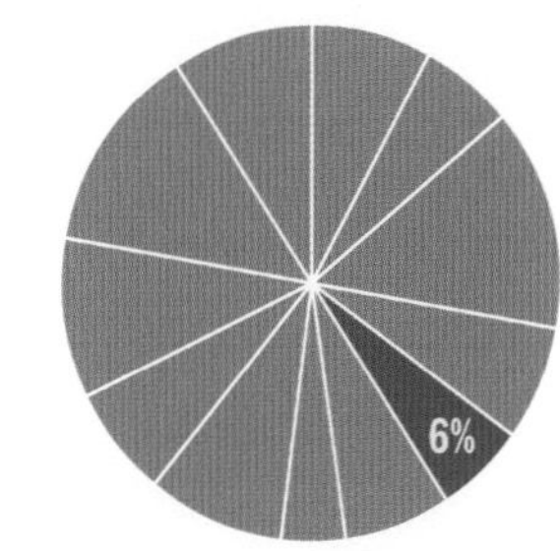

The content in this chapter should be relevant to about 6% of all questions about the behavioral sciences on the MCAT.

This chapter covers material from the following AAMC content categories:

6C: Responding to the world

7A: Individual influences on behavior

Introduction

A heart condition, substantial bone density decreases, esophageal and stomach ulcers, a loss of 25 percent of body mass—these are some of the possible consequences of anorexia nervosa. Sometimes those who develop this condition have been put under extreme pressure in their work or personal lives to look a certain way. Over time, this pressure can motivate weight loss in an unhealthy manner. Patients are sometimes motivated by feelings of disgust and guilt that go directly against the body's basic needs. A combination of stress from job and peers, intrinsic and extrinsic motivation, and negative emotions can result in what could be a life-threatening condition.

In this chapter, we'll discuss motivation, emotion, and stress. We will look at factors that influence motivation, the components of emotion, and the stressors that lead to the stress response. We will also look at theories used to explain these processes and associated behaviors. The physiological, cognitive, and behavioral elements will be examined in order to understand the role that these topics play in everyday life. Exploring these ideas will enable us to tackle any question related to these concepts on Test Day.

5.1 Motivation

LEARNING OBJECTIVES

After Chapter 5.1, you will be able to:

- Describe the sources of motivation posited by instinct theory, arousal theory, drive reduction theory, and the need-based theories
- Order Maslow's proposed hierarchy of needs from most to least essential
- Apply opponent-process theory to drug use and withdrawal

Motivation is the purpose, or driving force, behind our actions. The word derives from the Latin *movere*, meaning "to move." There are many examples of motivation in our everyday lives. As you sit, studying for the MCAT, you realize you are thirsty, so you reach for your water bottle. When you realize it is empty, the need to quench your thirst drives you to get up, walk to the kitchen, and fill the bottle with water. Thus, the physical state of thirst motivated an action. The desire to go to medical school and become a physician has motivated you to complete required undergraduate coursework, strive for a competitive GPA, participate in extracurricular activities, and dedicate your time to study for a standardized test. The goal of staying fit and healthy motivates many to spend hours in the gym, while the initial discomfort of physical activity might motivate others to stay sedentary. Motivation can be directed toward minimizing pain, maximizing pleasure, or it can be rooted in the desire, or **appetite**, to fullfil a physical need. Though the term appetite is commonly used to refer to a need for food, this term can more generally be applied to any need such as eating, drinking, sleeping, or social acceptance.

REAL WORLD

Intrinsic motivation can be reduced by introducing external reward into a scenario. In one study, children were given art supplies and told to draw. Some children were told they would receive a gold ribbon for their drawing, while others were not. Those who knew of the potential reward were driven by extrinsic motivation and spent less time drawing, acting only to receive the reward. Children who did not know about the reward drew for the sake of drawing—by intrinsic motivation—and spent more time drawing.

Motivation can manifest from external forces, such as rewards and punishments, or internal forces, where the behavior is personally gratifying. External forces, coming from outside oneself, create extrinsic motivation. **Extrinsic motivation** can include rewards for showing a desired behavior or avoiding punishment if the desired behavior is not achieved. Examples of such motivation include working hard at your job for praise from your boss, practicing regularly for a sport so that you will perform strongly in an upcoming game, or studying for months on end to achieve a high score on the MCAT. Each of these acts results in external, tangible rewards. Extrinsic motivation can also include doing chores to avoid punishment and working to avoid being fired. Competition is a strong form of external motivation because people are incentivized to beat others and not only to win, perform, or achieve for themselves. Motivation that comes from within oneself is referred to as **intrinsic motivation**. This can be driven by interest in a task or pure enjoyment. A student who takes interest in the subject matter at hand and has the goal of mastering the content is driven by intrinsic motivation, while the goal of achieving high grades is considered extrinsic.

The primary views of motivation focus on instincts that elicit natural behavior, the desire to maintain optimal levels of arousal, the drive to reduce uncomfortable states, and the goal of satisfying physiological and psychological needs.

Instinct Theory

Early attempts to understand the basis of motivation focused on **instincts**, which are innate, fixed patterns of behavior. For example, wolves are instinctively pack creatures that naturally follow the alpha male of their group. Additionally, they are highly territorial creatures, protecting areas that are much larger than needed to hunt and dwell. This protection includes scent-marking, howling, and direct aggressive attacks on intruders. Humans also have instinctive behavior; for example, thumb sucking is an instinctual response to stress in babies that is aimed at self-soothing. As discussed in Chapter 1 of *MCAT Behavioral Sciences Review*, primitive reflexes like the grasp reflex, shown in Figure 5.1, are also instinctual. Note that some instincts last for the entire lifetime, while others may appear or disappear with age.

Figure 5.1 The Grasp Reflex
Primitive reflexes are examples of instincts seen in infants that extinguish with age.

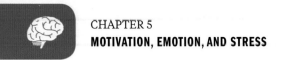

KEY CONCEPT

An instinct is an innate, fixed pattern of behavior. It may be consistent throughout life, or it may appear or disappear with time.

According to the **instinct theory** of motivation, certain behaviors are based on evolutionarily programmed instincts. This theory was one of the first to describe motivation and was derived from Darwin's theory of evolution. William James, the father of modern psychology, was one of the first to write about human instincts in his 1890 publication of *Principles of Psychology*. He stated that humans were motivated by many instincts, possibly more than any other animal studied. James suggested that human actions are derived from 20 physical instincts, including suckling and locomotion, and 17 mental instincts, including curiosity and fearfulness. However, he said that many of these instincts were in direct conflict with each other and could be overridden by experience. Arguably the greatest proponent of instinct theory was William McDougall, who proposed that humans were led to all thoughts and behaviors by 18 distinctive instincts, including flight and acquisition. James and McDougall postulated that the instincts of suckling and carrying food to the mouth result in naturally motivating one to eat.

Arousal Theory

Another factor that influences motivation is **arousal**, the psychological and physiological state of being awake and reactive to stimuli. Arousal involves the brainstem, autonomic nervous system, and endocrine system and plays a vital role in behavior and cognition.

Arousal theory states that people perform actions in order to maintain an optimal level of arousal: seeking to increase arousal when it falls below their optimal level, and to decrease arousal when it rises above their optimum level. Additionally, the **Yerkes–Dodson law** postulates a U-shaped function between the level of arousal and performance. This law states that performance is worst at extremely high and low levels of arousal and optimal at some intermediate level, as depicted in Figure 5.2. The optimal level of arousal varies between different types of tasks: lower levels are optimal for highly cognitive tasks, while higher levels are optimal for activities that require physical endurance and stamina. Further, simple tasks generally require slightly higher arousal than complex tasks.

REAL WORLD

While a moderate level of arousal is optimal for performance, certain individuals seek out higher levels of arousal. Such people may seek out dangerous activities, such as skydiving or bungee jumping. These individuals are sometimes found to have lower levels of monoamine oxidase (MAO), which breaks down catecholamines. Low levels of MAO result in higher neurotransmitter levels, which may lead to motivation to experience high levels of arousal.

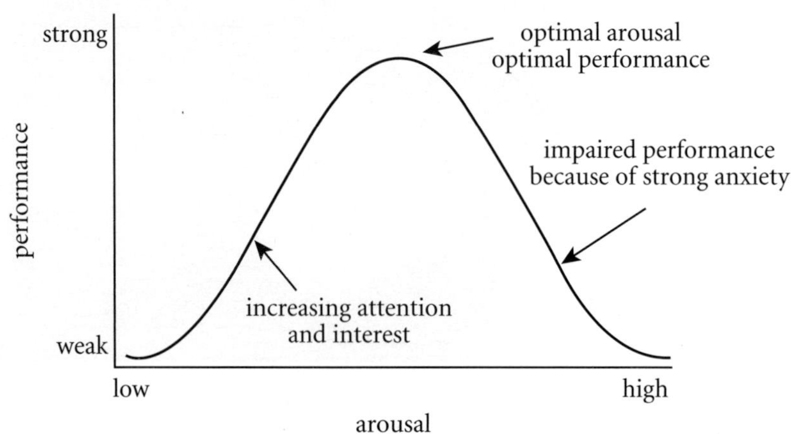

Figure 5.2 Yerkes-Dodson Law

Drive Reduction Theory

Drives are defined as internal states of tension that activate particular behaviors focused on goals. Drives are thought to originate within an individual without requiring any external factors to motivate behavior. In other words, drives help humans survive by creating an uncomfortable state, ensuring motivation to eliminate this state or to relieve the internal tension created by unmet needs. **Primary drives**, including the need for food, water, and warmth, motivate us to sustain bodily processes in homeostasis. **Homeostasis** is the regulation of the internal environment to maintain an optimal, stable set of conditions. In homeostatic regulation, external factors are encountered, and the system will react to push the system back to its optimal state.

Homeostasis is usually controlled by **negative feedback** loops. A common real-life example of a negative feedback loop is a thermostat. A thermostat is set to a desired temperature, and then sensors monitor the air temperature in relation to this desired temperature. If the air temperature gets too cold, the heater will turn on; if the temperature gets too warm, the heater will turn off. Negative feedback loops in the body operate the same way. Likewise when our bodies are lacking nutrients and energy, feedback systems release hormones like ghrelin that create hunger and motivate eating. After we consume food, feedback is sent to the brain to turn off the hunger drive through hormones like leptin. Hunger is a complex feedback system involving these hormones, receptors in the walls of the stomach, levels of glucose (maintained by the liver), and insulin and glucagon levels (released by the pancreas). The concentrations of many hormones of the endocrine system are regulated by three-organ "axes," such as the hypothalamic–pituitary–adrenal axis shown in Figure 5.3.

REAL WORLD

One well-characterized drive is the hunger (food) drive, which has been studied for more than 60 years. The hunger drive is a primary drive, in that food is necessary for life. Given the necessity of food to life, many species possess multiple processes that drive the individual to consume food. These biological processes include tie-ins to the sense of smell and taste, which were demonstrated by many studies, notably those of Janowitz and Grossman.

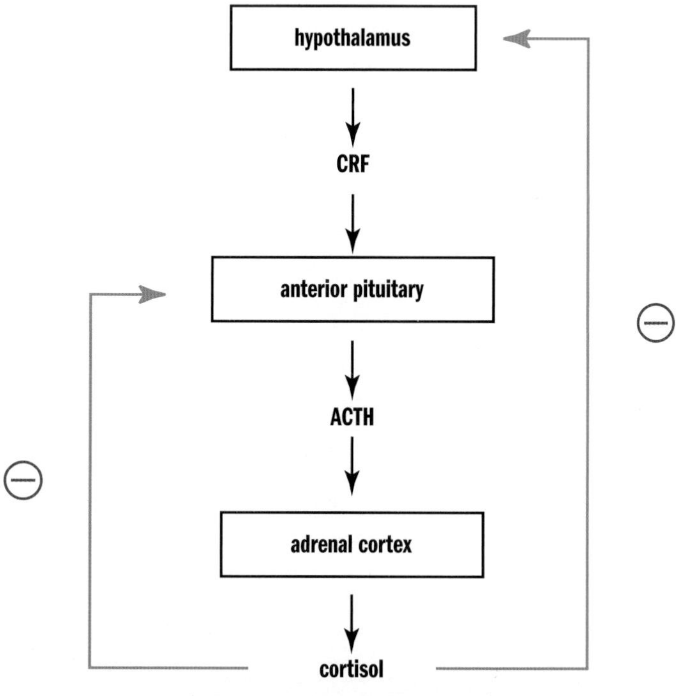

Figure 5.3 Negative Feedback in the Endocrine System

KEY CONCEPT

Primary drives are those that motivate us to sustain necessary biological processes. Secondary drives are those that motivate us to fulfill nonbiological, emotional, or "learned" desires.

BRIDGE

Drive reduction theory can be applied to motivation in terms of learning, and is commonly used to define motivational states within behavioral conditioning. Conditioning and learning are discussed in Chapter 3 of *MCAT Behavioral Sciences Review*.

MCAT EXPERTISE

Knowing the four primary factors that influence motivation is key for Test Day: instincts, arousal, drives, and needs. The MCAT will expect you to know the common theories for explaining motivation.

Additional drives that are not directly related to biological processes are called **secondary drives**. These drives are thought to stem from learning. The drive to matriculate into medical school and become a physician is an example of a secondary drive. Secondary drives also include certain emotions, such as the desire for nurturing, love, achievement, and aggression.

Drive reduction theory explains that motivation is based on the goal of eliminating uncomfortable states. Theorists hypothesize that certain physiological conditions result in a negative internal environment. This internal environment then drives motivation and seeks homeostasis in order to reduce the uncomfortable internal state.

Need-Based Theories

In need-based theories of motivation, energy and resources are allocated to best satisfy human needs. These needs may be **primary needs**, which are generally physiological needs such as the need for food, water, sleep, and shelter. Or these needs might be **secondary needs**, which are generally mental states, like a desire for power, achievement, or social belonging.

Abraham Maslow defined **needs** as relatively long-lasting feelings that require relief or satisfaction and tend to influence action. He observed that certain needs will yield a greater influence on our motivation and he established what is referred to as **Maslow's hierarchy of needs**. Maslow classified needs into five groups, and assigned different levels of priority to each group. The hierarchy is typically displayed as a pyramid, as shown in Figure 5.4, where the most primitive, essential, and important needs are at the base. The first four levels of the pyramid correspond to physiological needs, safety and security, love and belonging, and self-esteem. The highest level of the pyramid corresponds to **self-actualization**, or the need to realize one's fullest potential. Maslow theorized that if the lowest level of need is not met, motivation to meet that need will be the highest priority. Once the lowest level of needs is met, if additional needs exist, they will be satisfied based on priority. For example, a person's most basic motivation will be to satisfy physiological needs, followed by the need to establish a safe and secure environment.

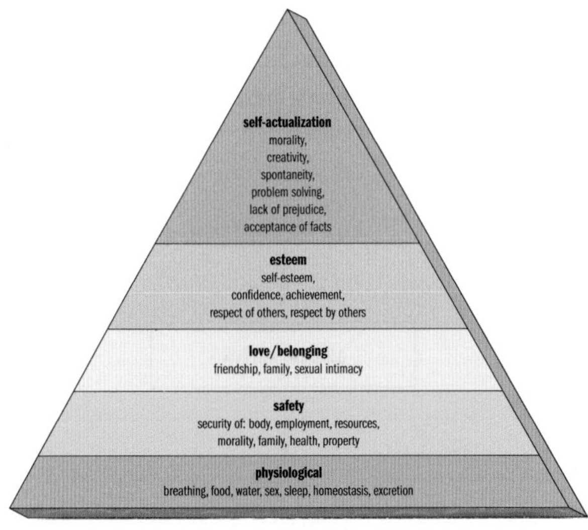

Figure 5.4 Maslow's Hierarchy of Needs

Another need-based motivational theory is the **self-determination theory** (**SDT**), which emphasizes the role of three universal needs: autonomy, the need to be in control of one's actions and ideas; competence, the need to complete and excel at difficult tasks; and relatedness, the need to feel accepted and wanted in relationships. Theorists explain that these three needs must be met in order to develop healthy relationships with oneself and others.

Additional Theories and Applications

There are a few other theories of motivation that you should know for the MCAT: opponent-process theory, sexual motivation, incentive theory, and expectancy-value theory. Opponent-process theory and sexual motivation will be discussed in more detail below. **Incentive theory** explains that behavior is motivated not by need or arousal, but by the desire to pursue rewards and to avoid punishments. **Expectancy-value theory** states that the amount of motivation needed to reach a goal is the result of both the individual's expectation of success in reaching the goal and the degree to which the individual values succeeding at the goal.

There are many motivations that stem from biology but that are impacted by additional psychological and sociocultural factors. One of the strongest natural motivations is hunger. However, people often eat for the sheer pleasure of the act, a motivation that may contribute to obesity occurring at alarming rates in the United States. Societal and cultural norms can determine what types of foods one eats and when. For example, some cultures have a traditional diet that is very high in fat and participate in many social activities involving food. At the other extreme, anorexia nervosa is also correlated to biological and cultural factors. It has been observed that those suffering from the disease are more likely to suffer from personality disorders as well. The prevalence of anorexia in the United States has increased significantly in the last several decades as the societal concept of beauty has changed from more full-bodied idols to extremely thin cultural icons.

Opponent-Process Theory

Motivations are considered destructive if they result in harm to oneself. For example, drug abusers can be motivated to take drugs by the pleasure experienced when taking the drug or by the removal of withdrawal symptoms. Most recreational drugs in the United States are psychoactive substances such as narcotics, sedatives, stimulants (e.g. caffeine and nicotine), hallucinogens, cannabis, and alcohol. A theory of motivation that explains continuous drug use is the **opponent-process theory**. This theory explains that when a drug is taken repeatedly, the body will attempt to counteract the effects of the drug by changing its physiology. For example, the body will counteract repeated use of alcohol, a depressant, by increasing arousal. The problem with this reaction is that it will last longer than the drug, resulting in withdrawal symptoms that are exactly opposite the effects of alcohol: sensations of anxiety, jitteriness, and irritability. The withdrawal created by this mechanism can create a physical dependence on the drug. Opponent-process theory can also explain **tolerance**, a decrease in perceived drug effect over time. Cultural and demographic factors also affect drug use. Young adults are the most likely age group to smoke, with a decline in smoking rates seen as the group ages. Smoking is also seen more commonly in disadvantaged socioeconomic groups. Across the globe, smoking rates are highest in Eastern Europe; this creates visibility, leading to additional pressure or desire to smoke in these communities.

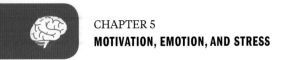

BEHAVIORAL SCIENCES GUIDED EXAMPLE WITH EXPERT THINKING

Researchers conducting a meta-analysis of the literature on intrinsic and extrinsic motivation examined the following two studies:

Intrinsic vs. extrinsic motivation

Study 1:

Undergraduate university students ($n = 27$) were recruited to perform a weekly filing task over four weeks at the university library. Participants were randomly divided into three groups: one group received $20 per week, one group received a general education credit for finishing all four weeks, and one group received no compensation. Participants completed a survey designed to measure their baseline engagement and enjoyment of the tasks they were asked to perform. At the end of the four weeks, the survey was administered again, and each participant was invited to continue the weekly tasks on a volunteer basis. No further compensation was offered. Results are shown in Figure 1.

This article has multiple studies that appear complex. I'll need to make sure to keep track of all variables.

IV: compensation ($)

DV: engagement in/enjoyment of task

2nd DV: 2nd survey + willingness to volunteer

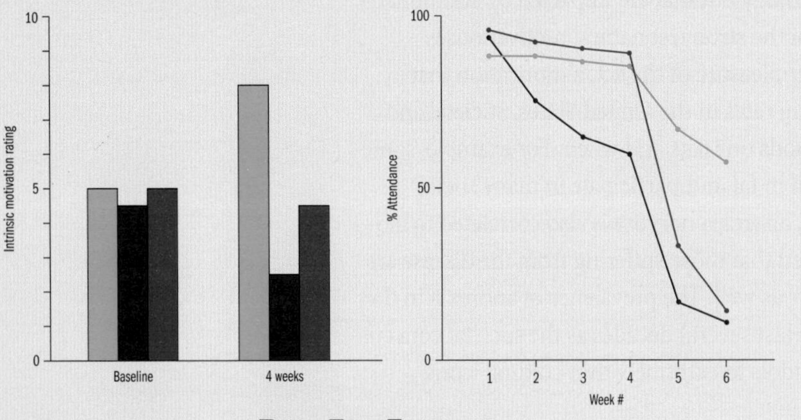

The no-compensation group seemed to enjoy the task much more than the other two groups compared to baseline, and more of them continued to volunteer after the study was over.

Figure 1 Enjoyment survey results and participation

Study 2:

A sample of men identified as being at high risk of heart disease ($n = 128$) were recruited for a multifaceted intervention program which required attending a clinic once per week for medication, physical exercise, and counseling. Participants came from both medium and low socioeconomic status (SES) households. All individuals in the low SES group experienced food insecurity. Participants in each SES group were divided into three conditions: monetary compensation for attendance, grocery credit for attendance, and no compensation for attendance. Compensation was offered for a six-month period. Participants completed an inventory designed to measure baseline intrinsic motivation to improve their health, with items such as "The reason I am participating in this course of treatment as prescribed is because participation in this treatment is consistent with my life goals." The inventory was administered again at six months and at twelve months. Results are shown in Figure 2.

IVs: SES and compensation for attendance

DV: intrinsic motivation, measured by survey

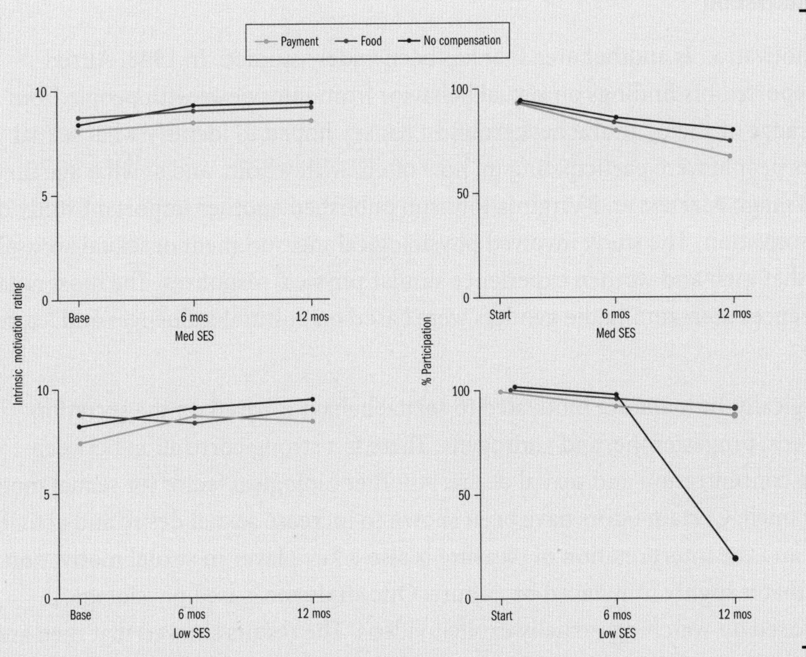

Motivation is high for all groups, as is participation, with the exception of the no-compensation low SES group once compensation ends.

Figure 2 Motivation and participation over 12 months

Adapted from: Czaicki NL, Dow WH, Njau PF, McCoy SI (2018) Do incentives undermine intrinsic motivation? Increases in intrinsic motivation within an incentive-based intervention for people living with HIV in Tanzania. *PLoS ONE* 13(6): e0196616. https://doi.org/10.1371/journal.pone.0196616

What do the results of these two studies suggest about the role of intrinsic and extrinsic factors in motivation?

This question is asking for a fairly high-level analysis of the three different conditions across the two studies. To answer, first we will need to consider what was being tested and the results from each study. Identifying the variables as we read the studies will make this task much more manageable and, as we have seen in guided examples from other chapters, is a generally good approach to reading any study. In Study 1, the only difference between groups is the offer of compensation, which is an extrinsic motivator. The results are consistent with conventional wisdom (and our outside content knowledge) that when an extrinsic motivator is introduced, intrinsic motivation decreases. Therefore, when the compensation is removed, participation declines. We can specifically see this within the data: groups receiving compensation see a substantial decline in intrinsic motivation and attendance as compared to the no-compensation group. Additionally, the lack of an extrinsic motivator increases intrinsic motivation, as demonstrated by the volunteerism rate in the no-compensation group.

Given the results of Study 1, Study 2 may seem surprising. Based on the results in the figure for Study 2, for the medium SES and low SES groups receiving compensation, intrinsic motivation remained consistent throughout the study. For the low SES group, however, participation declined sharply for the non-compensated group.

Fortunately, on Test Day we will never be asked to write an essay explaining these results, so we are not responsible for coming up with an explanation on our own. Among the answer choices will be an explanation that is consistent with both the scientific content we studied and the analysis of the results we conducted.

If a question like this shows up on the real MCAT, take a moment to consider what we know and what we've learned from the passage, and make a general prediction that targets the *kind* of information a correct answer could provide. In this case we are looking for a difference between the low and medium SES groups that might explain the observed difference in the no-compensation group. We might consider the implications of SES generally, but the passage provided a potential clue when it mentioned that the low SES group had food insecurity. Given what we know about Maslow's hierarchy of needs we might anticipate an answer that implies motivation to obtain food is more important than reducing the risk of heart disease.

Sexual Motivation

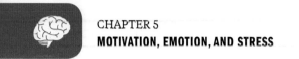

Sexual motivation is another area that has been widely studied. In 1948, Alfred Kinsey reported his findings on sexual behavior from interviews with people from a broad range of sociocultural backgrounds. Kinsey hoped to identify what sexual behaviors people were participating in, how often, with whom, and at what age they began. William Masters and Virginia Johnson published another important study of sexual motivation. The study involved physiological measurement of sexual arousal, proving that men and women experience similar physical responses. The most notable differences seen among the genders were based on cultural influences and learned behavior.

Physiologically, humans are motivated to sexual behavior based on the secretion of estrogens, progesterone, and androgens. There is a strong correlation between hormone concentration and sexual desire. Another biological factor for sexual motivation is smell. Certain odors have been shown to increase sexual desire and activity. Pleasure and the interpretation of pleasure is also a key player in sexual motivation and one that is highly influenced by culture. One study measured physiological arousal based on watching sexually explicit videos. The results showed that men and women experienced the same levels of arousal, but women more often reported being unaroused or having feelings of disgust based on subjective interviews. This study demonstrated that cognition plays a role in sexual motivation. Additionally, culture and society influence what is deemed appropriate sexual behavior, the age at which it is deemed appropriate, and with whom. Cultural norms and conditioning influence the desire for sexual interaction, or lack thereof.

MCAT CONCEPT CHECK 5.1

Before you move on, assess your understanding of the material with these questions.

1. For each of the theories listed below, what creates motivation?

Theory	Factor for Motivation
Instinct theory	
Arousal theory	
Drive reduction theory	
Need-based theories	

2. List Maslow's hierarchy of needs in decreasing priority:

    •

    •

•

•

•

---

3. Based on opponent-process theory, what clinical features would be expected with withdrawal from cocaine use?

5.2 Emotion

High-Yield

LEARNING OBJECTIVES

After Chapter 5.2, you will be able to:

- Describe the three elements of emotion: psychological response, behavioral response, and cognitive response
- Recall the seven universal facially expressed emotions
- Compare and contrast the James–Lange, Cannon–Bard, and Schachter–Singer theories of emotion
- Identify the names and functions of the parts of the limbic system

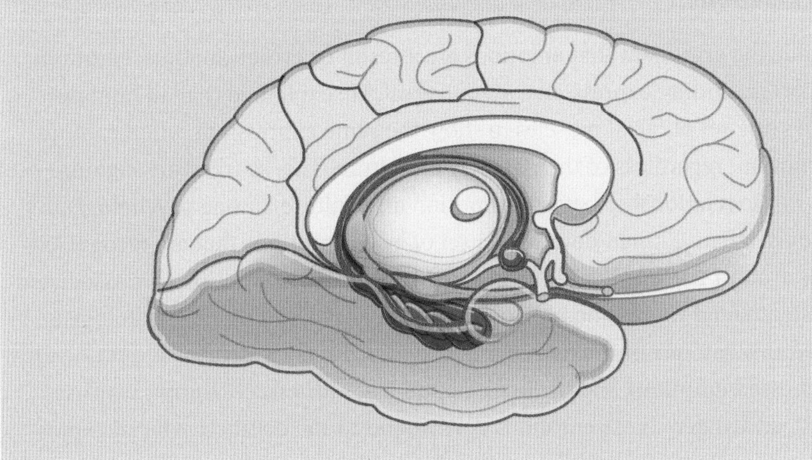

Emotion is a natural instinctive state of mind derived from one's circumstances, mood, or relationships with others. The word emotion is derived from the same Latin word as motivation.

Three Elements of Emotion

There are three elements of an emotion: the physiological response, the behavioral response, and the cognitive response.

Physiological Response

When a feeling is first experienced, arousal is stimulated by the autonomic nervous system. The physiological component includes changes in heart rate, breathing rate, skin temperature, and blood pressure. While it may be hard to recognize these changes and associate them with an emotion in everyday life, these changes have been detected in laboratory settings. Some emotions, such as fear, aggression, and embarrassment, are associated with more pronounced physiological changes than others.

Behavioral Response

The behavioral component of an emotion includes facial expressions and body language. For example, a smile, a friendly hand gesture, or even a subtle head tilt toward someone are commonly recognized as warm and happy signals. On the other hand, a frown, slumping of the shoulders, and looking downward are recognized as sad or downtrodden signals.

Cognitive Response

Finally, the cognitive component of emotion is the subjective interpretation of the feeling being experienced. Determination of one's emotion is an evaluative process largely based on memories of past experiences and perception of the cause of the emotion.

MCAT EXPERTISE

The AAMC lists fear, anger, happiness, surprise, joy, disgust, and sadness as universal emotions. Given the lack of consensus in the scientific community, the MCAT is more likely to test the topic conceptually than specifically.

Universal Emotions

Darwin made the argument that emotions are a result of evolution; thus, emotions and their corresponding expressions are universal. He explained that all humans evolved the same set of facial muscles to show the same expressions when communicating emotion, regardless of their society or culture. This sparked an ongoing discussion of the relationship between emotion and culture among psychologists and sociologists. Paul Ekman described a set of basic emotions that are recognized by societies around the world and further identified that six emotions are associated with consistent facial expressions across cultures. Ekman and other psychologists have revised this list; one of the more well-recognized sets consists of seven universal facially expressed emotions and includes: happiness, sadness, contempt, surprise, fear, disgust, and anger. These emotions correspond to the distinctive facial expressions explained in Table 5.1 and depicted in Figure 5.5.

EMOTION	FACIAL EXPRESSION CUES
Happiness	Smile, wrinkling around eyes, raised cheeks
Sadness	Frown, inner eyebrows pulled up and together
Contempt	One corner of the mouth pulled upwards
Surprise	Eyes widen, eyebrows pulled up and curved, jaw opens
Fear	Eyes widen, eyebrows pulled up and together, lips pulled toward ears
Disgust	Nose wrinkling and/or raising of upper lip
Anger	Glaring, eyebrows pulled down and together, lips pressed together

Table 5.1 Seven Universal Emotions

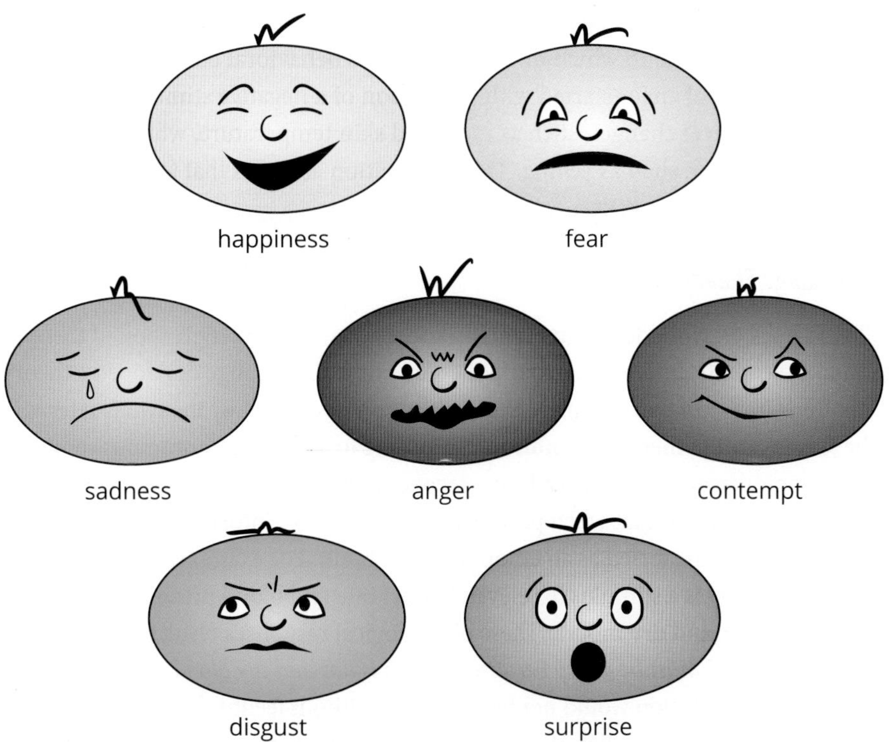

Figure 5.5 Seven Universal Emotions

While emotions are experienced universally, it is argued that they can be affected greatly by culture. Cultural dissimilarities in emotion include varying reactions to similar events, differences in the emotional experience itself, the behavior exhibited in response to an emotion, and the perception of that emotion by others within the society.

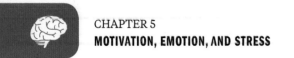

Adaptive Role of Emotion

In accordance with Darwin's thoughts on universal emotion, the **evolutionary perspective** states that everything we do, think, and feel is based on specialized functional programs designed for any problem we encounter. These programs are functionally coordinated in order to produce a cohesive response. Emotions are thought to be evolutionary adaptations due to situations encountered over the evolutionary history of the human species that guide sensory processing, physiological response, and behavior. Further, different emotions are thought to have evolved during different periods in history. Among the earliest to develop were primal emotions, such as fear; other, more evolutionarily progressive emotions include social emotions, such as guilt and pride.

Theories of Emotion

Early psychologists believed that the cognitive component of emotion led to the physiological component, which then produced the behavioral component. In other words, the feeling of anger started with perception of a negative stimulus, which caused physiological changes, such as increased skin temperature, which then resulted in behavior, such as yelling. This explanation assumes that feeling precedes arousal, which precedes action.

James–Lange Theory

William James, the founder of functionalist theory, viewed the progression of these emotional elements differently. Around the same time, Carl Lange developed a theory of emotion similar to that of James. The explanation developed by the two is referred to as the **James–Lange theory of emotion**. According to the theory, a stimulus results first in physiological arousal, which leads to a secondary response in which the emotion is labeled. James believed that when peripheral organs receive information and respond, that response is then labeled as an emotion by the brain. For example, a car cutting you off on the highway is a stimulus for elevated heart rate and blood pressure, increased skin temperature, and dry mouth. These physiological responses result in the cognitive labeling of anger: *I must be angry because my skin is hot and my blood pressure is high.* By extension, an emotion would not be processed without feedback from the peripheral organs; this theory predicts that individuals who cannot mount a sympathetic response, like patients with spinal cord injuries, should show decreased levels of emotion. Subsequent studies have proven this claim to be false; spinal cord injury subjects continue to show the same level of emotion after their injuries as before.

Cannon–Bard Theory

Walter Cannon and Philip Bard developed another scheme for explaining emotional components, referred to as the Cannon–Bard theory of emotion. In an attempt to test the James–Lange theory, Cannon studied the expression of emotion and its relationship to feedback from the sympathetic nervous system using cats whose afferent nerves had been severed. He hypothesized that physiological arousal and feeling an emotion occur at the same time, not in sequence. Thus, severing the feedback should not alter the emotion experienced. In this theory, a person will respond with action after experiencing the emotion both mentally and physically. Bard, a student of Cannon's, further explained that when exposed to a stimulus, sensory information is received and sent to both the cortex and the sympathetic nervous system simultaneously by the thalamus. Thus, the **Cannon–Bard theory of emotion**, depicted in Figure 5.6, states that the conscious experience of emotion and physiological arousal occur simultaneously, and then the behavioral component of emotion (i.e., action) follows: *I see a snake, so I feel afraid and my heart is racing . . . Let me out of here!*

While critics of the James–Lange theory cite the severed afferent nerve study as support for the Cannon–Bard theory, there are also weaknesses in this theory. The Cannon–Bard theory fails to explain the vagus nerve, a cranial nerve that functions as a feedback system, conveying information from the peripheral organs back to the central nervous system.

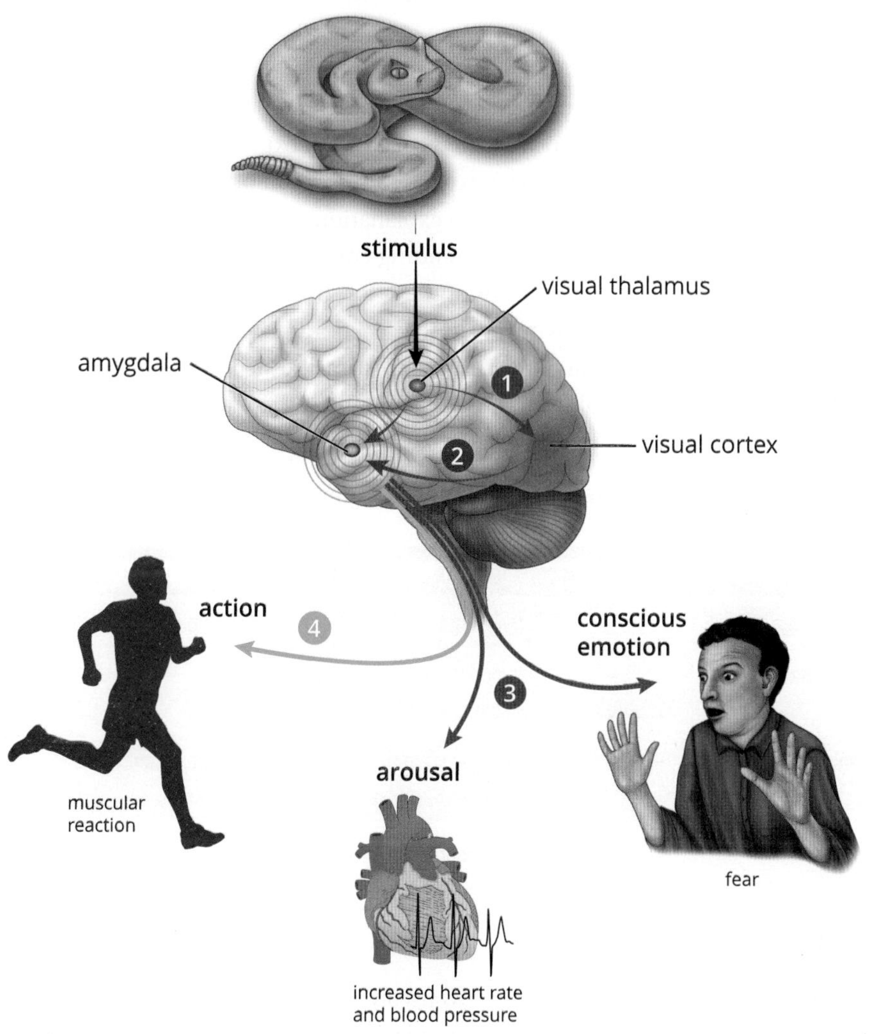

Figure 5.6 Cannon–Bard Theory of Emotion
Visual stimuli pass through the thalamus, and rough information is sent to the amygdala (fear) and the sympathetic nervous system (arousal). Action (muscle contraction) quickly follows. The visual cortex can either strengthen or quell this fear response once it has identified the stimulus.

MCAT EXPERTISE

The MCAT will expect that you have a solid understanding of the differences in emotional processing described by each of the three theories of emotion. It is best to mentally sort these theories by "first response" and "second response" to the stimulus.

Schachter–Singer Theory

A third theory is the **Schachter–Singer theory of emotion**, also termed the **cognitive arousal theory** or the **two-factor theory**, which states that two factors (physiological arousal and a cognitive label) are needed to experience emotion. According to this theory, physiological arousal alone is insufficient to elicit an emotional response. To feel an emotion, the mind must also identify the environmental stimulus causing that physiological arousal: *I am excited because my heart is racing and everyone else is happy.*

What is unique to the Schachter–Singer theory is this aspect of cognitive appraisal: to feel an emotion, one must consciously analyze the environment in relation to nervous system arousal. To study this phenomenon, Stanley Schachter and Jerome Singer gave injections of epinephrine or placebo to groups of subjects that were either informed, ignorant, or misinformed. They also manipulated external cues in the study by having an actor act either happy or angry. They observed that epinephrine did result in increased physiological arousal; however, they also discovered that the environment and cognitive processing affected the emotion experienced by the subjects. The groups who were misinformed or ignorant experienced the highest levels of emotion. Schachter and Singer explained this result by stating that subjects experiencing physiological arousal with no explanation or with a misleading explanation will attribute that arousal to the surrounding environment, and label themselves as happy or angry based on the behavior of the actor. In other words, the presence of unexpected arousal plus an environment that encourages a particular emotion is sufficient to create that emotion in the subject. Contrarily, the informed group knew to expect physiological arousal from the drug, and thus attributed their feelings to side effects of the epinephrine, rather than to emotions.

The three theories of emotion discussed in this section are summarized in Table 5.2.

THEORY		FIRST RESPONSE	SECOND RESPONSE
James-Lange	**Stimulus**	Nervous system arousal	Conscious emotion
Cannon-Bard	**Stimulus**	Nervous system arousal and conscious emotion	Action
Schachter-Singer	**Stimulus**	Nervous system arousal and cognitive appraisal	Conscious emotion

Table 5.2 Theories of Emotion

The Limbic System

Experiencing emotion is a complex process involving many parts of the brain. The most notable of these circuits is the **limbic system**, a complex set of structures that reside below the cerebrum on either side of the thalamus, as shown in Figure 5.7. The system is made up of the amygdala, thalamus, hypothalamus, hippocampus and fornix, septal nuclei, and parts of the cerebral cortex; it plays a large role in both motivation and emotion.

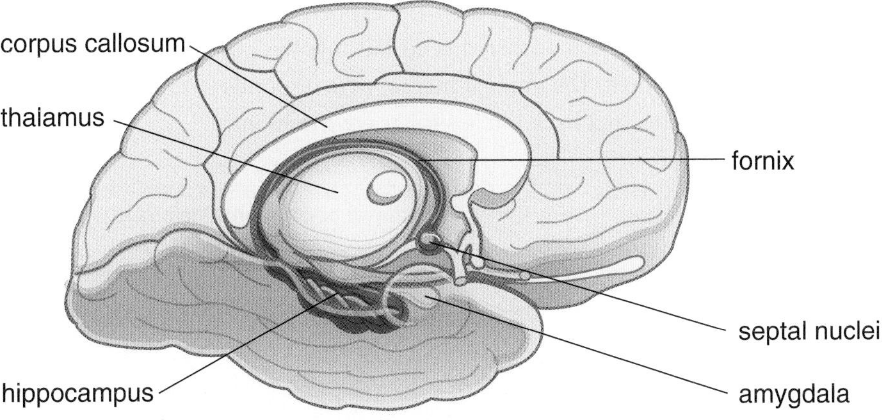

Figure 5.7 The Limbic System

REAL WORLD

When specific areas of the limbic system are stimulated during brain surgery, patients have described sensations such as happiness, contentment, and pleasure; in other areas, they report fear and anxiety.

The **amygdala** is a small round structure that signals the cortex about stimuli related to attention and emotions. The amygdala processes the environment, detects external cues, and learns from the person's surroundings in order to produce emotion. This region is associated with fear and also plays a role in human emotion through interpretation of facial expressions.

REAL WORLD

Research on rats has shown that when the amygdalae are damaged, the rats can no longer be classically conditioned to establish new fears. Similar effects have also been seen in humans wherein damage to this region prevents fear conditioning. This has been measured by autonomic nervous system responses as well as functional imaging.

The **thalamus** functions as a preliminary sensory processing station and routes information to the cortex and other appropriate areas of the brain. The **hypothalamus**, located below the thalamus, synthesizes and releases a variety of neurotransmitters. It serves many homeostatic functions, and is involved in modulating emotion. Indeed, by controlling the neurotransmitters that affect mood and arousal, the hypothalamus largely dictates emotional states.

The **hippocampus**, within the temporal lobe, is primarily involved in creating long-term memories. Along with the functions of the amygdala and hypothalamus, the storage and retrieval of emotional memories is key in producing an emotional response. The hippocampus also aids in creating context for stimuli to lead to an emotional experience. As described in Chapter 3 of *MCAT Behavioral Sciences Review*, memory systems can be divided into two categories: explicit and implicit. When an emotion is experienced, sensory systems transmit this information into both the explicit memory system, primarily controlled by the hippocampus in the medial temporal lobe, and the implicit memory system, controlled by the amygdala. Both memory systems are used for both the formation and retrieval of emotional memories, as shown in Figure 5.8. The conscious (explicit) memory is the memory of experiencing the actual emotion: remembering that you were happy at your high school graduation or that you were sad when you lost a loved one is explicit memory. Note that these are episodic memories: they are more properly considered memories *about* emotions than stored emotions. The unconscious (implicit) memory is referred to as **emotional memory**; this is the storage of the actual feelings of emotion associated with an event. When experiencing a similar event later on, these emotions may be retrieved. Thus, explicit memory of the emotion produces a conscious memory of the experience, and implicit memory determines the expression of past emotions. This distinction can be further identified when looking at individuals with posttraumatic stress disorder (PTSD). The explicit memory is the "story" of the event: what happened, where it occurred, who was involved, the fact that the scenario was traumatic, and so forth. The implicit memory corresponds to the sensations of unease and anxiety when put back into a similar environment.

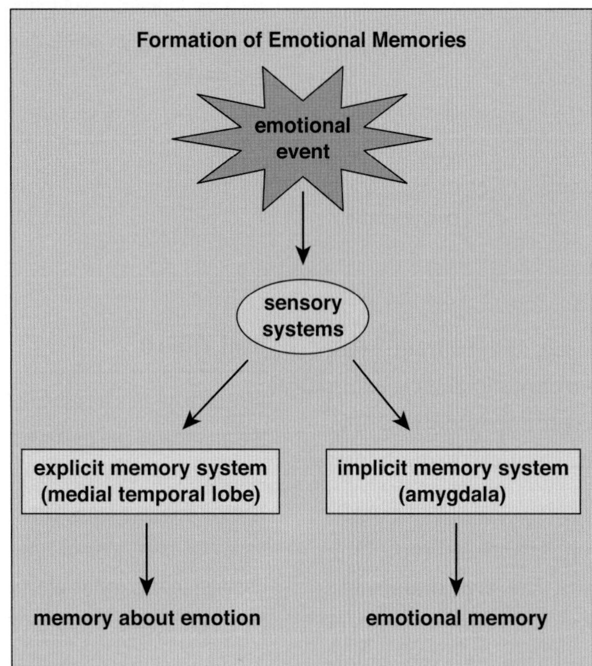

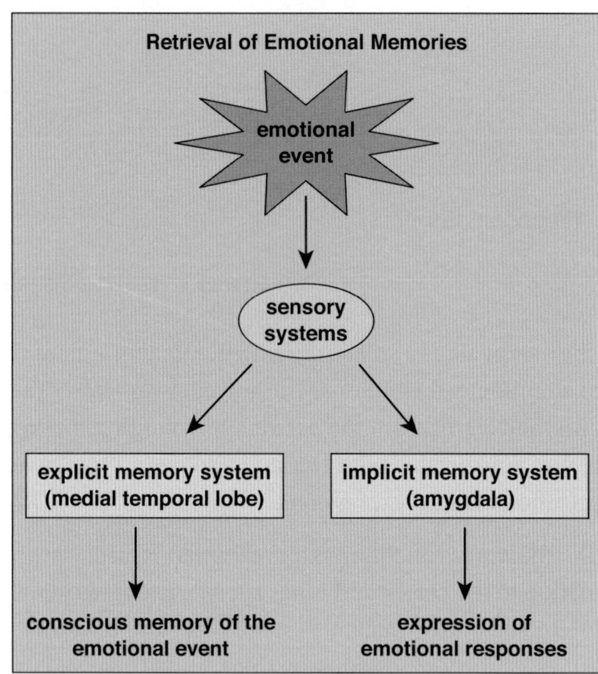

Figure 5.8 Formation and Retrieval of Emotional Memories

The ability to distinguish and interpret others' facial expressions is primarily controlled by the temporal lobe, with some input from the occipital lobe. This function is lateralized: the right hemisphere is more active when discerning facial expressions than the left. There are also gender differences: women demonstrate more activation of these brain areas than men. This ability is present but weak in children and develops into adulthood; adults are much more effective at identifying both positive and negative emotions.

REAL WORLD

One of the most notable studies on prefrontal cortex function is that of Phineas Gage. Gage was involved in an accident in which a metal rod pierced his brain, destroying the left frontal lobe. Gage's memory, speech, and motor skills were unaffected, but his personality was dramatically altered. Post-accident, Gage displayed irritable and impatient behavior, which inhibited his ability to complete simple tasks.

The **prefrontal cortex** is the anterior portion of the frontal lobes and is associated with planning intricate cognitive functions, expressing personality, and making decisions. The prefrontal cortex also receives arousal input from the brainstem, coordinating arousal and cognitive states. It has been demonstrated that the left prefrontal cortex is associated with positive emotions and the right prefrontal cortex with negative emotions. The **dorsal prefrontal cortex** is associated with attention and cognition, while the **ventral prefrontal cortex** connects with regions of the brain responsible for experiencing emotion. Specifically, the **ventromedial prefrontal cortex**, shown in Figure 5.9, is thought to play a substantial role in decision making and controlling emotional responses from the amygdala.

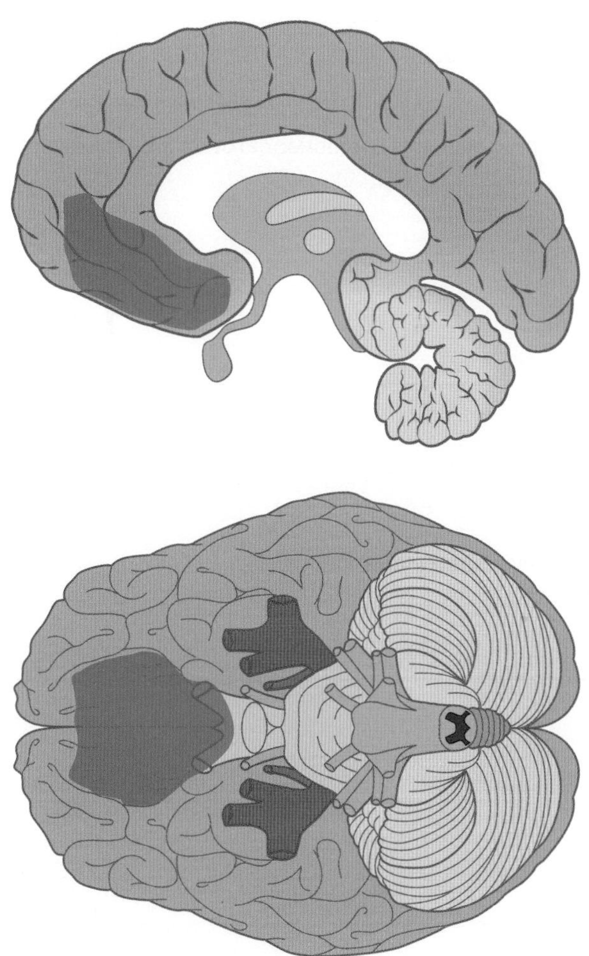

Figure 5.9 Ventromedial Prefrontal Cortex
The ventromedial prefrontal cortex is highlighted in blue.

As described earlier, the **autonomic nervous system** is also related to emotion; specific physiological reactions are associated with specific emotions. Skin temperature, heart rate, breathing rate, and blood pressure are all affected when experiencing emotion. Decreased skin temperature is detected in subjects experiencing fear, while increased skin temperature is associated with anger. Increased heart rate is observed in subjects experiencing both anger and fear, while decreased heart rate is observed in subjects who are feeling happiness. Heart rate variability is another factor used to determine emotion. Decreased heart rate variability is associated with stress, frustration, and anger. Blood pulse volume increases with anger or stress and decreases with sadness or relaxation. Skin conductivity is directly correlated with sympathetic arousal; however, a specific emotion cannot be identified by skin response. Diastolic blood pressure is increased to the greatest degree by anger, followed by fear, sadness, and happiness.

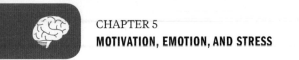

MCAT CONCEPT CHECK 5.2

Before you move on, assess your understanding of the material with these questions.

1. What are the three elements of emotion? Provide a brief description of each.
   - ___
   - ___
   - ___

2. What are the seven universal emotions?
   - ___
   - ___
   - ___
   - ___
   - ___
   - ___
   - ___

3. Compare and contrast the James–Lange, Cannon–Bard, and Schachter–Singer theories of emotion:

James–Lange Theory	Cannon–Bard Theory	Schachter–Singer Theory

4. What is the function of each part of the limbic system listed below?
   - Amygdala:

   ____

   - Thalamus:

   ____

   - Hypothalamus:

   ____

   - Hippocampus:

   ____

   - Ventromedial prefrontal cortex:

   ____

5.3 Stress

LEARNING OBJECTIVES

After Chapter 5.3, you will be able to:

- Distinguish between primary and secondary appraisals of stress
- Recall the three stages of general adaptation syndrome and the physiological changes associated with each stage
- Recognize common stressors and effective techniques for management of stress

In all aspects of life, at all times of day, we must make decisions, overcome challenges, and continue forward. While some of these decisions are small, others require planning and adaptation to new circumstances. Behavior of others and the perception of our surroundings affect our behavior and mental state, at times in a negative manner. It is our response to challenging events, be they physical, emotional, cognitive, or behavioral, that defines **stress**.

Cognitive Appraisal of Stress

Cognitive appraisal is the subjective evaluation of a situation that induces stress. This process consists of two stages. Stage 1, or **primary appraisal**, is the initial evaluation of the environment and the associated threat. This appraisal can be identified as irrelevant, benign–positive, or stressful. If primary appraisal reveals a threat, stage 2 appraisal begins. **Secondary appraisal** is directed at evaluating whether the organism can cope with the stress. This appraisal involves the evaluation of three things: harm,

MCAT EXPERTISE

The MCAT will expect you to know the two stages of stress appraisal: primary and secondary. Primary appraisal is the initial examination, which results in the identification of the stress as irrelevant, benign-positive, or stressful. If identified as a threat, secondary appraisal is an evaluation of one's ability to cope with the stress.

or damage caused by the event; threat, or the potential for future damage caused by the event; and challenge, or the potential to overcome and possibly benefit from the event. Individuals who perceive themselves as having the ability to cope with the event experience less stress than those who don't. In general, appraisal and stress level are personal, as individuals have different skills, abilities, and coping mechanisms. For example, while a spider might incite fear and stress in some, it would result in irrelevant appraisal in others. Some situations require ongoing monitoring through constant **reappraisal**, such as the perception of being followed.

Types of Stressors

A **stressor** is a biological element, external condition, or event that leads to a stress response. The severity of stressors can range from minimal or irritating hassles, like temporarily lost keys, to catastrophic scenarios, such as an impending natural disaster. Common stressors include:

- Environmental factors: uncomfortable temperature, loud sounds, inclement weather
- Daily events: running late, losing items, unexpected occurrences
- Workplace or academic setting: assignments, hierarchical interactions, time management
- Social expectations: demands placed on oneself by society, family, and friends
- Chemical and biological stressors: diet, alcohol, drugs, viruses, allergies, medications, medical conditions

Stressors are classified as either causing distress or causing eustress. **Distress** occurs when a stressor is perceived as unpleasant (e.g., a threat), whereas **eustress** is the result of a positively-perceived stressor (e.g., a challenge). Eustress can include life events such as graduating from college, studying to achieve a high score on the MCAT, getting married, or buying a house. While these events are largely positive, any event requiring individuals to change or adapt their lifestyle leads to stress. Stress level can be measured in "life change units" in a system called the **social readjustment rating scale**.

Stressors can also be psychological. Pressure, control, predictability, frustration, and conflict are all forms of psychological stress. Pressure is experienced when expectations or demands are put in place from external sources; this produces a feeling of urgency to complete tasks, perform actions, or display particular behaviors. The ability to control one's surroundings typically reduces stress levels; the inability to control a situation or event increases stress. In a study of nursing home patients, it was observed that those who had the most control of their daily environment displayed more active, positive, and social behavior. Predictability also plays a role in stress levels. For example, firefighters and police officers who cannot predict their daily scenarios experience higher levels of stress on the job. Frustration, which occurs when attaining a goal or need is prevented, increases stress. These frustration stresses can be external, such as not getting a raise, or internal, such as a disability interfering with everyday life. Finally, conflict stresses arise from the need to make a choice.

Approach–approach conflict refers to the need to choose between two desirable options. Avoidance–avoidance conflicts are choices between two negative options. Approach–avoidance conflicts deal with only one choice, goal, or event, but the outcome could have both positive and negative elements. For instance, while a job promotion might mean more money or status, it also comes with increased responsibility, potential for longer working hours, and increased pressure.

Physiological Response to Stressors

When subjected to stress, the body initially responds via the sympathetic nervous system. The "fight-or-flight" response initiates an increase in heart rate and decrease in digestion, with all available energy being reserved for reacting to the stressful event. The sequence of physiological responses developed by Hans Selye is called the **general adaptation syndrome** and consists of three distinct stages, as shown in Figure 5.10.

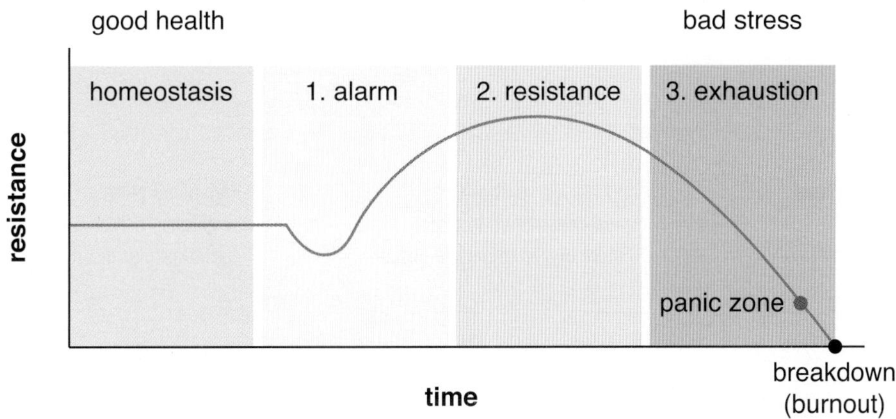

Figure 5.10 Three Stages of Stress Response

First is **alarm**, or the initial reaction to a stressor and the activation of the sympathetic nervous system. Shortly thereafter, the hypothalamus stimulates the pituitary to secrete adrenocorticotropic hormone (ACTH). This hormone stimulates the adrenal glands to produce cortisol, which maintains the steady supply of blood sugar needed to respond to stressful events. The hypothalamus also activates the adrenal medulla, which secretes epinephrine and norepinephrine to activate the sympathetic nervous system. The next stage is **resistance**, in which the continuous release of hormones allows the sympathetic nervous system to remain engaged to fight the stressor. Last, a person will experience **exhaustion** when the body can no longer maintain an elevated response with sympathetic nervous system activity. At this point, individuals become more susceptible to illnesses and medical conditions (such as ulcers and high blood pressure), organ systems can begin to deteriorate (with effects including heart disease), and in extreme cases, death can result. Some of the positive and negative effects of stress are shown in Figure 5.11.

Organ System	Acute Stress	Chronic Stress
Nervous system	Increased alertness, decreased pain perception	Memory problems, heightened risk of depressive disorders
Endocrine system	Adrenals secrete cortisol and other hormones to increase available energy	Slowed recovery from stress response due to high levels of cortisol and other hormones
Immune system	Thymus and other glands and tissues primed for immune response	Diminished immune response
Circulatory system	Increased heart rate, vasoconstriction to direct more oxygen to muscle tissue	High blood pressure, heightened risk of cardiovascular disorders
Reproductive system	Temporary suppression of functions	Heightened risk of infertility or miscarriage

Figure 5.11 Positive and Negative Effects of Stress

Emotional and Behavioral Responses to Stress

Beyond the effects on the human body, stress also takes a psychological toll on people who are unable to reduce their stress levels. On the emotional level, elevated stress can result in individuals feeling irritable, moody, tense, fearful, and helpless. They may also have difficulties with concentration and memory. Negative behavior responses to stress include withdrawing from others, difficulties at work or at school, substance use, aggression, and suicide. Additionally, chronic stress can lead to mental health disorders, such as anxiety and depression.

Coping and Stress Management

Strategies for coping with stress fall into two groups. Problem-focused strategies involve working to overcome a stressor, such as reaching out to family and friends for social support, confronting the issue head-on, and creating and following a plan of problem-solving actions. Emotionally focused strategies center on changing one's feelings about a stressor. They include taking responsibility for the issue, engaging in self-control, distancing oneself from the issue, engaging in wishful thinking, and using positive reappraisal to focus on positive outcomes instead of the stressor. Some coping strategies are adaptive, and reduce stress in a healthy way. A person who is feeling stressed could, for example, reach out to a loved one for help as a support-seeking coping strategy. However, coping strategies may also be maladaptive and include detrimental tactics, such as turning to drugs or alcohol.

Individuals can also engage in stress management to reduce their stress levels. Exercise is a powerful stress management tool that not only improves health and well-being, but also enhances mood. Exercise releases endorphins, opioid neuropeptides that act as "feel-good" neurotransmitters. Relaxation techniques, including meditation, diaphragmatic breathing, and progressive muscle relaxation have also been found to reduce stress. Additionally, studies have shown that engaging in a spiritual practice helps to manage stress.

BRIDGE

Maladaptiveness is a key factor in determining whether a behavior should be considered a disorder. Maladaptive behaviors are discussed in more detail in Chapter 7 of *MCAT Behavioral Sciences Review*.

MCAT EXPERTISE

Use these coping and stress management techniques to boost your performance (and mood) on Test Day!

MCAT CONCEPT CHECK 5.3

Before you move on, assess your understanding of the material with these questions.

1. What are the key features of primary and secondary cognitive appraisal of stress?
   - Primary appraisal:

   ____

   - Secondary appraisal:

   ____

2. What are the three stages of the general adaptation syndrome? What physiological changes are evident in each stage?

Stage	Physiological Changes

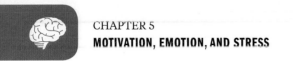

3. What are some common stressors? What are some effective techniques for managing stress?
   - Common stressors:

   ____

   - Stress management techniques:

   ____

   ____

   ____

Conclusion

The ability to strive for our goals and desires, be it for internal or external reasons, is an important aspect of psychology and behavior. Motivation is the mechanism used to meet our needs, act toward an end goal, and ultimately survive. While there are many factors that influence motivation, including instincts, arousal, drives, and needs, they all result in action to obtain perceived rewards, fulfill needs, or avoid perceived punishments. Emotion is a complex process resulting in physiological, cognitive, and behavioral elements, described in different fashions by the James–Lange, Cannon–Bard, and Schachter–Singer theories of emotion. Many components of the nervous system play a role in experiencing emotions, including the seven universal emotions. The response of the body and mind to challenges defines stress. Stress appraisal has phases that identify and allow the body to respond to the stressor encountered. The physical and mental response to stress can be severe, but there are many management and coping mechanisms commonly used to reduce the level of stress experienced.

Hopefully this chapter has left you motivated to keep working toward that goal of an excellent MCAT score and becoming the doctor you deserve to be. Studying for the MCAT certainly introduces a significant stress, but effective stress management techniques and a solid foundation in MCAT content and strategy will turn Test Day into eustress. Just keep your eyes on that white coat, an important garment that will someday be part of your identity—a topic we'll explore in the next chapter.

You've reviewed the content, now test your knowledge and critical thinking skills by completing a test-like passage set in your online resources!

GO ONLINE

CONCEPT SUMMARY

Motivation

- **Motivation** is the purpose, or driving force, behind our actions.
- Motivation can be **extrinsic**, based on external circumstances; or **intrinsic**, based on internal drive or perception.
- The primary factors that influence motivation are instincts, arousal, drives, and needs.
    - **Instincts** are innate, fixed patterns of behavior. In the **instinct theory** of motivation, people perform certain behaviors because of these evolutionarily programmed instincts.
    - In the **arousal theory**, people perform actions to maintain **arousal**, the state of being awake and reactive to stimuli, at an optimal level. The **Yerkes–Dodson law** shows that performance is optimal at a medium level of arousal.
    - **Drives** are internal states of tension that beget particular behaviors focused on goals. Primary drives are related to bodily processes; secondary drives stem from learning and include accomplishments and emotions. **Drive reduction theory** states that motivation arises from the desire to eliminate drives, which create uncomfortable internal states.
    - Satisfying **needs** may also motivate. **Maslow's hierarchy of needs** prioritizes needs into five categories: physiological needs (highest priority), safety and security, love and belonging, self-esteem, and self-actualization (lowest priority).
    - **Self-determination theory** emphasizes the role of three universal needs: autonomy, competence, and relatedness.
- **Incentive theory** explains motivation as the desire to pursue rewards and avoid punishments.
- **Expectancy–value theory** states that the amount of motivation for a task is based on the individual's expectation of success and the amount that success is valued.
- **Opponent-process theory** explains motivation for drug use: as drug use increases, the body counteracts its effects, leading to tolerance and uncomfortable withdrawal symptoms.
- Sexual motivation is related to hormones as well as cultural and social factors.

Emotion

- **Emotion** is a state of mind, or feeling, that is subjectively experienced based on circumstances, mood, and relationships.
- The three components of emotion are **cognitive** (subjective), **behavioral** (facial expressions and body language), and **physiological** (changes in the autonomic nervous system).

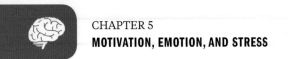

- The **seven universal emotions** are happiness, sadness, contempt, surprise, fear, disgust, and anger.
- There are multiple theories of emotion, based on the interactions of the three components of emotion.
  - In the **James–Lange theory**, nervous system arousal leads to an emotional experience.
  - In the **Cannon–Bard theory**, arousal of the nervous system and the experience of emotion occur simultaneously.
  - In the **Schachter–Singer theory**, nervous system arousal is combined with cognition to create the experience of emotion.
- The **limbic system** is the primary nervous system component involved in experiencing emotion.
  - The **amygdala** is involved with attention and fear, helps interpret facial expressions, and is part of the intrinsic memory system for emotional memory.
  - The **thalamus** is a sensory processing station.
  - The **hypothalamus** releases neurotransmitters that affect mood and arousal.
  - The **hippocampus** creates long-term explicit (episodic) memories.
  - The **prefrontal cortex** is involved with planning, expressing personality, and making decisions. The **ventral prefrontal cortex** is critical for experiencing emotion; the **ventromedial prefrontal cortex**, specifically, is involved in controlling emotional responses from the amygdala and decision making.

Stress

- The physiological and cognitive response to challenges or life changes is defined as **stress**.
- Stress appraisal has two stages:
  - **Primary appraisal** is classifying a potential stressor as irrelevant, benign–positive, or stressful.
  - **Secondary appraisal** is directed at evaluating if the organism can cope with the stress, based on harm, threat, and challenge.
- A **stressor** is anything that leads to a stress response and can include environment, daily events, workplace or academic settings, social expectations, chemicals, and biological stressors. Psychological stressors include pressure, control, predictability, frustration, and conflict.
- Stressors can lead to **distress** or **eustress**.
- The three stages of the **general adaptation syndrome** are alarm, resistance, and exhaustion.
- Stress management can include psychological, behavioral, and spiritual aspects.

ANSWERS TO CONCEPT CHECKS

5.1

1.

Theory	Factor for Motivation
Instinct theory	Instincts: innate, fixed patterns of behavior in response to stimuli
Arousal theory	Maintaining a constant level of arousal, the psychological and physiological state of being awake and reactive to stimuli
Drive reduction theory	Drives: internal states of tension or discomfort that can be relieved with a particular action
Need-based theories	Needs: factors necessary for physiological function or emotional fulfillment

2. Physiological needs, safety and security, love and belonging, self-esteem, self-actualization
3. Cocaine is a stimulant, causing euphoria, restlessness, increased heart rate, increased temperature, and anxiety. According to opponent-process theory, cocaine withdrawal should be the opposite: depressed mood, fatigue, decreased heart rate, decreased temperature, and apathy.

5.2

1. The three elements of emotion are as follows:
    - Physiological response (autonomic nervous system): heart rate, breathing rate, skin temperature, blood pressure
    - Behavioral response: facial expressions, body language
    - Cognitive response: subjective interpretation, memories of past experiences, perception of cause of emotion
2. The seven universal emotions are happiness, sadness, contempt, surprise, fear, disgust, and anger.

3.

James–Lange Theory	Cannon–Bard Theory	Schachter–Singer Theory
• Stimulus leads to physiological arousal • Arousal leads to the conscious experience of emotion • *My skin is hot and my blood pressure is high so I must be angry* • Requires connection between sympathetic nervous system and brain	• Stimulus leads to physiological arousal and feeling of emotion • Thalamus processes sensory information, sends it to cortex and sympathetic nervous system • Action is secondary response to stimulus • *I see a snake, so I am afraid and my heart is racing…Let me out of here!* • Does not explain vagus nerve	• Both arousal and labeling based on environment are required to experience an emotion • *I am excited because my heart is racing and everyone else is happy*

4. The amygdala is involved with attention and emotions (specifically fear), helps interpret facial expressions, and is part of the intrinsic memory system for emotional memory. The thalamus is a sensory processing station. The hypothalamus releases neurotransmitters that affect mood and arousal. The hippocampus creates long-term explicit memories (episodic memories). The ventromedial prefrontal cortex is involved in decision making and controlling emotional responses from the amygdala.

5.3

1. Primary appraisal is categorizing the stressor as irrelevant, benign–positive, or stressful. Secondary appraisal is the evaluation of the ability of the organism to cope with that stress.

2.

Stage	Physiological Changes
Alarm	Activation of sympathetic nervous system, release of ACTH and cortisol, stimulation of adrenal medulla to secrete epinephrine and norepinephrine
Resistance	Continuous release of hormones activates sympathetic nervous system
Exhaustion	Can no longer maintain elevated sympathetic nervous system activity, more susceptible to illness and medical conditions, organ systems deteriorate, death

3. Common stressors include environmental or physical discomfort, daily events, workplace or academic setting, social expectations, and chemical and biological stressors. Effective stress management techniques include exercise, relaxation techniques (meditation, diaphragmatic breathing, progressive muscle relaxation), spiritual practice, and many more.

SCIENCE MASTERY ASSESSMENT EXPLANATIONS

1. **A**

Due to the competitive nature of the motivation, this is considered extrinsic motivation. Extrinsic motivation is based on external conditions, including perceived reward or fear of punishment. In this case, the reward is beating the sibling. There is no suggestion of an uncomfortable internal state or tension, which is an aspect of drives, eliminating **(C)** and **(D)**.

2. **D**

The Yerkes–Dodson law states that there is an optimal level of arousal necessary to perform. If levels of arousal are too high, poor performance can result. In the case of this student performing at a recital, arousal level is very high as a result of nervousness and anxiety, resulting in a poor performance.

3. **A**

Drive reduction theory is the theory that one will act to eliminate uncomfortable internal states known as drives. The body will push toward equilibrium, or homeostasis.

4. **A**

The seven universal emotions are happiness, sadness, contempt, surprise, fear, disgust, and anger.

5. **C**

The three components of emotion are the physiological (changes in the autonomic nervous system), cognitive (subjective interpretation of an emotion), and behavioral (facial expressions and body language) responses.

6. **D**

Experiencing a physiological reaction to a stimulus and then labeling that response as emotion is in line with the James–Lange theory of emotion. In the statement, seeing the bear is the stimulus, an increase in heart rate and breathing rate is the physiological reaction, and identifying this as fear is the emotion experienced.

7. **C**

According to Darwin's theory of evolution, all species have instincts that help them survive. The instinct theory of motivation states that people are motivated to act based on instincts that they are programmed to exhibit.

8. **D**

The Cannon–Bard theory of emotion is based on the premise that conscious feelings and physiological components of emotion are experienced at exactly the same time. In this theory, this combination then leads to action. This is commonly confused with the Schachter–Singer theory, **(A)**, in which nervous system arousal occurs and then is labeled based on the context provided by the environment.

9. **C**

The amygdala, prefrontal cortex, and thalamus all play a role in the experience of emotions. The basal ganglia are involved in smooth movement and are not primarily responsible for the experience of emotions.

10. **A**

The left frontal lobe is associated with positive feelings, corresponding with joy and happiness. The right frontal lobe is associated with negative feelings, such as sadness and disgust, **(B)** and **(D)**.

11. **B**

Secondary appraisal of stress is the stage at which the ability of the organism to cope with the stressor is evaluated. This is based on the harm, threat, and challenge of the stressor, which are all correlated with its intensity. Primary appraisal is simply the initial determination of whether there is a negative association at all, not its intensity.

12. **D**

A positive stressor creates eustress. Because working out is used to relax, it is considered a eustress. Hassle, **(A)**, and frustration, **(B)**, are both types of distress, **(C)**, or negative stressors.

13. **A**

Approach–approach conflict is one in which both results are good outcomes. While one must be chosen, neither choice results in a negative outcome: for example, choosing between two desserts. Avoidance–escape conflict, **(D)**, is not a recognized form of conflict; these two terms are related to types of negative reinforcers.

14. **A**

The initial reaction to stress, which is activation of the sympathetic nervous system, is the alarm stage of stress response.

15. **C**

Progressive muscle relaxation is a relaxation technique demonstrated to help reduce stress in a manner that is beneficial to the body and psyche. The other methods described here, including avoidance of the stressor, **(D)**, serve to increase stress or merely change the source of the stress.

Consult your online resources for additional practice.

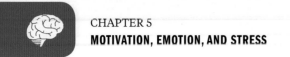

SHARED CONCEPTS

Behavioral Sciences Chapter 1
Biology and Behavior

Behavioral Sciences Chapter 3
Learning and Memory

Behavioral Sciences Chapter 7
Psychological Disorders

Biology Chapter 4
The Nervous System

Biology Chapter 5
The Endocrine System

Biology Chapter 10
Homeostasis

CHAPTER 6

IDENTITY AND PERSONALITY

SCIENCE MASTERY ASSESSMENT

Every pre-med knows this feeling: there is so much content I have to know for the MCAT! How do I know what to do first or what's important?

While the high-yield badges throughout this book will help you identify the most important topics, this Science Mastery Assessment is another tool in your MCAT prep arsenal. This quiz (which can also be taken in your online resources) and the guidance below will help ensure that you are spending the appropriate amount of time on this chapter based on your personal strengths and weaknesses. Don't worry though—skipping something now does not mean you'll never study it. Later on in your prep, as you complete full-length tests, you'll uncover specific pieces of content that you need to review and can come back to these chapters as appropriate.

How to Use This Assessment

If you answer 0–7 questions correctly:

Spend about 1 hour to read this chapter in full and take limited notes throughout. Follow up by reviewing **all** quiz questions to ensure that you now understand how to solve each one.

If you answer 8–11 questions correctly:

Spend 20–40 minutes reviewing the quiz questions. Beginning with the questions you missed, read and take notes on the corresponding subchapters. For questions you answered correctly, ensure your thinking matches that of the explanation and you understand why each choice was correct or incorrect.

If you answer 12–15 questions correctly:

Spend less than 20 minutes reviewing all questions from the quiz. If you missed any, then include a quick read-through of the corresponding subchapters, or even just the relevant content within a subchapter, as part of your question review. For questions you answered correctly, ensure your thinking matches that of the explanation and review the Concept Summary at the end of the chapter.

1. Each of the following is considered a part of a person's self-concept EXCEPT:
   A. the past self.
   B. the ought self.
   C. the future self.
   D. self-schemata.

2. As a gender identity, androgyny is defined as:
   A. low femininity, low masculinity.
   B. high femininity, low masculinity.
   C. low femininity, high masculinity.
   D. high femininity, high masculinity.

3. A high school student struggles consistently with math and feels that no amount of studying will help because "I just don't get it." Which of the following is the most likely short-term result with respect to the student's ability to do math?
   A. Low self-esteem
   B. Low self-efficacy
   C. Learned helplessness
   D. An external locus of control

4. A district attorney with an internal locus of control wins an important court trial. Which of the following best represents the lawyer's attribution of the events?
   A. "I won because I made great arguments and had more experience than the defense."
   B. "I won because the jury was on my side from the beginning and believed my arguments."
   C. "I won because the defense did not adequately present their side of the case."
   D. "I shouldn't have won because I don't deserve to be successful."

5. A person keeps an extremely tidy desk and becomes very nervous whenever things are disorganized or out of place. In which of the following stages would a psychodynamic therapist say this person had become fixated?
   A. The oral stage
   B. The anal stage
   C. The phallic stage
   D. The genital stage

6. According to Erikson's stages of psychosocial development, which of the following would be the most important for a recent college graduate to accomplish?
   A. Figuring out what identities are most important personally
   B. Feeling like a contributing member of society
   C. Forming an intimate relationship with a significant other
   D. Finding a feeling of accomplishment in life

7. Matt and Cati discuss the reasons why they avoid driving above the speed limit. Matt cites wanting to avoid a traffic fine, while Cati says that speeding is dangerous and, if everyone did it, there would be more accidents and people would get hurt. According to Kohlberg, which of the following describes the phases of moral reasoning demonstrated by Matt and Cati, respectively?
   A. Preconventional; conventional
   B. Preconventional; postconventional
   C. Conventional; preconventional
   D. Postconventional; conventional

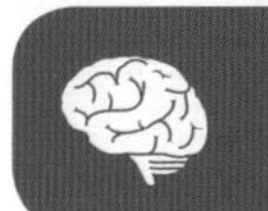

8. A child cannot make an origami swan alone, but is able to do so when observing and being assisted by an adult. This scenario is described in the ideas of which of the following theorists?

A. Albert Bandura
B. Alfred Adler
C. B. F. Skinner
D. Lev Vygotsky

9. Which of the following is a conclusion that can be made from research in role-taking and observational learning?

A. Young children will only model actions performed by their parents.
B. Celebrities and athletes are an adolescent's most important role models.
C. Children who role-take identities that are not gender typical are more likely to take on those roles later in life.
D. A female child is more likely to model the behavior of another female than a male.

10. A person feels extremely guilty after having an extramarital affair. According to the psychodynamic perspective, which of the following is responsible for this anxiety?

A. The id
B. The ego
C. The superego
D. The libido

11. A woman advances through the ranks of a company, eventually becoming the CEO. Which of the following Jungian archetypes reflects this woman's drive to be successful within the company?

A. The persona
B. The anima
C. The animus
D. The shadow

12. Researchers discover that polymorphisms in the *DRD2* gene can be associated with thrill-seeking behavior, and that individuals with certain forms of the gene are more likely to become extreme athletes and have more dangerous hobbies. Which of the following theories is supported by this discovery?

I. The social cognitive perspective
II. The behavioral perspective
III. The biological perspective

A. I only
B. III only
C. I and III only
D. II and III only

13. A person who works as an EMT identifies as a bit of a rebel, but is highly sociable and is able to keep calm in an emergency. This person would likely score in the lower range of which of the following traits?

A. Psychoticism
B. Neuroticism
C. Extraversion
D. Conscientiousness

14. Stockholm syndrome is a phenomenon in which people who are kidnapped or held hostage may begin to identify with or even feel affection for their captors. A psychoanalyst might explain Stockholm syndrome by citing which of the following defense mechanisms?

A. Reaction formation
B. Regression
C. Projection
D. Displacement

15. Having struggled for years through an economic recession, a young professional begins to buy lottery tickets every Friday. "If I won the lottery," the individual reasons, "I'd finally have the life I've always wanted. All my stress would go away and I could live comfortably." These thoughts regarding winning the lottery are most representative of:

A. a cardinal trait.
B. fictional finalism.
C. functional autonomy.
D. unconditional positive regard.

Answer Key

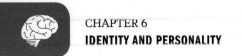

1. **B**
2. **D**
3. **B**
4. **A**
5. **B**
6. **C**
7. **A**
8. **D**
9. **D**
10. **C**
11. **C**
12. **C**
13. **B**
14. **A**
15. **B**

Detailed explanations can be found at the end of the chapter.

CHAPTER 6

IDENTITY AND PERSONALITY

In This Chapter

CHAPTER PROFILE

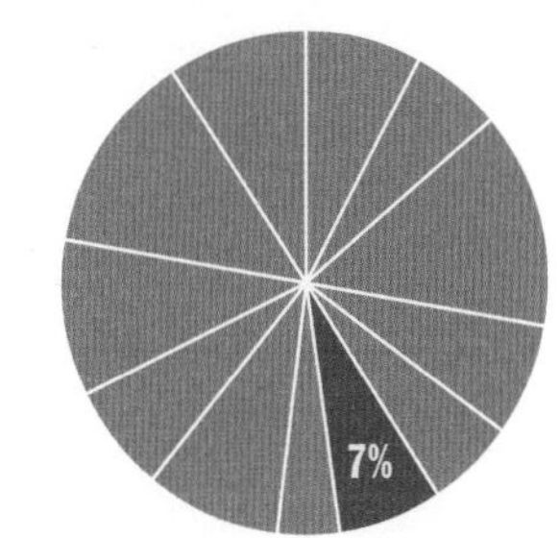

The content in this chapter should be relevant to about 7% of all questions about the behavioral sciences on the MCAT.

This chapter covers material from the following AAMC content categories:

7A: Individual influences on behavior

8A: Self-identity

Introduction

Social psychologists are concerned with how our social lives influence the ways in which we perceive ourselves. Specifically, researchers have focused on the influence that other people's views, our social roles, and our group memberships have on our perceptions of who we are.

Who are you? If you're like most people, you could probably answer that question in many different ways. You might list your physical characteristics, your family relationships, your emotional tendencies, or your skills and talents. In fact, many introductory psychology courses include an exercise in which students are asked to make a list of answers to the question *Who am I?* Completing this list gives each student a glimpse into their identity and personality. These ideas form the core of the study of psychology, in which the central goal is explaining our thoughts and behaviors. In this chapter, we'll discuss this and review the key theorists and their approaches to answering the question of who we are.

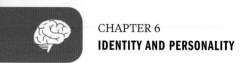

6.1 Self-Concept and Identity

LEARNING OBJECTIVES

After Chapter 6.1, you will be able to:

- Distinguish between self-concept and identity
- Recall the factors that contribute to different types of identity, such as gender and ethnicity
- Predict how a person would explain an event to themselves based on their locus of control

When you look in the mirror, whom do you see? If you're studying to take the MCAT, chances are some descriptors that come to mind include *student, intelligent, future doctor*, and so on. Our awareness of ourselves as distinct from others and our own internal list of answers to the question *Who am I?* form our **self-concept**. Many of the ways in which we define ourselves fall under the classification of a **self-schema**; that is, a self-given label that carries with it a set of qualities. For example, the *athlete* self-schema usually carries the qualities of youth, physical fitness, and dressing and acting in certain ways, although these qualities may change depending on culture, socioeconomic status, and personal beliefs. The idea of self-concept goes beyond these self-schemata; it also includes our appraisal of who we used to be and who we will become.

Sometimes the terms *self-concept* and *identity* are used interchangeably, but psychologists generally use them to refer to two different but closely related ideas. Social scientists define **identity** as the individual components of our self-concept related to the groups to which we belong. Whereas we have one self-concept, we have multiple identities that define who we are and how we should behave within any given context. Religious affiliation, sexual orientation, personal relationships, and membership in social groups are just a few of the identities that sum to create our self-concept. In fact, our individual identities do not always need to be compatible. Are you the same person when interacting with your friends as you are when you interact with coworkers or family? For most people the answer is *no*; they take on particular identities in different social situations.

Types of Identity

While there are many different types of identity, the MCAT—for historical or social reasons—tends to focus on some forms of identity more than others.

Gender Identity

Gender identity describes people's appraisals of themselves on scales of masculinity and femininity. While these concepts were long thought to be two extremes on a single continuum, theorists have reasoned that they must be two separate dimensions because an individual can achieve high scores on scales of both masculinity and femininity. **Androgyny** is defined as the state of being simultaneously very masculine and very feminine, while those who achieve low scores on both scales are referred to as **undifferentiated**. An initial sense of gender identity is usually established by age three, although it may morph and change over time. Some theories, such as the theory of **gender schema**, hold that key components of gender identity are transmitted through cultural and societal means.

Keep in mind that gender identity is not necessarily tied to biological sex or sexual orientation, although in many cultures these concepts are seen as closely related. While it is typical of many cultures to view gender as a strictly binary concept, some cultures recognize three or more genders. For example, some peoples of Samoa refer to androgynous but phenotypically male individuals as *fa'afafine*. To these Samoans, the *fa'afafine* are seen as an important social caste and are accepted as equals, although this is not always the case for all genders across all cultures.

REAL WORLD

Individuals who are transgender are those for whom gender identity does not match sex assigned at birth. They have been a heavily stigmatized group in American culture. In fact, it was not until the publication of the DSM-5 in 2013 that gender identity disorder was formally removed as a diagnosis. The DSM-5 includes the diagnosis gender dysphoria, which is given only to individuals for whom gender identity causes significant psychological stress.

Ethnic and National Identity

Ethnic identity refers to the part of one's identity associated with membership in a particular racial/ethnic group. Members in a given ethnic group often share a common ancestry, cultural heritage, and language. Many social psychologists study the ways in which our ethnic identity influences our perspectives of ourselves. In a 1947 study, Kenneth and Mamie Clark explored ethnic self-concepts among children who were ethnically White and Black using a doll preference task: the experimenter showed each child a Black doll and a White doll and asked the child a series of questions about how the child felt about the dolls. The majority of children, both White and Black, preferred the White doll. This study was important because it highlighted the negative effects of racism and underrepresented group status on the self-concept of Black children at the time. However, subsequent research using improved methodology (for example, randomizing the ethnicity of the experimenter), has shown that children who are Black hold more positive views of their own ethnicity; this may also represent societal changes at large.

While ethnicity is largely an identity into which we are born, **nationality** is based on political borders. National identity is the result of shared history, media, cuisine, and national symbols such as a country's flag. Nationality need not be tied to one's ethnicity or even to legal citizenship. Symbols play an important role in both ethnic and national identity. Some examples of symbols, representing American nationality, are shown in Figure 6.1.

Figure 6.1 Examples of National Symbols
Symbols commonly associated with residents of the United States of America

Other Types of Identity

Of course, there are many more categories through which we evaluate our identity. We compare ourselves to others in terms of age, class, religious affiliation, sexual orientation, and so on. Aspects of these other identities are explored in other parts of *MCAT Behavioral Sciences Review*.

It is important to know that there are several factors that determine which identity will be enacted in particular situations. It is believed that our identities are organized according to a **hierarchy of salience**, such that we let the situation dictate which identity holds the most importance for us at any given moment. For instance, male and female college students in same-gender groups are less likely to list gender in their self-descriptions than students in mixed-gender groups. Furthermore, researchers have found that the more salient the identity, the more we conform to the role expectations of the identities. Salience is determined by a number of factors, including the amount of work we have invested into the identity, the rewards and gratification associated with the identity, and the amount of self-esteem we have associated with the identity.

Self-Evaluation

BRIDGE

Remember that *esteem* is one of Maslow's hierarchy of needs (#4 in priority). This model is discussed in Chapter 5 of *MCAT Behavioral Sciences Review.*

Our individual self-concept plays a very important role in the way we evaluate and feel about ourselves. **Self-discrepancy theory** maintains that each of us has three selves and that perceived differences between these selves lead to negative feelings. Our self-concept makes up our **actual self**, the way we see ourselves as we currently are. Our **ideal self** is the person we would like to be, and our **ought self** is our representation of the way others think we should be. Generally, the closer these three selves are to one another, the higher our **self-esteem** or self-worth will be.

Those with low self-esteem don't necessarily view themselves as worthless, but they will be far more critical of themselves. As a result, they take criticism from others poorly and typically believe that people will only accept them if they are successful. Research also shows that they are more likely to use drugs, to be pessimistic, and to give up when facing frustration than their counterparts with high self-esteem.

While self-esteem is the measure of how we feel about ourselves, **self-efficacy** is our belief in our ability to succeed. Self-efficacy can vary by activity for individuals; we all can think of situations in which we hold the belief that we are able to be effective and, conversely, those in which we feel powerless. Of course, we are more motivated to pursue those tasks for which our self-efficacy is high, but we can get into trouble when it is too high. **Overconfidence** can lead us to take on tasks for which we are not ready, leading to frustration, humiliation, or sometimes even personal injury. On the other hand, self-efficacy can also be depressed; an individual can develop a perceived lack of control over the outcome of a situation, a phenomenon called **learned helplessness**. Learned helplessness has been found to be strongly related to clinical depression.

Locus of control is another core self-evaluation that is closely related to self-concept. **Locus of control** refers to the way we characterize the influences in our lives. People with an internal locus of control view themselves as controlling their own fate, whereas those with an external locus of control feel that the events in their lives are caused by luck or outside influences. For example, a runner who loses a race may attribute the cause of the loss internally (*I didn't train hard enough*) or externally (*My shoes didn't fit and the track was wet*).

All of these ideas work hand-in-hand to influence the way we feel about ourselves. The happiest among us are those who have high self-esteem, view themselves as effective people, feel that they are in control of their destinies, and see themselves as living up to their own expectations of who they would like to be.

BRIDGE

Locus of control and cognitive dissonance are integral to attribution theory. In order to preserve self-esteem, we often see our successes as a direct result of our efforts and our failures as the result of uncontrollable outside influences. Attribution theory is discussed in Chapter 10 of *MCAT Behavioral Sciences Review.*

MCAT EXPERTISE

Effective MCAT students review full-length exams with an internal locus of control: *What can I do to prepare myself better for the next practice test?* An external locus of control prevents students from actually gaining anything from their practice: *Oh, that was just a stupid question.*

MCAT CONCEPT CHECK 6.1

Before you move on, assess your understanding of the material with these questions.

1. What is the difference between self-concept and identity?

   ____

2. List three factors that contribute to a person's ethnic identity. How are these factors different from those that determine national identity?

   1. ______________________________
   2. ______________________________
   3. ______________________________

   - National identity:

   ____

3. A high school student fails a history test. How might a student with an internal locus of control interpret this event? What about a student with an external locus of control?

   - Internal:

   ____

   - External:

   ____

6.2 Formation of Identity

High-Yield

LEARNING OBJECTIVES

After Chapter 6.2, you will be able to:

- Identify the three major phases of Kohlberg's theory of moral development and their central tenets
- Explain, and distinguish between, the theories of identity of Erikson, Freud, Vygotsky, and Kohlberg

Psychologists generally agree that we are not born with our self-concept and identity in place and fully developed. As young children, our identities are largely defined by our relationship to our caregivers. As we move into adolescence, we begin to develop into unique individuals, deciding who we want to be when on our own. Several theorists have proposed stages through which we develop. They vary in scope with respect to both the aspects of our identity they describe and their time span, but they all have one thing in common: the MCAT loves to test your mastery of these theories!

Freud: Psychosexual Development

Sigmund Freud was a pioneer in charting personality and emotional growth. For Freud, human psychology and human sexuality were inextricably linked. In fact, Freud made the assertion that far from lying dormant until puberty, the **libido** (sex drive) is present at birth. Freud believed that libidinal energy and the drive to reduce libidinal tension were the underlying dynamic forces that accounted for human psychological processes.

Freud hypothesized five distinct stages of psychosexual development, summarized in Table 6.1 at the end of this section. In each stage, children are faced with a conflict between societal demands and the desire to reduce the libidinal tension associated with different erogenous zones of the body. Each stage differs in the manner in which libidinal energy is manifested and the way in which the libidinal drive is met. **Fixation** occurs when a child is overindulged or overly frustrated during a stage of development. In response to the anxiety caused by fixation, the child forms a personality pattern based on that particular stage, which persists into adulthood as a functional mental disorder known as a **neurosis.**

The first stage is the **oral stage**, spanning from 0 to 1 year of age. During this stage, gratification is obtained primarily through putting objects into the mouth, biting, and sucking. Libidinal energy is centered on the mouth. An adult who is orally fixated would be expected to exhibit excessive dependency.

Next is the **anal stage**, from 1 to 3 years, during which the libido is centered on the anus and gratification is gained through the elimination and retention of waste materials. Toilet training occurs during this stage. Fixation during this stage would lead to either excessive orderliness (*anal-retentiveness*) or sloppiness in the adult.

The **phallic stage**, sometimes known as the **Oedipal stage**, is the third of Freud's stages of psychosexual development. Generally, children aged 3 to 5 years are in this developmental stage. This stage centers on resolution of the **Oedipal conflict** for male children or the analogous **Electra conflict** for female children. In Freud's view, the male child envies his father's intimate relationship with his mother and fears castration

MCAT EXPERTISE

Though revolutionary at the time, Freud's theories have largely been discredited. They do, however, provide an interesting window into the history of personality theory. Expect the MCAT to test these ideas from a historical context, usually prefaced by *According to Freud's theory of psychosexual development . . .*

BRIDGE

Freud's stages of psychosexual development are based on the same principles as drive reduction theory, as discussed in Chapter 5 of *MCAT Behavioral Sciences Review:* libidinal energy creates internal tension, which we aim to reduce through certain behaviors.

at his father's hands. He wishes to eliminate his father and possess his mother, but the child feels guilty about these wishes. To successfully resolve the conflict, he deals with his guilty feelings by identifying with his father, establishing his sexual identity, and internalizing moral values. Also, the child to a large extent de-eroticizes, or **sublimates** his libidinal energy. This may be expressed through collecting objects or focusing on schoolwork. Freud did not elaborate much on the Electra conflict, although he theorized a similar desire. Because females cannot have castration fear (instead, they are theorized to have **penis envy**), female children are expected to exhibit less stereotypically female behavior and to be less morally developed in this theory.

Once the libido is sublimated, the child has entered the **latency stage**, which lasts until puberty is reached.

For Freud, the final stage is the **genital stage**, beginning in puberty and lasting through adulthood. According to Freud, if prior development has proceeded correctly, the person should enter into healthy heterosexual relationships at this point. Freud claimed, however, that if sexual traumas of childhood have not been resolved, such behaviors as homosexuality, asexuality, or fetishism may result.

STAGE	DESCRIPTION
Oral	Libidinal energy centered on the mouth; fixation can lead to excessive dependency
Anal	Toilet training occurs during this time; fixation can lead to excessive orderliness or messiness
Phallic	Oedipal or Electra conflict is resolved during this stage
Latency	Libido is largely sublimated during this stage
Genital	Begins at puberty; in theory, if previous stages have been successfully resolved, the person will enter into heterosexual relationships

Table 6.1 Freud's Stages of Psychosexual Development

Erikson: Psychosocial Development

Erik **Erikson's theory of psychosocial development** theorizes that personality development is driven by the successful resolution of a series of social and emotional conflicts. For example, the first such conflict is that of **trust** *vs.* **mistrust**, which occurs during the first year of life. Newborn humans are quite helpless and unsure of their environment. A newborn depends on their caregivers for support. So, the psychosocial conflict that a newborn faces is whether or not to trust caregivers to reliably provide that support. If caregivers do reliably care for the newborn, then the newborn will learn trust, which is a social skill. In different circumstances, the newborn could fail to learn to trust. However, according to Erikson's theory, such an individual may nevertheless move on to the next stage of psychosocial development, and may even learn trust later in life.

This example illustrates three key features of Erikson's theory. First, the conflicts that Erikson describes arise because an individual lacks some critical social or emotional skill. Each conflict therefore represents an opportunity to learn a new social or emotional skill, which, according to Erikson, is the mechanism for psychosocial development. Second, each conflict has either a positive or negative resolution. For example, a newborn can learn to be mistrustful. This outcome does represent a resolution of the trust *vs.* mistrust conflict. However, this outcome would be a negative outcome and, in Erikson's view, would represent a failure to develop. Psychosocial development means not only resolving each conflict, but obtaining a positive resolution. However, the third key idea is that, in Erikson's theory, an individual who fails to obtain a positive resolution at one stage can still advance to later stages and, later in life, may even learn the skill that they failed to learn during the developmental conflict.

The second conflict is **autonomy** *vs.* **shame and doubt** (1 to 3 years), where children begin to explore their surroundings and develop their interests. The favorable outcome here is feeling able to exert control over the world and to exercise choice as well as self-restraint. However, if children are overly controlled and criticized, the unfavorable outcome is a sense of doubt and a persistent external locus of control.

The next conflict confronted is **initiative** *vs.* **guilt** (3 to 6 years), in which children learn basic cause and effect principles in physics, and starting and finishing out tasks for a purpose. Favorable outcomes include a sense of purpose, the ability to initiate activities, and the ability to enjoy accomplishment. If guilt wins out, children will be so overcome by the fear of punishment that they may either unduly restrict themselves or may overcompensate by showing off.

Next is the conflict of **industry** *vs.* **inferiority** (6 to 12 years), where pre-adolescents are becoming aware of themselves as individuals. If resolved favorably, children will feel competent, be able to exercise their abilities and intelligence in the world, and be able to affect the world in the way that they desire. Unfavorable resolution results in a sense of inadequacy, a sense of inability to act in a competent manner, and low self-esteem.

During adolescence (12 to 20 years), individuals experience **identity** *vs.* **role confusion**. During this conflict, adolescents explore their independence to determine who they are and what their purpose is in society. At this stage, individuals either form a single identity or become unsure about their place in society. The favorable outcome is fidelity, the ability to see oneself as a unique and integrated person with sustained loyalties. Unfavorable outcomes are confusion about one's identity and an amorphous personality that shifts from day to day.

The main crisis of young adulthood (20 to 40 years) is **intimacy** *vs.* **isolation**, where people focus on creating long-lasting bonds with others. Favorable outcomes are love, the ability to have intimate relationships with others, and the ability to commit oneself to another person and to one's own goals. If this crisis is not favorably resolved, there will be an avoidance of commitment, alienation, and distancing of oneself from others and one's ideals. Isolated individuals are either withdrawn or capable of only superficial relationships with others.

The conflict of middle age (40 to 65 years) is **generativity** *vs.* **stagnation**, where the focus is on advancing present and future society. The successful resolution of this conflict results in an individual capable of being a productive, caring, and contributing member of society. If this crisis is not overcome, one acquires a sense of stagnation and may become self-indulgent, bored, and self-centered with little care for others.

REAL WORLD

The conflict of identity *vs.* role confusion has some positive effects: teenagers identifying their interests, gravitating toward friends who share these interests, and creating a sense of who they want to be. On the other hand, this conflict can lead to the formation of cliques, bullying, and significant peer pressure. The increase of online and in-person bullying among adolescents has led to a number of programs to ease this crisis, such as *StopBullying.gov* and the *It Gets Better* campaign.

Finally, old age (above 65 years) brings about the crisis of **integrity** *vs.* **despair**, where the focus tends to be reflective and contemplative. If favorably resolved, we will see wisdom, which Erikson defined as detached concern with life itself, with assurance in the meaning of life, dignity, and an acceptance of the fact that one's life has been worthwhile, along with a readiness to face death. If not resolved favorably, there will be feelings of bitterness about one's life, a feeling that life has been worthless, and at the same time, fear over one's own impending death.

ERIKSON'S STAGE (CRISIS)	AGE	EXISTENTIAL QUESTION
Trust *vs.* mistrust	0 to 1 year	Can I trust the world?
Autonomy *vs.* shame and doubt	1 to 3 years	Is it okay to be me?
Initiative *vs.* guilt	3 to 6 years	Is it okay for me to do, move, and act?
Industry *vs.* inferiority	6 to 12 years	Can I make it in the world of people and things?
Identity *vs.* role confusion	12 to 20 years	Who am I? What can I be?
Intimacy *vs.* isolation	20 to 40 years	Can I love?
Generativity *vs.* stagnation	40 to 65 years	Can I make my life count?
Integrity *vs.* despair	65 years to death	Is it okay to have been me?

Table 6.2 Erikson's Stages of Psychosocial Development

Kohlberg: Moral Reasoning

Lawrence Kohlberg's theory of personality development focuses not on urges or on resolving conflicts, but rather on the development of moral thinking. For this reason, this theory is often called **Kohlberg's theory of moral reasoning**. Kohlberg reasoned that, as our cognitive abilities grow, we are able to think about the world in more complex and nuanced ways, and this directly affects the ways in which we resolve moral dilemmas and perceive the notion of right and wrong.

Kohlberg's observations about moral reasoning were based on responses of subjects to hypothetical moral dilemmas. One often-cited example is the *Heinz dilemma*. In this scenario, a man named Heinz has a wife who is dying of a rare disease. There is a druggist in the town who invented a drug that could cure the disease. It costs him $200 to produce, yet he sells it for $2000. Heinz cannot afford this price, so he goes to the druggist and asks him if he would lower the price, a request that the druggist refuses. Desperate to save his wife, Heinz breaks into the druggist's office one night and steals the medication. Kohlberg presented dilemmas such as this one to volunteers and asked them to explain whether the characters in the story acted morally and why or why not. Kohlberg wasn't interested in the participants' appraisal of the actions as right or wrong, as he believed either answer could be justified. Instead, he was far more interested in the reasoning behind the appraisal. Based on the participants' responses, Kohlberg organized moral reasoning into six distinct stages ranging from the concrete to the abstract. He then organized these stages into three phases consisting of two stages each. Kohlberg's stages are summarized in Table 6.3.

Preconventional morality, the first of these phases, is typical of preadolescent thinking and places an emphasis on the consequences of the moral choice. Stage one (**obedience**) is concerned with avoiding punishment (*If I steal the drug, I'll go to jail*), while stage two (**self-interest**) is about gaining rewards (*I need to save my wife because I want to spend more of my life with her*). Stage two is often called the **instrumental relativist stage** because it is based on the concepts of reciprocity and sharing: *I'll scratch your back, you scratch mine.*

KEY CONCEPT

Conventional morality corresponds to average adult moral reasoning. Preconventional is therefore expected in children, and postconventional is expected in a smaller subset of adults with more advanced moral reasoning skills than the average population.

The second phase is **conventional morality**, which begins to develop in early adolescence when individuals begin to see themselves in terms of their relationships to others. This phase is based on understanding and accepting social rules. Stage three (**conformity**) places emphasis on the "nice person" orientation in which an individual seeks the approval of others (*I should not steal the drug because stealing is wrong*). Stage four (**law and order**) maintains the social order in the highest regard (*If everyone stole things they couldn't afford, people who produce those items would not be able to continue their business*).

The third phase is **postconventional morality**, which describes a level of reasoning that Kohlberg claimed not everyone was capable of and is based on social mores, which may conflict with laws. Stage five (**social contract**) views moral rules as conventions that are designed to ensure the greater good, with reasoning focused on individual rights (*Everyone has a right to live; businesses have a right to profit from their products*). Finally, stage six (**universal human ethics**) reasons that decisions should be made in consideration of abstract principles (*It is wrong for one person to hold another's life for ransom*).

Kohlberg viewed these stages as a progression in which each stage is adopted and then abandoned for the next as the individual progresses. In other words, we all begin in stage one and progress to varying degrees as our thinking matures.

Kohlberg is not without his critics. Some argue that postconventional morality describes views that are more prevalent in individualistic societies and is therefore biased against collectivist cultures. Similarly, Kohlberg's research was only performed using male subjects, which may cloud differences in reasoning patterns that may exist between genders.

PHASE	AGE	STAGES
Preconventional morality	Preadolescence	1: Obedience 2: Self-interest
Conventional morality	Adolescence to adulthood	3: Conformity 4: Law and order
Postconventional morality	Adulthood (if at all)	5: Social contract 6: Universal human ethics

Table 6.3 Kohlberg's Stages of Moral Development

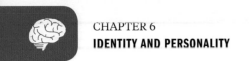

Vygotsky: Cultural and Biosocial Development

BRIDGE

Kohlberg and Vygotsky's theories of cognitive development were both heavily influenced by Piaget's work in this area, discussed in Chapter 4 of *MCAT Behavioral Sciences Review.*

Like Kohlberg, Lev Vygotsky's work was focused on understanding cognitive development. For Vygotsky, the engine driving cognitive development was the child's internalization of various aspects of culture: rules, symbols, language, and so on. As the child internalized these various interpersonal and cultural rules, cognitive activity developed accordingly.

Vygotsky is known for his concept of the **zone of proximal development**, referring to those skills and abilities that have not yet fully developed but are in the process of development. Gaining these skills successfully requires the help of a "more knowledgeable other," typically an adult. For example, children may struggle to ride a bicycle on their own, but with the help and guidance of an adult may be successful. Vygotsky would say that this skill is currently within the children's zone of proximal development.

The Influence of Others on Identity

Our personalities do not form in a vacuum; we are as much a product of those around us as a product of our own internal growth and development. Albert Bandura, who was also the psychologist behind the Bobo doll experiment described in Chapter 3 of *MCAT Behavioral Sciences Review*, claimed that observational learning contributes greatly to our future behaviors.

Young children observe and encode the behaviors they see in others, and may later imitate these behaviors. Children are more likely to imitate behaviors performed by someone who is like them: for example, young children will reliably mimic behaviors performed by their same-gender siblings. Children's first models are their caregivers, but as children grow and form more relationships, other role models emerge. Siblings, teachers, and the media all play an important role in modeling behavior for a developing child, but by adolescence, peers become the most important role models in a person's life.

As children grow, they become more able to see the identities of others as different from their own. They might experiment with other identities by taking on the roles of others, such as when children play *house* or *school*. Such **role-taking** is good practice for later in life, when a child begins to understand the perspectives and roles of others. Eventually, children become able to see how others perceive them and to imagine themselves from the outside. The ability to sense how another's mind works—for example, understanding how a friend is interpreting a story while you tell it—is referred to as **theory of mind**. Once we develop a theory of mind, we begin to recognize and react to how others think about us. We become aware of judgments from the outside world and react to these judgments. Our reactions to how others perceive us can be varied—maintaining, modifying, downplaying, or accentuating different aspects of our personality. Our understanding of how others see us, which relies on perceiving a reflection of ourselves based on the words and actions of others, is appropriately called the **looking-glass self**.

A related concept is a **reference group**, the group that we use as a standard to evaluate ourselves. Our self-concept often depends on whom we are comparing ourselves to. For example, in 2012, the average annual salary for a physician in the United States was about $200,000. Compared to the national median household salary (approximately $50,000), these individuals were quite well off. However, only 11 percent considered themselves "rich." Why? Many physicians live in higher socioeconomic areas and regularly interact with other physicians, and their responses may thus be biased by comparison to those around them.

MCAT CONCEPT CHECK 6.2

Before you move on, assess your understanding of the material with these questions.

1. Each of the following theorists evaluates an individual and determines that the person has failed in completing one of the theorist's developmental stages. What would each say is the most likely outcome for this person?
   - Freud:

   ____

   - Erikson:

   ____

   - Kohlberg:

   ____

2. Name and briefly describe the three major phases of Kohlberg's theory of moral development.

Phase	Description

3. How could Vygotsky's concept of zone of proximal development be applied to standardized test preparation?

____

6.3 Personality

LEARNING OBJECTIVES

After Chapter 6.3, you will be able to:

- Describe how personality is defined by the psychoanalytic, humanistic, type, behaviorist, social cognitive, biological, and trait perspectives
- List the traits described by Eysenck's PEN theory
- List the Big Five personality traits
- Recall the roles of the id, ego, and superego in the psychoanalytic perspective of personality

We've seen that identity is the way we define ourselves. **Personality**, while similar, describes the set of thoughts, feelings, traits, and behaviors that are characteristic of an individual across time and location. In a way, identity describes who we are, while personality describes how we act and react to the world around us. There are many different theories of personality, and different theorists within each category espouse sometimes conflicting views in an attempt to describe behavior. Like the various theories of development discussed earlier in this chapter, some of these ideas have been discredited, and so will only be tested on the MCAT from a historical perspective.

We can categorize theories of personality into four areas: psychoanalytic (psychodynamic), humanistic (phenomenological), type and trait, and behaviorist. There are great differences between and within these divisions in how personality is defined and how abnormal personalities are explained.

MCAT EXPERTISE

The psychoanalytic perspective, much like other Freudian and Jungian theories within psychology, is not backed by more modern understandings of personality and has fallen from use in professional circles, though it remains prevalent in popular culture. However, the psychoanalytic perspective is useful as a contrast to other theories of personality, and is included within testable materials as laid out by the AAMC.

The Psychoanalytic Perspective

The **psychoanalytic** or **psychodynamic theories of personality** contain some of the most widely varying perspectives on behavior, but they all have in common the assumption of unconscious internal states that motivate the overt actions of individuals and determine personality. The most noteworthy supporter of the psychoanalytic theory is Freud.

Sigmund Freud

Freud's contribution to the study of personality was his structural model, which involved three major entities: the id, ego, and superego, illustrated in Figure 6.2.

REAL WORLD

If a person is hungry and food is unavailable, wish fulfillment—fantasizing or daydreaming about food—helps relieve some of the tension created by the pleasure principle.

The **id** consists of all the basic, primal, inborn urges to survive and reproduce. It functions according to the **pleasure principle**, in which the aim is to achieve immediate gratification to relieve any pent-up tension. The **primary process** is the id's response to frustration based on the pleasure principle: *obtain satisfaction now, not later*. Mental imagery, such as daydreaming or fantasy, that fulfills this need for satisfaction is termed **wish fulfillment**.

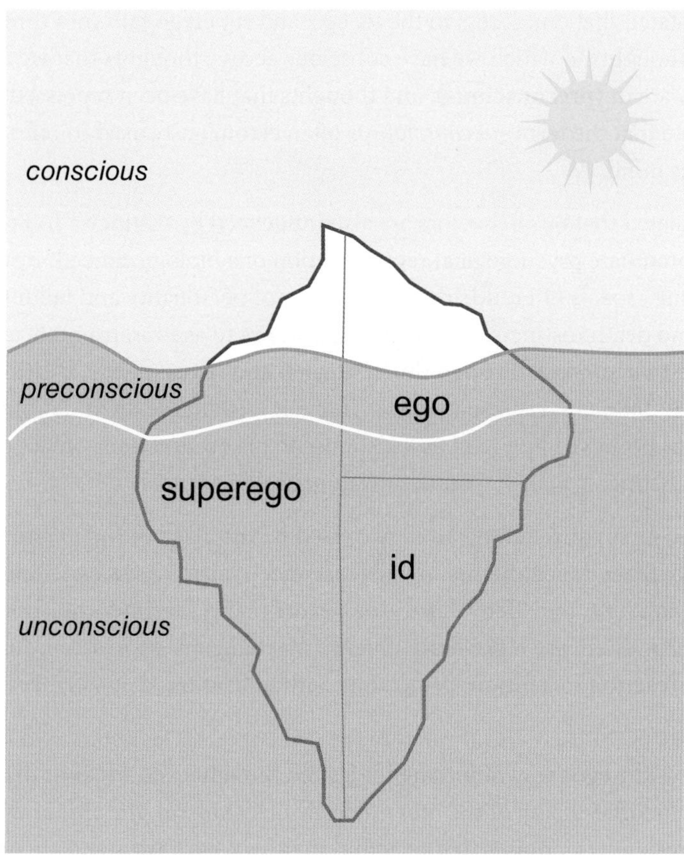

Figure 6.2 Freud's Topographic Model of the Mind

Because this mental image cannot effectively reduce tension on a permanent basis, the **ego** comes into play. The ego operates according to the **reality principle**, taking into account objective reality as it guides or inhibits the activity of the id and the id's pleasure principle. This guidance is referred to as the **secondary process**. The aim of the reality principle is to postpone the pleasure principle until satisfaction can actually be obtained. It must be emphasized that while the ego suspends the workings of the primary process, it does so only to meet the demands of objective reality. The mutual give and take of the ego and reality promotes the growth of perception, memory, problem solving, thinking, and reality testing. The ego can be understood to be the organizer of the mind: it receives its power from—and can never be fully independent of—the id.

The ego is also responsible for moderating the desires of the superego. Whereas the id's desires are basic needs, those of the superego are refined and focused on the ideal self. The **superego** is the personality's perfectionist, judging our actions and responding with pride at our accomplishments and guilt at our failures. The superego can be divided into two subsystems, both of which are a reflection of the morals taught to children by their caregivers. The **conscience** is a collection of the improper actions for which a child is punished, and the **ego-ideal** consists of those proper actions for which a child is rewarded. Ultimately, a system of right and wrong substitutes for caregivers' rewards and punishments.

REAL WORLD

When stuck in traffic, our id may desire to honk loudly at the cars in front of us, or to even pull over to the shoulder of the highway and drive recklessly past the congestion. Our ego knows that this would be unwise, and may advise us to breathe deeply and change the radio station to something calming instead.

Freud also stated that our access to the id, ego, and superego falls into three main categories: thoughts to which we have conscious access, thoughts that we aren't currently aware of (**preconscious**), and thoughts that have been repressed (**unconscious**). Note that the term *subconscious* is often erroneously used to refer to Freud's unconscious mind.

Freud postulated that our behaviors are also influenced by instincts. To Freud, an **instinct** is an innate psychological representation of a biological need. Instincts are the propelling aspects of Freud's dynamic theory of personality and fall into two types: life and death instincts. Life instincts, referred to as *Eros*, promote an individual's quest for survival through thirst, hunger, and sexual needs. Death instincts, referred to as *Thanatos*, represent an unconscious wish for death and destruction. *Thanatos* was proposed by Freud as a response to his observations of victims of trauma reenacting or focusing on their traumatic experiences.

The ego's recourse for relieving anxiety caused by the clash of the id and superego is through **defense mechanisms**. All defense mechanisms have two common characteristics: first, they deny, falsify, or distort reality; second, they operate unconsciously. There are eight main defense mechanisms: repression, suppression, regression, reaction formation, projection, rationalization, displacement, and sublimation.

KEY CONCEPT

While repression is unconscious forgetting (such as that which may occur after traumatic events), suppression is a conscious form of forgetting: *I'm not going to think about that right now.*

Repression is the ego's way of forcing undesired thoughts and urges to the unconscious and underlies many of the other defense mechanisms, the aim of which is to disguise threatening impulses that may find their way back from the unconscious. While repression is mostly an unconscious forgetting, **suppression** is a more deliberate, conscious form of forgetting.

Regression is reversion to an earlier developmental state. Faced with stress, older children may return to earlier behaviors such as thumb sucking, throwing temper tantrums, or clinging to their caregivers.

Reaction formation occurs when an individual suppresses urges by unconsciously converting these urges into their exact opposites. For example, a person pining after an inaccessible celebrity may outwardly express hatred for the celebrity as a way of reducing the stress caused by these unrequited feelings.

Projection is the defense mechanism by which individuals attribute their undesired feelings to others. *I hate my parents* might, for example, turn into *My parents hate me.* Projection is an important part of personality analysis. Tests that make use of projection to gain insight into a client's mind are common in psychoanalytic therapy. For example, the **Rorschach inkblot test**, shown in Figure 6.3, relies on the assumption that clients project their unconscious feelings onto the shape.

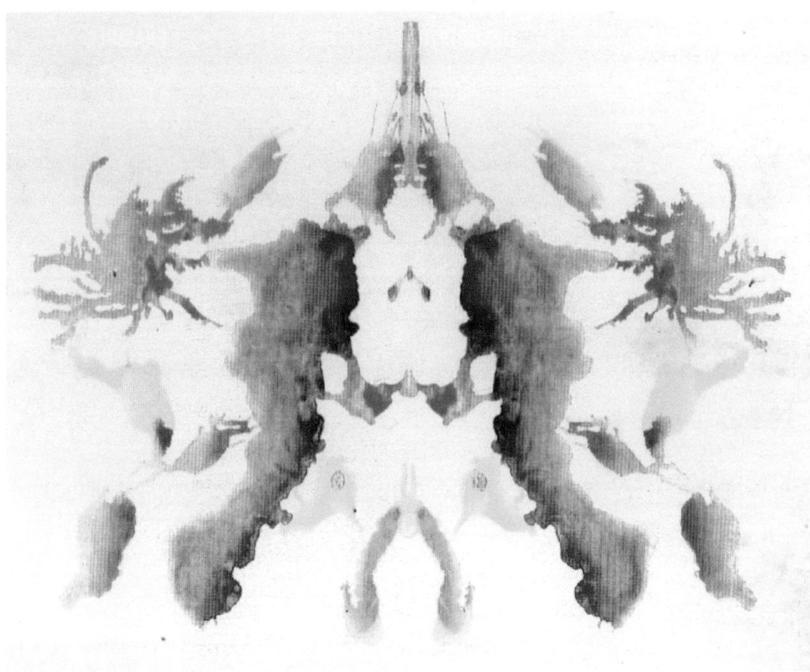

Figure 6.3 Card #10 from the Rorschach Inkblot Test

Similarly, the **thematic apperception test** consists of a series of pictures that are presented to the client, who is asked to make up a story about each one. The story, presumably, will elucidate the client's own unconscious thoughts and feelings.

Rationalization is the justification of behaviors in a manner that is acceptable to the self and society. Drivers who engage in reckless feats such as the Cannonball Run (a cross-country race from Los Angeles to New York) might justify their dangerous pursuits by saying, both to themselves and others: *I'm in complete control, and besides, there are plenty of dangerous drivers on the road. What difference will one more make?*

Displacement describes the transference of an undesired urge from one person or object to another. Someone angry at a supervisor at work may remain quiet there but snap at a family member at home.

Finally, **sublimation** is the transformation of unacceptable urges into socially acceptable behaviors. Freud might say that pent-up sexual urges may be sublimated into a drive for business success or artistic creativity.

The descriptions of the most commonly tested defense mechanisms, as well as examples, are provided in Table 6.4.

DEFENSE MECHANISM	DESCRIPTION	EXAMPLE
Repression	Unconsciously removing an idea or feeling from consciousness	A person who survived six months in a concentration camp cannot recall anything about life during that time period
Suppression	Consciously removing an idea or feeling from consciousness	A terminally ill cancer patient puts aside feelings of anxiety to enjoy a family gathering
Regression	Returning to an earlier stage of development	A person speaks to a significant other in "baby talk" when relaying bad news
Reaction formation	An unacceptable impulse is transformed into its opposite	Two coworkers fight all the time because they are actually very attracted to each other
Projection	Attribution of wishes, desires, thoughts, or emotions to someone else	A person who has committed adultery is convinced the spouse is cheating, despite a lack of evidence
Rationalization	Justification of attitudes, beliefs, or behaviors	A murderer who claims that, while killing is wrong, this particular victim "deserved it"
Displacement	Changing the target of an emotion, while the feelings remain the same	When sent to their room as punishment, children begin to punch and kick their pillows
Sublimation	Channeling of an unacceptable impulse in a socially acceptable direction	A boss who is attracted to an employee becomes that employee's mentor and advisor

Table 6.4 Commonly Tested Defense Mechanisms

Carl Jung

Later psychoanalytic theories have given more emphasis to interpersonal, sociological, and cultural influences, while maintaining their link with the psychoanalytic tradition. Carl Jung preferred to think of libido as psychic energy in general, not just psychic energy rooted in sexuality. Jung identified the ego as the conscious mind, and he divided the unconscious into two parts: the **personal unconscious**, similar to Freud's notion of the unconscious, and the collective unconscious. The **collective unconscious** is a powerful system that is shared among all humans and considered to be a residue of the experiences of our early ancestors. Its building blocks are images of common experiences, such as having caregivers as children. These images invariably have an emotional element, and are referred to as **archetypes** in Jung's theory. You can see an example of two archetypal images in Figure 6.4: God and the Devil.

Figure 6.4 Jungian Archetypes: God and the Devil
Archetypes are underlying forms or concepts that give rise to archetypal images, which may differ somewhat between cultures.

There are several important Jungian archetypes. The **persona** is likened to a mask that we wear in public, and is the part of our personality that we present to the world. Like our identity, Jung described the persona as adaptive to our social interactions, emphasizing those qualities that improve our social standing and suppressing our other, less desirable qualities. The **anima** (feminine) and the **animus** (masculine) describe gender-inappropriate qualities—in other words, feminine behaviors in males and masculine behaviors in females. For example, in Jung's theory, the anima is the suppressed female quality in males that explains emotional behavior (described by Jung as a *man's inner woman*), while the animus is the analogous male quality of females that explains power-seeking behavior (*a woman's inner man*). The **shadow** archetype is responsible for the appearance of unpleasant and socially reprehensible thoughts, feelings, and actions experienced in the unconscious mind.

The **self**, to Jung, was the point of intersection between the collective unconscious, the personal unconscious, and the conscious mind. The self strives for unity. Jung symbolized the self as a *mandala* (Sanskrit: "circle"), shown in Figure 6.5. Jung saw the mandala, a symbol of the universe in Buddhism and Hinduism, as the mythic expression of the self: the reconciler of opposites and the promoter of harmony. Jung also developed **word association testing** to assess how unconscious elements may be influencing the conscious mind and thus the self. In word association testing, patients respond to a single word with the first word that comes to mind. Jung believed that patient responses, in combination with evaluating mood and speed of response, would reveal elements of the unconscious.

KEY CONCEPT

Important Jungian archetypes:

- Persona—the aspect of our personality we present to the world
- Anima—a "man's inner woman"
- Animus—a "woman's inner man"
- Shadow—unpleasant and socially reprehensible thoughts, feelings, and actions experienced in the unconscious mind

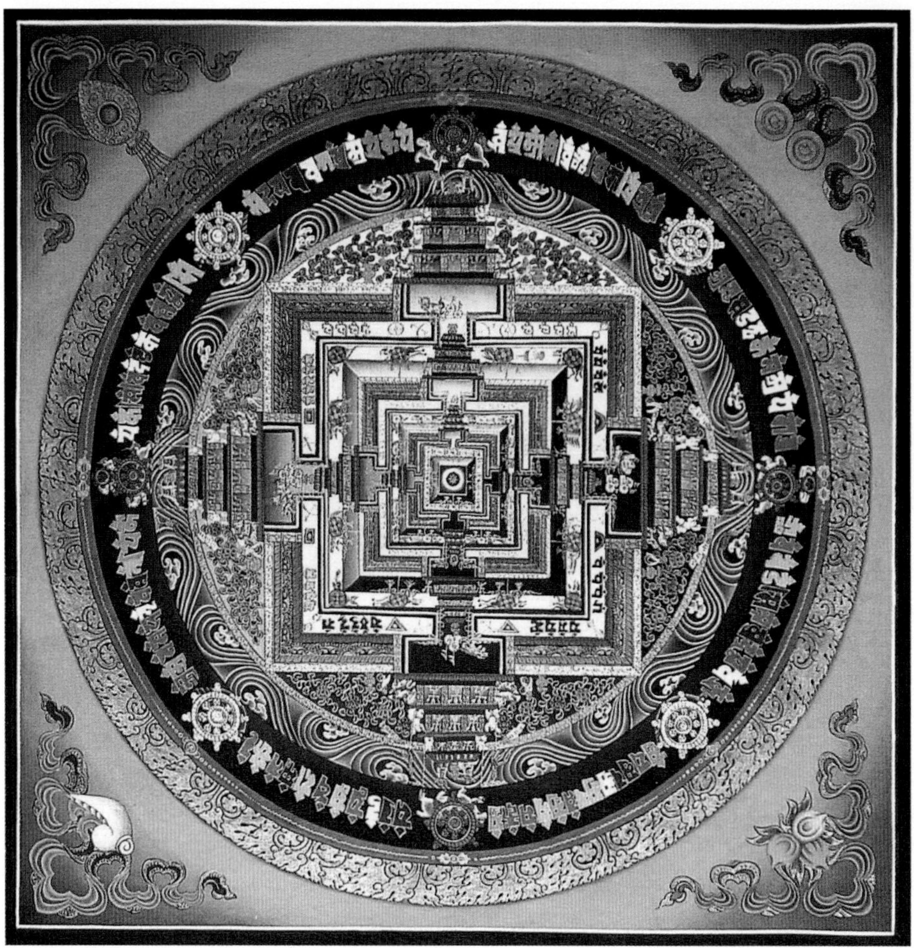

Figure 6.5 Tibetan Mandala

Jung saw the self as a mandala: the promoter of unity, balance, and harmony between the conscious mind, personal unconscious, and collective unconscious.

Jung described three dichotomies of personality:

- Extraversion (E, orientation toward the external world) *vs.* introversion (I, orientation toward the inner, personal world)
- Sensing (S, obtaining objective information about the world) *vs.* intuiting (N, working with information abstractly)
- Thinking (T, using logic and reason) *vs.* feeling (F, using a value system or personal beliefs)

In most individuals, both sides of each dichotomy are present to some degree, but one tends to dominate. Jung's work laid the groundwork for creation of the **Myers–Briggs Type Inventory (MBTI)**, a classic personality test. Each of Jung's three dichotomies, and a fourth—judging (J, preferring orderliness) *vs.* perceiving (P, preferring spontaneity)—is labeled as a specific personality type, as shown in Figure 6.6.

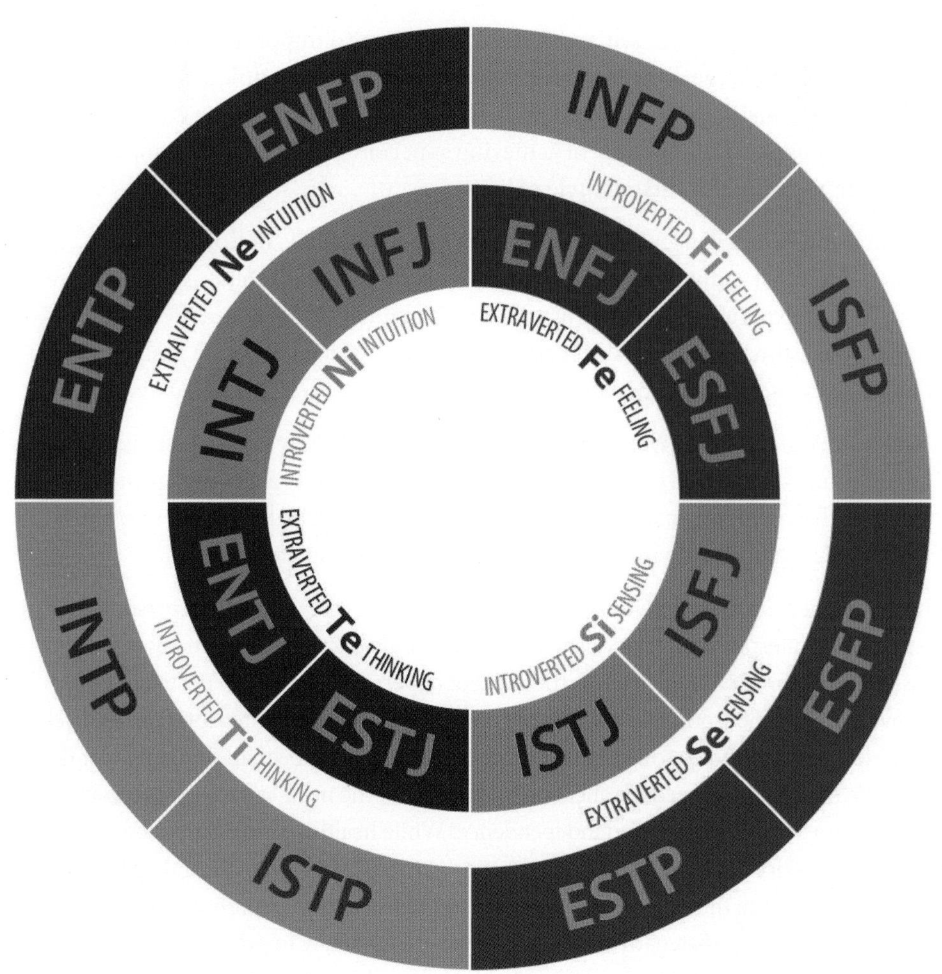

Figure 6.6 Myers-Briggs Type Inventory Personality Types

Other Psychoanalysts

As with most psychological movements, psychoanalysis gained a following of theorists who learned from and often disagreed with its original tenets. In opposition to many of Freud's key ideas, later psychoanalysts often emphasized social rather than sexual motivations for behavior. Jung can be counted among these, as can Alfred Adler, whose theory focused on the immediate social imperatives of family and society and their effects on unconscious factors.

Adler was the originator of the concept of the **inferiority complex**: an individual's sense of incompleteness, imperfection, and inferiority both physically and socially. According to Adler, striving for superiority drives the personality. This striving enhances the personality when it is oriented toward benefiting society, but yields disorder when it is selfish.

The notions of the creative self and style of life were also important to Adler's theory. The **creative self** is the force by which individuals shape their uniqueness and establish their personality. **Style of life** represents the manifestation of the creative self and describes a person's unique way of achieving superiority. The family environment is crucial in molding the person's style of life.

Another important concept in Adler's theory of personality is **fictional finalism**. This is the notion that individuals are motivated more by their expectations of the future than by past experiences. According to Adler, human goals are based on the subjective or fictional estimate of life's values rather than objective data from the past. Fictional finalism can often be summed up by the phrase *Life would be perfect if only*...

Notice the difference between Freud, Jung, and Adler. Whereas Freud's major assumption is that behavior is motivated by inborn instincts and Jung's principal axiom is that a person's conduct is governed by inborn archetypes, Adler assumes that people are primarily motivated by striving for superiority.

Karen Horney, another dissenting student of Freud's, likewise argued that personality is a result of interpersonal relationships, and adamantly disagreed with many of Freud's assumptions about women, such as the concept of penis envy. Horney postulated that individuals with neurotic personalities are governed by one of ten **neurotic needs**. Each of these needs is directed toward making life and interactions bearable. Examples of these neurotic needs are the need for affection and approval, the need to exploit others, and the need for self-sufficiency and independence. While healthy people have these needs to some degree, Horney emphasized that these needs become problematic if they fit at least one of four criteria: they are disproportionate in intensity, they are indiscriminate in application, they partially disregard reality, or they have a tendency to provoke intense anxiety. For instance, someone with a neurotic need for self-sufficiency and independence would go to great extremes to avoid being obligated to someone else in any way. As the central focus of the person's life, it would be a neurotic need and not a healthy one.

Horney's primary concept is that of basic anxiety. This is based on the premise that children's early perception of self is important and stems from their relationship with their caregivers. Inadequate caregiving can cause vulnerability and helplessness, which Horney termed **basic anxiety**, while neglect and rejection cause anger known as **basic hostility**. To overcome basic anxiety or basic hostility and attain a degree of security, children use three strategies in their relationships with others: moving toward people to obtain the goodwill of people who provide security; moving against people, or fighting them to obtain the upper hand; and moving away, or withdrawing, from people. These three strategies are the general headings under which the ten neurotic needs fall. Healthy people use all three strategies, depending on the situation. However, the highly threatened child will use one of these strategies rigidly and exclusively, and carries this strategy into adulthood.

Object relations theory also falls under the realm of psychodynamic theories of personality. In this context, *object* refers to the representation of caregivers based on subjective experiences during early infancy. These objects then persist into adulthood and impact our interactions with others, including the social bonds we create and our predictions of others' behavior.

The Humanistic Perspective

In direct contrast to the psychoanalysts, who focus on "sick" individuals and their troubling urges, **humanistic** or **phenomenological theorists** focus on the value of individuals and take a more person-centered approach, describing those ways in which healthy people strive toward self-realization. Humanism is often associated with **Gestalt therapy**, in which practitioners tend to take a holistic view of the self, seeing each individual as a complete person rather than reducing the person to individual behaviors or drives. For the humanists, our personality is the result of the conscious feelings we have for ourselves as we attempt to attain our needs and goals.

Kurt Lewin's **force field theory** puts very little stock in constraints on personalities such as fixed traits, habits, or structures such as the id, ego, and superego. Further, Lewin focused little on an individual's past or future, focusing instead on situations in the present. Lewin defined the field as one's current state of mind, which was simply the sum of the forces (influences) on the individual at that time. If the focus of humanistic psychology is exploring how an individual reaches self-realization, then these forces could be divided into two large groups: those assisting in our attainment of goals and those blocking the path to them.

Abraham Maslow, whose hierarchy of needs is discussed in Chapter 5 of *MCAT Behavioral Sciences Review*, was a humanist who studied the lives of individuals such as Ludwig van Beethoven, Albert Einstein, and Eleanor Roosevelt, who he felt were self-actualizers and had lived rich and productive lives. He identified several characteristics that these people had in common, including a nonhostile sense of humor, originality, creativity, spontaneity, and a need for some privacy. According to Maslow, self-actualized people are more likely than people who are not self-actualized to have what he called **peak experiences**: profound and deeply moving experiences in a person's life that have important and lasting effects on the individual.

George Kelly used himself as a model to theorize about human nature, and set aside the traditional concepts of motivation, unconscious emotion, and reinforcement in his descriptions of **personal construct psychology**. Kelly thought of the individual as a scientist, a person who devises and tests predictions about the behavior of significant people in the individual's life. The individual constructs a scheme of anticipation of what others will do, based on knowledge, perception, and relationships with these other people. Thus, the anxious person, rather than being the victim of inner conflicts and pent-up energy (as in psychodynamic theory), is one who is having difficulty constructing and understanding the variables in the environment. According to Kelly, psychotherapy is a process of insight whereby people acquire new constructs that will allow them to successfully predict troublesome events. Then, the individual will be able to integrate these new constructs into already existing ones.

Carl Rogers is most known for his psychotherapy technique known as **client-centered**, **person-centered**, or **nondirective** therapy. Rogers believed that people have the freedom to control their own behavior, and are neither slaves to the unconscious (as the psychoanalysts would suggest), nor subjects of faulty learning (as the behaviorists would say). Rather than providing solutions or diagnoses, the person-centered therapist helps clients reflect on problems, make choices, generate solutions, take positive action, and determine their own destiny. Rogers was the originator of the concepts of the real and ideal self discussed earlier in the chapter, and his therapeutic techniques aimed to help clients reconcile the differences between the various selves and reduce stress-inducing incongruence. Rogers also pioneered the concept of **unconditional positive regard**, a therapeutic technique by which the therapist accepts the client completely and expresses empathy in order to promote a positive therapeutic environment.

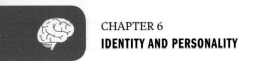

The Type and Trait Perspectives

The type and trait theorists were also borne out of dissatisfaction with psychoanalysis. **Type theorists** attempt to create a taxonomy of personality types, while **trait theorists** prefer to describe individual personality as the sum of a person's characteristic behaviors. For our purposes, we will consider them together.

Early attempts at personality types are generally discredited today. The ancient Greeks, for example, devised **personality types** based on **humors** or body fluids, an imbalance of which could lead to various personality disorders, as shown in Figure 6.7.

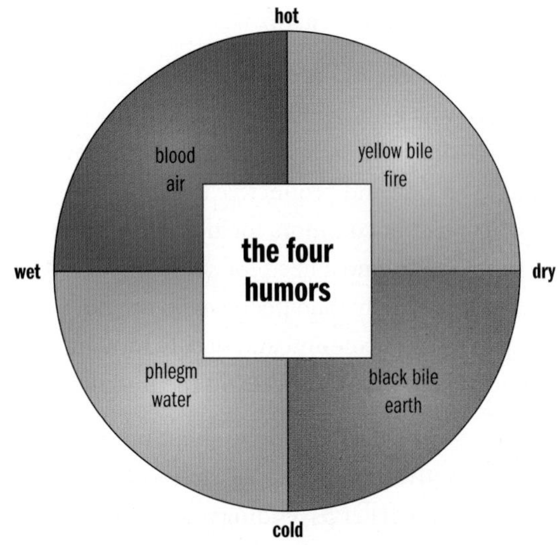

Figure 6.7 The Four Humors

Each humor was correlated with an element, an imbalance of which could lead to different personalities: blood (sanguine; impulsive and charismatic), bile (choleric; aggressive and dominant), black bile (melancholic; depressive and cautious), and phlegm (phlegmatic; relaxed and affectionate).

In the early 20th century, William Sheldon proposed personality types based on body type called **somatotypes**. Sheldon presumed that all short, stocky people were jolly, all tall people were high-strung and aloof, and people in between were strong and well-adjusted. One well-known type theory divides personalities into Types A and B. Individuals with **Type A** personalities are characterized by behavior that tends to be competitive and compulsive, while someone described as **Type B** is generally laid-back and relaxed. Not surprisingly, people with Type A personalities are more prone to heart disease than those with Type B personalities, although there is not much evidence to suggest that people with Type A personalities have a higher mortality rate.

The Myers–Briggs Type Inventory, described earlier in this chapter, also stands as a well-known example of a type theory.

Trait theorists instead use clusters of behaviors to describe individuals. Hans and Sybil Eysenck used factor analysis to group behaviors that typically occur together and assigned labels to those groups. These groups of behaviors are often also called **traits**. For example, people who are more reserved and less outspoken in groups also tend to enjoy solitary activities and avoid overstimulation. These behaviors fall under the label of introversion. The Eysencks described three traits in the **PEN model**: **Psychoticism** is a measure of nonconformity or social deviance. **Extraversion** is a measure of tolerance for social interaction and stimulation. **Neuroticism** is a measure of emotional arousal in stressful situations.

Negative affect, though not directly included in the PEN model, is related to neuroticism and describes how a person thinks of themselves and experiences negative emotions. High negative affect corresponds with neuroticism and anxiety, and is associated with several mood disorders. The Eysencks reasoned that people could be distinguished from one another based on where they fell in each of these three dimensions. More recently, the PEN theory has been expanded to what is known as the Five Factor Model, the **Big Five**, which as the name would suggest, uses dimensions of five traits: openness, conscientiousness, extraversion, agreeableness, and neuroticism. In this theory, **openness** describes openness to experience, or willingness to engage with the world and desire to try new things. Low openness is associated with persevering in tasks and difficulty with abstraction. **Conscientiousness** is in some ways analogous to self control, with high conscientiousness associated with high impulse control and low conscientiousness associated with spontaneity. **Agreeableness** refers to the degree to which a person is concerned about maintaining peace and harmony in their interactions with others.

MNEMONIC

The Big Five Traits of Personality: **OCEAN**

- **O**penness
- **C**onscientiousness
- **E**xtraversion
- **A**greeableness
- **N**euroticism

Gordon Allport, primarily a trait theorist, listed three basic types of traits or dispositions: cardinal, central, and secondary. **Cardinal traits** are traits around which people organize their lives. For instance, Mother Teresa's cardinal trait may have been self-sacrifice. While not everyone develops a cardinal trait, everyone does have central and secondary traits. **Central traits** represent major characteristics of the personality that are easy to infer, such as honesty or charisma. **Secondary traits** are other personal characteristics that are more limited in occurrence: aspects of one's personality that only appear in close groups or specific social situations. A major part of Allport's theory is the concept of **functional autonomy**, in which a behavior continues despite satisfaction of the drive that originally created the behavior. A hunter, for example, may have originally hunted to obtain food to eat. However, the hunter may continue even after there is enough food simply for the enjoyment of the hunt: that which began as a means to obtain a goal became the goal itself.

David McClelland identified a personality trait that is referred to as the need for achievement (N-Ach). People who are rated high in N-Ach tend to be concerned with achievement and have pride in their accomplishments. These individuals avoid high risks (to avoid failing) and low risks (because easy tasks will not generate a sense of achievement). Additionally, they set realistic goals, and stop striving toward a goal if success is unlikely.

BEHAVIORAL SCIENCES GUIDED EXAMPLE WITH EXPERT THINKING

Flow is a psychological state marked by high but subjectively effortless attention to a task that usually is accompanied by feelings of complete immersion, enjoyment, and loss of a sense of time. Feelings of flow are maximal in cases in which the task provides a significant challenge, but the individual's skill allows them to meet that challenge.

New term: flow. Challenge and task tell me this may be related to Yerkes–Dodson. I'll be on the lookout for more relationships between these concepts.

Flow proneness, the likelihood that an individual will enter a flow state given an appropriate compatibility between skill and demand, correlates positively with psychological well-being, self-esteem, life satisfaction, coping strategies, and conscientiousness. It has also been shown that flow proneness correlates negatively with traits related to neuroticism, such as trait anxiety.

New term: flow proneness. Self-esteem stands out here as a component of identity. I also see conscientiousness and neuroticism, which opens the door for the MCAT to ask about the Big Five.

Behavioral inhibition is a temperamental trait associated with fear and avoidance of novel experience. Because behavioral inhibition is also associated with anxiety and neurotic introversion, it would be reasonable to expect that low behavioral inhibition correlates with flow proneness.

Yet another new term: behavioral inhibition (BI). I want to keep these relationships straight. BI is related to the same OCEAN traits that correspond with low flow proneness (FP), so FP should relate to low BI.

Since the feelings of happiness and enjoyment related to flow derive from intrinsic motivation and rewards, it has also been shown that flow proneness is also correlated with an internal locus of control. Indeed, those with an internal locus of control are more sensitive to differences between individual skill and task challenge but are more likely to experience flow when performing tasks in which skill and challenge are compatible. Locus of control has also been shown to relate to high conscientiousness and low neuroticism.

More relationships: locus of control (LoC) is related in several ways to FP, and one of those ways involves the OCEAN traits again. High FP = Internal LoC = Low BI

Researchers analyzed data from a web-based survey of approximately 11,000 twins registered with the Swedish Twin Registry. This survey included a flow proneness questionnaire, a locus of control scale, and a measure of behavioral inhibition.

Twin study = genetics.
IV: genetic relationship
DV: survey results

These results confirmed a low but statistically significant correlation between flow proneness, low behavioral inhibition, and internal locus of control.

Analyzing relationship between dependent variables (survey results). Confirms relationship that correlation is low but significant.

Analysis also showed a strong relationship between specific genetic markers and all three measures, indicating that a specific set of genes account for a significant amount of the variance in these traits between individuals, possibly through dopaminergic pathways. The data support a dominant inheritance pattern.

Genetics confirmed. Summary: a few genes related to dopamine might be behind incidence of all of these traits.

Adapted from: Mosing MA, Pedersen NL, Cesarini D, Johannesson M, Magnusson PKE, Nakamura J, et al. (2012) Genetic and Environmental Influences on the Relationship between Flow Proneness, Locus of Control and Behavioral Inhibition. *PLoS ONE* 7(11): e47958. https://doi.org/10.1371/journal.pone.0047958

Which theory of personality is most supported by this experiment, and in what way do concepts from the study of both identity and personality contribute to this support?

The question demands that we integrate the data from the study with our outside knowledge of theories of personality. This passage blends quite a few topics together, so we'll want to read strategically and identify where and when new topics are introduced and explained; any time the passage alludes to content that we've studied we want to be ready to answer at least one question about that topic, so it's worth taking a few seconds to remind ourselves of what we know. Trait theory is definitely included, since the passage makes explicit reference to two of the Big Five: conscientiousness, which is the tendency to be organized, self-disciplined, and achievement focused; and neuroticism, the tendency to be prone to psychological stress and quickness to experience unpleasant emotions such as anger and anxiety. Two more Big Five traits are hinted at in the discussion of behavioral inhibition, which mentions avoidance of novel experience (low openness to experience) and introversion (the opposite of extraversion). If we're careful to look for these types of links as we read the first time, it will save us quite a lot of time on Test Day.

The passage also mentions self-esteem and locus of control, two ideas related to self-concept, and states that they are related to the traits discussed. With all this talk of traits, it would be reasonable to think that the passage is supporting trait theory, but the phrasing of the question stem is key here. We are asked about the experiment specifically, which is only located in the last paragraph of the study. In fact, while we've included a lot of analysis for the sake of completeness, a particularly astute test taker might even avoid some of this analysis and focus solely on analyzing the experiment to answer this question. By tying genetics to personality traits, the experiment in the passage most strongly supports the biological perspective. Overall, the results demonstrate a correlation between the genes regulating dopamine and reward pathways and the traits relating to rewards such as discipline, motivation, and locus of control. Then those traits are linked to the tendency in question in the study, flow proneness.

In summary, while the passage discusses other theories of personality and does provide some support for trait theory, the experiment and data provided most directly support the biological perspective of personality theory.

Other Theories of Personality

Of course, entire textbooks can be (and in fact are) devoted to personality theorists and their ideas. The MCAT tests only the key ideas of each theory, or the concepts that overlap heavily with other topics in this text.

The **behaviorist** perspective, championed by B. F. Skinner, is based heavily on the concepts of operant conditioning, discussed in Chapter 3 of *MCAT Behavioral Sciences Review*. Skinner reasoned that personality is simply a reflection of behaviors that have been reinforced over time. Therapy, then, should focus on learning skills and changing behaviors through operant conditioning techniques. **Token economies**, for example, are often used in inpatient therapeutic settings: positive behavior is rewarded with tokens that can be exchanged for privileges, treats, or other reinforcers.

The **social cognitive** perspective takes behaviorism one step further, focusing not just on how our environment influences our behavior, but also on how we interact with that environment. Albert Bandura's concept of reciprocal determinism is a central idea to this perspective. **Reciprocal determinism** refers to the idea that our thoughts, feelings, behaviors, and environment all interact with each other to determine our actions in a given situation. People choose environments that suit their personalities, and their personalities determine how they will feel about and react to events in those environments. Locus of control is another important concept in the social cognitive perspective: some people feel more in control of their environment while others feel that their environment controls them. For a social cognitive theorist, the best predictor of future behavior is past behavior in similar situations.

On the other end of the spectrum lies the **biological** perspective, which holds that personality can be explained as a result of genetic expression in the brain. The biological and trait perspectives are closely linked, as biological theorists maintain that many traits can be shown to result from genes or differences in brain anatomy.

The dichotomy presented by the social cognitive and biological perspectives of personality is similar to another debate in psychology: whether behavior is primarily determined by an individual's personality (the **dispositional approach**) or by the environment and context (the **situational approach**). This division is investigated in depth in the section on attribution theory in Chapter 10 of *MCAT Behavioral Sciences Review*.

MCAT CONCEPT CHECK 6.3

Before you move on, assess your understanding of the material with these questions.

1. For each of the following perspectives, briefly describe how each would define personality.
   - Psychoanalytic:
   - Humanistic:
   - Type:
   - Trait:
   - Behaviorist:
   - Social cognitive:
   - Biological:
2. What are the roles of the id, ego, and superego, according to the psychoanalytic perspective?
   - Id:
   - Ego:
   - Superego:

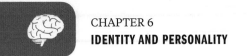

3. What are the traits described by the Eysencks' PEN theory, and what does each describe?
   - P:

   - E:

   - N:

   ---

4. What are the Big Five personality traits?
   1. ______________________
   2. ______________________
   3. ______________________
   4. ______________________
   5. ______________________

Conclusion

In this chapter, we discussed two concepts that are central to the study of psychology: identity, which describes who we are, and personality, which describes the set of thoughts, feelings, traits, and behaviors that are characteristic of each of us across time and different locations. We discussed a variety of approaches to both, each with their own theorists and critics. The ideas presented in this chapter are ripe for the MCAT to test; expect questions on Test Day to ask you to identify the various perspectives and the differences between them as they apply to behavior and research.

Many of the theories of personality focus on problems in everyday life: how we cope with stresses, anxiety, and depression. In medical school, your focus will be primarily on these pathologic states of mind, or psychological disorders. It is to this extremely important topic—abnormal psychology—that we turn our attention in the next chapter.

You've reviewed the content, now test your knowledge and critical thinking skills by completing a test-like passage set in your online resources!

GO ONLINE

CONCEPT SUMMARY

Self-Concept and Identity

- **Self-concept** is the sum of the ways in which we describe ourselves: in the present, who we used to be, and who we might be in the future.
- Our **identities** are individual components of our self-concept related to the groups to which we belong. Religious affiliation, sexual orientation, and ethnic and national affiliations are examples of identities.
- **Self-esteem** describes our evaluation of ourselves. Generally, the closer our **actual self** is to our **ideal self** (who we want to be) and our **ought self** (who others want us to be), the higher our self-esteem will be.
- **Self-efficacy** is the degree to which we see ourselves as being capable at a given skill or in a given situation. When placed in a consistently hopeless scenario, self-efficacy can be diminished to the point where **learned helplessness** results.
- **Locus of control** is a self-evaluation that refers to the way we characterize the influences in our lives. People with an internal locus of control see their successes and failures as a result of their own characteristics and actions, while those with an external locus of control perceive outside factors as having more of an influence in their lives.

Formation of Identity

- Freud's psychosexual stages of personality development are based on the tensions caused by the **libido**. Failure at any given stage leads to **fixation** that causes personality disorders. Freud's phases (oral, anal, phallic [Oedipal], latent, and genital) are based on the erogenous zones that are the focus of each phase of development.
- Erikson's stages of psychosocial development stem from conflicts that occur throughout life (trust *vs.* mistrust, autonomy *vs.* shame and doubt, initiative *vs.* guilt, industry *vs.* inferiority, identity *vs.* role confusion, intimacy *vs.* isolation, generativity *vs.* stagnation, integrity *vs.* despair). These conflicts are the result of decisions we are forced to make about ourselves and the environment around us at each phase of our lives.
- Kohlberg's stages of moral development describe the approaches of individuals to resolving moral dilemmas. Kohlberg believed that we progress through six stages divided into three main phases: **preconventional**, **conventional**, and **postconventional**.
- Vygotsky described development of language, culture, and skills. He proposed the idea of the **zone of proximal development**, which describes those skills that a child has not yet mastered and require a **more knowledgeable other** to accomplish.

- **Imitation** and **role-taking** are common ways children learn from others. Children first reproduce the behaviors of role models, and later learn to see the perspectives of others and practice taking on new roles.
- Our self-concept depends in part on our **reference group**, or the group to which we compare ourselves. Two individuals with the same qualities might see themselves differently depending on how those qualities compare to their reference groups.

Personality

- The **psychoanalytic** perspective views personality as resulting from unconscious urges and desires.
  - Freud's theories are based on the **id** (base urges of survival and reproduction), the **superego** (the idealist and perfectionist), and the **ego** (the mediator between the two and the conscious mind). The ego makes use of **defense mechanisms** to reduce stress caused by the urges of the id and the superego.
  - Jung assumed a **collective unconscious** that links all humans together. He viewed the personality as being influenced by **archetypes**.
  - Other psychoanalysts such as Adler and Horney have distanced themselves from Freud's theories, claiming that the unconscious is motivated by social rather than sexual urges.
- The **humanistic** perspective emphasizes the internal feelings of healthy individuals as they strive toward happiness and self-realization. Maslow's **hierarchy of needs** and Rogers's therapeutic approach of **unconditional positive regard** flow from the humanistic view of personality.
- **Type** and **trait** theorists believe that personality can be described as a number of identifiable traits that carry characteristic behaviors.
  - Type theories of personality include the ancient Greek notion of humors, Sheldon's **somatotypes**, division into **Types A** and **B**, and the **Myers–Briggs Type Inventory**.
  - The Eysencks identified three major traits which could be used to describe all individuals. The acronym for these traits is PEN: **psychoticism** (nonconformity), **extraversion** (tolerance for social interaction and stimulation), and **neuroticism** (arousal in stressful situations). Later trait theorists expanded these traits to the **Big Five**: openness, conscientiousness, extraversion, agreeableness, and neuroticism.
  - Allport identified three basic types of traits: cardinal, central, and secondary. **Cardinal traits** are the traits around which people organize their lives; not everyone develops a cardinal trait. **Central traits** represent major characteristics of the personality and **secondary traits** are more personal characteristics and are limited in occurrence.
  - McClelland identified the personality trait of the need for achievement (N-Ach).

- The **social cognitive** perspective holds that individuals interact with their environment in a cycle called **reciprocal determinism**. People mold their environments according to their personalities, and those environments in turn shape our thoughts, feelings, and behaviors.
- The **behaviorist** perspective, based on the concept of operant conditioning, holds that personality can be described as the behaviors one has learned from prior rewards and punishments.
- **Biological** theorists claim that behavior can be explained as a result of genetic expression.

ANSWERS TO CONCEPT CHECKS

6.1

1. Self-concept describes the sum of all of the phrases that come to mind when we think of who we are, who we used to be, and who we may become in the future. Identity, on the other hand, describes a set of behaviors and labels we take on when in a specific group.
2. Ethnic identity is determined by common ancestry, cultural heritage, and language, among other similarities. Rather than being determined by birth, national identity is determined by the political borders of where one lives and the cultural identity of that nation.
3. A student with an internal locus of control will look for internal factors, such as not having studied hard enough. A student with an external locus of control will blame external factors such as bad luck or the test being too difficult.

6.2

1. Freud would say that the individual has become fixated in that stage and will display the personality traits of that fixation for life. Erikson would say that the individual will still move through subsequent phases, but will be lacking the skills and virtues granted by successful resolution of that stage. Kohlberg would say that the individual was incapable of reasoning at the level of failure, and that the individual would use the reasoning described in previous stages to resolve moral dilemmas.
2.

Phase	Description
Preconventional	Reasoning is based on individual rewards and punishments
Conventional	Reasoning is based on the relationship of the individual to society
Postconventional	Reasoning is based on abstract principles

3. Zone of proximal development does not just apply to children, but rather the acquisition of new skills and abilities at any age. Standardized tests require students to utilize many skills that they may not yet have fully developed, but are in the process of development. This is why results can be improved by getting the help of those with more knowledge about the exam, such as expert instructors and authors.

6.3

1. Psychoanalytic: Personality is the result of unconscious urges and desires.

   Humanistic: Personality comes from conscious feelings about oneself resulting from healthy striving for self-realization.

   Type: Personalities are sets of distinct qualities and dispositions into which people can be grouped.

   Trait: Personalities are assembled from having different degrees of certain qualities and dispositions.

   Behaviorist: Personality is the result of behavioral responses to stimuli based on prior rewards and punishments.

   Social cognitive: Personality comes from the interactions between individuals and their environment.

   Biological: Personality is based on genetic influences and brain anatomy.
2. The id is the sum of our basic urges to reproduce and survive, while the superego is our sense of perfectionism and idealism. The ego mediates the anxieties caused by the actions of the id and superego by using defense mechanisms.
3. Psychoticism: nonconformity or social deviance

   Extraversion: tolerance for social interaction and stimulation

   Neuroticism: emotional arousal in stressful situations
4. The Big Five personality traits are openness, conscientiousness, extraversion, agreeableness, and neuroticism.

SCIENCE MASTERY ASSESSMENT EXPLANATIONS

1. **B**

Self-concept is defined as the sum of all of the ways in which we see ourselves, including who we are, as in **(D)**, who we were in the past, **(A)**, and who we may become in the future, **(C)**. The ought self, while closely related to self-esteem, is our appraisal of how others see us and is not a part of our self-concept.

2. **D**

Androgyny is defined as scoring highly on scales of both femininity and masculinity. Achieving a low score on both scales, **(A)**, would be considered undifferentiated, while **(B)** and **(C)** would be described as feminine and masculine, respectively.

3. **B**

Because there is nothing in the question stem to suggest that this situation will fundamentally change this student's attitudes in the short term, **(A)** and **(D)** can be eliminated. **(C)** is unlikely in the short term, as learned helplessness requires a repeated inability to have any effect on a situation over a long period of time and is much more severe, usually manifesting as depression. It is far more likely that the student will simply feel ineffective when it comes to math, which is low self-efficacy.

4. **A**

Because we know the lawyer has an internal locus of control, we expect belief in being in control of the events happening in life. Both **(B)** and **(C)** attribute success to outside factors, representing an external locus of control. While **(D)** perhaps represents an attribution that could correlate to low self-esteem, it is not indicative of locus of control.

5. **B**

Both excessive organization and excessive sloppiness are indicative of fixation in the anal stage of psychosexual development—what is commonly referred to as anal-retentiveness.

6. **C**

As a postadolescent young adult, this person would be described by Erikson as experiencing the conflict of intimacy *vs.* isolation, and so forming significant relationships with others would be a primary goal. **(B)** and **(D)** represent the next two stages in life (generativity *vs.* stagnation and integrity *vs.* despair, respectively), while **(A)** is the conflict that Erikson would say should have been resolved in adolescence (identity *vs.* role confusion).

7. **A**

Matt's reasoning reflects a desire to avoid punishment, which reflects stage one in Kohlberg's preconventional phase (obedience). Cati's reasoning takes into account social order, reflecting stage four in the conventional phase (law and order).

8. **D**

This situation is best described by Lev Vygotsky's zone of proximal development theory, which holds that children are often unable to perform tasks by themselves, but can complete the task with the help of a more knowledgeable other.

9. **D**

(C) has no support from role-taking research and can be eliminated. **(A)** and **(B)** are both inaccurate; the Bobo doll experiment shows young children modeling behavior not performed by their parents, and teens are most influenced by their peers, not celebrities and athletes. The research does suggest, however, that children are more likely to engage in behavior modeled by individuals who are like themselves; thus, a female child is more likely to imitate behavior by another female.

10. **C**

The superego is responsible for moral guilt when we do not live up to our ideals. While the id and the libido, **(A)** and **(D)**, may be responsible for the urge to have an affair, the superego is responsible for the anxiety one feels afterward.

11. **C**

Jung saw the drive for power and success as typically male traits, so Jung would say this woman is exercising her "inner man." The animus is the archetype that most closely reflects this quality.

12. **C**

This research supports a link between genetic expression and behavior, which is a central tenet of the biological perspective. The social cognitive perspective also holds that people's behaviors and traits shape their environments, which in turn have an effect on their identity, so the discovery also supports this perspective. Behaviorism is not supported, as the discovery is not related to rewards and punishments.

13. **B**

As a rebel and a sociable person, this individual would score highly on both psychoticism and extraversion, respectively. Neuroticism is associated with high emotional arousal in stressful situations, so being able to keep calm in an emergency is a sign of low neuroticism. Conscientiousness, a trait associated with being hardworking and organized instead of impulsive, is not described by the question stem.

14. **A**

Reaction formation is a defense mechanism that converts unwanted feelings into their exact opposite. A psychodynamic theorist would say that the terror and hatred victims feel toward their captors might be unconsciously turned into affection in an effort to reduce the stress of the situation.

15. **B**

Fictional finalism is comprised of internal, idealistic beliefs about the future. The assumption that winning the lottery will solve all of life's problems is representative of this form of thinking. Cardinal traits, **(A),** are the traits around which individuals organize their entire life. Functional autonomy, **(C)**, is when a behavior continues after the drive behind the behavior has ceased; for example, if this young professional continued purchasing lottery tickets after winning simply for enjoyment, then this behavior would have gained functional autonomy. Unconditional positive regard, **(D)**, is used in some forms of humanistic therapy in which the therapist believes in the internal good of the client and does not judge the client negatively for any words or actions.

Consult your online resources for additional practice.

SHARED CONCEPTS

Behavioral Sciences Chapter 3
Learning and Memory

Behavioral Sciences Chapter 4
Cognition, Consciousness, and Language

Behavioral Sciences Chapter 5
Motivation, Emotion, and Stress

Behavioral Sciences Chapter 7
Psychological Disorders

Behavioral Sciences Chapter 9
Social Interaction

Behavioral Sciences Chapter 10
Social Thinking

CHAPTER 7

PSYCHOLOGICAL DISORDERS

SCIENCE MASTERY ASSESSMENT

Every pre-med knows this feeling: there is so much content I have to know for the MCAT! How do I know what to do first or what's important?

While the high-yield badges throughout this book will help you identify the most important topics, this Science Mastery Assessment is another tool in your MCAT prep arsenal. This quiz (which can also be taken in your online resources) and the guidance below will help ensure that you are spending the appropriate amount of time on this chapter based on your personal strengths and weaknesses. Don't worry though—skipping something now does not mean you'll never study it. Later on in your prep, as you complete full-length tests, you'll uncover specific pieces of content that you need to review and can come back to these chapters as appropriate.

How to Use This Assessment

If you answer 0–7 questions correctly:

Spend about 1 hour to read this chapter in full and take limited notes throughout. Follow up by reviewing **all** quiz questions to ensure that you now understand how to solve each one.

If you answer 8–11 questions correctly:

Spend 20–40 minutes reviewing the quiz questions. Beginning with the questions you missed, read and take notes on the corresponding subchapters. For questions you answered correctly, ensure your thinking matches that of the explanation and you understand why each choice was correct or incorrect.

If you answer 12–15 questions correctly:

Spend less than 20 minutes reviewing all questions from the quiz. If you missed any, then include a quick read-through of the corresponding subchapters, or even just the relevant content within a subchapter, as part of your question review. For questions you answered correctly, ensure your thinking matches that of the explanation and review the Concept Summary at the end of the chapter.

1. Which of the following is an example of a negative symptom seen in schizophrenia?
   A. Auditory hallucinations
   B. Disorganized behavior
   C. Disturbance of affect
   D. Delusions

2. During an interview with a schizophrenic patient, a psychiatrist notices that the patient keeps repeating what the psychiatrist says. This phenomenon is known as:
   A. echolalia.
   B. echopraxia.
   C. loosening of associations.
   D. neologisms.

3. A 42-year-old patient has always been extremely neat and tidy. This person works as a secretary and stays long after normal working hours to check the punctuation and spelling of letters prepared during the day. The patient was referred for counseling by a supervisor after repeatedly getting into fights with coworkers. "They don't take the job to heart," the patient says. "They just joke around all day." The most likely preliminary diagnosis for this patient is:
   A. obsessive–compulsive personality disorder.
   B. antisocial personality disorder.
   C. narcissistic personality disorder.
   D. borderline personality disorder.

4. Which of the following is true with regard to a major depressive episode?
   A. It may last less than two weeks.
   B. It must involve thoughts of suicide or a suicide attempt.
   C. It may involve a decrease in sleep.
   D. It must involve feelings of sadness.

5. A young person moves to a new city to start work on an accelerated degree program. After a few months this individual visits the doctor complaining of stress and isolation and is diagnosed with depression. The doctor ascribes the cause to low levels of serotonin and prescribes an SSRI for treatment. Which of the following best describes the physician's approach to diagnosis and treatment?
   A. Biomedical approach with indirect therapy
   B. Biomedical approach with direct therapy
   C. Biopsychosocial approach with indirect therapy
   D. Biopsychosocial approach with direct therapy

6. A young person of unknown age is brought by the Philadelphia police to the local emergency department for evaluation after being found wandering in a park. The patient carries no identification and is unable to provide a first or last name or any life details, except that the name Phoenix seems familiar. The police in Arizona are contacted and find a missing persons report matching the patient's description. Based on this information, the most likely diagnosis for this patient is:
   A. depersonalization/derealization disorder.
   B. dissociative identity disorder.
   C. somatic symptom disorder.
   D. dissociative amnesia with dissociative fugue.

7. In addition to being a freestanding diagnosis, agoraphobia is most often seen in association with which other psychiatric diagnosis?
   A. Obsessive–compulsive disorder
   B. Avoidant personality disorder
   C. Generalized anxiety disorder
   D. Panic disorder

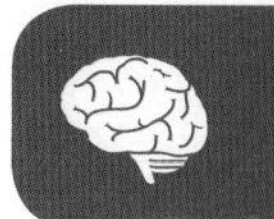

8. A researcher is interested in studying fear responses to a variety of different stimuli. To rule out potential confounding variables, the research excludes individuals with phobias from the study. In a sample of 50 participants, how many will likely be excluded from the study, assuming that the sample is representative of the overall population?
   - **A.** 45
   - **B.** 15
   - **C.** 10
   - **D.** 5

Questions 9–10 refer to the scenario described below.

A physician is attempting to diagnose a patient's mental disorder based on a set of symptoms. The confirmed symptoms currently include appetite disturbance, substantial weight change, decreased energy, a feeling of worthlessness, and excessive guilt.

9. What two disorders could these symptoms indicate?
   - **A.** Major depressive and bipolar disorders
   - **B.** Dissociative amnesia and depersonalization/derealization disorder
   - **C.** Alzheimer's disease and Parkinson's disease
   - **D.** Specific phobia and panic disorder

10. What should the physician ask about to distinguish between the two possible disorders affecting that patient?
   - **A.** Whether the patient has amnesia
   - **B.** Whether the patient has also had manic episodes
   - **C.** Whether the patient is irrationally afraid of anything
   - **D.** Whether the patient has experienced difficulty performing familiar tasks

11. All of the following are risk factors or diagnostic criteria for Alzheimer's disease EXCEPT:
   - **A.** Extra copies of the β-amyloid precursor protein gene.
   - **B.** Decreased abundance of serotonin and norepinephrine.
   - **C.** Neurofibrillary tangles of hyperphosphorylated tau protein.
   - **D.** Deficient blood flow to the parietal lobes.

12. Splitting is a defense mechanism commonly seen with which personality disorder?
   - **A.** Antisocial personality disorder
   - **B.** Borderline personality disorder
   - **C.** Histrionic personality disorder
   - **D.** Narcissistic personality disorder

13. A patient comes to the doctor with a two-week history of complete paralysis of the left arm. The patient has had no injury to the extremity, and full neurological workup fails to demonstrate any underlying cause. The patient seems surprisingly unconcerned about the paralysis, and seems more worried about an argument that happened one month ago in which the patient hit a family member. Based on this information, the patient's most likely diagnosis is:
   - **A.** conversion disorder.
   - **B.** generalized anxiety disorder.
   - **C.** illness anxiety disorder.
   - **D.** histrionic personality disorder.

14. A patient is taken to the doctor after starting to move the fingers in such a way that it looks like rolling something, despite nothing actually being there. The patient also exhibits slowed movement and a shuffling gait. Which neurotransmitter is likely to be present in decreased levels in the patient's brain?
   - **A.** Epinephrine
   - **B.** Histamine
   - **C.** Dopamine
   - **D.** Serotonin

15. Which of the following is/are true regarding bipolar disorders?
   - **I.** They have little, if any, genetic heritability.
   - **II.** They are associated with increased levels of serotonin in the brain.
   - **III.** They all require at least one depressive episode for diagnosis.

   - **A.** I only
   - **B.** II only
   - **C.** I and III only
   - **D.** II and III only

Answer Key

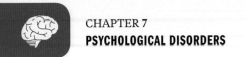

1. **C**
2. **A**
3. **A**
4. **C**
5. **B**
6. **D**
7. **D**
8. **D**
9. **A**
10. **B**
11. **B**
12. **B**
13. **A**
14. **C**
15. **B**

Detailed explanations can be found at the end of the chapter.

CHAPTER 7

PSYCHOLOGICAL DISORDERS

In This Chapter

CHAPTER PROFILE

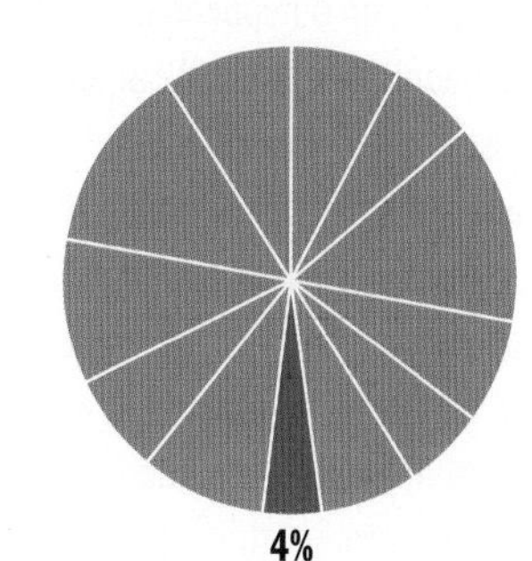

The content in this chapter should be relevant to about 4% of all questions about the behavioral sciences on the MCAT.

This chapter covers material from the following AAMC content category:

7A: Individual influences on behavior

Introduction

The progress in our understanding of hysteria has come largely through the elaboration of the so-called mechanisms by which the symptoms arise. These mechanisms have been declared to reside or to have their origin in the subconsciousness or coconsciousness. The mechanisms range all the way from the conception of Janet that the personality is disintegrated owing to lowering of the psychical tension to that of Freud, who conceives all hysterical symptoms as a result of dissociation arising through conflicts between repressed sexual desires and experiences and the various censors organized by the social life.

The above is an excerpt from the *Journal of Abnormal Psychology* in 1915. Merely a century ago, our understanding of psychological disorders was in its infancy. Hysteria—the antiquated name for conversion disorder—was thought to result from marital discord and repressed sexual desires. We are now beginning to understand the underlying psychological and biological factors at play in a number of mental illnesses. In this chapter, we will focus on several different types of psychological disorders, their classification, causes, and frequencies.

MCAT EXPERTISE

If you've taken a glance at the rest of this book, you might have noticed that this chapter has a low percentage of relevant content as compared to the rest of the chapters within *MCAT Behavioral Sciences Review*. While Chapter 7 materials will be tested less often than the other materials in this book, there are so many psychology questions on the MCAT that even lower-yield topics within the behavioral sciences will get you points!

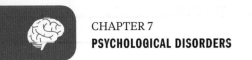

7.1 Understanding Psychological Disorders

LEARNING OBJECTIVES

After Chapter 7.1, you will be able to:

- Compare and contrast the biomedical and the biopsychosocial models of psychological disorders
- Recall the most common psychological disorders in the United States

REAL WORLD

Thomas Szasz, an outspoken critic of labeling people "mentally ill," argues that most of the disorders treated by clinicians are not really illnesses. Rather, they are traits or behaviors that differ from the cultural norm. Szasz argues that labeling people as mentally ill is a way to force them to change and conform to societal norms rather than allowing them to attack the societal causes of their problems.

Psychological disorders are characteristic sets of thoughts, feelings, or actions that cause noticeable distress to the sufferer, are considered deviant by the individual's culture, or cause **maladaptive** functioning in society, meaning that some aspect of the individual's behavior negatively impacts others or leads to self-defeating outcomes. Many disorders can be treated once diagnosed. The process of defining these disorders varies, and there are two main classification systems you'll need to know for the MCAT.

Biomedical *vs.* Biopsychosocial Approaches

The first classification system to know for the MCAT is the **biomedical approach** to psychological disorders. Biomedical therapy emphasizes interventions that rally around symptom reduction of psychological disorders. In other words, this approach assumes that any disorder has roots in biomedical disturbances, and thus the solution should also be of a biomedical nature. This view is thought of as narrower than other approaches because it fails to take into account many of the other sources of disorders, such as lifestyle and socioeconomic status. For example, heart disease clearly has roots within the mechanisms of the cardiac muscle, but the causes of these malfunctions have as much to do with biomedical causes (such as genetics) as they do with lifestyle causes (such as a diet rich in salty, fatty foods; smoking; and alcohol use). Similarly, this biomedical approach can miss some underlying sources of psychological disorders and is often more effective when supplemented with a broader approach to diagnosis and treatment.

MCAT EXPERTISE

The biopsychosocial model was originally theorized in a 1977 *Science* article and has grown in breadth, depth, and applicability since it was first described. In fact, the increased recognition of psychological and social factors on patient care was one of the primary drivers for the creation of the *Psychological, Social, and Biological Foundations of Behavior* section of the MCAT!

A broader classification system commonly used for these psychological disorders is the **biopsychosocial approach**. This method assumes that there are biological, psychological, and social components to an individual's disorder. The biological component of a disorder is something in the body, like having a particular genetic syndrome. The psychological component of a disorder stems from the individual's thoughts, emotions, or behaviors. Finally, the disorder's social component results from the individual's surroundings and can include issues of perceived class in society and even discrimination or stigmatization. All three of these aspects of a disorder are considered in the biopsychosocial approach for both diagnosis and treatment.

To better understand the biopsychosocial approach, consider depression. Certain genetic factors can make an individual more or less susceptible to depressive tendencies, showing a purely biological influence on the disorder. However, from a psychological perspective, the levels of stress that the individual experiences can also contribute to the severity of the depression experienced. Finally, the social environment may provide additional stressors or support from one's career, family, and friends. Accordingly, in the biopsychosocial model, the goal is often to provide not only **direct therapy**—treatment that acts directly on the individual, such as medication or periodic meetings with a psychologist—but also **indirect therapy**, which aims to increase social support by educating and empowering family and friends of the affected individual.

Classifying Psychological Disorders

To aid clinicians in considering these factors, the ***Diagnostic and Statistical Manual of Mental Disorders*** (**DSM**) was created. Originally, the manual was written to collect statistical data in the United States. It is now used as a diagnostic tool in the United States and various other countries. The manual is currently in its fifth edition, which was published in May 2013, so the common abbreviation seen is DSM-5. This manual is a compilation of many known psychological disorders. The DSM-5's classification scheme is not based on theories of etiology (cause) or treatments of different disorders. Rather, it is based on descriptions of symptoms. It is used by clinicians to fit lists of compiled symptoms from a patient into a category and thus to diagnose that patient. The DSM-5 has 20 diagnostic classes of mental disorders; those that will be tested on the MCAT are discussed in this chapter.

Rates of Psychological Disorders

Suffering from a mental disorder can be a lonely experience because the disorder usually occurs only in the mind of the patient. However, the rates of these psychological disorders are higher than this experience would otherwise suggest. Table 7.1 covers these rates in detail.

REAL WORLD

David Rosenhan studied whether it was possible to be judged sane if you are in an "insane place" (a psychiatric hospital). Rosenhan and seven other "sane" people were admitted into psychiatric hospitals by reporting auditory hallucinations. Each of these pseudopatients was diagnosed to have either schizophrenia or bipolar disorder, and each was admitted. Once admitted, they acted completely normal—but it still took an average of three weeks to be discharged, and each was still given the diagnosis of schizophrenia in remission. Once labeled, it is very hard to distance oneself from the diagnosis of mental illness.

MCAT EXPERTISE

The MCAT tests many, but not all, categories of mental disorder described within the DSM-5. Neurodevelopmental disorders, eating disorders, impulse control disorders, sleeping disorders, and others are not listed within the AAMC's guide to MCAT content, and as such are not included within this text.

DISORDER	PERCENTAGE AFFECTED	NUMBER AFFECTED (IN MILLIONS)
Any mental disorder	18.3	44.7
Specific phobia	9.1	22.2
Social anxiety disorder	7.1	17.3
Major depressive disorder	6.7	16.4
Posttraumatic stress disorder	3.6	8.8
Bipolar disorder	2.8	6.8
Generalized anxiety disorder	2.7	6.6
Panic disorder	2.7	6.6
Borderline personality disorder	1.4	3.4
Obsessive-compulsive disorder	1.2	2.9
Agoraphobia	0.9	2.2
Anorexia nervosa	0.6	1.5
Schizophrenia	0.6	1.5
All cancers*	6.1	15.1
Diabetes*	9.4	23.0

*Note: These nonpsychological conditions are included for comparison.

All data from this website: https://www.nimh.nih.gov/health/statistics/index.shtml

Table 7.1 One-Year Prevalence Rates for Psychological Disorders in the United States

MCAT CONCEPT CHECK 7.1

Before you move on, assess your understanding of the material with these questions.

1. What is the difference between the biomedical and biopsychosocial models of psychological disorders?

2. Name three psychological disorders with greater than 2% one-year prevalence in the United States (affecting more than 1 in 50 people per year). Refer to Table 7.1 if you get stuck.

 •

 •

 •

7.2 Types of Psychological Disorders

High-Yield

LEARNING OBJECTIVES

After Chapter 7.2, you will be able to:

- List the major positive symptoms and major negative symptoms of schizophrenia and psychotic disorders
- Recall the features of major depressive episodes, manic and hypomanic episodes
- Distinguish between the testable mood disorders
- Relate obsessions and compulsions to the symptoms of obsessive–compulsive disorder
- Describe and explain the symptoms of posttraumatic stress disorder (PTSD)
- Describe and distinguish dissociative and somatic symptom disorders
- Describe the features and individual disorders that fall under cluster A, B, and C personality disorders

As mentioned earlier, the DSM-5 categorizes common symptoms into 20 diagnostic classes. Many of these classes represent significant revisions from the DSM-5's immediate predecessor, the DSM-IV-TR. The most heavily tested diagnostic classes on the MCAT are schizophrenia spectrum and other psychotic disorders, bipolar and related disorders, depressive disorders, anxiety disorders, obsessive–compulsive and related disorders, trauma- and stressor-related disorders, dissociative disorders, somatic symptom and related disorders, and personality disorders.

Schizophrenia and Other Psychotic Disorders

According to the DSM-5, individuals with a **psychotic disorder** present with one or more of the following symptoms: delusions, hallucinations, disorganized thought, disorganized behavior, catatonia, and negative symptoms. Like most psychological categories, psychotic disorders are on a spectrum. To delineate the psychotic disorders as described in the DSM-5, psychotic symptoms must be understood.

REAL WORLD

The term schizophrenia is a relatively recent term, coined in 1911 by Eugen Bleuler. Before Bleuler, schizophrenia was called *dementia praecox*. Schizophrenia literally means "split mind" because the disorder is characterized by distortions of reality and disturbances in the content and form of thought, perception, and affect. Unfortunately, this has led to confusion with dissociative identity disorder (formerly multiple personality disorder). By *split mind*, Bleuler did not mean that the mind is split into different personalities, but that the mind is split from reality.

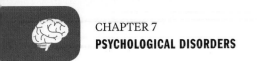

Psychotic symptoms are divided into positive and negative types. **Positive symptoms** are behaviors, thoughts, or feelings added to normal behavior. In other words, positive symptoms are features that are experienced in individuals with psychotic disorders that are not present in the general population. Examples include delusions and hallucinations, disorganized thought, and disorganized or catatonic behavior. Positive symptoms are considered by some to be two distinct dimensions—the psychotic dimension (delusions and hallucinations) and the disorganized dimension (disorganized thought and behavior)—perhaps with different underlying causes. **Negative symptoms** are those that involve the absence of normal or desired behavior, such as disturbance of affect and avolition.

Positive Symptoms

MCAT EXPERTISE

The fact that delusions must be considered deviant from the society in which an individual lives provides an excellent opportunity for the MCAT to integrate mental illness and sociology. For example, a belief in shamanism–which is common in the Caribbean, Central and South America, Africa, and in some American Indian tribes–would not be considered a delusion within societies that endorse shamanic medicine.

Delusions are false beliefs discordant with reality and not shared by others in the individual's culture. These delusions are maintained often in spite of strong evidence to the contrary. Common delusions include delusions of reference, persecution, and grandeur. **Delusions of reference** involve the belief that common elements in the environment are directed toward the individual. For example, people with delusions of reference may believe that characters in a TV show are talking to them directly. **Delusions of persecution** involve the belief that the person is being deliberately interfered with, discriminated against, plotted against, or threatened. **Delusions of grandeur**, also common in bipolar I disorder, involve the belief that the person is remarkable in some significant way, such as being an inventor, historical figure, or religious icon. Other common delusions involve the concept of **thought broadcasting**, which is the belief that one's thoughts are broadcast directly from one's head to the external world, **thought withdrawal**, the belief that thoughts are being removed from one's head, and **thought insertion**, the belief that thoughts are being placed in one's head.

Hallucinations are perceptions that are not due to external stimuli but which nevertheless seem real to the person perceiving them. The most common form of hallucination is auditory, involving voices that are perceived as coming from inside or outside the patient's head. Visual and tactile hallucinations are less common, but may be seen in drug use or withdrawal.

BRIDGE

Word salad can be seen in severe schizophrenia as well as Wernicke's (receptive) aphasia. Patients will string together unrelated words, although the prosody of the speech (its rhythm, stress, and intonation) remains intact. Aphasias are discussed in Chapter 3 of *MCAT Behavioral Sciences Review*.

Disorganized thought is characterized by loosening of associations. This may be exhibited as speech in which ideas shift from one subject to another in such a way that a listener would be unable to follow the train of thought. A patient's speech may be so disorganized that it seems to have no structure—as though it were just words thrown together incomprehensibly. This is sometimes called **word salad**. In fact, a person with schizophrenia may even invent new words, called **neologisms**.

Disorganized behavior refers to an inability to carry out activities of daily living, such as paying bills, maintaining hygiene, and keeping appointments. **Catatonia** refers to certain motor behaviors characteristic of some people with schizophrenia. The patient's spontaneous movement and activity may be greatly reduced or the patient may maintain a rigid posture, refusing to be moved. At the other extreme, catatonic behavior may include useless and bizarre movements not caused by any external stimuli, **echolalia** (repeating another's words), or **echopraxia** (imitating another's actions).

Negative Symptoms

The classic negative symptoms of schizophrenia and related psychotic disorders are disturbance of affect and avolition. **Affect** refers to the experience and display of emotion, so disturbance of affect is any disruption to these abilities. Affective symptoms may include **blunting**, in which there is a severe reduction in the intensity of affect expression; **emotional flattening** (**flat affect**), in which there are virtually no signs of emotional expression; or **inappropriate affect**, in which the affect is clearly discordant with the content of the individual's speech. For example, a patient with inappropriate affect may begin to laugh hysterically while describing a parent's death. Interestingly, it has become more difficult to assess the affective aspects of schizophrenia because the antipsychotic medications used in treatment frequently blunt and flatten affect as well. Finally, **avolition** is marked by decreased engagement in purposeful, goal-directed actions.

Schizophrenia

Schizophrenia is the prototypical psychotic disorder in this category of disorders. **Schizophrenia** is characterized by a break between an individual and reality. In fact, the term *schizophrenia* literally means "split mind." Eugen Bleuler coined the term in reference to the splitting of one's mind from reality. To be given the diagnosis of schizophrenia, an individual must show continuous signs of the disturbance for at least six months, and this six-month period must include at least one month of positive symptoms (delusions, hallucinations, or disorganized speech).

MCAT EXPERTISE

When the MCAT tests schizophrenia, it is likely to include a connection to sociology through the *downward drift hypothesis*, which states that schizophrenia causes a decline in socioeconomic status, leading to worsening symptoms, which sets up a negative spiral for the patient toward poverty and psychosis. This is why rates of schizophrenia are much, much higher among homeless and indigent people.

Phases of Schizophrenia

The diagnosis and course of schizophrenia typically follows a specific path, termed the phases of schizophrenia. Before schizophrenia is diagnosed, a patient often goes through a phase characterized by poor adjustment. This phase is called the **prodromal phase**. The prodromal phase is exemplified by clear evidence of deterioration, social withdrawal, role functioning impairment, peculiar behavior, inappropriate affect, and unusual experiences. This is followed by the **active phase** in which pronounced psychotic symptoms are displayed. If schizophrenia development is slow, correct diagnosis is difficult and the prognosis is especially poor. If the onset of symptoms is intense and sudden, the diagnosis is readily made and the prognosis is better. Diagnosis usually occurs during the active phase. The **residual phase,** also called the **recovery phase,** occurs after an active episode and is characterized by mental clarity often resulting in concern or depression as the individual becomes aware of previous behavior.

Other Psychotic Disorders

Other psychotic disorders differ from schizophrenia by the presence, severity, and duration of psychotic symptoms. As a general trend, the *other psychotic disorders* present symptoms to a lesser degree in comparison to schizophrenia.

- **Schizotypal Personality Disorder**: Include both personality disorder and psychotic symptoms, with the personality symptoms having been already established before psychotic symptoms present. This condition is covered in greater detail in *Personality Disorders.*

- **Delusion Disorder**: Psychotic symptoms are limited to delusions and are present for at least a month.
- **Brief Psychotic Disorder**: Positive psychotic symptoms are present for at least a day, but less than a month.
- **Schizophreniform Disorder**: Same diagnostic criteria as schizophrenia except in duration; the required duration for this diagnosis is only 1 month.
- **Schizoaffective Disorder**: Major mood episodes (major depressive episodes and manic episodes) while also presenting psychotic symptoms.

Depressive Disorders

Sadness is a natural part of life, especially in response to stressful life events like the death of a loved one. During periods of sadness, people might call themselves depressed. However, periodic sadness in response to life events is not a mental disorder. **Depressive disorders**, in contrast, are conditions characterized by feelings of sadness that are severe enough, in both magnitude and duration, to meet specific diagnostic criteria.

To understand the DSM-5's categorization of the spectrum of depressive disorders, we must first discuss the 9 **depressive symptoms** defined in the DSM-5. These symptoms can be recalled with the mnemonic **sadness + SIG E. CAPS**:

- **Sadness**: Depressed mood, feelings of sadness and emptiness
- Sleep: Insomnia or hypersomnia
- Interest: Loss of interest and pleasure in activities that previously sparked joy, termed **anhedonia**
- Guilt: A feeling of inappropriate guilt or worthlessness
- Energy: Lower levels of energy throughout the day
- Concentration: Decrease in ability to concentrate (self described, or observed by others)
- Appetite: Pronounced change in appetite (increase or decrease) resulting in a significant change (5%+) in weight.
- Psychomotor symptoms: Psychomotor retardation (slowed thoughts and physical movements) and psychomotor agitation (restlessness resulting in undesired movement)
- Suicidal thoughts: Recurrent suicidal thoughts

In addition to depressive symptoms, the DSM-5 also categorizes depressive disorders based on duration, timing, and cause of depressive symptoms.

Major Depressive Disorder

The key diagnostic feature of **major depressive disorder (MDD)** is the presence of major depressive episodes. A **major depressive episode** is defined as a 2-week (or longer) period in which 5 of the 9 defined depressive symptoms are encountered, which must include either depressed mood or **anhedonia** (inability to feel and anticipate pleasure). In addition, the symptoms must be severe enough to impair one's daily social- or work-related activities.

MNEMONIC

Symptoms of a major depressive episode:
SIG E. CAPS

Sadness +

- **S**leep
- **I**nterest
- **G**uilt
- **E**nergy
- **C**oncentration
- **A**ppetite
- **P**sychomotor symptoms
- **S**uicidal thoughts

Persistent Depressive Disorder

Considering the difference in naming between major depressive disorder and persistent depressive disorder, it may seem reasonable to assume that persistent depressive disorder is a lesser form of major depressive disorder. However, this is not the case. In fact, major depressive episodes can coincide with persistent depressive disorder. A diagnosis of **persistent depressive disorder (PDD)**, also known as **dysthymia**, is given when an individual experiences a period, lasting at least 2 years, in which they experience a depressed mood on the majority of days. With the primary diagnostic feature of PDD being time, a patient can receive both the PDD and MDD diagnosis if they meet both the duration and severity requirements of both disorders.

BRIDGE

The most common first-line treatment for depression is the class of medications called selective serotonin reuptake inhibitors (SSRIs). These block the reuptake of serotonin by the presynaptic neuron, resulting in higher levels of serotonin in the synapse and relief of symptoms. The nervous system is outlined in Chapter 1 of *MCAT Behavioral Sciences Review* and Chapter 4 of *MCAT Biology Review*.

Other Depressive Disorders

Whereas major depressive disorder and persistent depressive disorder are characterized by severity and duration of depressive symptoms, other depressive disorders can be characterized by their age of incidence and apparent cause.

Children often exhibit more dramatic emotional responses than adults and, in previous editions of the DSM, this led to the potential overdiagnosis of bipolar disorders in children. To address this concern the DSM-5 includes **disruptive mood dysregulation disorder**, which is typically diagnosed between the ages of 6 and 10, and has the key diagnostic feature of persistent and recurrent emotional irritability in multiple environments (school, home, etc.).

Depressive symptoms can also arise in response to specific times and situations; if these symptoms meet certain diagnostic criteria then they are considered disorders. **Premenstrual dysphoric disorder** is characterized by mood changes, often depressed mood, occurring a few days before menses and resolving after menses onset.

Although not freestanding diagnoses in the DSM-5, both seasonal affective disorder and postpartum depression are conditions that have an apparent cause. In **seasonal affective disorder (SAD)**, the dark winter months are believed to be the source of depressive symptoms and thus the disorder is best categorized as major depressive disorder with seasonal onset, while in **postpartum depression** the rapid change in hormone levels just after giving birth is the cause of the depressive symptoms. In the case of seasonal affective disorder, depressive symptoms are present only in the winter months. This disorder may be related to abnormal melatonin metabolism; it is often treated with **bright light therapy**, where the patient is exposed to a bright light for a specified amount of time each day, as demonstrated with a plant in Figure 7.1.

Figure 7.1 Bright Light Therapy for Seasonal Affective Disorder

Bipolar and Related Disorders

This category of disorders is characterized by the presence of manic and depressive symptoms, which if severe and persistent enough can be labelled as episodes. **Manic symptoms** are associated with an exaggerated elevation in mood, accompanied by an increase in goal-directed activity and energy. Put simply, manic symptoms can be thought of as the prolonged and exaggerated emotion of happiness or joy. According to the DSM-5, there are 7 manic symptoms. These symptoms can be recalled with the mnemonic **DIG FAST**:

- **D**istractibility: Inability to remain focused on an activity
- **I**rresponsibility: Engaging in risky activities without considering future consequences
- **G**randiosity: Exaggerated and unrealistic increase in self-esteem
- **F**light of thoughts: Racing thoughts, self-reported or revealed through rapid speech
- **A**ctivity or agitation: Increase in goal-oriented work or social activities
- **S**leep: Decreased need for sleep, e.g. sleeping for only a couple hours but feeling rested
- **T**alkative: Exaggerated desire to speak

The presence of manic symptoms are considered a **hypomanic episode** if the symptoms are present for at least 4 days and include at least 3 or more of the 7 defined manic symptoms, yet the symptoms are not severe enough to impair the person's social or work activities. However, the diagnosis progresses to a **manic episode** if the manic symptoms (3 or more of the defined 7) are severe enough to impair a person's social or work activities and persist for at least 7 days.

In addition to manic symptoms and their associated episodes, the presence or absence of depressive symptoms and their associated episodes are also used to differentiate bipolar and related disorders. Specifically, these disorders are classified by the presence or absence of manic, hypomanic, and major depressive episodes. Depressive symptoms were covered in *Depressive Disorders*.

Bipolar I Disorder

When manic episodes are present, a diagnosis of **bipolar I disorder** is likely to be made, as the key diagnostic feature of this disorder is the presence of manic episodes. While most diagnoses of bipolar I disorder also include depressive symptoms, often major depressive episodes, they are not a requirement. To illustrate this point consider two hypothetical patients: Patient A only experiences manic episodes, while Patient B regularly experiences both manic and major depressive episodes, cycling between the two regularly. Despite both patients presenting very differently, both fit the categorization of bipolar I disorder.

Bipolar II Disorder

The key feature of a **bipolar II disorder** diagnosis is the presence of both a major depressive episode and an accompanying hypomanic episode, but not a manic episode. To avoid confusion, it is worth noting that if a patient has experienced both major depressive episodes and manic episodes, a diagnosis of bipolar I disorder will likely be made. In addition, if a person experiences only major depressive symptoms (absence of hypomanic and manic episodes), then a diagnosis of major depressive disorder is likely to be made. Thus, the diagnosis of bipolar II only captures individuals who experience major depressive episodes and the lesser, hypomanic episodes.

Cyclothymic Disorder

The diagnostic features of cyclothymic disorder are the presence of both manic and depressive symptoms that are not severe enough to be considered episodes. In other words, the patient has not experienced major depressive, manic, or hypomanic episodes. Or, more specifically, the patient has never experienced *3 or more of the 7 manic symptoms in a 4 day period* (diagnostic criteria for hypomanic episode) and has never experienced *5 or more of the 9 depressive symptoms in a 2-week period* (diagnostic criteria for a major depressive episode). Considering the relatively low threshold of symptom requirements, it may seem that everyone would be diagnosed with cyclothymic disorder. However despite the relatively low symptom requirements, the duration requirements for this disorder are high. For a diagnosis of **cyclothymic disorder** to be made, a person must have experienced numerous periods of manic and depressive symptoms for the majority of time over a 2-year (or longer) period.

Before moving on to anxiety disorders, a brief discussion of proposed neurological etiologies of mood disorders is warranted. The most common explanation revolves around the neurotransmitters norepinephrine and serotonin. These two are often linked together into what is called the **monoamine** or **catecholamine theory of depression**. This theory holds that too much norepinephrine and serotonin in the

REAL WORLD

Depressive and manic episodes are essentially two sides of the same coin: Depression is associated with low norepinephrine and serotonin levels, and manic episodes are associated with high levels of these neurotransmitters. When patients are put on treatment for depression, they must be watched for signs of mania because antidepressant medications may trigger manic symptoms or episodes.

synapse leads to mania, while too little leads to depression. Although more recent research has shown that it is not that simple, you should be aware of this theory for the MCAT.

Anxiety Disorders

BRIDGE

For all anxiety disorders, clinicians must rule out hyperthyroidism—excessive levels of the thyroid hormones triiodothyronine (T_3) and thyroxine (T_4)—because increasing the whole body's metabolic rate will create anxiety-like symptoms. Thyroid function is discussed in Chapter 5 of *MCAT Biology Review.*

From an evolutionary perspective, emotions served to direct and modulate behavior based on environmental stimuli. As seen in bipolar and depressive disorders, when the regulation of emotions, such as happiness or sadness, are insufficient, then symptoms arise. In the case of anxiety, fear is the associated emotion. Fear is often defined as an emotional response to an immediate threat, while **anxiety** can be viewed as fear of an upcoming or future event. Like fear, anxiety is healthy and important in one's life. It is only considered an **anxiety disorder** when irrational and excessive fear or anxiety affects an individual's daily functioning.

There are more than 10 disorders listed in the anxiety disorders portion of the DSM-5. These disorders are categorized by the situation or stimulus that induces anxiety.

Specific Phobias

The most common type of anxiety disorder is a phobia. A **phobia** is an irrational fear of something that results in a compelling desire to avoid it. Most of the phobias that you are likely familiar with are what the DSM-5 calls specific phobias. A **specific phobia** is one in which fear and anxiety are produced by a specific object or situation. Unlike other sources of anxiety, specific phobias lack a specific ideation or thought pattern and instead present as an immediate and irrational fear response to the specific object or situation. For example, *claustrophobia* is an irrational fear of closed places, *acrophobia* is an irrational fear of heights, and *arachnophobia* is an irrational fear of spiders.

Figure 7.2 Specific Phobia
Arachnophobia, the fear of spiders, is a common example of a specific phobia.

Separation Anxiety Disorder

Separation anxiety is the excessive fear of being separated from one's caregivers or home environment. Although some separation anxiety is common and to be expected in young children, when this anxiety is excessive and persists beyond the age where it is deemed developmentally appropriate, the person may be diagnosed with **separation anxiety disorder**. This diagnosis is accompanied by the ideation that when separated, the caregiver or the individual themselves will be harmed (e.g. kidnapping, getting sick). These persistent beliefs may result in avoidant behaviors such as refusal to leave the home, shadowing the caregiver, etc.

Social Anxiety Disorder

On the surface, social anxiety disorder can be viewed as a social phobia, that is, fear and anxiety towards social situations and encounters. However, unlike previously discussed specified phobias, social anxiety disorder has an accompanying ideation in which individuals think that they will be perceived negatively by others. Thus, the key diagnostic feature of **social anxiety disorder** is fear or anxiety towards social situations with the belief that the individual will be exposed, embarrassed, or simply negatively perceived by others.

Like other anxiety disorders, avoidant behaviors are often conditioned as a means to reduce the associated anxiety. In the case of social anxiety, this can be as broad as avoiding social situations entirely or as narrow as avoiding handshakes out of fear of sweaty palms. Avoidant behavior to the point of social or occupational impairment is necessary for a social anxiety disorder diagnosis.

Selective Mutism

Although categorized as a separate anxiety disorder, **selective mutism** is heavily associated with social anxiety disorder and characterized by the consistent inability to speak in situations where speaking is expected. However, in situations that are more relaxed or when communication is not expected, speaking is unaffected. From this perspective, selective mutism may be conceptualized as a patient's fear of being negatively evaluated for what the patient might say.

Panic Disorder

The key diagnostic feature of **panic disorder** is the recurrence of unexpected panic attacks. To understand panic disorder, we must first cover panic attacks. From a physiological perspective, a panic attack is the misfiring of the sympathetic nervous system resulting in an unwanted *fight or flight* response. From a psychological perspective, a panic attack includes the associated emotions that accompany the sympathetic response, such as intense fear and *a sense of impending doom/danger*. Combining these two perspectives, a **panic attack** is the sudden surge of fear in which individuals feel that they are losing control of their body and/or that they are dying. The occurrence of an individual's attacks may be associated with specific triggers, in which case the attacks are termed **expected panic attacks**. If there is no clear trigger and the panic attacks are seemingly random, they are termed **unexpected panic attacks**.

BRIDGE

Notice that a large number of the symptoms of panic disorder are caused by excess activation of the sympathetic nervous system (autonomic overdrive). These include trembling, sweating, hyperventilation, shortness of breath, a racing heart rate, and palpitations. The autonomic nervous system is discussed in Chapter 1 of *MCAT Behavioral Sciences Review* and Chapter 4 of *MCAT Biology Review*.

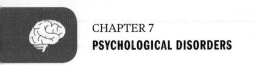

The diagnosis of panic disorder requires the recurrence of *unexpected* panic attacks. The unexpected panic attacks themselves can produce an associated anxiety. In other words, an individual may become anxious at the thought of having an unexpected panic attack. If this anxiety impairs one's daily functions and persists for at least a month, then the diagnosis of panic disorder is made.

It is worth noting that panic attacks themselves are not considered a psychological disorder. They may occur in the absence of physiological disorders or may be associated with anxiety disorders in which there is a clear trigger. For instance, an individual with arachnophobia may experience expected panic attacks when encountering a spider.

Agoraphobia

Agoraphobia is an anxiety disorder characterized by a fear of being in places or situations where it might be difficult for an individual to escape. This fear may stem from the thought that the individual may experience a panic attack or similar event in which they would need to escape to avoid embarrassment. People who have agoraphobia tend to be uncomfortable leaving their homes, using public transport, being in open spaces, waiting in lines, or simply being in crowds. Due to agoraphobia's association with panic attacks and fear of being negatively evaluated by others, it is often comorbid with panic disorder, social anxiety disorder, and specific phobias.

Generalized Anxiety Disorder

As previously mentioned, the DSM-5 categorizes anxiety disorders based on the stimulus that induces fear or anxiety. For instance, anxiety towards social interactions is termed social anxiety; anxiety at the thought of separation from one's caregivers is termed separation anxiety. On the other hand, some individuals have more anxious temperaments, making them susceptible to anxiety triggered by a multitude of stimuli. It is for this reason that specific anxiety disorders are often comorbid with one another, resulting in patients having multiple diagnoses. However, a better diagnosis for some patients with many triggers for anxiety might be generalized anxiety disorder.

Generalized anxiety disorder (GAD) is defined as a disproportionate and persistent worry about many different things—making mortgage payments, doing a good job at work, returning emails, political issues, and so on—for at least six months. In addition, the worrying is difficult to control, even in cases where the individual knows that their worrying and fear is irrational. These individuals often have physical symptoms like fatigue, muscle tension, and sleep problems that accompany the worry. General anxiety disorder is relatively common in the US population, with approximately 3% of the population experiencing GAD in a 12-month period. Furthermore, over the course of a lifetime, individuals have a 1 in 10 chance of meeting the diagnostic criteria for general anxiety disorder.

Obsessive–Compulsive and Related Disorders

Formerly classified under anxiety and somatic symptom disorders, the disorders in this group were relabeled as obsessive–compulsive and related disorders in the DSM-5.

The reason for this organizational change reveals the common feature among these conditions. Across all of the following disorders, individuals perceive a particular need and respond to the need by completing a particular action. Disorders in this category are differentiated by the compulsiveness of the need to be met as well as the nature of the action.

Obsessive-Compulsive Disorder

Obsessive–compulsive disorder (OCD) is characterized by obsessions (persistent, intrusive thoughts and impulses), which produce tension, and compulsions (repetitive tasks) that relieve tension but cause significant impairment in a person's life. The relationship between the two is key: obsessions raise the individual's stress level, and the compulsions relieve this stress. Obsessions and their compulsions are ego-dystonic, meaning that the individual knows that their behavior is irrational, but the anxiety that arises when compulsions are not performed cannot be ignored.

Obsessions are perceived needs with the accompanying ideation that if a particular need is not met, then disastrous events will follow. Actions paired with obsessions are termed **compulsions**. As individuals with OCD attempt to satisfy their obsessions, rituals or sets of rules are developed for how their compulsions must be performed. For example, individuals may need to wash their hands for a specific length of time or else the intrusive thought of getting sick occurs. Alternatively, individuals may need to check if their door is locked a specific number of times or else worry obsessively about getting robbed. To be diagnosed with OCD, the compulsions must impair one's daily activities, for instance by taking up a lot of time during the day.

Body Dysmorphic Disorder

In **body dysmorphic disorder**, a person has an unrealistic negative evaluation of personal appearance and attractiveness, usually directed toward a certain body part. This is known as a **preoccupation**, a type of worry which lacks the disastrous ideation that accompanies obsessions. Patients with this disorder see their nose, skin, or stomach as ugly or even horrific when actually ordinary in appearance. This body–focused preoccupation also disrupts day-to-day life, and the sufferer may seek multiple plastic surgeries or other extreme interventions. A common association with this disorder is **muscle dysmorphia**, in which individuals believe that their body is too small or unmuscular (a preoccupation) and respond through working out. Like body dysmorphic disorder in general, this belief persists even with clear evidence to the contrary.

Hoarding Disorder

Hoarding disorder presents as a need to save or keep items and is often paired with excessive acquisition of objects. This behavior stems from several possible sources, ranging from the belief that kept items will eventually be useful to the feeling that the patient has a responsibility to care for the items. As a result, individuals with hoarding disorder often fill their homes with seemingly useless items even past the point where the accumulation of belongings impairs daily life.

Obsessive–Compulsive and Related Disorders with Body-Focused Repetitive Behaviors

Trichotillomania and excoriation are two obsessive–compulsive and related disorders that both present with body-focused compulsions. In the case of **trichotillomania**, individuals are compelled to pull out their hair, while with **excoriation disorder,** individuals are compelled to pick at their skin. A required diagnostic feature in both of these disorders is that patients have previously attempted to stop their body-focused compulsions but have so far failed.

Trauma- and Stressor-Related Disorders

This category captures disorders where a traumatic event is the source of the symptoms and thus is a diagnostic requirement in these disorders. The typical response to traumatic events includes fear, helplessness, and perhaps anxiety. In trauma and stressor related disorders, however, individuals also present with maladaptive symptoms like anhedonia, dysphoria (generalized dissatisfaction with life), aggression, or dissociation.

By far, the most notable disorder in this category is **posttraumatic stress disorder** (**PTSD**). PTSD occurs after experiencing or witnessing a traumatic event, such as war, a home invasion, rape, or a natural disaster, and consists of intrusion symptoms, arousal symptoms, avoidance symptoms, and negative cognitive symptoms.

- **Intrusion symptoms** include recurrent reliving of the event, flashbacks, nightmares, and prolonged distress.
- **Arousal symptoms** include an increased startle response, irritability, anxiety, self-destructive or reckless behavior, and sleep disturbances.
- **Avoidance symptoms** include deliberate attempts to avoid the memories, people, places, activities, and objects associated with the trauma.
- **Negative cognitive symptoms** include an inability to recall key features of the event, negative mood or emotions, feeling distanced from others, and a persistent negative view of the world.

To meet the criteria of PTSD, a particular number of these symptoms must be present for at least one month. If the same symptoms last for less than one month (but more than three days), it may be called **acute stress disorder**.

From a behaviorist perspective, symptoms of PTSD can be explained by the traumatic event and one's reaction to it. *Intrusion* and *arousal symptoms* can be explained by associative learning, specifically classical conditioning, in which the event has become associated with traumatic triggers and has generalized to include everyday stimuli. *Avoidance symptoms* can be explained through operant conditioning, specifically avoidance learning, in which an individual learns behavior to avoid unpleasant stimuli, or involuntary responses in the case of PTSD. Finally, negative cognitive symptoms can be viewed as a form of dissociation, which is a defense mechanism to avoid unpleasant stimuli. Dissociation will be covered in greater detail in *Dissociative Disorders.*

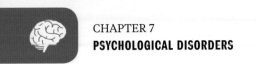

Dissociative Disorders

Patients with **dissociative disorders** avoid stress by escaping from parts of their identity. Such patients otherwise still have an intact sense of reality. Examples of dissociative disorders include dissociative amnesia, dissociative identity disorder (formerly multiple personality disorder), and depersonalization/derealization disorder.

Dissociative Amnesia

Dissociative amnesia is characterized by an inability to recall past experiences. The qualifier *dissociative* simply means that the amnesia is not due to a neurological disorder. This disorder is often linked to trauma. Some individuals with this disorder may also experience **dissociative fugue**: a sudden, unexpected move or purposeless wandering away from one's home or location of usual daily activities. Individuals in a fugue state are confused about their identity and can even assume a new identity. Significantly, they may actually believe that they are someone else, with a complete backstory.

Dissociative Identity Disorder

In **dissociative identity disorder** (**DID**, formerly multiple personality disorder), there are two or more personalities that recurrently take control of the patient's behavior, as represented in Figure 7.3. This disorder results when the components of identity fail to integrate. In most cases, patients have suffered severe physical or sexual abuse as young children. After much therapy, the personalities can sometimes be integrated into one. The existence of dissociative identity disorder is justifiably debated within the medical community, but its characteristics are still important to recognize on Test Day.

REAL WORLD

One of the first and most famous cases of dissociative identity disorder in the media is Shirley Ardell Mason, also known as "Sybil," who claimed to have at least 13 separate personalities. Mason underwent years of therapy in an attempt to combine her personalities into a single one. Two separate TV movies, both called *Sybil*, have been produced to tell the story of Sybil's struggle with this disorder.

Figure 7.3 Dissociative Identity Disorder (DID)
One artist's interpretation of many personalities seen in DID.

Depersonalization/Derealization Disorder

In **depersonalization/derealization disorder**, individuals feel detached from their own minds and bodies (**depersonalization**) or from their surroundings (**derealization**). This often presents as a feeling of automation, and can include symptoms like a failure to recognize one's reflection. An out-of-body experience is an example of depersonalization. Derealization is often described as giving the world a dreamlike or insubstantial quality. Such patients may also experience depersonalization and derealization simultaneously. These feelings cause significant impairment of regular activities. However, even during these times, such patients do not display psychotic symptoms like delusions or hallucinations.

Somatic Symptom and Related Disorders

Diagnoses in this category are marked by somatic (bodily) symptoms that cause significant stress or impairment.

Somatic Symptom Disorder

Individuals with **somatic symptom disorder** have at least one somatic symptom, which may or may not be linked to an underlying medical condition, and that is accompanied by disproportionate concerns about its seriousness, devotion of an excessive amount of time and energy to it, or elevated levels of anxiety.

Illness Anxiety Disorder

Illness anxiety disorder is characterized by being consumed with thoughts about having or developing a serious medical condition. Individuals with this disorder are quick to become alarmed about their health, and either excessively check themselves for signs of illness or avoid medical appointments altogether. Most patients classified under hypochondriasis in the DSM-IV-TR now fit into either somatic symptom disorder if somatic symptoms are present, or illness anxiety disorder if they are not.

Conversion Disorder

Conversion disorder, also known as **functional neurological symptom disorder**, is characterized by symptoms affecting voluntary motor or sensory functions that are incompatible with the patient's neurophysiological condition. The symptoms generally begin soon after the individual experiences high levels of stress or a traumatic event, but may not develop until some time has passed after the initiating experience. Examples include paralysis or blindness without evidence of neurological damage. The patient may be surprisingly unconcerned by the symptom—what is called ***la belle indifférence***. Conversion disorder was historically called *hysteria*. The symptoms seen in conversion disorder may sometimes be connected with the inciting event in a literal or poetic way; for example, a parent going blind shortly after watching a child die tragically.

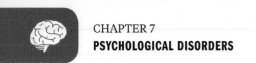

Personality Disorders

A **personality disorder** is a pattern of behavior that is inflexible and maladaptive, causing distress or impaired functioning in at least two of the following: cognition, emotions, interpersonal functioning, or impulse control. Personality disorders are considered **ego-syntonic**, meaning that individuals perceive their behavior as correct, normal, or in harmony with their goals. This is in contrast to the other disorders covered in this chapter that are **ego-dystonic**, meaning that individuals see the illness as something thrust upon them that is intrusive and bothersome. In addition to **general personality disorder** (which fits the diagnostic criteria described above), there are ten personality disorders grouped into three clusters: cluster A (paranoid, schizotypal, and schizoid), cluster B (antisocial, borderline, histrionic, and narcissistic), and cluster C (avoidant, dependent, and obsessive–compulsive). Personality disorder criteria will continue changing over time; the DSM-5 includes a section specifically devoted to research models for redefining personality disorders.

REAL WORLD

The distinction between **ego-syntonic** and **ego-dystonic** symptoms is a key feature in differential diagnosis of disorders in the DSM-5. For instance, social anxiety disorder shares many of the same symptoms as avoidant personality disorder, such as anxiety directed towards social interactions and maladaptive avoidance behavior. The distinction between these disorders is that individuals with social anxiety disorder often know that their fear of being ridiculed is irrational (ego-dystonic), while individuals with avoidant personality disorder actually believe they are inferior and that their fear of ridicule is valid (ego-syntonic).

Cluster A (Paranoid, Schizotypal, and Schizoid Personality Disorders)

The **cluster A personality disorders** are all marked by behavior that is labeled as odd or eccentric by others. Its three examples include paranoid, schizotypal, and schizoid personality disorders.

Paranoid personality disorder is marked by a pervasive distrust of others and suspicion regarding their motives. In some cases, these patients may actually be in the prodromal phase of schizophrenia and are termed premorbid.

Schizotypal personality disorder refers to a pattern of odd or eccentric thinking. These individuals may have ideas of reference (similar to delusions of reference, but not as extreme in intensity) as well as magical thinking, such as superstitiousness or a belief in clairvoyance.

Finally, **schizoid personality disorder** is a pervasive pattern of detachment from social relationships and a restricted range of emotional expression. People with this disorder show little desire for social interactions, have few or no close friends, and have poor social skills. It should be noted that neither schizotypal nor schizoid personality disorder are the same as schizophrenia.

Cluster B (Antisocial, Borderline, Histrionic, and Narcissistic Personality Disorders)

The **cluster B personality disorders** are all marked by behavior that is labeled as dramatic, emotional, or erratic by others. Its four examples include antisocial, borderline, histrionic, and narcissistic personality disorders.

Antisocial personality disorder is three times more common in males than in females. The essential feature of the disorder is a pattern of disregard for and violations of the rights of others. This is evidenced by repeated illegal acts, deceitfulness, aggressiveness, or a lack of remorse for said actions. Many serial killers and career criminals who show no guilt for their actions have this disorder. Additionally, people with this disorder comprise about 20 to 40 percent of prison populations.

Borderline personality disorder is two times more common in females than in males. In this disorder, there is pervasive instability in interpersonal behavior, mood, and self-image. Interpersonal relationships are often intense and unstable. There may be profound identity disturbance with uncertainty about self-image, sexual identity, long-term goals, or values. There is often intense fear of abandonment. Individuals with borderline personality disorder may use **splitting** as a defense mechanism, in which they view others as either all good or all bad (an *angel* vs. *devil* mentality). Suicide attempts and self-mutilation (cutting or burning) are common.

Histrionic personality disorder is characterized by constant attention-seeking behavior. These individuals often wear colorful clothing, are dramatic, and are exceptionally extroverted. They may also use seductive behavior to gain attention.

In **narcissistic personality disorder**, the patient has a grandiose sense of self-importance or uniqueness, preoccupation with fantasies of success, a need for constant admiration and attention, and characteristic disturbances in interpersonal relationships such as feelings of entitlement. As used in everyday language, narcissism refers to those who like themselves too much. However, people with narcissistic personality disorder have very fragile self-esteem and are constantly concerned with how others view them. There may be marked feelings of rage, inferiority, shame, humiliation, or emptiness when these individuals are not viewed favorably by others.

Cluster C (Avoidant, Dependent, and Obsessive–Compulsive Personality Disorders)

The **cluster C personality disorders** are all marked by behavior that is labeled as anxious or fearful by others. Its three examples include avoidant, dependent, and obsessive–compulsive personality disorders.

In **avoidant personality disorder**, the affected individual has extreme shyness and fear of rejection. Individuals who have this disorder will see themselves as socially inept and are often socially isolated, despite an intense desire for social affection and acceptance. These individuals tend to stay in the same jobs, life situations, and relationships despite wanting to change.

Dependent personality disorder is characterized by a continuous need for reassurance. Individuals with dependent personality disorder tend to remain dependent on one specific person, such as a caregiver or significant other, to take actions and make decisions.

In **obsessive–compulsive personality disorder** (**OCPD**), the individual is perfectionistic and inflexible, tending to like rules and order. Other characteristics may include an inability to discard worn-out objects, lack of desire to change, excessive stubbornness, lack of a sense of humor, and maintenance of careful routines. Note that obsessive–compulsive personality disorder is not the same as obsessive–compulsive disorder. Whereas OCD has obsessions and compulsions that are focal and acquired, OCPD is lifelong. OCD is also ego-dystonic (*I can't stop washing my hands because of the germs!*), whereas OCPD is ego-syntonic (*I just like rules and order!*).

KEY CONCEPT

Obsessive-compulsive disorder (OCD) and obsessive-compulsive personality disorder (OCPD) are not synonymous. OCD is marked by obsessions (intrusive thoughts causing tension) and compulsions (repetitive tasks that relieve this tension but cause significant impairment). OCPD is a personality disorder in which individuals are perfectionistic and inflexible.

BEHAVIORAL SCIENCES GUIDED EXAMPLE WITH EXPERT THINKING

The following individuals are patients at an inpatient mental health facility.

Patient A's hospitalization is the result of an intense argument with a family member that involved threats of violence, which prompted the police to be called. For the first two weeks of her stay, Patient A spoke often about plans to start several online business ventures, saying that she felt she was a "business genius" and that she would be a billionaire by the end of the year. She slept very little and was irritable, often becoming angry with clinic staff when they tried to reason with her about the soundness of her plans. In the following weeks of Patient A's stay, her mood leveled and she expressed regret over her treatment of her family.

A: Elevated mood, irritable, rapid speech. Lasts 2 weeks.

Patient B arrived at the clinic as a result of complaints from his neighbors. When police entered his apartment, they found it in complete disrepair, and it was clear that he had stopped attending to his personal hygiene long ago. During the first few weeks of his stay, Patient B made no effort to speak to staff or other patients. When he did respond to questions, his answers were short phrases, sometimes unrelated to the question, and sometimes simply a repetition of a few words from the question. He spent his days isolated, often sitting, immobile and unresponsive to occurrences around him.

B: Not taking care of himself, not talking or reacting to others.

Based only on the information provided, what diagnosis from the DSM-5 is most likely for each of these patients?

If you're asked to make a diagnosis for a hypothetical patient, make a checklist of the symptoms described in the passage or question, and then match them to what you know about the disorders that are within the scope of the MCAT. It is worth taking a moment to consider the severity of the symptoms as well, since it can help to differentiate between similar disorders (depression and dysthymic disorder, for example).

Patient A exhibits grandiose self-esteem, rapid speech, a lack of need for sleep, and irritability, which are all symptoms of mania and together are sufficient to apply the label of manic episode. The described transition out of mania into a different mood state suggests bipolar disorder. Specifically, the presence of a full manic episode rules out bipolar II and would lead to a diagnosis of bipolar I, which does not require a depressive episode to follow.

Patient B is trickier, but consider that the writers of the MCAT know that you are not a trained psychiatrist and so will not require a nuanced diagnosis from you. The symptoms that they present will be straightforward and should add up to a description of a disorder that you are familiar with. For patient B, those symptoms are avolition (an inability to perform basic goal-directed activities), flat or blunted affect (lack of emotional expression), and alogia (reduction in speaking). These are all negative symptoms of schizophrenia. The description of patient B also includes echolalia (repetition of words or short phrases), which is a positive symptom and is a signal to you as a test taker that this is indeed schizophrenia, rather than a case of severe depression. While there is still the possibility that Patient B has another, related disorder, the MCAT would not present you with a choice that would require you to distinguish between, say, schizophrenia and schizoid personality disorder without substantial additional information allowing that decision to be made.

In summary, patient A is most likely experiencing bipolar I disorder, and patient B is most likely experiencing schizophrenia or a related disorder.

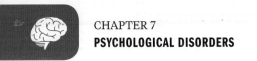

MCAT CONCEPT CHECK 7.2

Before you move on, assess your understanding of the material with these questions.

1. A schizophrenic patient is started on the atypical antipsychotic *risperidone*, which is effective for treatment of the positive symptoms of schizophrenia, but not the negative symptoms. Which of the patient's symptoms are likely to improve, and which are not?

____

____

____

2. What are the features of a major depressive episode? Of a manic episode?
   - Major depressive episode:

   ____

   - Manic episode:

   ____

3. For each of the following disorders, briefly describe their makeup with respect to depressive episodes, manic episodes, and other mood disturbances:
   - Major depressive disorder:

   ____

   - Bipolar I disorder:

   ____

   - Bipolar II disorder:

   ____

   - Cyclothymic disorder:

   ____

4. A patient with obsessive–compulsive disorder believes that the latch on the apartment door must be checked five times before it is okay to go to bed. Without checking the latch five times, the patient cannot sleep for fear that someone will break into the apartment. Identify the patient's obsession, the patient's compulsion, and how they are related in obsessive–compulsive disorder.
   - Obsession:

   - Compulsion:

   - Relationship:

   ---

5. What features describe each cluster of personality disorders? Which personality disorders fall into each cluster?

Cluster	Features	Personality Disorders
A		
B		
C		

7.3 Biological Basis of Nervous System Disorders

LEARNING OBJECTIVES

After Chapter 7.3, you will be able to:

- Describe the impact of depression on hormone and neurotransmitter levels
- Recall the general features and risk factors for Alzheimer's disease
- Explain the role of dopamine in schizophrenia and Parkinson's disease

In addition to knowing the psychological and sociological components of these diagnoses, the MCAT also expects you to know the biological basis of a few mental disorders. These disorders include schizophrenia, depression, Alzheimer's disease, and Parkinson's disease. Research into how to stop the progression of the biological component of these diseases is widespread and will also be something to stay apprised of as a medical student and a physician.

Schizophrenia

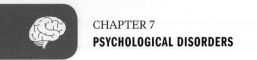

Schizophrenia is an area of active research, though some biological factors have been isolated. Most potential causes are genetic, but trauma at birth, especially hypoxemia (low oxygen concentrations in the blood), is also considered to be a risk factor. Other exposures may also play a role; excessive marijuana use in adolescence is associated with increased risk. There is significant data to indicate that schizophrenia is partially inherited. If a person has this disorder, the risk that first-degree relatives will also have the disorder is ten times that of an unrelated person in the general population; this measurement controls for environmental effects.

Schizophrenia may be associated with structural changes in the brain, but more research is needed to determine their significance and prevalence within the affected population. Schizophrenia is highly associated with an excess of dopamine in the brain; many medications used to treat schizophrenia, such as neuroleptics, block dopamine receptors. The term **neuroleptic** means that these medications depress nerve function. Neuroleptics are also known as **antipsychotics**.

Depressive and Bipolar Disorders

There are a host of markers associated with depression:

- Abnormally high glucose metabolism in the amygdala
- Hippocampal atrophy after a long duration of illness
- Abnormally high levels of glucocorticoids (cortisol)
- Decreased norepinephrine, serotonin, and dopamine (monoamine theory of depression)

It has been found that both these neurotransmitters and their metabolites are decreased in depressed patients, meaning that their actual production is decreased (rather than production staying the same and their degradation increasing).

For bipolar disorders, there exists a different set of biological factors and genetic corollaries that contribute to the disease:

- Increased norepinephrine and serotonin (monoamine theory)
- Higher risk if parent has bipolar disorder
- Higher risk for persons with multiple sclerosis

Alzheimer's Disease

Alzheimer's disease is a type of dementia characterized by gradual memory loss, disorientation to time and place, problems with abstract thought, and a tendency to misplace things. Later stages of the disease are associated with changes in mood or behavior, changes in personality, difficulty with procedural memory, poor judgment, and loss of initiative. Now, each of these symptoms alone doesn't necessarily point to Alzheimer's; however, when all or almost all of these symptoms are seen in one person, and especially when the symptoms end up inhibiting normal daily function, this points to Alzheimer's disease. This disease is most common in patients older than 65, and women are at greater risk than men. Family history is a significant risk factor and, interestingly, there is a lower risk of developing the disease with higher levels of education.

There is a genetic component to Alzheimer's disease. Research shows that mutations in the *presenilin* genes on chromosomes 1 and 14 contribute to having the disease, and mutations in the *apolipoprotein E* gene on chromosome 19 can also alter the likelihood of acquiring the disease. Finally, the *β-amyloid precursor protein* gene on chromosome 21 is known to contribute to Alzheimer's disease, explaining the much higher risk of Alzheimer's in individuals with Down syndrome.

While the precise biological cause of Alzheimer's disease is unknown, there are many biological markers that are found in patients with the disease. Don't worry about understanding each of these markers in depth, but rather be able to recognize these factors if you see them on the MCAT:

- Diffuse atrophy of the brain on CT or MRI
- Flattened sulci in the cerebral cortex
- Enlarged cerebral ventricles, shown in Figure 7.4(a)
- Deficient blood flow in parietal lobes, which is correlated with cognitive decline
- Reduction in levels of acetylcholine
- Reduction in *choline acetyltransferase* (ChAT), the enzyme that produces acetylcholine
- Reduced metabolism in temporal and parietal lobes
- Senile plaques of **β-amyloid** (a misfolded protein in β-pleated sheet form), shown in Figure 7.4(b)
- Neurofibrillary tangles of hyperphosphorylated tau protein, shown in Figure 7.4(c)

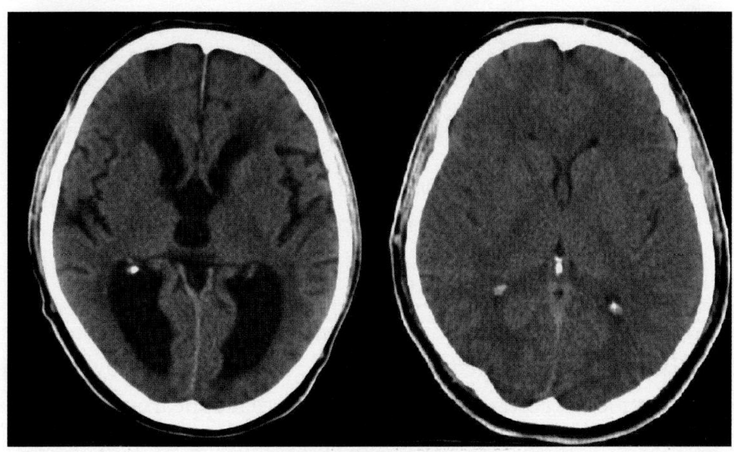

Figure 7.4(a) Symptoms of Alzheimer's Disease
Enlarged cerebral ventricles (left) vs. *normal cerebral ventricles (right)*

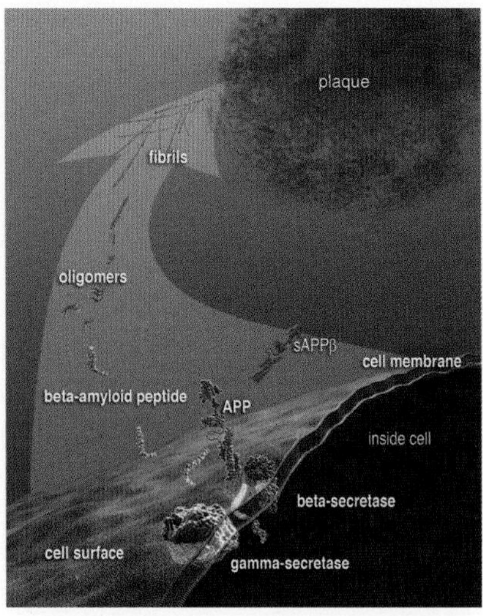

Figure 7.4(b) Symptoms of Alzheimer's Disease
Formation of senile plaques of β-amyloid

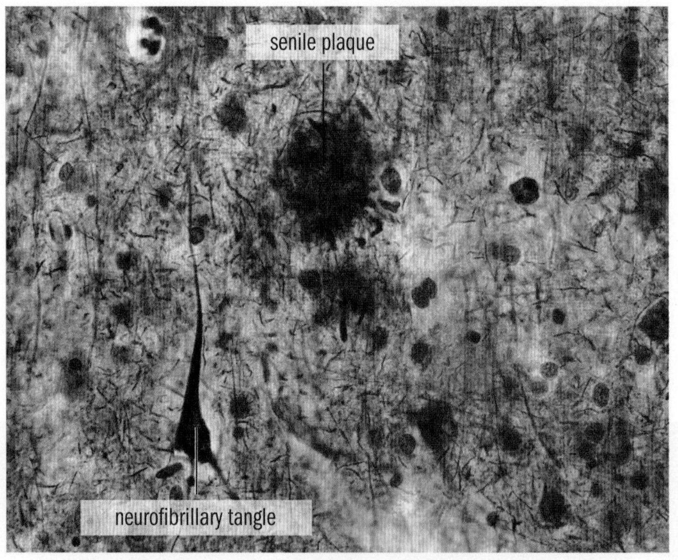

Figure 7.4(c) Symptoms of Alzheimer's Disease
Neurofibrillary tangles of hyperphosphorylated tau protein (magenta)

Parkinson's Disease

Parkinson's disease, demonstrated in Figure 7.5, is characterized by **bradykinesia** (slowness in movement), **resting tremor** (a tremor that appears when muscles are not being used), **pill-rolling tremor** (flexing and extending the fingers while moving the thumb back and forth, as if rolling something in the fingers), **masklike facies** (static and expressionless facial features, staring eyes, and a partially open mouth), **cogwheel rigidity** (muscle tension that intermittently halts movement as an examiner attempts to manipulate a limb), and a **shuffling gait** with stooped posture.
A common but not characteristic symptom is depression. Dementia is also common in Parkinson's disease.

Figure 7.5 Findings of Parkinson's Disease
Note the masklike facies, shuffling gait, and stooped posture.

The biological basis of this disease is decreased dopamine production in the **substantia nigra**, a layer of cells in the brain that functions to produce dopamine to permit proper functioning of the **basal ganglia**, as shown in Figure 7.6. The basal ganglia are critical for initiating and terminating movements, as well as sustaining repetitive motor tasks and smoothening motions; thus, the symptoms of Parkinson's disease flow logically from its underlying cause. This condition can be partially managed, therefore, with **L-DOPA**, a precursor that is converted to dopamine once in the brain, replacing that which is lost due to Parkinson's disease. There have also been attempts to regenerate dopaminergic neurons in the substantia nigra using stem cells placed into the central nervous system. Similar stem cell-based therapies have been used in other contexts, such as after a spinal cord injury or stroke to attempt to regenerate function in the central nervous system, with limited results.

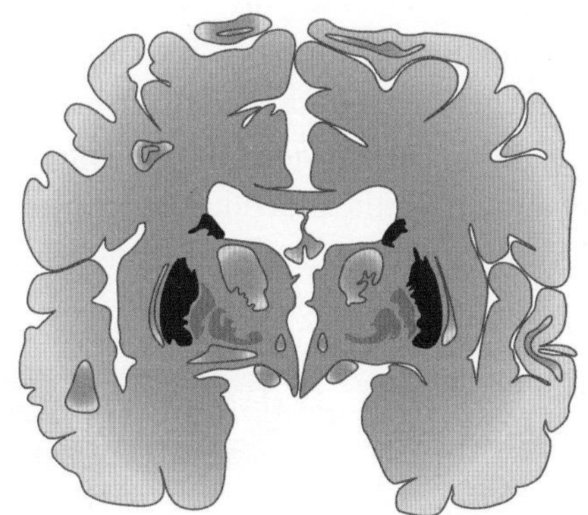

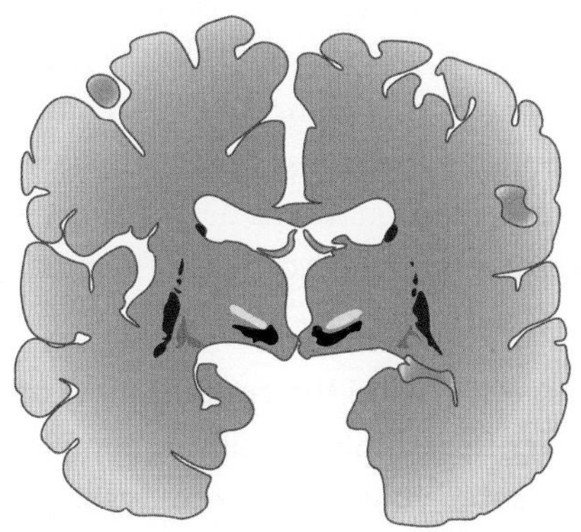

Figure 7.6 The Basal Ganglia
The substantia nigra in the midbrain (black) releases dopamine to activate the other regions of the basal ganglia (blue, red, and yellow).

REAL WORLD

Note the connection between schizophrenia and psychosis (caused by an excess of dopamine) and Parkinson's disease (caused by a deficit of dopamine). Antipsychotic medications often lead to "parkinsonian" side effects, like muscle rigidity and flattened affect. Medications used in Parkinson's disease often lead to psychotic side effects, such as hallucinations and delusions.

MCAT CONCEPT CHECK 7.3

Before you move on, assess your understanding of the material with these questions.

1. Which hormone and neurotransmitter concentrations are elevated in depression? Which ones are reduced?
   - Elevated:

   ____

   - Reduced:

   ____

2. Provide an example of a genetic factor that appears to increase risk of Alzheimer's disease.

   ____

3. How are dopamine levels related in schizophrenia and Parkinson's disease?

   ____

Conclusion

The content covered in this chapter will allow you to score more points on the MCAT—and to prepare for your clinical clerkships in psychiatry. This chapter is unique in that it covers not how the mind normally works, as we see in the other chapters in this book, but rather how the mind works when it is functioning abnormally. The MCAT tests critical thinking; one common way to do this is to ask what happens when a system—like the mind—is not functioning normally. Thus, this chapter covered high-yield information that is likely to appear on the MCAT because it connects all three subjects of the *Psychological, Social, and Biological Foundations of Behavior* section. In the next chapter, we move away from the individual as we begin to explore social psychology; from there, we'll continue expanding outward as we move into sociology.

You've reviewed the content, now test your knowledge and critical thinking skills by completing a test-like passage set in your online resources!

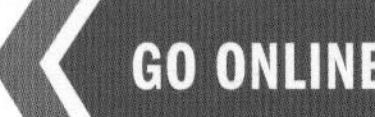

CONCEPT SUMMARY

Understanding Psychological Disorders

- The **biomedical approach** to psychological disorders takes into account only the physical and medical causes of a psychological disorder. Thus, treatments in this approach are of a biomedical nature.
- The **biopsychosocial approach** considers the relative contributions of biological, psychological, and social components to an individual's disorder. Treatments also fall into these three areas.
- The ***Diagnostic and Statistical Manual of Mental Disorders*** is used to diagnose psychological disorders. Its current version is DSM-5 (published May 2013). It categorizes mental disorders based on symptom patterns.
- Psychological disorders, especially anxiety, depressive, and substance use disorders, are very common in the population.

Types of Psychological Disorders

- **Schizophrenia** is the prototypical disorder with psychosis as a feature. It contains positive and negative symptoms.
  - **Positive symptoms** add something to behavior, cognition, or affect and include delusions, hallucinations, disorganized speech, and disorganized behavior.
  - **Negative symptoms** are the loss of something from behavior, cognition, or affect and include disturbance of affect and avolition.
- **Depressive disorders** include major depressive disorder, dysthymia, and seasonal affective disorder.
  - **Major depressive disorder** contains at least one major depressive episode.
  - **Persistent depressive disorder (dysthymia)** is the presence of depressive symptoms for at least two years that do not meet criteria for major depressive disorder.
- **Bipolar and related disorders** have manic or hypomanic episodes.
  - **Bipolar I disorder** contains at least one manic episode.
  - **Bipolar II disorder** contains at least one major depressive episode and least one hypomanic episode.
  - **Cyclothymic disorder** describes periods of manic and depressive symptoms that are not severe enough to be labeled an episode. These symptoms must persist for at least 2 years and be present the majority of that time.
- **Anxiety disorders** capture conditions in which excessive fear or anxiety impairs one's daily functions. Anxiety disorders are differentiated by the stimuli that induces anxiety.
  - **Specific phobias** are irrational fears of specific objects or situations.

- **Separation anxiety disorder** is anxiety due to separation from one's caregivers, often with the ideation that if separated, either the caregiver or the patient will be harmed.
- **Social anxiety disorder** is anxiety due to social or performance situations with the ideation that the patient will be negatively evaluated.
- **Selective mutism** disorder is the impairment of speech in situations where speaking is expected.
- **Panic disorder** is marked by recurrent panic attacks: intense, overwhelming fear and sympathetic nervous system activity with no clear stimulus. It may lead to agoraphobia.
- **Agoraphobia** is a fear of places or situations where it is hard for an individual to escape.
- **Generalized anxiety disorder** is a disproportionate and persistent worry about many different things for at least six months.

- **Obsessive–compulsive disorder and related disorders** are characterized by perceived needs (obsessions or preoccupations) and paired actions to meet those needs (compulsions).
  - **Obsessive–compulsive disorder** is characterized by **obsessions** (persistent, intrusive thoughts and impulses) and **compulsions** (repetitive tasks that relieve tension but cause significant impairment in a person's life).
  - **Body dysmorphic disorder** is characterized by an unrealistic negative evaluation of one's appearance or a specific body part. The individual often takes extreme measures to correct the perceived imperfection.
  - **Hoarding disorder** is characterized by the reluctance of giving up one's physical possessions. Often this behavior is associated with excessive acquisition of physical items.
- **Posttraumatic stress disorder** (**PTSD**) is characterized by intrusion symptoms (reliving the event, flashbacks, nightmares), avoidance symptoms (avoidance of people, places, objects associated with trauma), negative cognitive symptoms (amnesia, negative mood and emotions), and arousal symptoms (increased startle response, irritability, anxiety). These symptoms can be explained from the behaviorist perspective.
- **Dissociative disorders** include dissociative amnesia, dissociative identity disorder, and depersonalization/derealization disorder.
  - **Dissociative amnesia** is an inability to recall past experience without an underlying neurological disorder. In severe forms, it may involve **dissociative fugue**, a sudden change in location that may involve the assumption of a new identity.
  - **Dissociative identity disorder** is the occurrence of two or more personalities that take control of a person's behavior.
  - **Depersonalization/derealization disorder** involves feelings of detachment from the mind and body or from the environment.

- **Somatic symptom and related disorders** involve significant bodily symptoms.
  - **Somatic symptom disorder** involves at least one somatic symptom, which may or may not be linked to an underlying medical condition, that causes disproportionate concern.
  - **Illness anxiety disorder** is a preoccupation with thoughts about having, or coming down with, a serious medical condition.
  - **Conversion disorder** involves unexplained symptoms affecting motor or sensory function and is associated with prior trauma.
- **Personality disorders (PD)** are patterns of inflexible, maladaptive behavior that cause distress or impaired functioning in at least two of the following: cognition, emotions, interpersonal functioning, or impulse control. They occur in three **clusters**: **A** (odd, eccentric), **B** (dramatic, emotional, erratic), and **C** (anxious, fearful).
  - Cluster A includes paranoid, schizotypal, and schizoid PDs. Cluster B includes antisocial, borderline, histrionic, and narcissistic PDs. Cluster C includes avoidant, dependent, and obsessive–compulsive PDs.
  - **Paranoid PD** involves a pervasive distrust and suspicion of others.
  - **Schizotypal PD** involves ideas of reference, magical thinking, and eccentricity.
  - **Schizoid PD** involves detachment from social relationships and limited emotion.
  - **Antisocial PD** involves a disregard for the rights of others.
  - **Borderline PD** involves instability in relationships, mood, and self-image. **Splitting** is characteristic, as are recurrent suicide attempts.
  - **Histrionic PD** involves constant attention-seeking behavior.
  - **Narcissistic PD** involves a grandiose sense of self-importance and need for admiration.
  - **Avoidant PD** involves extreme shyness and fear of rejection.
  - **Dependent PD** involves a continuous need for reassurance.
  - **Obsessive–compulsive PD** involves perfectionism, inflexibility, and preoccupation with rules.

Biological Basis of Nervous System Disorders

- Schizophrenia may be associated with genetic factors, birth trauma, adolescent marijuana use, and family history. There are high levels of dopaminergic transmission.
- Depression is accompanied by high levels of glucocorticoids and low levels of norepinephrine, serotonin, and dopamine.

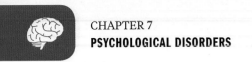

- Bipolar disorders are accompanied by high levels of norepinephrine and serotonin. They are also highly heritable.
- Alzheimer's disease is associated with genetic factors, brain atrophy, decreases in acetylcholine, senile plaques of **β-amyloid**, and **neurofibrillary tangles** of hyperphosphorylated tau protein.
- **Parkinson's disease** is associated with **bradykinesia**, **resting tremor**, **pill-rolling tremor**, **masklike facies**, **cogwheel rigidity**, and a **shuffling gait**. There is decreased production of dopamine by cells in the **substantia nigra**.

ANSWERS TO CONCEPT CHECKS

7.1

1. Whereas the biomedical model considers only the physical, pathological mechanisms that underlie mental illness, the biopsychosocial model considers the contributions of these biological factors along with psychology (thoughts, emotions, or behaviors) and social situation (environment, social class, discrimination, or stigmatization).
2. The following disorders occur in greater than 2 percent of the United States population per year: specific phobia, social anxiety disorder, major depressive disorder, posttraumatic stress disorder, bipolar disorder, generalized anxiety disorder, and panic disorder.

7.2

1. Positive symptoms of schizophrenia, including delusions, hallucinations (usually auditory), disorganized thought, and disorganized behavior, are likely to improve from treatment with an antipsychotic medication. Negative symptoms, including disturbance of affect and avolition, are largely unaffected by antipsychotic medications.
2. Major depressive episodes include a two-week duration of at least five of the following symptoms: depressed mood, loss of interest (anhedonia), sleep disturbance, feelings of guilt, lack of energy, difficulty concentrating, changes in appetite, psychomotor symptoms, and suicidal thoughts. At least one of the symptoms must be depressed mood or anhedonia. Manic episodes include a one-week duration of at least three of the following symptoms: elevated or expansive mood, distractibility, decreased need for sleep, grandiosity, flight of ideas or racing thoughts, agitation, pressured speech, and engagement in risky behavior.
3. Major depressive disorder contains at least one major depressive episode with no manic episodes. Bipolar I disorder has at least one manic episode with or without depressive episodes. Bipolar II disorder has at least one hypomanic episode with at least one major depressive episode. Cyclothymic disorder has hypomanic episodes and dysthymia that is not severe enough to be a major depressive episode.
4. Obsessions are persistent, intrusive thoughts and impulses that produce tension. In this case, the obsession is the patient's thought that someone will break into the apartment. Compulsions are repetitive tasks that relieve tension but cause significant impairment in a person's life. This patient's compulsion is having to check the latch on the apartment door five times before going to bed. Their relationship is that obsessions raise tension while compulsions relieve that tension.

5.

Cluster	Features	Personality Disorders
A	Odd or eccentric	Paranoid, schizotypal, schizoid
B	Dramatic, emotional, or erratic	Antisocial, borderline, histrionic, narcissistic
C	Anxious or fearful	Avoidant, dependent, obsessive–compulsive

7.3

1. In depression, levels of cortisol are increased. Many neurotransmitter levels are reduced, including norepinephrine, serotonin, and dopamine.
2. Mutations in the *presenilin* genes (chromosomes 1 and 14) and β-amyloid precursor protein gene (chromosome 21) are associated with increased risk for Alzheimer's disease.
3. Dopamine levels are elevated in schizophrenia and reduced in Parkinson's disease. Thus, treatments for one disorder may cause symptoms similar to those of the other.

SCIENCE MASTERY ASSESSMENT EXPLANATIONS

1. **C**

Negative symptoms are the absence of normal or desired behavior, which include disturbance of affect and avolition. Positive symptoms are the addition of abnormal behavior, including hallucinations, **(A)**, disorganized behavior, **(B)**, and delusions, **(D)**.

2. **A**

Echolalia is an involuntary repetition of others' words and utterances and may be seen in schizophrenia. Echopraxia, **(B)**, is imitation of others' actions. Loosening of associations, **(C)**, is a type of disordered thought in which the patient moves between remotely related ideas. Neologisms, **(D)**, are newly invented words.

3. **A**

Focusing on details, loving routine, having a sense that there is only one right way to do things, and lack of humor suggests an obsessive–compulsive personality disorder.

4. **C**

Depression is marked by a period of at least two weeks in which the patient has five of nine cardinal symptoms, one of which must be depressed mood or lack of interest (anhedonia). While decreased need for sleep is commonly seen in manic episodes, it may also appear in depression, as sleep disturbance is one of the nine cardinal symptoms. Not all depressed individuals are suicidal, as in **(B)**. In older individuals, depression may often manifest as anhedonia without feelings of sadness, invalidating **(D)**.

5. **B**

The doctor is using the biomedical approach since the focus is on only the biological cause of the patient's symptoms. Furthermore, since the treatment centers only on the patient, the doctor is using direct therapy. This combination of approach and therapy matches **(B)** as the correct answer. By contrast, a doctor using the biopsychosocial approach would likely consider the patient's recent move, stresses associated with the accelerated program, and lack of support structure as potential causes of depression. Additionally, treatment using the biopsychosocial approach includes indirect therapy which focuses on increasing social support.

6. **D**

Dissociative fugue is characterized by sudden travel or change in normal day-to-day activities and occurs in some cases of dissociative amnesia. Symptoms include an inability to recall one's past or confusion about one's identity. Thus, the described patient is most likely to be suffering dissociative amnesia accompanied by dissociative fugue.

7. **D**

Agoraphobia, or a fear of places or situations in which it would be difficult to escape, is commonly seen in panic disorder. Concern about having a panic attack in public may make these individuals fearful of leaving their home.

8. **D**

Phobias affect approximately 9% of the population, meaning that on average 9 in 100 individuals will have some sort of phobia. In a sample of 50 people, 4 or 5 people will need to be excluded, assuming that the sample is representative of the larger population. Thus, **(D)** is correct.

9. **A**

The symptoms listed indicate a major depressive episode. However, depressive episodes can be a part of bipolar disorders, which also contain manic episodes. Thus, if manic episodes have not yet been asked about, one cannot distinguish between depression or bipolar disorder as the correct diagnosis yet.

10. **B**

To determine if this patient has major depressive disorder or a bipolar disorder, the presence of manic (or hypomanic) episodes should be confirmed. Bipolar disorders contain manic (or hypomanic) episodes, while major depressive disorder does not.

11. **B**

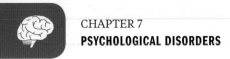

The correct answer should be a statement that is not relevant to Alzheimer's disease. Choice **(B)**, a decreased abundance of serotonin and norepinephrine, describes the monoamine theory of depression. These neurochemical changes are not known risk factors or diagnostic criteria for Alzheimer's disease, making **(B)** the correct answer. The remaining choices are either risk factors or diagnostic criteria for Alzheimer's disease. Choice **(A)** describes the increased risk of Alzheimer's disease in individuals with Down syndrome, as the gene for β-amyloid precursor is on chromosome 21. Choices **(C)** and **(D)** both describe biological markers of Alzheimer's disease.

12. **B**

Splitting, the consideration of others as either "all good" or "all bad," is characteristic of borderline personality disorder.

13. **A**

Conversion disorder is marked by a motor or sensory symptom in the absence of an underlying physical or neurological cause. It is associated with an inciting event that, in this case, may have been the argument with the family member. The patient's lack of concern over the deficit is referred to as *la belle indifférence*.

14. **C**

The symptoms indicate that the patient likely has Parkinson's disease. This disease is caused by decreased dopamine production in the substantia nigra.

15. **B**

Bipolar disorders have been shown to be highly heritable and are associated with increased levels of norepinephrine and serotonin in the brain. Bipolar I disorder can be diagnosed with a single manic episode and does not require a major depressive episode. Bipolar II disorder requires at least one hypomanic episode and one major depressive episode. Cyclothymic disorder contains at least one hypomanic episode and dysthymia.

Consult your online resources for additional practice. **GO ONLINE**

SHARED CONCEPTS

Behavioral Sciences Chapter 1
Biology and Behavior

Behavioral Sciences Chapter 3
Learning and Memory

Behavioral Sciences Chapter 12
Social Stratification

Biology Chapter 4
The Nervous System

Biology Chapter 5
The Endocrine System

Biology Chapter 12
Genetics and Evolution

CHAPTER 8

SOCIAL PROCESSES, ATTITUDES, AND BEHAVIOR

SCIENCE MASTERY ASSESSMENT

Every pre-med knows this feeling: there is so much content I have to know for the MCAT! How do I know what to do first or what's important?

While the high-yield badges throughout this book will help you identify the most important topics, this Science Mastery Assessment is another tool in your MCAT prep arsenal. This quiz (which can also be taken in your online resources) and the guidance below will help ensure that you are spending the appropriate amount of time on this chapter based on your personal strengths and weaknesses. Don't worry though—skipping something now does not mean you'll never study it. Later on in your prep, as you complete full-length tests, you'll uncover specific pieces of content that you need to review and can come back to these chapters as appropriate.

How to Use This Assessment

If you answer 0–7 questions correctly:

Spend about 1 hour to read this chapter in full and take limited notes throughout. Follow up by reviewing **all** quiz questions to ensure that you now understand how to solve each one.

If you answer 8–11 questions correctly:

Spend 20–40 minutes reviewing the quiz questions. Beginning with the questions you missed, read and take notes on the corresponding subchapters. For questions you answered correctly, ensure your thinking matches that of the explanation and you understand why each choice was correct or incorrect.

If you answer 12–15 questions correctly:

Spend less than 20 minutes reviewing all questions from the quiz. If you missed any, then include a quick read-through of the corresponding subchapters, or even just the relevant content within a subchapter, as part of your question review. For questions you answered correctly, ensure your thinking matches that of the explanation and review the Concept Summary at the end of the chapter.

1. The behavior of the individuals in the Stanford prison experiment is best explained by which of the following terms?

I. Bystander effect
II. Deindividuation
III. Internalization
IV. Social loafing

A. I only
B. III only
C. II and III only
D. II and IV only

2. A jury member who initially feels that a strict penalty should be placed on the defendant votes for an even stricter penalty after deliberation with the other jury members. This behavior is best described by which social phenomenon?

A. Social facilitation
B. Group polarization
C. Assimilation
D. Socialization

3. Which of the following would decrease the likelihood of a bystander lending aid to a victim?

A. Increasing the number of people in the room
B. Increasing the degree of danger experienced by the victim
C. Making the victim an acquaintance instead of a stranger
D. Being alone in the room with the victim

4. During groupthink, members of the group do all of the following EXCEPT:

A. stereotype members outside of the group.
B. withhold opposing views.
C. ignore warnings against the ideas of the group.
D. create a sense of negativity against risk taking.

5. Adult prison systems may attempt to change the behavior of inmates through all of the following mechanisms of socialization EXCEPT:

A. primary socialization.
B. secondary socialization.
C. anticipatory socialization.
D. resocialization.

6. Your neighbors ask you to collect their mail while they are out of town and you agree. Later that day, they ask you to water their plants as well. What technique for compliance are they using in this scenario?

A. Lowball technique
B. That's-not-all technique
C. Foot-in-the-door technique
D. Door-in-the-face technique

7. Which of the following statements represents the affective component of an attitude?

A. "I love action movies."
B. "I'm going to see a new action movie at the theater."
C. "Action movies are much better than comedies."
D. "Tomorrow, I'm going to rent an action movie."

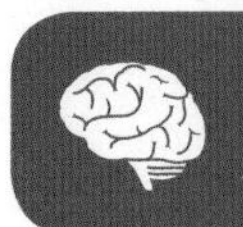

8. After sitting in a lecture, determining that a professor is a bad teacher based on the professor's unprofessional attire and monotone speech is an example of which type of processing?
   A. Knowledge route processing
   B. Adaptive route processing
   C. Central route processing
   D. Peripheral route processing

9. In the Milgram shock experiment, many subjects were willing to give the maximal voltage shock because they were influenced by which psychological principle?
   A. Deviance
   B. Obedience
   C. Conformity
   D. Compliance

10. Each individual in a group of teenagers is asked to estimate the height of a tree. One individual estimates the height to be 25 feet, but after discussing with the group is convinced that the height is likely closer to 40 feet. Which type of conformity is seen here?
   A. Normative
   B. Identification
   C. Internalization
   D. Compliance

11. Which of the following is NOT a component of the functional attitudes theory?
   A. Knowledge
   B. Acceptance
   C. Ego defense
   D. Ego expression

12. The swimming times for all members of a swim team are tracked over a six-month period in team-only practices and at public meets. For 14 of the 16 members, top times were clocked at the meets. What social phenomenon does this evidence support?
   A. Social facilitation
   B. Peer pressure
   C. Identification
   D. Group polarization

13. An 18-year-old is completing the final months of high school and begins to wake up early each day to run five miles in preparation for joining the Army. What type of socialization is this young person experiencing?
   A. Normative socialization
   B. Informative socialization
   C. Resocialization
   D. Anticipatory socialization

14. Which of the following best reflects the difference between social action and social interaction?
   A. Social action refers to positive changes one makes in their society; social interaction refers to the route by which these changes occur.
   B. Social action refers to the effects of a group on an individual's behavior; social interaction refers to the effects that multiple individuals all have on each other.
   C. Social action refers to changes in behavior caused by internal factors; social interaction refers to changes in behavior caused by external factors.
   D. Social action refers to changes in behaviors that benefit only the individual; social interaction refers to changes in behavior that benefit others.

15. In the group setting, the mentality of "If you aren't with us, you're against us" is most representative of which factor of groupthink?
   A. Illusion of invulnerability
   B. Illusion of morality
   C. Pressure for conformity
   D. Self-censorship

Answer Key

1. **C**
2. **B**
3. **A**
4. **D**
5. **A**
6. **C**
7. **A**
8. **D**
9. **B**
10. **C**
11. **B**
12. **A**
13. **D**
14. **B**
15. **C**

Detailed explanations can be found at the end of the chapter.

CHAPTER 8

SOCIAL PROCESSES, ATTITUDES, AND BEHAVIOR

In This Chapter

CHAPTER PROFILE

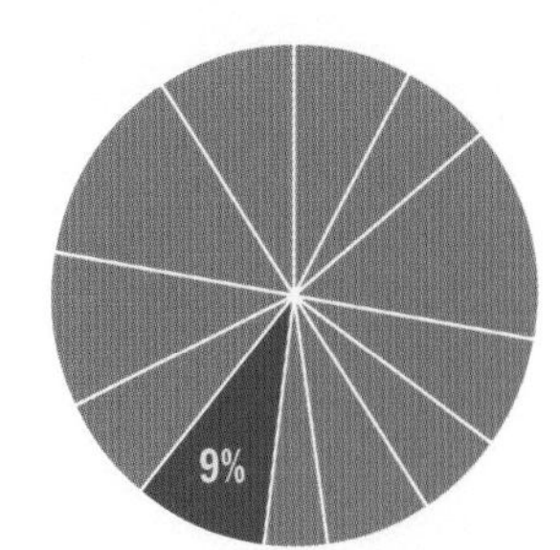

The content in this chapter should be relevant to about 9% of all questions about the behavioral sciences on the MCAT.

This chapter covers material from the following AAMC content categories:

7A: Individual influences on behavior

7B: Social processes that influence human behavior

7C: Attitude and behavior change

Introduction

The renowned Italian painter and sculptor Michelangelo Buonarroti stated that a sculptor simply releases and uncovers the ideal figures that are hidden within stone. This idea has led psychologists and sociologists to describe what is known as the **Michelangelo phenomenon**. The concept of self is made up of both the intrapersonal self, the ideas that individuals have regarding their own abilities, traits, and beliefs, and the interpersonal self, the manner in which others influence creation of the ideal self. In the Michaelangelo phenomenon, a close relationship between two individuals 'sculpts' both individuals' skills and traits. This development occurs because of each individual's perceptions of the other, and their behaviors in response to one another.

In this chapter, you will learn about the social processes and interactions that develop this self. The behavior and attitudes of individuals are highly influenced by the people with whom they interact, the society in which they live, and the culture in which they are immersed. Humans, being naturally social creatures, learn how to behave and react based on their relationships and experiences. The following pages will give us an in-depth look at the ways behavior is affected by the presence of others, group processes, culture, and socialization, as well as how attitudes are formed and how they impact behavior.

8.1 Group Psychology

LEARNING OBJECTIVES

After Chapter 8.1, you will be able to:

- Describe social facilitation, deindividuation, the bystander effect, social loafing, and peer pressure
- Compare and contrast group polarization and groupthink
- Distinguish between assimilation and multiculturalism

Understanding social processes and interaction has long been a goal of sociologists, notably Max Weber, who was one of the first sociologists to study this interaction. Weber attempted to understand and describe **social action**, which he defined as actions and behaviors that individuals are performing or modulating because others are around. The idea is that humans will behave in different ways based on their social environment and how their behavior will affect those around them. If individuals predict a negative reaction from those around them, they will often modify their behavior.

Social Action

Social action should be contrasted with social interaction. Social action considers just the individual that is surrounded by others. When examining social interaction, we will look at the behavior and actions of two or more individuals who take one another into account.

Social Facilitation

It has been observed that people tend to perform better on simple tasks when in the presence of others. This tendency is known as **social facilitation**, and it supports the idea that people naturally exhibit a performance response when they know they are being watched. Although being in the presence of others does not necessarily constitute an evaluation, the theory suggests that performance sparks a perceived evaluation in the individual performing. According to the **Yerkes–Dodson law of social facilitation**, being in the presence of others will significantly raise arousal, which enhances the ability to perform tasks one is already good at (or **simple tasks**), and hinders the performance of less familiar tasks (or **complex tasks**). For example, an expert pianist may perform better in concert than when alone in practice sessions. However, someone with very limited knowledge of music would perform worse in a social setting than when alone. This is demonstrated in Figure 8.1.

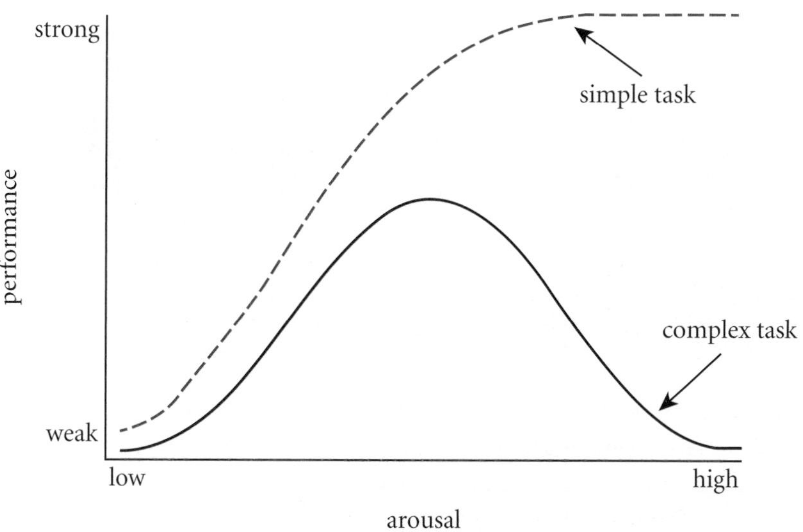

Figure 8.1 Yerkes-Dodson Law

Note the difference between Figure 5.2 (in Chapter 5 of *MCAT Behavioral Sciences Review*) and Figure 8.1. Whereas Figure 5.2 focuses solely on new or less-familiar tasks, Figure 8.1 also includes simple tasks, which are not adversely affected by heightened arousal. Social facilitation reflects the idea that performance is not solely influenced by individual ability, but also by social environment and awareness of that environment.

BRIDGE

The Yerkes-Dodson law is also used to describe the relationship between stress or sympathetic arousal and performance. Just as social facilitation can enhance the ability to perform tasks, so can moderate levels of arousal. Arousal can also be an effect of being surrounded by others and feeling pressure to perform: if there's too much pressure, performance drops. Motivation and stress are discussed in Chapter 5 of *MCAT Behavioral Sciences Review.*

Deindividuation

Deindividuation describes the loss of one's self-awareness in a group setting and the associated adoption of a more group oriented identity. This phenomenon is sometimes called mob mentality, as the individuals of the group lose their own sense of morals and judgment and follow along with the actions of the group. To explain this phenomenon social psychologists have cited group cohesion and individual anonymity as factors impacting deindividuation. If an individual can relate to the group, perhaps through a sports team or political affiliation, then the likelihood of that person adopting the group identity increases. In addition, group settings increase anonymity and thus diffuse the sense of individual responsibility for the actions of the group. Thus, as group anonymity increases, so does the strength of deindividuation. Applied to the real world, deindividuation often leads to **antinormative behavior**, that is, behavior not socially acceptable in most social circumstances; behavior "against the norm." A commonly cited example of antinormative behavior in the context of deindividuation is the violence seen in crowds and riots. Large, homogenous groups are postulated to increase deindividuation effects, particularly when relative anonymity is a factor due to the group's size. These factors can be further enhanced when a group is in uniform or masked, as shown in Figure 8.2.

Figure 8.2 Deindividuation
Being masked or in uniform facilitates anonymity in a crowd.

REAL WORLD

The violent 1964 murder of Kitty Genovese outside her home in Queens created interest in the bystander effect after her murder and a lack of response by neighbors were reported in the newspaper. Kitty reportedly cried out while being attacked in her apartment parking lot. One neighbor called out the window for the attacker to leave her alone. The attacker left, but returned ten minutes later and found Kitty barely conscious just outside the back door. Genovese was attacked again for over half an hour and ultimately died en route to the hospital. Of the 38 witnesses (bystanders), it was reported that no one had called the police. Later investigation determined this not to be true, but the misreported incident was nevertheless influential for promoting the idea of the bystander effect.

Bystander Effect

The **bystander effect** is another observed phenomenon that occurs in social groups wherein individuals do not intervene to assist those who are in perceived need when other people are present. The likelihood and timeliness of response is inversely related to the number of bystanders. In other words, the more people standing by, the less likely any one of those people is to help. There are several factors at play in the bystander effect. First, when in groups, people are less likely to notice danger or anything out of the ordinary. Additionally, when in groups, humans take cues from others. If other people are not responding to a situation, an individual is less likely to perceive the situation as a threat or emergency. The degree of emergency or the perceived danger plays a role in response. In low-danger scenarios, bystanders are less likely to provide aid; in high-danger scenarios, bystanders are more likely to intervene. Another factor is the degree of responsibility felt by the bystanders. This is determined by the competency of the bystanders, their relationship to the at risk individual(s), and whether they consider considers the person(s) at risk to be deserving of aid. Finally, cohesiveness of the group has been shown to influence responsiveness. In groups made up of strangers, the likelihood of a response, and the speed of that response, is much slower than in a group of well-acquainted individuals.

Social Loafing

Social loafing refers to the tendency of individuals to reduce effort when in a group setting. This phenomenon may apply in many contexts: physical effort, such as carrying a heavy object; mental effort, such as working on a group project; or initiative, such as coming up with the solution to a problem.

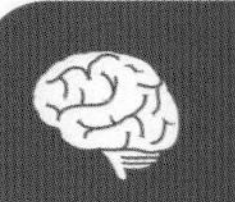

Peer Pressure

From a sociology perspective, **peers** are individuals who are regarded as equals within a social group. **Peer pressure** refers to the social influence placed on an individual by one's peers. Peer pressure exists at all ages. This pressure can come in many forms, including religious ideals, appearance, values, and sexual behavior. It can be positive or negative; certain types of peer pressure can benefit the individual experiencing the influence. In children, social acceptance is associated with being most like the social norm of the group, regardless of positive or negative connotations.

In adolescence, peers play an extremely important role in determining lifestyle, appearance, and social activities. While caregivers and other adults provide the foundation for development of beliefs and values, peers become very important as teenagers become independent from their parental figures. The pressure exerted by peers can cause changes in behavior, attitudes, or beliefs to conform to the norms of the group. Stress and the presence of peers can lead to poor choices and potentially facilitate **risky behaviors** such as binge eating, reckless driving, and violent activities.

Changes in beliefs or behavior due to peer pressure can be explained by the **identity shift effect**. When an individual's state of harmony is disrupted by a threat of social rejection, the individual will often conform to the norms of the group. Upon doing so, however, the individual will begin to experience internal conflict because the behavior is outside the normal character of the individual. To eliminate the sense of internal conflict, individuals experience an identify shift wherein they adopt the standards of the group as their own. The identity shift effect also highlights a larger theme in psychology: **cognitive dissonance**, the simultaneous presence of two opposing thoughts or opinions. This generally leads to an internal state of discomfort, which may manifest as anxiety, fear, anger, or confusion. Individuals will try to reduce this discomfort by changing, adding to, or minimizing one of these dissonant thoughts.

Solomon Asch's conformity experiment showed that individuals will often conform to an opinion held by the group. In this experiment, male college students participated in simple tasks of perception. The study was set up to have one individual who made observations in the presence of confederates, or actors who were pretending to be a part of the experiment. The point of the study was to examine if the behavior of the individual was influenced by the confederates. The participants were shown two cards like those in Figure 8.3. They were then asked to say aloud which line on the second card, labeled A, B, or C, matched the length of the line on the first card. Prior to the experiment, the confederates were secretly told to unanimously respond correctly or incorrectly to the question. When the confederates answered correctly, the error rate for the real participants was less than 1 percent. However, when the confederates answered incorrectly, it was seen that the real participants answered incorrectly up to one-third of the time. Thus, Asch concluded, individuals will sometimes provide answers they know to be untrue if it avoids going against the group: the urge toward conformity could outweigh the desire to provide the correct answer.

MCAT EXPERTISE

The MCAT can ask about psychological experiments by name, so you should be familiar with some of the landmark experiments in social psychology. For example, a question may refer to the Asch conformity experiment by name, or it may include a similar experiment without a direct reference and ask you to draw conclusions using your content knowledge.

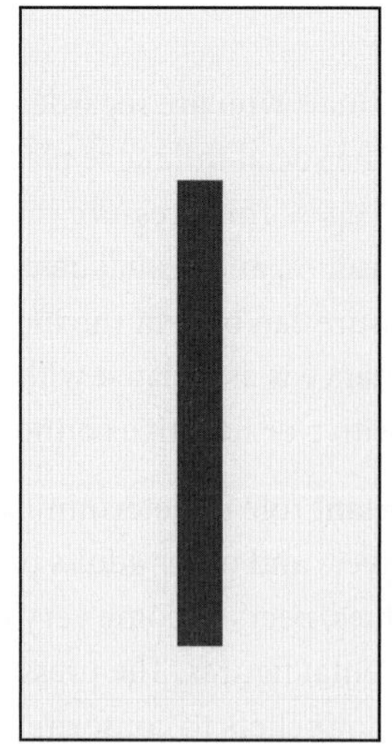

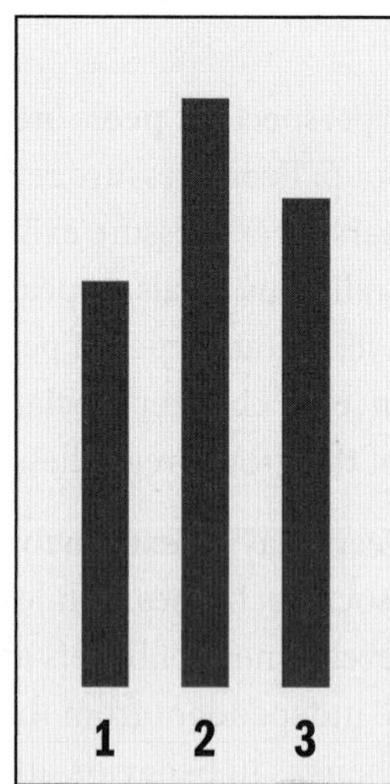

Figure 8.3 Example of Cards Used in the Asch Conformity Experiment

Group Processes

In contrast to social action, **social interaction** explores the ways in which two or more individuals can both shape each other's behavior. These include group processes and establishment of culture.

Group Polarization

Group polarization describes the tendency for groups to collaboratively make decisions that are more extreme than the individual ideas and inclinations of the members within the group. Thus, polarization can lead to riskier or more cautious decisions based on the initial tendencies of the group members toward risk or caution. This phenomenon has shown that individuals in groups will form opinions that are more extreme than the opinions they would reach in isolation. The hypothesis underlying polarization is that initial ideas tend not to be extreme, but that through discussion within the group, these ideas tend to become more and more extreme. This concept was originally termed **risky shift** because it was noted that groups tended to make riskier decisions than individuals. However, when psychologists began to realize that groups could also shift toward caution, the term became **choice shift**. Choice shift and group polarization refer to very similar concepts. However, choice shift refers specifically to measured changes in decisions before and after group interaction, whereas group polarization refers more generally to the tendency of a group to move to more extreme conclusions and decisions as a result of interaction.

Group polarization explains many real-life scenarios, including policy making, violence, and terrorism. For example, members of the same political party may espouse the same ideals and opinions in the group setting, but may waver slightly on issues when alone. This kind of polarization is also seen in jury deliberation. In the case of punitive damages (monetary penalties for a certain behavior), jurors who initially favor a high punishment may deliberate and decide upon an even higher punishment

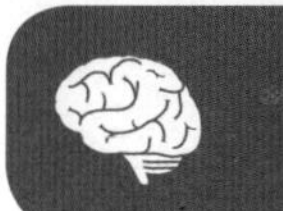

after discussion. As social media has exploded in recent decades, research has shown that the group does not necessarily need to be together physically in order for polarization to occur. Simply reading others' ideas on social media sites can result in more extreme ideas from individuals.

Groupthink

Groupthink refers to a social phenomenon in which desire for harmony or conformity results in a group of people coming to an incorrect or poor decision. In an attempt to eliminate or minimize conflict among the group members, consensus decisions are reached without alternative ideas being assessed. In these cases, the desire to agree with the group causes a loss of independent critical thinking. The group also begins to isolate and ignore external viewpoints, seeing its own ideas as correct without question.

Groupthink can have a large impact on group decision making and is influenced by a variety of factors, including group cohesiveness, group structure, leadership, and situational context. Irving Janis conducted the first research on the theory in the 1970s. Janis studied the effect of extreme stress on group cohesiveness and its resulting effect on groupthink. Janis further investigated the decision making of groups that had led to disastrous American foreign policy decisions, including the Bay of Pigs invasion. Janis specifically examined eight factors that are indicative of groupthink:

- **Illusion of invulnerability:** Members encourage risks, ignore possible pitfalls and are too optimistic.
- **Collective rationalization:** Members ignore expressed concerns about group approved ideas.
- **Illusion of morality:** Members believe ideas produced by the group are morally correct, disregarding evidence to the contrary.
- **Excessive stereotyping:** Members construct stereotypes of those expressing outside opinions.
- **Pressure for conformity:** Members feel pressured not to express opinions that disagree with the group, and view opposition as disloyal.
- **Self-censorship:** Members withhold ideas and opinions that disagree with the group.
- **Illusion of unanimity:** Members believe the decisions and judgments of the group to be without disagreement, even if it does exist.
- **Mindguards:** Some members may decide to take on a role protecting the group against opposing views.

Many of these factors, including illusion of morality, excessive stereotyping, pressure for conformity, and mindguards can be seen in Figure 8.4, a propaganda poster from the United States during the McCarthy era. The poster draws on antisemitic stereotypes and fear of Communist influence to argue against public health measures like water fluoridation, polio vaccines, and mental healthcare.

REAL WORLD

The Bay of Pigs Invasion and Cuban Missile Crisis were used by Janis as case studies. When JFK took over the White House, the administration inherited a CIA Cuban invasion plan, and it was accepted without critique. When Senator Fulbright and Secretary Schlesinger expressed objections, they were ignored by the Kennedy team. Over time, Fulbright and Schlesinger started to perform self-censorship. After the invasion, it was revealed that there were many inaccuracies in the CIA plan, including underestimation of the Cuban air force and the assumption that Castro would not have the ability to quell uprisings.

At the Sign of THE UNHOLY THREE

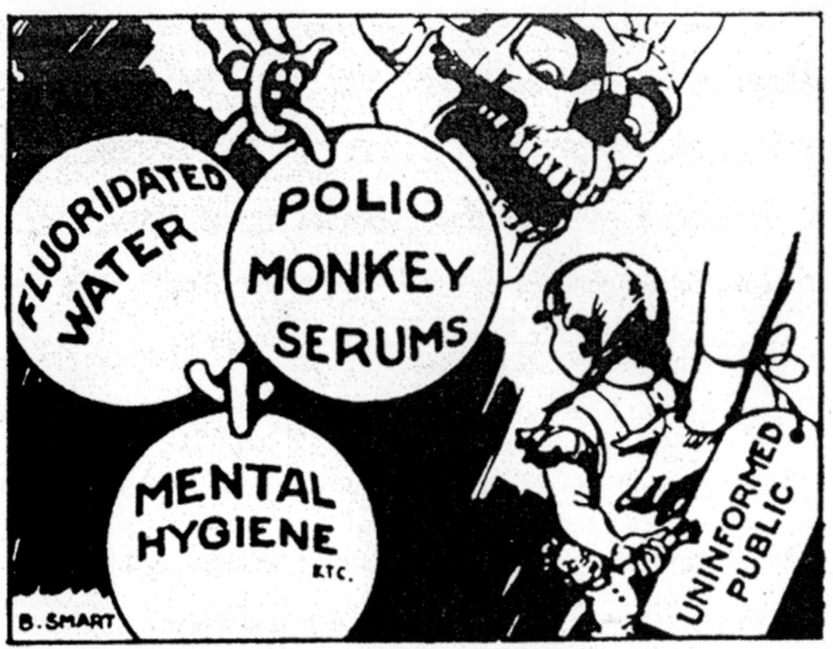

Are you willing to PUT IN PAWN to the UNHOLY THREE all of the material, mental and spiritual resources of this GREAT REPUBLIC?

FLUORIDATED WATER

1—Water containing Fluorine (rat poison—no antidote) is already the only water in many of our army camps, making it very easy for saboteurs to wipe out an entire camp personel. If this happens, every citizen will be at the mercy of the enemy—already within our gates.

POLIO SERUM

2—Polio Serum, it is reported, has already killed and maimed children; its future effect on minds and bodies cannot be guaged. This vacciné drive is the entering wedge for nation-wide socialized medicine, by the U. S. Public Health Service, (heavily infiltrated by Russian-born doctors, according to Congressman Clare Hoffman.) In enemy hands it can destroy a whole generation.

MENTAL HYGIENE

3—Mental Hygiene is a subtle and diabolical plan of the enemy to transform a free and intelligent people into a cringing horde of zombies.

Rabbi Spitz in the American Hebrew, March 1, 1946: "American Jews must come to grips with our contemporary anti-Semites; we must fill our insane asylums with anti-Semitic lunatics."

FIGHT COMMUNISTIC WORLD GOVERNMENT by destroying THE UNHOLY THREE ! ! ! It is later than you think!

KEEP AMERICA COMMITTEE
Box 3094, Los Angeles 54, Calif. H. W. Courtois, Secy. May 16, 1953

Figure 8.4 Groupthink as Seen in McCarthy-Era Propaganda

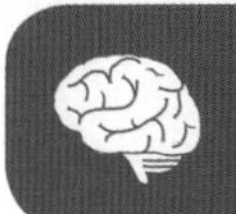

Similar patterns of thinking, in which a group arrives at a common (but often extreme) consensus also underlie many cultural phenomena, including riots, fads, and mass hysteria. Antinormative behavior in riots was described previously in the section on deindividuation. Still, like groupthink, a shared political or social motivation may urge groups to engage in potentially violent and destructive behavior. A **fad** is a behavior that is transiently viewed as popular and desirable by a large community. Fads can include owning certain objects (such as pet rocks in the 1970s, Rubik's cubes in the 1980s, and pogs in the 1990s) or engaging in certain behaviors (using catchphrases, altering clothing in some way, or engaging in particular types of media such as viral videos). Finally, **mass hysteria** refers to a shared, intense concern about the threats to society. In mass hysteria, many features of groupthink—collective rationalization, illusion of morality, excessive stereotyping, and pressure for conformity, in particular—lead to a shared delusion that is augmented by distrust, rumors, propaganda, and fear mongering. Perhaps the most notable historical case of mass hysteria was the Salem witch trials in colonial Massachusetts, which led to the execution of twenty individuals for fears of witchcraft.

Culture

Culture can be defined as the beliefs, behaviors, actions, and characteristics of a group or society of people. Culture is learned by living within a society, observing behaviors and traits, and adopting them. Culture is also passed down from generation to generation. While a "cultured" individual is often thought of as someone who has knowledge of the arts and expensive taste, sociology considers all people to be cultured by living within a society and participating in its culture. Culture is universal throughout humanity; while many animals exhibit purely instinctual behavior, humans show variable behaviors based on the cultures in which they reside. For example, while all wolf parents care for their pups in the same manner, human parents show vast differences in their caregiving. In some cultures, children are breastfed for years, while in others, infants are breastfed for mere months or not at all. Some groups have multiple caregivers who are not a parent, while others allow only a parent to care for the child. Even within a given culture, beliefs about the correct way to respond to infants crying can vary dramatically: some groups instantly comfort crying children and others let them "cry it out." The beliefs held by an individual are typically based on learned behavior, expectations, and pressure from the group one is in. Cultural differences include everything from typical jobs, common dwellings, and diet to what time of day one eats and where one travels on vacation, if at all. When traveling outside of one's own society, these cultural differences can seem quite dramatic and are often referred to as **culture shock**.

BRIDGE

A discussion of culture in the context of social structure is described in Chapter 11 of *MCAT Behavioral Sciences Review*.

Assimilation and Multiculturalism

Cultural **assimilation** is the process by which an individual's or group's behavior and culture begin to resemble that of another group. This can also mean that groups with different cultures begin to merge. Assimilation integrates new aspects of a society and culture with old ones, transforming the culture itself. While one society melds into another, it is typically not an even blend. One group will generally have more power and influence than the other, resulting in more traits of that culture being displayed after transformation. Four primary factors are sometimes used to assess the degree of assimilation in immigrant communities: socioeconomic status, geographic distribution, language attainment, and intermarriage.

One alternative to assimilation is the creation of **ethnic enclaves**, which are locations (usually neighborhoods) with a high concentration of one specific ethnicity, as shown in Figure 8.5. These are most common in urban areas and often have names like Chinatown or Little Italy.

Figure 8.5 An Ethnic Enclave
Entrance to Chinatown, Sydney, Australia

KEY CONCEPT

- Assimilation—(usually uneven) merging of cultures; a melting pot
- Multiculturalism—celebration of coexisting cultures; a cultural mosaic

Multiculturalism, also known as **cultural diversity**, refers to communities or societies containing multiple cultures or ethnic groups. From a sociology perspective, multiculturalism encourages, respects, and celebrates cultural differences, as shown in Figure 8.6. This view can enhance acceptance of cultures within society, which contrasts with the concept of assimilation. While multiculturalism is often described as a creating a *cultural mosaic*, or mixture of cultures and ethnic groups that coexist in society, assimilationism is described as creating a *melting pot*, or melting together of different elements of culture into one homogeneous culture.

Figure 8.6 Multiculturalism
Multiculturalism may be celebrated through holidays and festivals, such as Harmony Day in Australia, shown here.

Subcultures

Subcultures refer to groups of people within a culture that distinguish themselves from the primary culture to which they belong. When studying subcultures, symbolic attachment to things such as clothing or music can differentiate the group from the majority. Subcultures can be formed based on race, gender, ethnicity, sexuality, and other differentiating factors from the whole of society.

Subcultures can be perceived as negative when they subvert the majority culture's definitions of normalcy. In the case of **counterculture**, the subculture group gravitates toward an identity that is at odds with the majority culture and deliberately opposes the prevailing social mores.

MCAT CONCEPT CHECK 8.1

Before you move on, assess your understanding of the material with these questions.

1. Provide a brief definition for the following social phenomena:
    - Social facilitation:

      ___

    - Deindividuation:

      ___

    - Bystander effect:

      ___

    - Social loafing:

      ___

    - Peer pressure:

      ___

2. What are the similarities and differences between group polarization and groupthink?

   ___

3. What are the differences between assimilation and multiculturalism?

   ___

8.2 Socialization

High-Yield

LEARNING OBJECTIVES

After Chapter 8.2, you will be able to:

- Distinguish between conformity, compliance, and obedience
- Compare and contrast primary and secondary socialization
- Describe compliance techniques, such as foot-in-the-door, door-in-the-face, lowball, and that's-not-all

More than any other animal, humans use social experiences to learn acceptable behavior in the culture in which they live. Sociologists and psychologists use the term **socialization** when discussing the process of developing, inheriting, and spreading norms, customs, and beliefs. Individuals gain the knowledge, skills, habits,

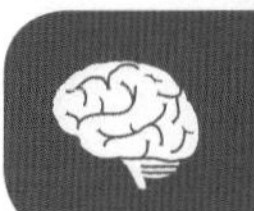

and behaviors that are necessary for inclusion in society. Widely held views in a society become the accepted viewpoints and are generally adopted by the majority of individuals within that society. Beliefs, customs, and cultural norms are often passed down from one generation to another within a society in a process called **cultural transmission** or **cultural learning**. Spread of norms, customs, and beliefs from one culture to another can also occur, and is called **cultural diffusion**.

Socialization can be further categorized. **Primary socialization** occurs during childhood when we initially learn acceptable actions and attitudes in our society, primarily through observation of our caregivers and other adults in close proximity. In children, this sets the stage for future socialization and provides the foundation for creating personal opinions. **Secondary socialization** is the process of learning appropriate behavior within smaller sections of the larger society. This type of socialization occurs outside of the home and is based on learning the rules of specific social environments. For example, the behavior necessary to thrive in school is different from that in the home setting, and also from that which is acceptable on a sports field or in a church. Secondary socialization is typically associated with adolescents and adults and includes smaller changes and refinements to behavior that were established in primary socialization. Secondary socialization can also occur when moving to a new region or changing schools or professions. **Anticipatory socialization** is the process by which a person prepares for future changes in occupations, living situations, or relationships. A premedical student shadowing physicians to assimilate and practice appropriate behaviors in expectation of one day becoming a doctor is an example of anticipatory socialization. **Resocialization** is another process by which one discards old behaviors in favor of new ones, typically through intensive retraining, and can have positive or negative connotations. The method by which members of the armed forces are trained to obey orders and commands without hesitation is a prime example of resocialization, but so is attracting and indoctrinating members into a cult.

REAL WORLD

Cults that have become a mainstay in media today are often "Doomsday cults." This term refers both to groups that prophesy catastrophe and apocalypse and to those who attempt to bring it about. In December of 2012, nearly 1000 members of the Chinese cult Church of Almighty God were arrested for broadcasting fears of apocalypse and encouraging the overthrow of the Communist Party.

Norms

Sociologists define **norms** as societal rules that define the boundaries of acceptable behavior. **Mores** are widely observed social norms. While norms are not laws, they do provide a mechanism for regulating the behavior of individuals and groups and thereby serve as a means of **social control**. Penalties for misconduct or rewards for appropriate behavior, called **sanctions**, can also be used to maintain social control. Negative sanctions punish behaviors that deviate from norms, while positive sanctions reward behaviors that comply with norms. Sanctions can also be categorized as formal or informal. **Formal sanctions** are enforced by formal social institutions like governments or employers and can include receiving a promotion (positive) or a jail sentence (negative). By contrast, **informal sanctions** are enforced by social groups. Informal sanctions might include being allowed to sit at a particular table in the school cafeteria (positive) or exclusion from a social group (negative).

Norms provide us with a sense of what is appropriate, what we should do, and what is considered **taboo**—socially unacceptable, disgusting, or reprehensible. Norms exist for behavior, speech, dress, home life, and more and can differ between groups within a society, and also between different cultures. For example, many Americans tend to be extraverted and talkative, even among strangers, while Japanese culture sometimes teaches that showing too much of oneself in a public setting is a sign of weakness. Thus, a very quiet person who does not make eye contact could seem odd in America but may fit in perfectly in Japan. **Folkways** are norms that refer to behavior that is considered polite in particular social interactions, such as shaking hands after a sports match, as seen in Figure 8.7.

Figure 8.7 Folkways
An act as simple as shaking hands after a sporting match is an example of a folkway.

Agents of Socialization

Any part of society that is important when learning social norms and values is called an **agent of socialization**. For children, the primary agents of socialization are caregivers, often parents or other family members. Direct family remains an important agent of socialization for adolescents, but social circles—including friends, peers, and teachers—become important agents as well. For adults, colleagues and bosses can also act as agents of socialization. Aside from personal relationships, the environment is another agent of socialization. For example, when entering college, some teenagers may experience a complete lifestyle change and are in nearly constant interaction with people of their own age. This new environment creates a shift in acceptable behavior that can include late nights out with friends or all-night study sessions. When entering the workforce, another change in environment leads to socialization within the organization. Ethnic background, religion, and government also play a role in learned behavior, and are therefore also important agents of socialization. And geography at the national, regional, and neighborhood levels also dictates norms of behavior: acceptable behavior in downtown Manhattan is not identical to acceptable behavior in rural Montana.

REAL WORLD

A study published in 2002 examined the link between television and violence. In the study, the researchers followed 707 subjects for 17 years. Study subjects who watched more than one hour of TV per day were approximately four times more likely to act aggressively toward others later in life than those who watched less than one hour of TV. Of those who watched more than three hours of TV per day, nearly 30% were involved in assaults, robberies, or other aggressive behaviors. While this is an interesting link, remember that it shows correlation—not causation!

Furthermore, the media are an important agent of socialization through their influence on what is accepted within a particular society. **Popular culture**, i.e. common

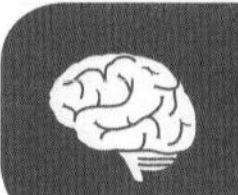

trends and beliefs prevalent at a given point in time, is heavily influenced by the media. The media can determine what is considered important in a particular society. Mass media is most commonly accessed through television, radio, newspapers, and the Internet. It delivers impersonalized communication to a vast audience, and can thereby establish trends in American or international popular culture. Many of the agents of socialization are summarized in Figure 8.8.

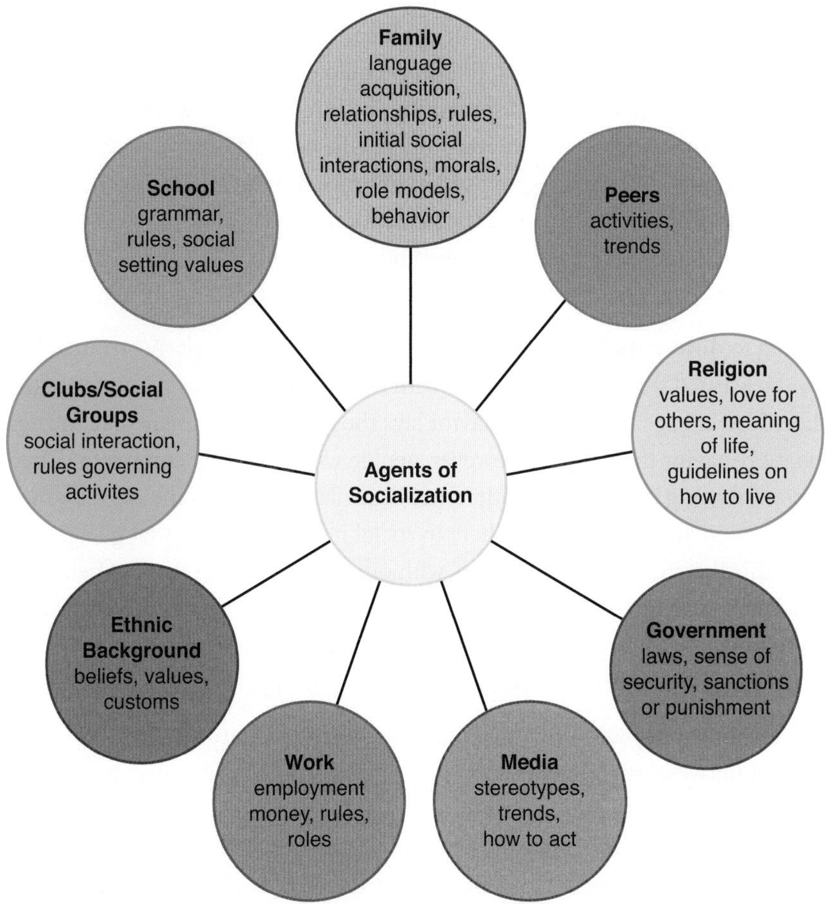

Figure 8.8 Agents of Socialization

Deviance and Stigma

Deviance refers to any violation of norms, rules, or expectations within a society. It is important to note that using the term *deviant* is often associated with strongly negative connotations; however, in the sociological context, it simply refers to any act that goes against societal norms. Deviance can vary in severity, from something as simple as jaywalking to something as serious as committing murder. Deviance also includes any act that meets with disapproval from the larger society.

Social stigma is the extreme disapproval or dislike of a person or group based on perceived differences from the rest of society. These deviations from the norm can include differences in beliefs, abilities, behaviors, and appearance. Certain medical conditions such as HIV, achondroplasia (dwarfism), and obesity can also be stigmatized. Stigma can

BRIDGE

Mental illness has long been stigmatized in American society. While this is slowly changing, the potential stigma associated with a mental health diagnosis continues to be a hurdle to many patients seeking out or receiving care. Many common psychological disorders are discussed in Chapter 7 of *MCAT Behavioral Sciences Review*.

also spread to affect others who are associated with a particular individual. For example, family members of an alleged or convicted murderer or rapist can be stigmatized. Stigma also evolves over time: whereas divorce was stigmatized in the early twentieth century, it no longer has such strong negative connotations.

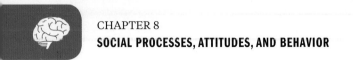

Deviance, stigmatization, and reputation are strongly linked with the **labeling theory**. This theory posits that the labels given to a person affect not only how others respond to that person, but also affect that person's self-image. Labels can lead to channeling of behavior into deviance or conformity. For example, if members of society label a person as a criminal, this label could either lead to further criminality or to a change in behavior toward something more in line with what is accepted in that society. In many instances, we resist being labeled, particularly with labels we perceive as negative. However, groups may embrace deviant labels. Biker gangs, for example, utilize labeling to enhance the perception of their own subgroup. Internalizing a label and assuming the role implied by the label may lead to the assumed role taking over a person's identity, a phenomenon known as **role engulfment**.

According to **differential association theory**, deviance, particularly criminal behavior, can be learned through interactions with others. In this theory, intimate exposure to others who engage in deviant behavior lays the groundwork for people to engage in deviant behavior themselves. However, people will also likely come into contact with norm-abiding individuals. Differential association, then, is the degree to which one is surrounded by ideals that adhere to social norms *vs.* ideals that go against them. In this theory, when associations with others engaging in deviant behavior are more numerous or intense than those engaging in normative behavior, individuals begins to gravitate toward deviant behavior themselves. In common language, we might describe someone as having "fallen into the wrong group."

Finally, **strain theory** attempts to explain deviance as a natural reaction to the disconnect between social goals and social structure. One common example in strain theory is the American dream, which refers to acquiring wealth and personal stability through achievement and hard work. The American dream is considered a desirable social goal, but the structure of society is unable to guarantee the education and opportunity needed to achieve this goal to all citizens. Therefore, deviant behavior such as theft may arise as an attempt to achieve the social goal *outside* of the limiting social structure.

While deviance is often associated with negative behavior such as crime, functional theorists argue that it is necessary for social order. These theorists argue that deviance provides a clear perception of social norms and acceptable boundaries, encourages unity within society, and can even promote social change.

Conformity, Compliance, and Obedience

While deviance is defined as going against societal norms, conformity, compliance, and obedience are manners of adhering to social expectations or others' requests.

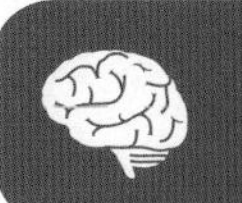

Conformity

Conformity is matching one's attitudes, beliefs, and behaviors to group or societal norms. The pressure to conform can be real or imagined: an actual pressure from others, or a perceived pressure or expectation. Conformity is also known as **majority influence**. The Asch experiments, described earlier, showed the strength of social influence on **normative conformity**, the desire to fit into a group because of fear of rejection.

There are distinct types of conformity, including internalization and identification. **Internalization** involves changing one's behavior to fit with a group while also privately agreeing with the ideas of the group. **Identification** refers to the outward acceptance of others' ideas without personally taking on these ideas.

A classic experiment looking at internalization was Philip Zimbardo's Stanford Prison Experiment. Zimbardo advertised for a role-playing experiment in which he recruited 21 male college students. The study participants were randomly assigned the role of prisoner or guard. The prisoners were arrested in their homes and taken to a "prison" created in the Stanford University psychology building. Guards were issued uniforms, including whistles, handcuffs, and dark glasses to prevent eye contact. The prisoners and guards quickly fell into their roles and displayed related behaviors almost immediately. Guards began to taunt and harass prisoners, appearing to enjoy their role. Prisoners also adopted their new role, taking the prison rules very seriously, and becoming more and more dependent on the guards. As the guards became more aggressive, the prisoners became more submissive, although they also attempted to mount a revolt. The study had to be ended after six days because the guards had begun to physically abuse the prisoners so severely that ethical concerns were raised. After the study ended, Zimbardo interviewed each participant. The guards and prisoners, who had internalized their roles, were both shocked by their behavior during the experiment.

The likelihood of conformity differs among cultures. For instance, individualist cultures tend to value independent thought and unique ideas and are thus less likely to conform; in collectivist cultures, group mentality often supersedes the individual. The latter type of society tends toward conformity.

KEY CONCEPT

Internalization and identification both deal with accepting others' ideas, but whereas internalization also reflects a change in internal thoughts to agree with the idea, identification is acceptance of the idea on the surface level without internalizing it.

REAL WORLD

In recent years, design flaws in the Stanford Prison Experiment have come to light. These include sampling biases, ethical concerns, lack of controls, and active participation in the experiment by Zimbardo himself.

Compliance

Compliance is a change in behavior based on a direct request. The person or group that asks the individual to make the change typically has no actual power or authority to command the individual, yet will ask the individual to change behavior. There are several notable techniques used to gain compliance of others, particularly within the marketing arena. The **foot-in-the-door** technique begins with a small request, and after gaining compliance, a larger request is made. An example of this scenario could be a fellow classmate asking to borrow your notes after missing class. You agree and offer to share the notes at the next class session. Later in the day, when you see the student again, you're asked if you would be willing to make copies of your notes too. Many people will still agree at this point, as the first request opened the door to continued compliance.

The next technique is called the **door-in-the-face technique**. This is the opposite of the foot-in-the-door technique, wherein a large request is made at first and, if refused, a second, smaller request is made. Often, this smaller request is the actual goal of the requester. Using this technique, a fellow student might ask you to make a copy of your notes from class and bring them to the next class. If you deny the request, the student might follow up with a smaller request, asking to borrow your notes to make personal copies. The second, more reasonable request may be granted.

Another common method of achieving compliance is the **lowball technique**. In this technique, the requestor will get an initial commitment from an individual, and then raise the cost of the commitment. It is important to note that cost need not only include money, but can also include effort and time. An example of this technique is a scenario in which you are asked by your boss to head a committee with a time commitment of five hours per month of meetings. You agree to head the committee, but later discover that the commitment also includes written reports from each meeting and a quarterly presentation.

Yet another technique used to gain compliance is the **that's-not-all technique**. In this method, an individual is made an offer, but before making a decision, is told the deal is even better than expected. This method is frequently seen in infomercials:
We can offer you these earrings for the stunningly low price of $19.99. But wait! If you buy them, you'll also receive our matching necklace, normal retail value $49.99, absolutely free.

Obedience

While compliance deals with requests made by people without actual authority over an individual, **obedience** is changing one's behavior in response to a direct order or expectation expressed by an authority figure. While a classmate has no authority to demand notes from you, an authority figure has social power over other individuals. For instance, if a teacher demands that you hand over your notes from class, you would be obeying rather than complying. People are far more likely to obey than comply due to the real or perceived social power of the individual.

One of the most notable obedience experiment series was conducted by Stanley Milgram. In this classic set of studies, Milgram claimed to be recruiting participants for a study to test the effects of punishment on learning behavior. Participants were told they would be randomly assigned to be the "teacher" or "learner"; however, the "learner" was actually a paid actor (confederate). The teachers were told that they would be controlling an electrical panel that would administer shocks to the learners if they made mistakes. Prior to giving the first shock, the teachers were given a sample 45 V shock to make them aware of what they would be doing to the learners. The teachers were then told that they would need to increase the voltage by 15 V each time an incorrect response was given. The learners, who received no actual shock, were provided with scripts telling them to show pain, ask to stop the experiment, scream, and even feign passing out. As the learners acted more and more uncomfortable, the teachers became less willing to increase the shock voltage. However, by using increasingly demanding language (from *Please continue* to *You have no other choice, you must go on*), the researchers were able to get 65 percent of the participants to administer shocks to the maximum of 450 V, even if they showed discomfort in doing so. Milgram and other researchers were surprised at the level of obedience the participants showed during the experiment. This type of experiment has been repeated many times and has consistently shown that more than 60 percent of people will obey even if they do not wish to continue.

MCAT CONCEPT CHECK 8.2

Before you move on, assess your understanding of the material with these questions.

1. What is the difference between primary and secondary socialization?

____

2. What are conformity, compliance, and obedience?
   - Conformity:

____

   - Compliance:

____

   - Obedience:

____

3. For each of the compliance techniques listed below, provide a brief description:
   - Foot-in-the-door:

____

   - Door-in-the-face:

____

   - Lowball:

____

   - That's-not-all:

____

8.3 Attitudes and Behavior

High-Yield

LEARNING OBJECTIVES

After Chapter 8.3, you will be able to:

- Recall the three components of attitude
- Describe the four functional areas of the functional attitude theory
- Identify the roles of central route and peripheral route processing in the elaboration likelihood model
- Recall the three interactive factors of Bandura's triadic reciprocal causation

Social cognition focuses on the ways in which people think about others and how these ideas impact behavior. Our attitudes—the ways in which we perceive others—impact the ways we behave toward them.

Components of Attitudes

An **attitude** is the expression of positive or negative feeling toward a person, place, thing, or scenario. Attitudes develop from experiences with others who affect our opinions and behaviors. Even prior to meeting someone, past experiences and information from others can influence your attitude toward a person.

MNEMONIC

Components of Attitude: **ABC**

- **A**ffective
- **B**ehavioral
- **C**ognitive

There are three primary components of attitude: affective, behavioral, and cognitive. The **affective** component of attitude refers to the way a person feels toward something, and is the emotional component of attitude. *Snakes scare me* and *I love my family* are both affective expressions of attitude. The **behavioral** component of attitude is the way a person acts with respect to something. For example, avoiding snakes and spending time with one's family would reflect the behavioral component of the attitudes described earlier. Finally, the **cognitive** component of attitude is the way an individual thinks about something, which is usually the justification for the other two components. In the snake example above, knowing that snakes can be dangerous (and sometimes venomous) provides a reason to be afraid of snakes and to avoid them.

Theories of Attitudes

The **functional attitudes theory** states that attitudes serve four functions: knowledge, ego expression, adaptation, and ego defense. The **knowledge function** can be summarized as follows: attitudes help provide organization to thoughts and experiences, and knowing the attitudes of others helps to predict their behavior. For example, one would predict that an individual who cares about political action would vote in an upcoming election. Attitudes facilitate being **ego-expressive**, allowing us to communicate and solidify our self-identity. For instance, if a person strongly identifies with a sports team, that person might wear a team hat to identify as having a positive attitude towards that team. **Adaptive** attitude is the idea that expressed socially acceptable attitudes will lead to acceptance. For example, a person declaring to a social group that they enjoyed a popular movie can help to build social bonds.

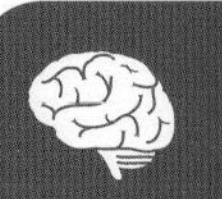

Lastly, attitudes are **ego-defensive** if they protect our self-esteem or justify actions that we know are wrong. For example, a child who has difficulty doing math may develop a negative attitude toward the subject.

Learning theory posits that attitudes are developed through different forms of learning. Direct contact with the object of an attitude can influence attitude towards that object. For example, children form a positive attitude toward sweets almost immediately after tasting them. Direct instruction from others can also influence attitudes. For instance, a child who is taught by caregivers not to use curse words can form a negative attitude toward curse words and, indirectly, a negative attitude toward those who use curse words. Our attitudes can also be influenced by others' attitudes. For example, teenagers may begin to have a positive attitude toward smoking if they notice that all of their friends smoke. Finally, attitudes may be formed through classical conditioning, operant conditioning, or observational learning, all of which are discussed in Chapter 3 of *MCAT Behavioral Sciences Review*.

The **elaboration likelihood model** is a theory of attitude formation and attitude change that separates individuals on a continuum based on how they process persuasive information. At one extreme are those who elaborate extensively, that is, those who think deeply about information, scrutinize its meaning and purpose, and draw conclusions or make decisions based on this analysis. Deep thinking in this manner is referred to as **central route processing**. When an attempt to influence attitudes uses information that appeals to central route processing, this attempt is said to be using the **central route to persuasion**. A scientific paper would be one example of an attempt to influence attitudes that uses the central route to processing. At the other extreme are those who do not elaborate and focus instead on superficial details such as the appearance of the person delivering the argument, catchphrases and slogans, and credibility. This type of processing is known as **peripheral route processing**. When attempts to influence attitudes appeal to peripheral route processing, these attempts are said to be using the **peripheral route to persuasion**. An advertisement with just a logo that contains a visually appealing image is one example of an attempt to influence attitudes that uses the peripheral route to persuasion. To contrast these two types of processing, consider two voters watching a well-informed and charismatic politician speak: One voter might be swayed by the cogent arguments made by the politician, and this illustrates high elaboration, central route processing. The other voter might be swayed by the perception that the speaker is likable and a good person, illustrating low elaboration, peripheral route processing. Most individuals fall in the middle of this continuum, and the degree to which we elaborate on information can vary depending on the specific situation.

KEY CONCEPT

- Central route processing (high elaboration)—scrutinizing and analyzing the content of persuasive information
- Peripheral route processing (low elaboration)—focusing on superficial details of persuasive information, such as appearances, catchphrases and slogans, and credibility

Social cognitive theory postulates that people learn how to behave and shape attitudes by observing the behaviors of others. According to this theory, behavior is not learned by trial-and-error, but develops through direct observation and replication of the actions of others. This learning is influenced by personal factors (such as thoughts about the behavior) and the environment in which the behavior is observed. These three factors—behavior, personal factors, and environment—are not independent concepts, but influence each other, as shown in Bandura's triadic

BEHAVIORAL SCIENCES GUIDED EXAMPLE WITH EXPERT THINKING

Corporate researchers conducted a longitudinal study of gym members at ten locations in the second week of January for five consecutive years. Participants engaged in a short series of physical activities and measurements assessing physical fitness, then responded to questions about attitudes and fitness. Participants who were present for at least three of the five years and received consistent, low scores in physical fitness measures were associated with the following: lower self-reported gym attendance, lower perceived value of exercise, and low expressed regard for those who attended the gym with a frequency greater than five visits/week. These participants also reported high levels of seasonal motivation (New Year's resolutions).

This is giving me context on a specific set of beliefs involving consistent behavior over time. With a consistent social behavior (high motivation and low adherence) I'm expecting questions about social psychology.

In response to these data, gym owners established a social media group devoted to creating small local workout groups meeting 3–4 days per week with regular check-ins for their members and invited gym attendees within this demographic set to join.

Social psychology: it sounds like the gym owner is trying for positive peer pressure.

How might the group who received consistent low scores in physical fitness measures respond to the new initiative if they choose to participate? How do the results of the five year study inform predictions of their adherence to the 3–4 day per week workout plan?

The writers of the MCAT will hint at a concept in the passage or question stem and then ask us a question about recognizing or applying that concept to the situation described. We'll want to be familiar enough with the content that when the question tells a story or mentions some specific example, we can say to ourselves "Oh, that's ________!"

In order to answer this question, we're going to have to dive into the scenario and identify factors that will influence behavioral outcomes. In the first paragraph, these gym members have maintained the same low level of fitness and low gym attendance for five years, which should remind us of attribution theory: consistent behaviors tend to have dispositional rather than situational explanations. Despite the finding of high seasonal motivation to engage with exercise, it's likely that this group of individuals aren't already in shape because of their attitudes towards themselves or working out ("I'm nervous people will judge me at the gym" or "I just don't enjoy lifting weights"), rather than situational barriers ("My preferred gym is being renovated" or "I'm just super busy this year"). In other words, such attributions are about "who they are" instead of "what has happened to them." We also get a hint in this direction with the finding about low regard for people who attend the gym very regularly. The cause/effect relationship for this set of findings and past behaviors is difficult to assess with the information provided, and may differ between participants. It could be that for some, the thoughts on gym enthusiasts influence their avoidance of the gym in an effort not to adopt similar behaviors. But for others, it could also be that their inability to get to the gym regularly might be influencing their attitudes about those who do in order to protect their egos.

Things are different this year, thanks to the creation of the new fitness group by the gym owners. Some subset of this group of participants have now joined a group of people who will be going to the gym together and will hold each other accountable. We should identify a specific social phenomenon as soon as we read that sentence: peer pressure!

Now that we've identified the important information in the scenario, we should return to the question itself. The question asks how the information we already have about this group of participants might predict their behaviors, which are the three factors of social-cognitive theory. Social-cognitive theory says that all of these things influence one another. If we put together everything we've learned, we see that the attitudes expressed in this group make it less likely that they'll stick to working out, but they've also taken steps to change their environment by opting to join a new peer group for themselves, which makes it more likely that their behaviors may change. While it's impossible to tell which of these influences will win out over the other, it is possible that putting themselves into this new environment and changing their behaviors will in fact change their attitudes over time to be more accepting of those people (now including themselves!) who go to the gym often.

In short, our answer is: the data on this group indicates a possible disposition that makes it less likely they will consistently go and work out at the gym. However, there is now a competing influence to go to the gym that applies to them based on peer pressure from their new group.

reciprocal causation in Figure 8.9. For example, the work ethic of employees in a company (behavior) is affected by how hard their colleagues work, their previous attitudes toward hard work (personal), and the systems and infrastructure of the company (environment). Reciprocally, this behavior may create a change in the employees' attitude toward work (personal) and the systems within the company (environment).

MCAT EXPERTISE

MCAT passages tend to describe an experiment or a scenario and drop in sentences or even short phrases that hint at related scientific content that is then used in questions. The better you are at recognizing this content, the more ready you will be to answer these questions quickly and correctly. But don't feel like you need to scour each passage for every single concept that could appear in the question set.

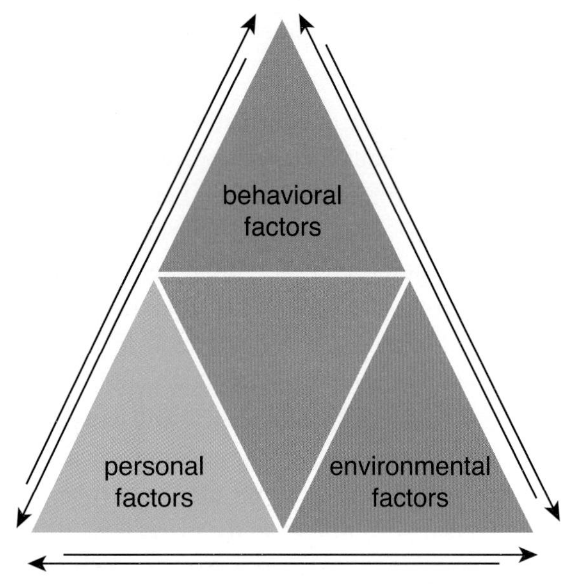

Figure 8.9 Bandura's Triadic Reciprocal Causation

MCAT CONCEPT CHECK 8.3

Before you move on, assess your understanding of the material with these questions.

1. What are the three components of attitude?

    -
    -
    -

2. What are the four functional areas of the functional attitudes theory?

    -
    -
    -
    -

3. What are the routes of processing used to explain the elaboration likelihood model? Which is associated with high elaboration?

    -
    -

4. What are the three interactive factors of Bandura's triadic reciprocal causation?
   -
   -
   -

Conclusion

Human behavior is learned and influenced by those around us. Simply being around others can lead to changes in behavior from how we act when alone. Pressure from others can also lead to changes in behavior when people crave acceptance: inclusion in a group can lead to changes in thought processes and decision making. Social norms, which are learned through experience and observation, can be rejected by individuals, leading to deviance and stigmatization. However, the need to fit in can lead to conformity, compliance, and obedience. Attitudes are also developed through observations, experiences, and interactions with others, and there are multiple theories explaining their specific methods of formation.

This chapter focused primarily on how groups influence an individual's behavior. In the next chapter, we'll look at the structure of these groups and how we present ourselves to the larger society. We'll explore methods of communication between individuals—both verbal and nonverbal—and consider how we encourage others to gain certain impressions about us.

You've reviewed the content, now test your knowledge and critical thinking skills by completing a test-like passage set in your online resources!

CONCEPT SUMMARY

Group Psychology

- **Social facilitation** describes the tendency of people to perform at a different level based on the fact that others are around.
- **Deindividuation** is a loss of self-awareness in large groups, which can lead to drastic changes in behavior.
- The **bystander effect** describes the observation that when in a group, individuals are less likely to respond to a person in need.
- **Peer pressure** refers to the social influence placed on individuals by others they consider equals.
- Group decision making may differ from individual decision making.
  - Group **polarization** is the tendency toward making decisions in a group that are more extreme than the thoughts of the individual group members.
  - **Groupthink** is the tendency for groups to make decisions based on ideas and solutions that arise within the group without considering outside ideas. Ethics may be disturbed as pressure is created to conform and remain loyal to the group.
- **Culture** describes the beliefs, ideas, behaviors, actions, and characteristics of a group or society of people.
  - **Assimilation** is the process by which a group or individual's culture begins to melt into another culture.
  - **Multiculturalism** refers to the encouragement of multiple cultures within a community to enhance diversity.
  - **Subcultures** refer to a group of people within a culture that distinguish themselves from the primary culture to which they belong.

Socialization

- **Socialization** is the process of developing and spreading norms, customs, and beliefs.
- **Norms** are what determine the boundaries of acceptable behavior within society.
- Agents of socialization include family, peers, school, religious affiliation, and other groups that promote socialization.
- **Stigma** is the extreme disapproval or dislike of a person or group based on perceived differences from the rest of society.
- **Deviance** refers to any violation of norms, rules, or expectations within a society.
- **Conformity** is changing beliefs or behaviors in order to fit into a group or society.

- **Compliance** occurs when individuals change their behavior based on the requests of others. Methods of gaining compliance include the foot-in-the-door technique, door-in-the-face technique, lowball technique, and that's-not-all technique, among others.
- **Obedience** is a change in behavior based on a command from someone seen as an authority figure.

Attitudes and Behavior

- **Attitudes** are tendencies toward expression of positive or negative feelings or evaluations of something.
- There are affective, behavioral, and cognitive components to attitudes.
- The **functional attitudes theory** states that there are four functional areas of attitudes that serve individuals in life: knowledge, ego expression, adaptability, and ego defense.
- The **learning theory** states that attitudes are developed through forms of learning: direct contact, direct interaction, direct instruction, and conditioning.
- The **elaboration likelihood model** states that attitudes are formed and changed through different routes of information processing based on the degree of elaboration (**central route processing**, **peripheral route processing**).
- The **social cognitive theory** states that attitudes are formed through observation of behavior, personal factors, and environment.

ANSWERS TO CONCEPT CHECKS

8.1

1. Social facilitation describes the tendency of people to perform at a different level based on the fact that others are around. Deindividuation is the idea that people will lose a sense of self-awareness and can act dramatically different because of the influence of a group. The bystander effect describes the observation that individuals are less likely to respond to a person in need when in a group. Social loafing refers to a decrease in effort seen when individuals are in a group. Peer pressure refers to the social influence placed on individuals by others they consider their equals.
2. Group polarization and groupthink are both social processes that occur when groups make decisions. Group polarization is the tendency toward extreme decisions in a group. Groupthink is the tendency for groups to make decisions based on ideas and solutions that arise within the group without considering outside ideas, given the pressure to conform and remain loyal to the group.
3. Societies that contain multiple cultures can exhibit multiculturalism or assimilation. Assimilation is the process by which multiple cultures begin to merge into one, typically with an unequal blending of ideas and beliefs. Multiculturalism refers to the idea that multiple cultures should be encouraged and respected without one culture becoming dominant overall.

8.2

1. Primary socialization refers to the initial learning of acceptable behaviors and societal norms during childhood, which is facilitated mostly by caregivers and other adults. Secondary socialization refers to learning the norms of specific subgroups or situations during adolescence and adulthood.
2. Conformity is changing beliefs or behaviors in order to fit into a group or society. Compliance occurs when individuals change their behavior based on the request of others who do not wield authority over the individual. Obedience is a change in behavior because of a request from an authority figure.
3. The foot-in-the-door technique refers to asking for favors that increase in size with each subsequent request. The door-in-the-face technique refers to making a large request and then, if refused, making a smaller request. The lowball technique refers to gaining compliance without revealing the full cost (money, effort, or time) of the favor. The that's-not-all technique refers to increasing the reward for a request before an individual has the chance to make a decision.

8.3

1. The three components of attitude are affective, behavioral, and cognitive.
2. The four functional areas of the functional attitudes theory are knowledge, ego expression, adaptation, and ego defense.
3. The routes of processing used to explain the elaboration likelihood model are central route processing and peripheral route processing. Central route processing is associated with high elaboration.
4. The three interactive factors of Bandura's triadic reciprocal causation are behavior, personal factors, and environment.

SCIENCE MASTERY ASSESSMENT EXPLANATIONS

1. **C**

When fulfilling particular roles, an individual's behavior can be very out of character. The changing of one's behavior (and internal ideas) to match a group is called internalization conformity. This was a key part of the experiment. The experiment also involved deindividuation, the loss of self-identity in the group setting that can lead to antinormative or violent behavior.

2. **B**

The fact that individual opinions became more extreme during group discussion is explained by group polarization. The jury member initially felt that a strict penalty should be given, but this opinion became more extreme after conversation with the rest of the group.

3. **A**

It has been observed that increasing the number of bystanders decreases the likelihood that any of them will aid a victim. Increasing the degree of danger experienced by the victim, **(B)**, making the victim an acquaintance instead of a stranger, **(C)**, and being alone in the room with the victim, **(D)**, would increase the likelihood that the bystander would help the victim.

4. **D**

With groupthink, a member would perform all of the actions described by the answer choices except create a sense of negativity against risk taking; in fact, there is optimism and encouragement toward risk taking in groupthink.

5. **A**

Primary socialization is the teaching of acceptable actions and attitudes during childhood, which would occur too early to be part of the adult prison system. Resocialization, **(D)**, is the process by which one changes behaviors by discarding old routines and patterns and transitions to new behaviors necessary for a life change. The prison environment is designed to change unlawful behavior into desired behavior, making this an incorrect choice. When entering prison, an inmate must also undergo secondary socialization, **(B)**, learning the rules of the specific social environment of the prison. Finally, if the inmate is not incarcerated for life, attempts at anticipatory socialization, **(C)**, must be made before releasing the inmate in preparation for life outside of the prison.

6. **C**

This is a prime example of the foot-in-the-door technique. The neighbors first ask for a small favor and, after receiving commitment, ask for a larger favor.

7. **A**

The affective component of attitude consists of feelings and emotions toward something.

8. **D**

Peripheral route processing deals with processing information that is not based on content, but instead on superficial parameters such as boring speech patterns or appearance of the speaker. Central route processing, **(C)**, is the processing of information through analysis of its content.

9. **B**

The Milgram shock experiment showed that individuals would obey orders from authority figures even if they were not comfortable with the task at hand. Conformity and compliance, **(C)** and **(D)**, also deal with changes in individual behavior, but are not based on the requests of an authority figure.

10. **C**

Internalization refers to the type of conformity in which individuals change their outward opinion to match the group and also personally agree with those ideas.

11. **B**

The four functional areas of the functional attitudes theory are knowledge, adaptability, ego expression, and ego defense. Acceptance into a group may influence attitudes or opinions; however, this is not a part of the functional attitudes theory.

12. **A**

For 14 out of the 16 members, the record times were obtained during public meets. The fact that the team members performed better when in front of a crowd supports the notion of social facilitation.

13. **D**

This young person is preparing to spend time in a new social setting. The process of preparing for future changes in environment is considered anticipatory socialization.

14. **B**

Social action is best described as the effects that a group has on individual behavior, including social facilitation, deindividuation, the bystander effect, social loafing, and peer pressure. Social interaction describes how two or more individuals influence each other's behavior, including group polarization and groupthink.

15. **C**

Placing spoken or unspoken expectations on individuals to agree with the ideas of the group is best described as pressure for conformity.

Consult your online resources for additional practice.

SHARED CONCEPTS

Behavioral Sciences Chapter 5
Motivation, Emotion, and Stress

Behavioral Sciences Chapter 6
Identity and Personality

Behavioral Sciences Chapter 9
Social Interaction

Behavioral Sciences Chapter 10
Social Thinking

Behavioral Sciences Chapter 11
Social Structure and Demographics

Physics and Math Chapter 11
Reasoning About the Design and Execution of Research

CHAPTER 9

SOCIAL INTERACTION

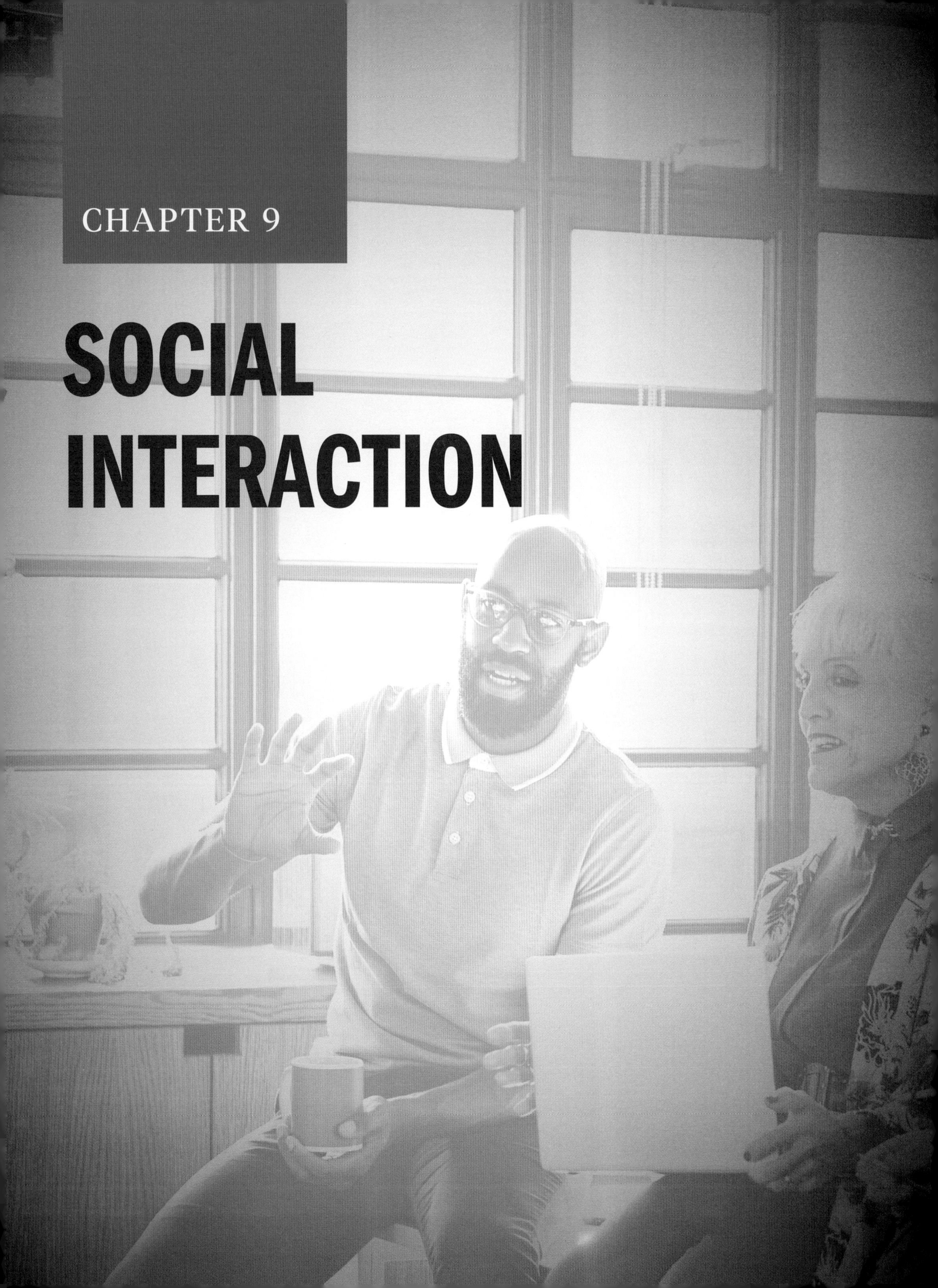

SCIENCE MASTERY ASSESSMENT

Every pre-med knows this feeling: there is so much content I have to know for the MCAT! How do I know what to do first or what's important?

While the high-yield badges throughout this book will help you identify the most important topics, this Science Mastery Assessment is another tool in your MCAT prep arsenal. This quiz (which can also be taken in your online resources) and the guidance below will help ensure that you are spending the appropriate amount of time on this chapter based on your personal strengths and weaknesses. Don't worry though—skipping something now does not mean you'll never study it. Later on in your prep, as you complete full-length tests, you'll uncover specific pieces of content that you need to review and can come back to these chapters as appropriate.

How to Use This Assessment

If you answer 0–7 questions correctly:

Spend about 1 hour to read this chapter in full and take limited notes throughout. Follow up by reviewing **all** quiz questions to ensure that you now understand how to solve each one.

If you answer 8–11 questions correctly:

Spend 20–40 minutes reviewing the quiz questions. Beginning with the questions you missed, read and take notes on the corresponding subchapters. For questions you answered correctly, ensure your thinking matches that of the explanation and you understand why each choice was correct or incorrect.

If you answer 12–15 questions correctly:

Spend less than 20 minutes reviewing all questions from the quiz. If you missed any, then include a quick read-through of the corresponding subchapters, or even just the relevant content within a subchapter, as part of your question review. For questions you answered correctly, ensure your thinking matches that of the explanation and review the Concept Summary at the end of the chapter.

1. Which of the following best describes the sociological definition of a status?
   - **A.** The emotional state of a social interaction
   - **B.** Expectations that are associated with a specific title in society
   - **C.** A position in society used to classify an individual
   - **D.** A means to describe one's peers

2. Becoming a college graduate requires hard work and diligence in academics. As such, being a college graduate could be considered a(n):
   - **A.** ascribed status.
   - **B.** achieved status.
   - **C.** master status.
   - **D.** cardinal status.

3. A bureaucracy is a specific example of a(n):
   - **A.** immediate network.
   - **B.** primary group.
   - **C.** organization.
   - **D.** reference group.

4. Which of the following is NOT characteristic of a bureaucracy?
   - **A.** Rigidly defined work procedures
   - **B.** Requirement for officials to hold an advanced degree
   - **C.** Regular salary increases
   - **D.** Election by constituents

5. Which of the following is a form of verbal communication?
   - **A.** Facial expressions
   - **B.** Eye contact
   - **C.** Written text
   - **D.** Body language

6. Which of the following best describes the impression management strategy of aligning actions?
   - **A.** Adhering to the behaviors that are expected for a given role in society
   - **B.** Relieving tension brought about by holding conflicting views in one's head
   - **C.** Providing socially acceptable reasons to explain unexpected behavior
   - **D.** Dictating that members of a group should follow similar practices to one another

7. While on the phone, a friend says: "A good friend would let me borrow the bike." This friend is using which impression management strategy?
   - **A.** Managing appearances
   - **B.** Alter-casting
   - **C.** Ingratiation
   - **D.** Self-disclosure

8. Which of the following is an example of a *Gesellschaft*?
   - **A.** A large corporation
   - **B.** A small rural neighborhood
   - **C.** Members of the same family
   - **D.** An ethnic enclave in a large city

9. In some cultures, it is considered taboo for one to show too much sadness at a funeral. In other cultures, wailing and crying loudly is expected. These cultures differ in their:
   - **A.** characteristic institutions.
   - **B.** display rules.
   - **C.** authentic selves.
   - **D.** peer groups.

10. Which of the following is NOT a dimension of the system for multiple level observation of groups (SYMLOG)?

- **A.** Friendliness *vs*. unfriendliness
- **B.** Dominance *vs*. submission
- **C.** Conformity *vs*. contrast
- **D.** Instrumentally controlled *vs*. emotionally expressive

11. Political campaign ads often focus on "exposing" an opposing candidate's negative characteristics. In the dramaturgical approach, one would describe this as:

- **A.** bringing the front stage self to the back stage.
- **B.** bringing the back stage self to the front stage.
- **C.** removing the front stage self.
- **D.** removing the back stage self.

12. In the context of impression management, which of the following selves is most similar to the ought self?

- **A.** The ideal self
- **B.** The tactical self
- **C.** The authentic self
- **D.** The presented self

13. The evolutionary role of emotions has been used as support for which model(s) of emotional expression?

- **A.** The basic model only
- **B.** The social construction model only
- **C.** Both the basic model and social construction model
- **D.** Neither the basic model nor the social construction model

14. Which of the following is an example of intraspecific animal communication?

- **A.** A dog who barks when a stranger enters the house
- **B.** An anglerfish that uses a bioluminescent appendage to attract prey
- **C.** Bats using echolocation to detect the surrounding environment
- **D.** A cat who uses scent glands to mark his territory for other cats

15. Primary groups differ from secondary groups in that:

- **A.** primary groups are shorter-lived than secondary groups.
- **B.** primary groups are larger than secondary groups.
- **C.** primary groups are formed of stronger bonds than secondary groups.
- **D.** primary groups are assigned while secondary groups are chosen.

Answer Key

1. **C**
2. **B**
3. **C**
4. **D**
5. **C**
6. **C**
7. **B**
8. **A**
9. **B**
10. **C**
11. **B**
12. **B**
13. **A**
14. **D**
15. **C**

Detailed explanations can be found at the end of the chapter.

CHAPTER 9

SOCIAL INTERACTION

In This Chapter

CHAPTER PROFILE

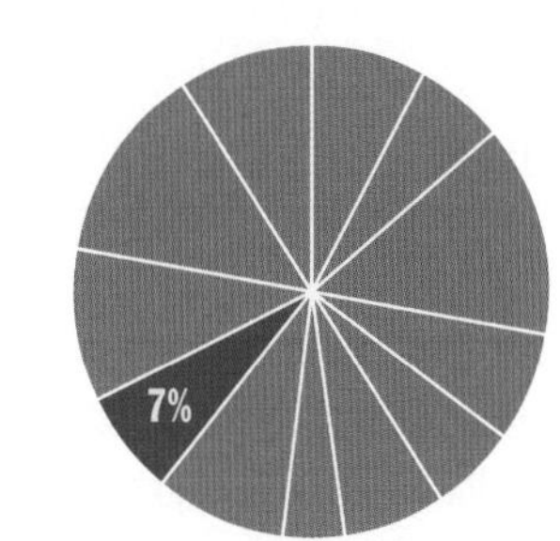

The content in this chapter should be relevant to about 7% of all questions about the behavioral sciences on the MCAT.

This chapter covers material from the following AAMC content categories:

7B: Social processes that influence human behavior

8C: Social interactions

Introduction

Every day, you present yourself to others and interact with society. You interact with others in a number of ways, using emotional expression, verbal communication, and cultural norms. You may also interact with nonhuman animals on a day-to-day basis. This interaction can be just as complex and meaningful to your emotional state and life experience. What shapes and molds your interactions? How do you know the appropriate way to connect with other members of society?

9.1 Elements of Social Interaction

High-Yield

LEARNING OBJECTIVES

After Chapter 9.1, you will be able to:

- Identify the meaning of social interaction terms such as status, role, group, network, and organization
- List the three types of status and provide an example of each

Society has developed out of necessity for human beings to survive and develop. Social interaction is the basis of social life and helps humans reach their full potential. Social interaction is facilitated by preexisting commonalities between individuals and shared understanding or experiences, such as a shared language. Through our social interactions, we develop culture.

Statuses

In most human societies, people do not view every individual as an equal. Instead, we create a hierarchical structure with inequalities of material goods, social opportunities, social acceptance, and skills. Some are wealthy, and some are impoverished; some are talented in sports, while others are not. Some are admired by others, most are liked, and some are disliked or even stigmatized. **Social statuses** are perceived positions in society that are used to classify individuals. Being a premed student, for example, is considered a status. Most statuses exist in relation to other statuses: being a premed student does not have meaning unless there are other statuses with which to compare it, such as *medical student* or *resident*. It is important to note that not all personal characteristics are considered to be social statuses. For example, being left-handed is not considered a status.

KEY CONCEPT

Types of statuses:

- Ascribed—given involuntarily, based on race, ethnicity, sex, family background, and so on
- Achieved—gained as a result of one's efforts or choices
- Master—status by which one is most identified; is pervasive in that person's life

There are three key types of statuses: ascribed, achieved, and master statuses. An **ascribed status** is one that is given involuntarily (usually at birth), due to such factors as race, ethnicity, sex, and family background. An **achieved status** is a status that is gained as a result of one's efforts or choices, such as being a doctor. A person can hold multiple statuses at the same time (collectively known as one's status set), but one's **master status** is the status by which a person is most identified. This status is typically the most important status the individual holds and affects all aspects of that person's life. It is also generally how people view themselves and often holds a symbolic value. Master statuses can also cause pigeonholing: we may view individuals only through the lens of their master status, without regard to any other personal characteristics (such as with a president or other major political figure).

Roles

Each status is associated with **roles**, or sets of beliefs, values, attitudes, and norms that define expectations for those who hold the status. **Role performance** is the carrying out of behaviors associated with a given role. Individuals can vary in how successful they are at performing a role. For example, part of a doctor's role is to translate medical information into language their patients can understand; however, some doctors are far better at this skill than others. Role performance can also change depending on the social situation and context of the interaction. When doctors interact with each other, the pertinent parts of their roles are quite different than when interacting with patients. Behaviors and expectations thus change as a result of the **role partner**—the person with whom one is interacting. Doctors have many role partners: patients, nurses, patients' relatives, other doctors, residents, and hospital administration. The various roles associated with a status are referred to as a **role set**.

REAL WORLD

A great example of role conflict is a single parent who also works a full-time job. Both of these roles carry a very large set of expectations, which are often at odds with each other.

Through our lives, we each take on numerous statuses, each of which may contain a variety of roles. Additionally, we are often playing several roles at one time. Due to the complex nature of statuses and role sets, it is not surprising that conflict, challenges, uncertainty, and ambivalence arise as we try to navigate the many expectations of day-to-day life. **Role conflict** is the difficulty in satisfying the requirements or expectations of multiple roles, whereas **role strain** is the difficulty in satisfying multiple requirements of the same role. **Role exit** is the dropping of one identity for another.

Groups

Another major component of social interaction involves groups. In sociological terms, a **group** (also known as a **social group**) consists of two or more people who share any number of similar characteristics as well as a sense of unity. The simplest of social groups is called a **dyad** (two people), followed by a **triad** (three people). As group size increases, the group trades intimacy for stability. Social groups are more complex than a group of individuals who happen to be in the same physical space. For example, people waiting to cross the street at a crosswalk do not constitute a social group. Common characteristics shared by social groups include values, interests, ethnicity, social background, family ties, and political representation. Many sociologists see social interaction as the most important characteristic that strengthens a social group.

We center most of our lives around social groups, from the camaraderie of teammates to the complexity of governments. Social groups also meet many of the needs we have; these groups provide an opportunity to belong and be accepted and they offer protection, safety, and support. Many people also learn, earn a living, and practice religion in groups. Groups can also be a source of conflict, including discrimination, persecution, oppression, and war. These conflicts sprout from the relationships within and between groups.

An **in-group** is a social group with which a person experiences a sense of belonging or identifies as a member. An **out-group**, on the other hand, refers to a social group with which an individual does not identify. An in-group can form based on a variety of identifying characteristics, including but not limited to race, culture, gender, religion, profession, or education. Out-groups can sometimes compete with or oppose in-groups, creating **group conflict**. Notably, negative feelings toward an out-group are not necessarily based on a sense of dislike toward the characteristics of the out-group; rather, they can be based on favoritism for the in-group and the absence of favoritism for the out-group.

A **peer group** is a group that consists of self-selected equals associated by similar interests, ages, or statuses. Peer groups provide an opportunity for friendship and feelings of belonging. A **family group**, by contrast, is not self-selected but determined by birth, adoption, and marriage. It joins members of various ages, genders, and generations through emotional ties. The family group can be filled with conflict at times; this is often true in adolescence when peer groups begin to compete with family groups for time and loyalty. Family groups may also struggle with cultural gaps and social differences between generations, such as speaking in different languages.

Another important type of group is a **reference group**. These are groups that individuals use as a standard for evaluating themselves. For example, to determine how strong a medical school applicant you are, you might compare yourself to the reference group of all medical school applicants.

Primary and Secondary Groups

Groups can also be categorized into primary and secondary groups. In a **primary group**, interactions between members of the group are direct, with close bonds providing warm, personal, and intimate relationships to members. These groups often last a long period of time and may include a core circle of friends, a tightly knit family, or members of a team. In a **secondary group**, the interactions are impersonal and businesslike, with few emotional bonds and with the goal of accomplishing a specific purpose. Secondary groups typically last for a short period of time, and they form and dissolve without any special significance to those involved, an example being students working together on a group project.

Community and Society

The German sociologist Ferdinand Tönnies distinguished two major types of groups. His theory is known as ***Gemeinschaft und Gesellschaft***, which translates to *community and society. Gemeinschaft* (community) refers to groups unified by feelings of togetherness due to shared beliefs, ancestry, or geography. Families and neighborhoods are examples of *Gemeinschaften. Gesellschaft* (society) refers to less personal groups that are formed out of mutual self-interests working together toward the same goal. Companies and countries are examples of *Gesellschaften*.

Observing and Analyzing Groups

Group size may vary; the smallest size a group can be is two people. Smaller group sizes, like dyads or triads, allow individuals to present more of themselves to the group. **Interaction process analysis** is a technique for observing, classifying, and measuring the interactions within small groups. In the 1970s, it was revised to the **system for multiple level observation of groups** (**SYMLOG**), which is based on the belief that there are three fundamental dimensions of interaction: dominance *vs.* submission, friendliness *vs.* unfriendliness, and instrumentally controlled *vs.* emotionally expressive.

Extensive research on groups has revealed that a group holds power over its members, creating group pressure that can ultimately shape members' behaviors. This is called **group conformity**; individuals are compliant with the group's goals, even when the group's goals may be in direct contrast to the individual's goal. Individuals conform in an attempt to fit in and be accepted by the group. Individuals will often participate in behaviors they normally would not.

Groupthink is related to group conformity and occurs when members focus on reaching a consensus at the cost of critical evaluation of relevant information. This can lead to groups not exploring all sides of an issue and may limit the group's options or views; further, group members may self-censor by not expressing their beliefs. A more extensive discussion of the effects of groups on individual behavior (social action) and group dynamics is explored in Chapter 8 of *MCAT Behavioral Sciences Review*.

Networks

The term **network** is used to describe the observable pattern of social relationships among individuals or groups. Patterns of relationship can be determined by mapping the interactions between individual units, the nature of which can be highly variable. For example, a sociologist may look at the patterns in the interactions between friends, family members, or societal institutions. Researchers often display networks with maps containing a series of points, with each point representing a unit in the network. They connect the points with lines to display the interactions between units, as shown in Figure 9.1. Not all contact points within a network are necessarily unique.

If there are overlapping connections with the same individual, it is referred to as **network redundancy**. Network analysis can be used to gain understanding of the actions of individuals and groups and to study the broader social structure.

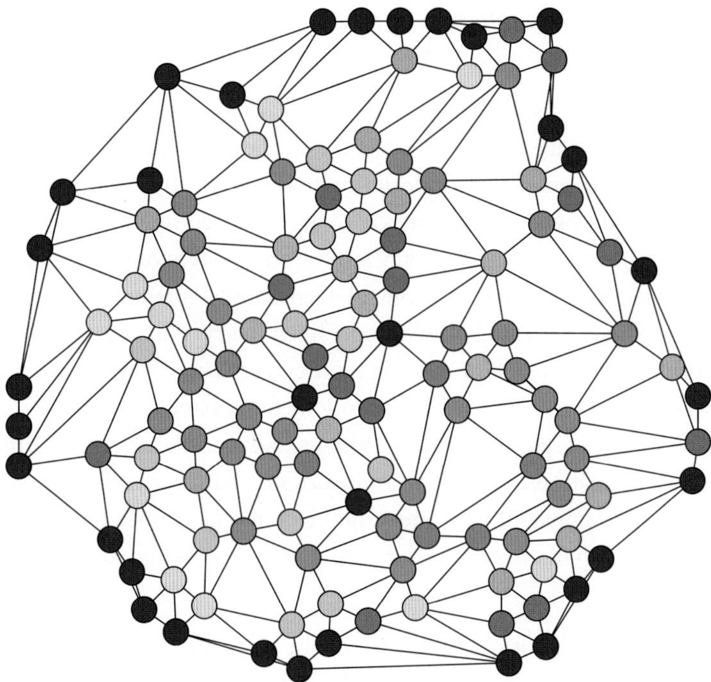

Figure 9.1 Example of a Social Network Diagram

Individuals in networks face the demands and expectations of other members, constraining what they are able to do. They also may have access to resources through the network. An example of a network is a university's alumni association: the members are held to certain standards and commitments, but also may reap the benefits of the network when searching for a job. **Immediate networks** are dense with strong ties, whereas **distant networks** are looser and contain weaker ties; immediate networks may be composed of friends, whereas distant networks may include acquaintances. The combination of immediate and distant networks provide the most benefit to individuals, which is augmented if the networks work complementarily to provide different resources.

BRIDGE

A genetic pedigree can be thought of as a specific type of network map, in which geneticists can track genetic patterns. The lines in genetic pedigrees represent mating patterns, parent-child relationships, and other familial structures. While pedigree analysis does not appear on the MCAT, the related topic of genetics is discussed in Chapter 12 of *MCAT Biology Review*.

Organizations

In sociology, **organizations** are complex secondary groups that are set up to achieve specific goals and are characterized by having a structure and a culture. We have all been members of multiple organizations, such as schools, companies, music groups, sports teams, fraternities and sororities, political organizations, community action committees, and so on. The study of organizations is at the heart of sociology because of the importance that organizations have throughout a person's life.

The modern **formal organization** developed during the Industrial Revolution as a way to maximize efficiency. The formal aspect derives from the explicit goals that guide the members and their activities. Furthermore, formal organizations have enforcement procedures that seek to control the activities of their members. Lastly, these organizations are characterized by the hierarchical allotment of formal roles or duties to members. Formal organizations can be quite large.

The basic organization of society is found in its **characteristic institution**. Throughout history this has changed. In prehistoric times, the characteristic institution was primarily the kin, clan, or sib. In modern times, as we have transformed our cities into urban centers of trade and commerce, we have moved to bureaucracy as the characteristic institution. A **bureaucracy** is a rational system of political organization, administration, discipline, and control. Generally, a bureaucracy has these six characteristics: paid, nonelected officials on a fixed salary; officials who are provided rights and privileges as a result of making their careers out of holding office; regular salary increases, seniority rights, and promotions upon passing exams or milestones; officials who enter the organization by holding an advanced degree or training; responsibilities, obligations, privileges, and work procedures rigidly defined by the organization; and responsibility for meeting the demands of one's position. Due to these characteristics, bureaucracies are often slow to change and less efficient than other organizations.

Bureaucracies have been criticized over time. The **iron law of oligarchy** states that democratic or bureaucratic systems naturally shift to being ruled by an elite group. This shift is due to a number of factors, including the necessity of a core body of individuals to carry out the day-to-day activities of the organization, increased need for specialization, and leadership characteristics of certain members of the group. Thus, even a group established with democratic principles and complete egalitarianism will ultimately centralize, placing power in the hands of a few key leaders.

McDonaldization is commonly used to refer to a shift in focus toward efficiency, predictability, calculability, and control in societal practices. While the original model for McDonaldization was, of course, the fast-food restaurant and its push towards efficiency, examples of these same characteristics can be seen in many other institutions. For example, 24-hour news channels, which feature running footers of the latest news stories as "bite-size" headlines, demonstrate efficient and predictable sources of information. Corporations may mine "big data" to make business decisions using controlled, standardized methods, allowing the business to focus on the calculable outcomes of a choice such as profit and loss analysis and market share.

MCAT CONCEPT CHECK 9.1

Before you move on, assess your understanding of the material with these questions.

1. List the three types of statuses and provide an example of each:

Status	Example

2. For each of the sociological terms below, provide a brief definition:
   - Status:
   - Role:
   - Group:
   - Network:
   - Organization:

9.2 Self-Presentation and Interacting with Others

LEARNING OBJECTIVES

After Chapter 9.2, you will be able to:

- Classify forms of communication as verbal or nonverbal
- Distinguish between front-stage self and back-stage self
- Identify examples of body language, facial expression, visual display, scent, and vocalization in communication

To Erving Goffman, the sociologist who developed the dramaturgical perspective, every interaction we have with other people is a theatrical performance in which we consciously or unconsciously use the "scene," our "costume," and the role that we "perform" to influence the way others think or feel. Whenever we try to influence others' perception with respect to a person, object, or event, we are engaging in impression management. One form of impression management is **self-presentation**, the process of displaying ourselves to society both visually (through clothing, grooming, etc.) and through our actions, often to make sure others see us in the best possible light.

Expressing and Detecting Emotions

Expressed emotions include both verbal and nonverbal behaviors that communicate internal states. We can express emotions with or without conscious awareness.

The **basic model of emotional expression** was first established by Charles Darwin. Darwin stated that emotional expression involves a number of components: facial expressions, behaviors, postures, vocal changes, and physiological changes. Darwin claimed that expression is consistent with his theories on evolution and should be similar across cultures. Darwin also stated that primates and animals exhibit rudimentary muscle actions that are similar to those used by humans for facial expressions. Since Darwin, many researchers have found that a number of basic human emotions are universally experienced and that their corresponding facial expressions are universally recognized. The **appraisal model** is closely related, and accepts that there are biologically predetermined expressions once an emotion is experienced, but that there is a cognitive antecedent to emotional expression.

BRIDGE

Paul Ekman's work with universal emotions, as detailed in Chapter 5 of *MCAT Behavioral Sciences Review*, was a key development in the basic model of emotional expression. Individuals knowledgeable about Ekman's work are capable of detecting very subtle and transient facial expressions that may indicate that an individual is trying to be deceptive about the emotions being conveyed.

Three of the primary models that describe individual emotion (James–Lange, Cannon–Bard, and Schachter–Singer) were discussed in Chapter 5 of *MCAT Behavioral Sciences Review*. In this chapter, we will look at how emotions are shaped by social context and culture.

The **social construction model** assumes that there is no biological basis for emotions. Instead, emotions are based on experiences and the situational context alone. It also suggests that certain emotions can only exist within social encounters and that emotions are expressed differently—and thus play different roles—across cultures. In this model, one must be familiar with social norms for a certain emotion to perform the corresponding emotional behaviors in a given social situation.

Culture provides the foundation to understand and interpret behaviors. Studies have suggested that cultural differences can lead to very different social consequences when emotions are expressed. Cultural expectations of emotions are often referred to as **display rules**. For example, in Utkuhikhalik Inuit society, anger is rarely expressed; individuals who demonstrate anger are considered social pariahs. Display rules govern which emotions can be expressed and to what degree. They may differ as a function of the culture, gender, or family background of an individual. Emotional expressions can be managed in several different ways: by simulating feelings one does not actually feel; by qualifying, amplifying, or deamplifying feelings; by masking an emotion with another emotion; or by neutralizing any emotional expression whatsoever.

A **cultural syndrome** is a shared set of beliefs, attitudes, norms, values, and behaviors among members of the same culture that are organized around a central theme. Cultural syndromes influence the rules for expressing or suppressing emotions and can even influence the ways emotions are experienced. For example, happiness is generally considered a positive emotion across cultures. However, in countries with more individualistic cultural syndromes, like the United States, happiness is viewed as infinite, attainable, and internally experienced. In contrast, in countries with a more collectivist cultural syndrome, such as Japan, happiness is a very rational emotion and generally applied to collective experiences more than to individual successes or experiences. This difference is illustrated in the contrast between the phrases *I am happy* and *I am sharing happiness with others*.

Gender can also play an important role in emotional expression. Research on the expression of emotion in the United States has shown that women are expected to express anger in public less often than men, while men are expected to repress the expression of sadness. Research also supports the conclusion that women are better at detecting subtle differences in emotional expression than men.

Impression Management

Impression management refers to our attempts to influence how others perceive us. This is done by regulating or controlling information we present about ourselves in social interactions. Impression management is often used synonymously with self-presentation. When describing impression management, theorists describe three "selves": the authentic self, the ideal self, and the tactical self. The **authentic self** describes who the person actually is, including both positive and negative attributes. The **ideal self**, as described in Chapter 6 of *MCAT Behavioral Sciences Review*, refers to who we would like to be under optimal circumstances. The **tactical self** refers to who we market ourselves to be when we adhere to others' expectations of us. This is similar to the ought self described in Chapter 6 of *MCAT Behavioral Sciences Review*.

People use a number of impression management strategies when in the presence of others. Some common strategies are summarized in Table 9.1, with examples of each.

STRATEGY	DEFINITION	EXAMPLE(S)
Self-disclosure	Giving information about oneself to establish an identity	Disclosing that you are a premedical student
Managing appearances	Using props, appearance, emotional expression, or associations with others to create a positive image	Wearing a white coat, keeping calm while dealing with a difficult patient, mentioning associations with important researchers during an interview
Ingratiation	Using flattery or conforming to expectations to win someone over	Blindly agreeing to someone else's opinion, complimenting a friend before asking for a favor
Aligning actions	Making questionable behavior acceptable through excuses	Justifications for missing deadlines, blaming a bad grade on too little sleep
Alter-casting	Imposing an identity onto another person	Any example in this course that says *As a good MCAT student, you should*...in which Kaplan is assigning *you* the role of *good MCAT student*

Table 9.1 Impression Management Strategies

Erving Goffman described impression management through the **dramaturgical approach**, using the metaphor of a theatrical performance to describe how individuals create images of themselves in various situations. In this analogy, Goffman relates a person's attempts to manage the impressions of others to an actor's performance in a play. A person's **front stage self** is the persona they present to an audience. A person will adapt their front stage self depending on the social situation, similar to an actor on stage in front of an audience performing according to the setting, role, and script of the play. In contrast, when an actor is back stage, the actor is hidden from the audience and is free to act in ways that may not be congruent with the actor's character in the play. According to Goffman, the **back stage self** is the persona adopted when not in a social situation and there is no concern about upholding the performance of a desired public image.

Another theory comes from George Herbert Mead, who described the self in two parts called the *Me* and the *I*. The part of self that is developed through interaction with society is the ***Me***. The development of the *Me* comes from considering the **generalized other**, which is based on a person's established perceptions of the expectations of society. Any time that a person tries to imagine what is expected of them in a social situation, they are taking on the perspective of the generalized other. And by considering the perspective of the generalized other and adapting one's behavior appropriately, the *Me* develops. By contrast, the ***I*** is the individual's own impulses. However, the *I* is not totally independent of the *Me*. Rather, a person's impulses are shaped by their interpretation of society's expectations. In short, the *Me* shapes the *I*.

MCAT EXPERTISE

Many of the sociological theories tested on the MCAT are far more extensive than the knowledge base the AAMC expects of test takers. The dramaturgical approach, for example, describes over twenty sociological concepts in theatrical terms; however, the MCAT only expects you to know front stage *vs.* back stage self.

BRIDGE

Mead's description of *Me* and *I* formed the foundation for the sociological theory of symbolic interactionism, which is described in Chapter 11 of *MCAT Behavioral Science Review.*

BEHAVIORAL SCIENCES GUIDED EXAMPLE WITH EXPERT THINKING

Researchers conducted two experiments in an effort to investigate social behavior.

Experiment 1:
Five-year-old children participated in one of four conditions. In all four conditions, the children played a game in which they received a sheet of paper with five symbols drawn on it. The children were also given a sheet of stickers, some of which matched the symbols, and were given 90 seconds to find the appropriate stickers and affix them to the sheet of paper over their matching symbols. Also on the table was a second set of materials, which the children were told was for another child who would be playing the game later.

Exp. 1: Sticker matching game

In the stealing condition, the participant's sticker sheet was missing one of the five symbols necessary to complete the game, while the sticker could be found on the second sheet. In the helping condition, the second sheet was conspicuously missing one of the stickers needed to complete the game, and the participant's sheet included an extra of those stickers, which participants were told they could keep or give to the next child. In some trials, the experimenter left the child alone while the child completed the task (the unobserved case), and in others, another child who was participating in the study but not part of the same experimental group was seated near the child during the task (the observed case).

Two conditions: children need to steal to complete their task; children can give to another child who is missing stickers.

IVs: steal or help condition; observation

DV: behavior

In the stealing condition, 4% of subjects stole in the observed case and 24% stole in the unobserved case ($p = .02$). In the helping condition, 28% of subjects helped in the observed case and 11% helped in the unobserved case ($p = .07$). Results are summarized in Figure 1.

The percentages are stated here, but are summarized in the figure. Missing from the figure are the p-values. Looks like the stealing condition was significant and the helping condition wasn't, but we could call it a trend.

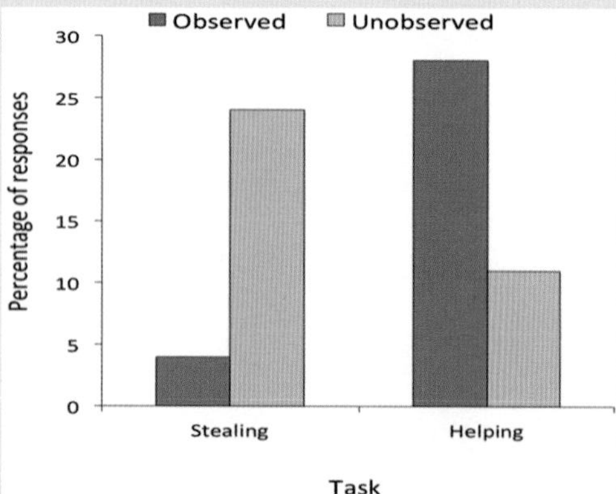

Figure 1 Stealing and helping behavior for children in their respective conditions.

Children helped when people were watching and stole when no one was.

Experiment 2:
Two chimpanzees were placed in separate cages. One of the chimps had access to a rope attached to a tray that contained food. In the stealing condition, the food was placed such that the other chimp could reach it, but pulling the rope moved the tray out of reach of both chimps. In the helping condition, the food was placed out of reach of both chimps, but pulling the rope moved the tray within reach of the second chimp. The subject chimps were not able to see the recipient chimps during the experiment, but were taught what the rope mechanism did prior to testing. Just as in Experiment 1, each of these conditions was carried out in one of two cases: either the chimps were alone in the experimental room, or a third dominant male chimp was present, observing the interaction. It is known that in chimpanzees, fitness improves with relationships to dominant group members. Results are summarized in Figure 2.

Similar conditions here. IVs: help or hurt, observation; DV: behavior.

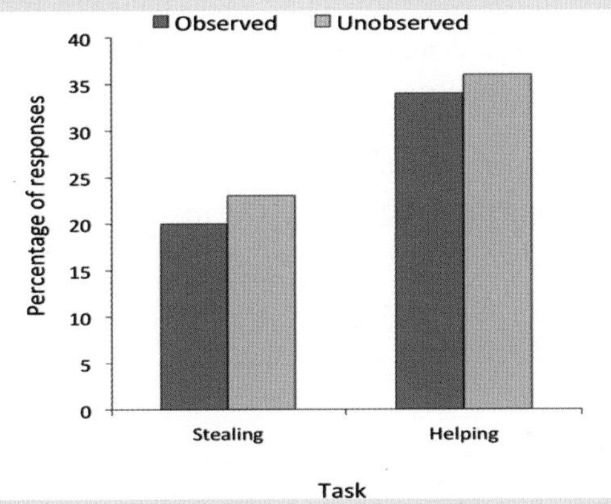

Figure 2 Stealing and helping behavior for chimps in their respective conditions.

This is different from the children. Observation didn't seem to matter.

Adapted from: Engelmann JM, Herrmann E, Tomasello M (2012) Five-Year Olds, but Not Chimpanzees, Attempt to Manage Their Reputations. *PLoS ONE* 7(10): e48433. https://doi.org/10.1371/journal.pone.0048433

What conclusion can be drawn based on these studies about the differences between human children and chimpanzees in social settings?

The question asks for a conclusion, so our plan should be to summarize the results of the studies and relate them to outside knowledge. Fortunately, once we understand the design of the two studies, the results aren't difficult to interpret. In Experiment 1, it looks like the children were more likely to steal when they were alone and more likely to help when they were being observed. From the p-values, it looks like that first result is statistically significant and the second result would count as merely a trend. Chimpanzees, on the other hand, showed no such concern for the presence of others. If this were a multiple-choice question, we would look for an answer that explained both results.

Knowing exactly what bit of content to tie these results to might be tricky without answer choices, but the passage does provide a clue. In Experiment 2, we're told that it is adaptive for chimpanzees to gain the favor of dominant group members, so these researchers are likely interested in the way individuals may change their behavior in order to gain the favor of others. This could be called self-presentation, reputation management, or impression management.

Applying this concept to the results of the study, we can conclude that children have at least some capacity for acting in a prosocial manner to manage impressions, but chimpanzees in the given situation are incapable of or unconcerned with doing the same.

Verbal and Nonverbal Communication

Communication is the ability to convey information by speech, writing, signals, or behavior. It is the foundation of social interaction and is often used to elicit changes, generate action, create understanding, share a point of view, or inform. Effective communication occurs when the desired message is received by the recipient.

Verbal communication is the transmission of information via the use of words, whether spoken, written, or signed. It is tied to nonverbal communication and is often dependent on nonverbal cues for the receiver to understand the sender's full meaning. While face-to-face conversations are rich with nonverbal communication, even phone conversations include nonverbal means of communication, such as pauses and changes in tone.

Nonverbal communication refers to how people communicate, intentionally or unintentionally, without words. Some examples of nonverbal communication are facial expressions, tone of voice, body position and movement, touches, and eye positioning. Gestures involving the hands and other parts of the body are often nonverbal, but the formal signs of a sign language count as verbal communication. Nonverbal cues serve a number of functions in communication, including expression of emotions, as shown in Figure 9.2, conveyance of attitudes and personality traits, and facilitation of verbal communication. Nonverbal communication is often dictated by culture. For example, in US culture, people can be suspicious of someone who does not make eye contact, as this is sometimes considered to be a sign of lying. However, in many other cultures, direct eye contact is used far less often than in the United States. For example, many children in Thailand are taught not to make eye contact with teachers and adults in order to show respect. Some types of verbal and nonverbal communication are listed in Table 9.2.

BRIDGE

Strong communication skills are tested everywhere on the MCAT, but are particularly important in the *Critical Analysis and Reasoning Skills* section. See Chapter 2 of *MCAT Critical Analysis and Reasoning Skills Review* for a discussion of analyzing rhetoric.

Figure 9.2 Human Body Language
Sadness is associated with drooping upper eyelids, staring into the distance, frowning, and slumping of shoulders, the last of which is seen here.

VERBAL	NONVERBAL
Spoken language	Facial expressions
Written language (print and electronic)	Body language (posture)
Sign languages (American Sign Language)	Gestures
Tactile languages (Braille alphabet)	Tone of voice (prosody)
	Eye contact
	Amount of personal space

Table 9.2 Verbal and Nonverbal Communication

Animal Signals and Communication

We not only communicate with other people, but also with other living creatures. **Animal communication** is defined as any behavior of one animal that affects the behavior of another.

Nonhuman animals communicate with one another in order to convey information such as emotions, intent, status, health, and the location of resources. They communicate with one another through a variety of nonverbal means, including body language, rudimentary facial expressions, visual displays, scents, and vocalizations.

The use of body language is common across a number of species. Body language can indicate that an animal is frightened, as shown in Figure 9.3, aggressive, relaxed, or even embarrassed; dogs often tuck their tails between their legs when scolded or fearful. Body language can also have significance for reproduction, as many animals will get into certain positions to signify readiness to mate.

Figure 9.3 Animal Body Language
When surprised or scared, cats will recoil, crouch, and remain relatively motionless.

While humans possess far finer motor control of the muscles of facial expression, many animals (especially mammals) use facial expressions to indicate similar emotions to body language. It is noteworthy, however, that facial expressions appear to be more highly conserved between species than body language. For example, baring teeth and lunging forward, as shown in Figure 9.4, are perceived almost universally as signs of aggression or readiness to attack.

Figure 9.4 Animal Facial Expressions
Baring teeth and lunging forward are recognized by many animals as signs that an attack is imminent.

Animals may also use visual displays for communication. This is common for sex discrimination in birds; females are often less colorful than males because it permits them a greater degree of camouflage and protection when caring for their young. However, this also serves as communication between birds, as sex is readily apparent from the bird's appearance. Other visual displays include bioluminescence (the production of light), colorful plumage (as in peacocks), and dancing. Bees are well-known for communicating through dancing, as shown in Figure 9.5.

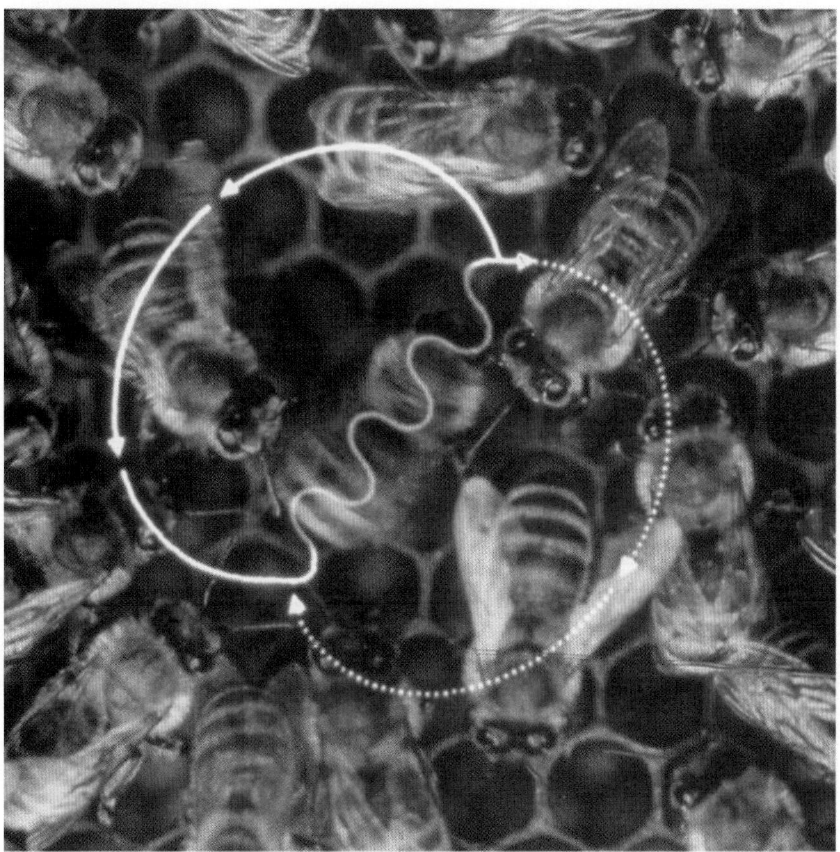

Figure 9.5 Bee Communication through Movement
The "waggle dance," illustrated here, indicates the location of food relative to the hive.

Many animals use scents to communicate both intraspecifically (between members of the same species) and interspecifically (between members of different species). Pheromones are a common example and are given off by members of a species to attract a mate. Scents can be used to mark an animal's territory or as a method of defense, such as in skunks.

Finally, animals also communicate through vocalizations with various levels of sophistication. For example, research has shown that prairie dogs have different "words" for specific predators, and can even create new words for novel objects. Bird calls are species specific and are used to attract a mate or warn of a threat.

In addition to interacting in the wild, humans use both verbal and nonverbal communication when interacting with domesticated animals, as is often seen between owners and their pets. Dog owners may use vocal commands to tell their pets to come, stay, or sit. Additionally, just as tone of voice can express joy or anger to a person, it can communicate the same information to a pet. Pets can be scolded with a look or a gesture. Communication works in the opposite direction as well, as a pet's body language and expressions convey information to its owner.

BRIDGE

It is debatable if pheromones actually have an effect on humans because we lack many of the genes necessary for function of the vomeronasal organ, an accessory olfactory organ seen in other animals. Olfaction and scent detection is discussed in Chapter 2 of *MCAT Behavioral Sciences Review*.

Communication between humans and animals is not confined strictly to pets. One of the most famous examples of animal communication is Koko, a gorilla who was able to communicate with humans through the use of American Sign Language. Koko's vocabulary included more than one thousand words.

MCAT CONCEPT CHECK 9.2

Before you move on, assess your understanding of the material with these questions.

1. Classify the following forms of communication as verbal or nonverbal:

• American Sign Language	Verbal	Nonverbal
• Turning your body away from another person	Verbal	Nonverbal
• Text messages	Verbal	Nonverbal
• Giving a "high five"	Verbal	Nonverbal
• Frowning	Verbal	Nonverbal

2. What is the front stage self? The back stage self?
   - Front stage self:

____

____

   - Back stage self:

____

____

3. For each of the methods of animal communication below, provide one example:

Method of Communication	Example
Body language	
Facial expressions	
Visual displays	
Scents	
Vocalizations	

Conclusion

Skunks are unique in how they communicate with other animals they perceive as threats. Their anal glands are capable of producing high concentrations of thiol-containing compounds, which create a distinctive malodorous scent. But it is noteworthy that skunks only carry five or six sprays' worth of material at a time—thus, they tend to use other forms of animal communication, such as body language, hissing, and foot stamping before resorting to spraying. The spray, however, is an ultimate defense: intense, caustic, and very sticky. Animals who are sprayed quickly learn that the skunk is not an animal to mess with.

Humans also use many methods of communication. While they may certainly not follow the same patterns as skunks, humans use combinations of vocalization, body language, facial expressions, and gestures to interact with each other socially. The field of sociology flows from these interactions as we create groups, networks, and organizations; organize our society into hierarchies with statuses; and fulfill the roles dictated by our statuses. We put much of our energy into controlling how we communicate with others, trying to create the optimal image of ourselves through impression management.

The content of this chapter plays a large role in your day-to-day life. Every day you interact with other people, and how you interact is largely determined by the culture and society in which you live. In the next chapter, we begin to analyze specific types of interactions, like attraction and altruism, and then examine the dark side of human society: bias, prejudice, discrimination, and stereotypes.

GO ONLINE

You've reviewed the content, now test your knowledge and critical thinking skills by completing a test-like passage set in your online resources!

CONCEPT SUMMARY

Elements of Social Interaction

- A **status** is a perceived position in society used to classify individuals.
  - An **ascribed status** is involuntarily assigned to an individual based on race, ethnicity, gender, family background, and so on.
  - An **achieved status** is voluntarily earned by an individual.
  - A **master status** is the status by which an individual is primarily identified.
- A **role** is a set of beliefs, values, and norms that define the expectations of a certain status in a social situation.
  - **Role performance** refers to carrying out the behaviors of a given role.
  - A **role partner** is a person with whom one is interacting who helps define the roles within the relationship.
  - A **role set** contains all of the different roles associated with a status.
  - **Role conflict** occurs when one has difficulty in satisfying the requirements of multiple roles simultaneously; **role strain** occurs when one has difficulty satisfying multiple requirements of the same role simultaneously.
- **Groups** are made up of two or more individuals with similar characteristics that share a sense of unity.
  - A **peer group** is a self-selected group formed around similar interests, ages, or statuses.
  - A **family group** is the group into which an individual is born, adopted, or married.
  - An **in-group** is a social group with which a person experiences a sense of belonging or identifies as a member.
  - An **out-group** is a social group with which an individual does not identify.
  - **Group conflict** occurs when an out-group competes with or opposes an in-group.
  - A **reference group** is a group to which individuals compare themselves.
  - **Primary groups** are those that contain strong, emotional bonds.
  - **Secondary groups** are often temporary and contain fewer emotional bonds and weaker bonds overall.
  - ***Gemeinschaft*** **(community)** is a group unified by feelings of togetherness due to shared beliefs, ancestry, or geography.
  - ***Gesellschaft*** **(society)** is a group unified by mutual self-interests in achieving a goal.
  - **Groupthink** occurs when members begin to conform to one another's views without critical evaluation.
- A **network** is an observable pattern of social relationships between individuals or groups.
- **Organizations** are bodies of people with a structure and culture designed to achieve specific goals.

Self-Presentation and Interacting with Others

- Various models have been proposed for how we express emotion in social situations.
  - The **basic model** states that there are universal emotions, along with corresponding expressions that can be understood across cultures.
  - The **social construction model** states that emotions are solely based on the situational context of social interactions.
- **Display rules** are unspoken rules that govern the expression of emotion.
- A **cultural syndrome** is a shared set of beliefs, norms, values, and behaviors organized around a central theme, as is found among people sharing the same language and geography.
- **Impression management** refers to the maintenance of a public image, which is accomplished through various strategies.
  - **Self-disclosure** is sharing factual information.
  - **Managing appearances** refers to using props, appearance, emotional expression, or associations to create a positive image.
  - **Ingratiation** is using flattery or conformity to win over someone else.
  - **Aligning actions** is the use of excuses to account for questionable behavior.
  - **Alter-casting** is imposing an identity onto another person.
- The **dramaturgical approach** says that individuals create images of themselves in the same way that actors perform a role in front of an audience.
  - The **front stage** is where individuals are seen by the audience and where they strive to preserve their desired image.
  - The **back stage** is where individuals are not in front of an audience and where they are free to act outside of their desired image.
- Communication includes both verbal and nonverbal elements.
  - **Verbal communication** is the conveyance of information through spoken, written, or signed words.
  - **Nonverbal communication** is the conveyance of information by means other than the use of words, such as body language, prosody, facial expressions, and gestures.
  - **Animal communication** takes place not only between nonhuman animals, but between humans and other animals as well. Animals use body language, rudimentary facial expressions, visual displays, scents, and vocalizations to communicate.

ANSWERS TO CONCEPT CHECKS

9.1

1.

Status	Example
Ascribed	Any status given involuntarily, due to factors such as race, ethnicity, gender, and family background
Achieved	Any status that is gained as a result of one's efforts or choices
Master	Any status by which a person would be most readily identified and that pervades all aspects of an individual's life

2. Statuses are perceived positions in society used to classify individuals. Roles are the behaviors and expectations associated with a status in a particular context. A group is a collection of at least two individuals. A network is a more formal illustration of the relationships between individuals, usually through graphic representation. An organization is a body with a specific set of goals, a structure, and a culture; organizations are complex secondary groups that are set up to achieve specific goals.

9.2

1. Verbal: American Sign Language, text messages

   Nonverbal: turning your body away (body language), giving a "high five" (gesture), frowning (facial expression)

2. The front stage self refers to when we are on stage and performing. This requires us to live up to the roles and expectations assumed by our status. The back stage self is when we are away from others and may include behaviors that would not be appropriate or consistent with the front stage self.

3. Examples may vary.

Method of Communication	Example
Body language	Dogs: tail between the legs
Facial expressions	Various animals: baring teeth
Visual displays	Peacocks: colorful plumage
Scents	Insects (and others): pheromones
Vocalizations	Birds: birdcalls

SCIENCE MASTERY ASSESSMENT EXPLANATIONS

1. **C**

A status is a position in society used to classify a person and exists in relation to other statuses. The specific behaviors associated with this status, **(B)**, best describe a role.

2. **B**

An achieved status is one that is acquired through personal efforts. This is in contrast to an ascribed status, **(A)**, in which the status is involuntarily given based on race, ethnicity, gender, family background, and so on. A master status, **(C)**, is one that influences all aspects of an individual's life. While being a college graduate is an important aspect of day-to-day life, it does not usually pervade every part of one's life.

3. **C**

A bureaucracy is an example of an organization, specifically one with the goal of performing complex tasks as efficiently as possible. Immediate networks and primary groups, **(A)** and **(B),** are characterized by strong, intimate bonds, which are not commonly seen in bureaucracies. Reference groups, **(D)**, are those groups to which we compare ourselves for various characteristics.

4. **D**

Generally, bureaucracies are marked by six characteristics: paid officials on a fixed salary; nonelected officials who are provided rights and privileges as a result of making their career out of holding office; regular salary increases, seniority rights, and promotions upon passing exams or milestones, **(C)**; officials who enter the organization by holding an advanced degree or training, **(B)**; responsibilities, obligations, privileges, and work procedures rigidly defined by the organization, **(A)**; and responsibility for meeting the obligations of the office one holds.

5. **C**

Verbal communication uses words (whether spoken, written, or signed). Nonverbal communication uses other means of signaling emotions or ideas, such as body language, **(D)**, facial expressions, **(A)**, eye contact, **(B)**, prosody, and personal space. Hand gestures and other bodily movements are also often nonverbal communication, but can be verbal when used as part of a sign language.

6. **C**

Aligning actions is an impression management technique in which one provides socially acceptable reasons for unexpected behavior. This may manifest as providing an excuse for poor performance or laughing off an inappropriate comment as a joke. Tension created from having conflicting thoughts or opinions, as mentioned in **(B)**, refers to cognitive dissonance.

7. **B**

Imposing a role on another person (in this case, "good friend") is the hallmark of alter-casting. This example is also the opposite of ingratiation, **(C)**, because the implication behind the statement is that not lending the bike makes one a "bad friend"; ingratiation is the use of flattery or conformity to win over someone else.

8. **A**

A *Gesellschaft* (society) is one in which individuals are working toward the same goal, such as a company or country. *Gemeinschaften* (communities), on the other hand, are those that are bonded together by beliefs, ancestry, or geography.

9. **B**

Display rules are those that dictate cultural expectations of emotion. In some cultures, sadness is considered personal and internal; in others, sadness is shared externally with the community.

10. **C**

SYMLOG is a method for analyzing group dynamics and considers groups along three dimensions: dominant *vs.* submissive, friendliness *vs.* unfriendliness, and instrumentally controlled *vs.* emotionally expressive.

11. **B**

If a candidate is "exposed," then personal characteristics that are usually shielded from public view have been brought in front of the public. This would be pulling aspects of the back stage self to the front stage. It would not be considered removing the front stage self, **(C)**, because the candidate still has a public image, even if it has been tarnished.

12. **B**

The ought self is who others think we should be: the expectations imposed by others on us. This is most similar to the tactical self, which is the self we present to others when we adhere to their expectations. The presented self, **(D)**, is a combination of the authentic, ideal, and tactical selves.

13. **A**

The basic model of emotion, as proposed by Charles Darwin, states that emotions serve an evolutionary purpose and thus are similar across cultures. The seven universal emotions have also been used as support for this theory. The social construction model states that emotions are always a product of the current social situation and does not posit any biological basis for emotions, implying a lack of a role for emotions in evolution.

14. **D**

Intraspecific communication refers to communication between members of the same species. Interspecific communication, on the other hand, refers to communication between members of different species. Echolocation, **(C)**, is not an example of intraspecific communication because the sender of the signal and the recipient are the same organism; this would be considered autocommunication.

15. **C**

Primary groups have direct and close bonds between members, providing warm, personal, and intimate relationships to its members. Secondary groups, in contrast, form superficial bonds and tend to last for a shorter period of time.

Consult your online resources for additional practice. GO ONLINE

SHARED CONCEPTS

Behavioral Sciences Chapter 3
Learning and Memory

Behavioral Sciences Chapter 5
Motivation, Emotion, and Stress

Behavioral Sciences Chapter 6
Identity and Personality

Behavioral Sciences Chapter 8
Social Processes, Attitudes, and Behavior

Behavioral Sciences Chapter 10
Social Thinking

Behavioral Sciences Chapter 11
Social Structure and Demographics

CHAPTER 10

SOCIAL THINKING

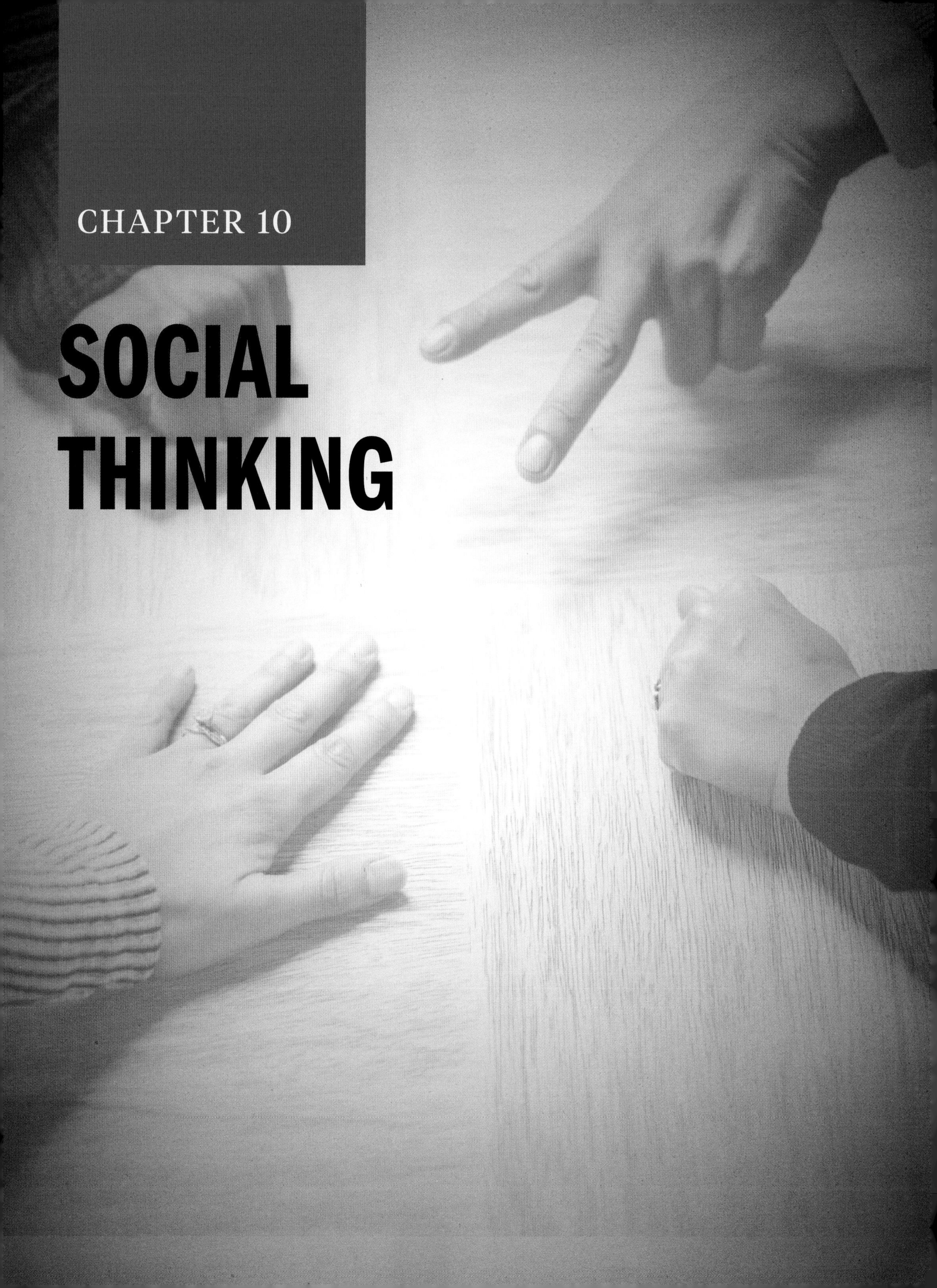

SCIENCE MASTERY ASSESSMENT

Every pre-med knows this feeling: there is so much content I have to know for the MCAT! How do I know what to do first or what's important?

While the high-yield badges throughout this book will help you identify the most important topics, this Science Mastery Assessment is another tool in your MCAT prep arsenal. This quiz (which can also be taken in your online resources) and the guidance below will help ensure that you are spending the appropriate amount of time on this chapter based on your personal strengths and weaknesses. Don't worry though—skipping something now does not mean you'll never study it. Later on in your prep, as you complete full-length tests, you'll uncover specific pieces of content that you need to review and can come back to these chapters as appropriate.

How to Use This Assessment

If you answer 0–7 questions correctly:

Spend about 1 hour to read this chapter in full and take limited notes throughout. Follow up by reviewing **all** quiz questions to ensure that you now understand how to solve each one.

If you answer 8–11 questions correctly:

Spend 20–40 minutes reviewing the quiz questions. Beginning with the questions you missed, read and take notes on the corresponding subchapters. For questions you answered correctly, ensure your thinking matches that of the explanation and you understand why each choice was correct or incorrect.

If you answer 12–15 questions correctly:

Spend less than 20 minutes reviewing all questions from the quiz. If you missed any, then include a quick read-through of the corresponding subchapters, or even just the relevant content within a subchapter, as part of your question review. For questions you answered correctly, ensure your thinking matches that of the explanation and review the Concept Summary at the end of the chapter.

1. The tendency to become close friends with neighbors rather than people in other neighborhoods is most strongly related to which of the following factors?
   - **A.** Proximity
   - **B.** Reciprocity
   - **C.** Self-disclosure
   - **D.** Similarity

2. Which of the following would be associated with high levels of aggression?
   - **I.** Increased amygdala activity
   - **II.** Decreased amygdala activity
   - **III.** Increased prefrontal cortex activity
   - **IV.** Decreased prefrontal cortex activity

   - **A.** I and III only
   - **B.** I and IV only
   - **C.** II and III only
   - **D.** II and IV only

3. A child who cries when a caregiver departs and smiles and runs to the caregiver upon return is displaying which type of attachment pattern?
   - **A.** Avoidant attachment
   - **B.** Ambivalent attachment
   - **C.** Disorganized attachment
   - **D.** Secure attachment

4. Elephant seal males mate with multiple females each mating season, while females only have one mate each. What type of mating system is this?
   - **A.** Polyandry
   - **B.** Polygyny
   - **C.** Monogamy
   - **D.** Promiscuity

5. A person with a ventromedial hypothalamus injury will likely show which behavior?
   - **A.** Increased empathy
   - **B.** Decreased empathy
   - **C.** Increased food intake
   - **D.** Decreased food intake

6. Female great reed warblers are attracted to males with larger song repertoires because they tend to produce offspring with higher viability. This is an example of which of the following?
   - **A.** Runaway selection
   - **B.** Sensory bias
   - **C.** Direct phenotypic benefits
   - **D.** Indirect phenotypic benefits

7. In several species of shrimp, the larger adults will sacrifice themselves to protect the younger, smaller shrimp. How is this behavior best explained?
   - **A.** Inclusive fitness
   - **B.** Direct benefit
   - **C.** Sensory bias
   - **D.** Foraging

8. Which of the following is NOT a component of social perception?
   - **A.** The target
   - **B.** The situation
   - **C.** The perceiver
   - **D.** The process

9. When you first meet Dustin, he is very rude to you. You run into him twice more and he is very friendly, but you still dislike him because of your first meeting. What impression bias does this describe?
   - **A.** Primacy effect
   - **B.** Recency effect
   - **C.** Reliance on central traits
   - **D.** Proximity

10. Mei brings cookies to work. Although you have not yet tasted them, you say to another coworker, "Mei is such a great person; I'm sure these cookies are fantastic!" What type of bias is this?
   - **A.** Reliance on central traits
   - **B.** Direct benefits
   - **C.** Halo effect
   - **D.** Similarity

11. A friend wins a tennis game and says, "I trained so hard—that was a great win!" After losing a subsequent match, the friend says, "My baby brother kept me up all night crying; I was tired for the match." These statements reflect which of the following principles?
   - **A.** Just-world hypothesis
   - **B.** Fundamental attribution error
   - **C.** Self-serving bias
   - **D.** Esteem bias

12. Carlos is always happy and smiling. Today, you notice he seems down and think something must have happened to upset him. What types of attribution are you making?
   - **I.** Internal
   - **II.** External
   - **III.** Situational
   - **IV.** Dispositional

   - **A.** I and III only
   - **B.** I and IV only
   - **C.** II and III only
   - **D.** II and IV only

13. A group of men and women are going to be rated on their driving abilities. The role of gender is emphasized in the experiment and the women perform worse than the men. In another experiment, the role of gender is not mentioned and the ratings are comparable between the two groups. Which principle do these results support?
   - **A.** Institutional discrimination
   - **B.** Stereotype threat
   - **C.** Prejudice
   - **D.** The just-world hypothesis

14. The behavior that accompanies the negative attitudes a person has toward a group or individual is referred to as:
   - **A.** stereotyping.
   - **B.** cultural relativism.
   - **C.** prejudice.
   - **D.** discrimination.

15. Game theory is designed to study:
   - **A.** reliance on central traits.
   - **B.** behavior attribution.
   - **C.** decision-making behavior.
   - **D.** self-enhancement.

Answer Key

1. **A**
2. **B**
3. **D**
4. **B**
5. **C**
6. **D**
7. **A**
8. **D**
9. **A**
10. **C**
11. **C**
12. **C**
13. **B**
14. **D**
15. **C**

Detailed explanations can be found at the end of the chapter.

CHAPTER 10

SOCIAL THINKING

In This Chapter

CHAPTER PROFILE

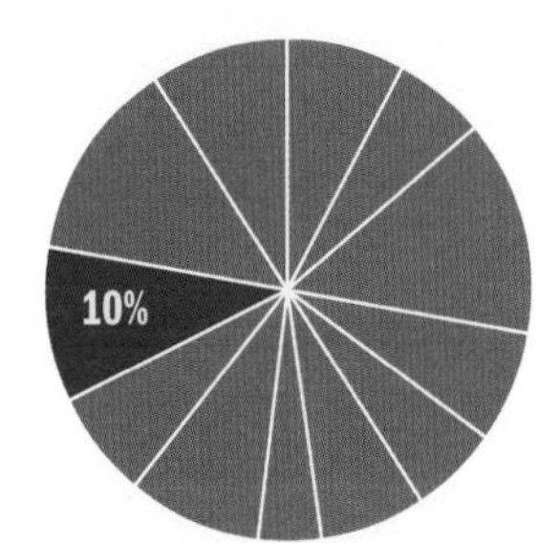

The content in this chapter should be relevant to about 10% of all questions about the behavioral sciences on the MCAT.

This chapter covers material from the following AAMC content categories:

8B: Social thinking

8C: Social interactions

Introduction

Social psychology is concerned with social behavior, including the ways people influence each other's attitudes and behavior. It looks at the impact that individuals have on one another, that social groups have on individual group members, that individual group members have upon the social group, and that social groups have on other social groups. In this chapter, we will continue our discussion of social psychology, highlighting its close relationship to sociology and the other fields within psychology, theoretical perspectives on human behavior within the social environment, and key concepts and classical studies in the field of social psychology. But whereas the last two chapters focused on how individuals are affected by groups and how individuals interact within groups, we will turn our attention in this chapter to specific behaviors seen across human beings, including attraction, aggression, attachment, and the need for social support. We'll also take a look at the dark side of social psychology as we look at patterns of blame in attribution theory and the misappropriation of social structure for prejudice and discrimination. This will be highly relevant for you as a physician, as many patients face prejudice based on their diagnoses or are discriminated against because of personal characteristics, including age, sex, race, ethnicity, socioeconomic status, sexual orientation, gender identity, and more.

10.1 Social Behavior

LEARNING OBJECTIVES

After Chapter 10.1, you will be able to:

- Describe interpersonal attraction, including the factors that influence interpersonal attraction
- Recall the meaning of the term *aggression*, including examples of aggressive behaviors
- List the four types of attachment and how they affect childhood behavior
- Identify the common types of social support
- Explain the relationship between altruism and inclusive fitness

Social behaviors involve interactions with others. These may flow from positive feelings, such as attraction or attachment, or they may flow from negative feelings, such as aggression.

Attraction

KEY CONCEPT

Interpersonal attraction is influenced by many factors, including physical characteristics, similarity, self-disclosure, reciprocity, and proximity.

Have you ever wondered what makes some people friends and others enemies? How second graders choose their best friends? Why you keep eyeing that attractive person in your physics class? Social psychologists call this phenomenon of individuals liking each other **interpersonal attraction**. Researchers have found several factors that affect attraction, including similarity, self-disclosure, reciprocity, and proximity. Outward appearance also plays a role; the more symmetric someone's face is, the more physically attractive people tend to perceive that person to be. Humans are also attracted to individuals with certain body proportions approximating the **golden ratio** (1.618:1).

We tend to be attracted to people who are similar to us in attitudes, intelligence, education, height, age, religion, appearance, and socioeconomic status. One reason for this may be convenience: it's easier to spend time together if you both want to go on a bike ride or if you both enjoy Thai food. Also, people are drawn to having their values and choices validated by another person. So why is there a cliché about opposites attracting? Social psychologists find that attraction also occurs if opposing qualities match up with each other; for example, a nurturer is attracted to someone who craves being nurtured. Notably, successful complementary relationships still have fundamental similarities in some attitudes that make the complementary aspects of the relationship work.

Another component of attraction lies in the opportunity for **self-disclosure**, or sharing one's fears, thoughts, and goals with another person and being met with nonjudgmental empathy. Engaging in this behavior deepens attraction and friendship. This must be a reciprocal behavior, however. Revealing one's innermost secrets creates a sense of vulnerability that, if not met by the other person, can be interpreted as being taken advantage of. Reciprocity is important in other aspects of interpersonal attraction as well. **Reciprocal liking** is the phenomenon whereby people like others better when they believe the other person likes them. Researchers have shown that even if we disagree with others on important issues, we will have increased interest in them if we have indications that they like us.

Finally, **proximity**, or just being physically close to someone, plays a factor in attraction to a person. Studies have shown that we are more likely to form friendships with people in the same dorm as us or with the people who sit closest to us in class. Part of this is convenience; it's easier to have conversations and make plans with people in the same area. Another explanation is the **mere exposure effect** or **familiarity effect**, the tendency for people to prefer stimuli that they have been exposed to more frequently. You may have observed this in your everyday life: Have you disliked a song the first time you heard it, only to find yourself singing along and saying, *I like this song!* after hearing it many more times? This principle is also used in marketing: the more people hear the name of a product, the more likely they are to be attracted to and purchase that product.

Aggression

Aggression is defined as a behavior that intends to cause harm or increase social dominance. Aggression can take the form of physical actions as well as verbal or nonverbal communication. Ethologists study aggression in terms of the interactions between animals in natural settings. Aggression in these settings can include bodily contact, as seen in Figure 10.1, but most displays of aggression are settled by threat and withdrawal without actual bodily harm. Threat displays are common in both animals and humans. Before a fight, someone might puff up the chest or pull back a fist to threaten another person. This display may or may not result in physical harm or violence. Other examples of aggression include a bully hurling insults at another child or a gang member making threatening gestures to a member of another gang.

Figure 10.1 Aggression Following Threat Displays of Elephant Seals
While threat displays may lead to violence, as seen here, threat displays commonly lead to withdrawal to prevent fights.

What is the purpose of aggressive behavior if it causes so much destruction? Evolutionarily, aggression offers protection against perceived and real threats. Aggression helped our ancestors fight off predators. It also helps organisms gain access to resources such as food, additional territory, or mates. In cases of limited resources, aggression could be the deciding factor that allows one to pass on genes.

BRIDGE

The prefrontal cortex is critically important to managing the limbic system, which is important in managing emotion and stress. These roles of the prefrontal cortex are discussed in Chapter 5 of *MCAT Behavioral Sciences Review.*

From a biological perspective, multiple parts of the brain contribute to violent behavior. The **amygdala** is the part of the brain responsible for associating stimuli and their corresponding rewards or punishments. In short, it is responsible for telling us whether or not something is a threat. If the amygdala is activated, this increases aggression. However, higher-order brain structures, such as the prefrontal cortex, can hit the brakes on a revved-up amygdala, reducing emotional reactivity and impulsiveness. Reduced activity in the prefrontal cortex has been linked to increased aggressive behavior.

REAL WORLD

Alcohol has been shown to increase aggressive behavior. Alcohol impairs judgment and limits one's ability to control aggressive reactions. It also makes one feel less inhibited by social mores that would normally restrict aggressive behavior.

Aggression is also under hormonal control. Higher levels of testosterone have been linked to more aggressive behavior in humans irrespective of sex or gender. Some speculate that the higher levels of testosterone in phenotypical males compared to phenotypical females may explain the trend that males are generally more aggressive than females across cultures and that males commit a disproportionate majority of violent crimes.

Beyond the biological contributions to aggressive behavior, studies have found many psychological and situational predictors of aggression. Do you find yourself snapping at people more when you're in pain? Have you ever gotten annoyed with a waiter when you were extremely hungry? Such responses are accounted for by the **cognitive neoassociation model**, which states that we are more likely to respond to others aggressively whenever we are feeling negative emotions, such as being tired, sick, frustrated, or in pain. This can also be seen on a large scale: riots are more likely to happen on hot days than cool ones; drivers without air conditioning are more likely to honk at other drivers than those with air conditioning.

Another factor that contributes to aggressive behavior is exposure to violent behavior. The effects on children of media portrayals of violence continue to be a hot topic. Research findings are mixed but tend to show that viewing violent behavior indeed correlates to an increase in aggressive behavior. The contribution of modeling to violence in children was also explored in Albert Bandura's Bobo doll experiment, described in Chapter 3 of *MCAT Behavioral Sciences Review.*

Attachment

KEY CONCEPT

In attachment, a secure base is a caregiver who is consistent, available, comforting, and responsive.

Attachment is an emotional bond between a caregiver and a child that begins to develop during infancy. While parental figures are most common, emotional bonds can occur with any caregiver who is sensitive and responsive during social interaction. After World War II, psychiatrist John Bowlby noticed the negative effects of isolation on social and emotional development in orphaned children and started the study of attachment. In the 1970s, psychologist Mary Ainsworth expounded on this theory, saying that infants need a secure base, in the form of a consistent caregiver during the first six months to two years of life, from which to explore the world and develop appropriately. Four main types of attachment styles have been described: secure, avoidant, ambivalent, and disorganized.

Secure Attachment

Secure attachment is seen when a child has a consistent caregiver and is able to go out and explore, knowing that there is a secure base to return to. The child will be upset at the departure of the caregiver and will be comforted by the return of the caregiver. The child trusts that the caregiver will be there for comfort, and while the child can be comforted by a stranger, the child will clearly prefer the caregiver. Having a secure attachment pattern is thought to be a vital aspect of a child's social development. Children with avoidant, ambivalent, or disorganized attachment can have deficits in social skills. Collectively, these attachment types are known as **insecure attachment**.

Avoidant Attachment

Avoidant attachment results when the caregiver has little or no response to a distressed child. Given the choice, these children will show no preference between a stranger and the caregiver. They show little or no distress when the caregiver leaves and little or no relief when the caregiver returns.

Ambivalent Attachment

Ambivalent attachment occurs when a caregiver has an inconsistent response to a child's distress, sometimes responding appropriately, sometimes neglectfully. As such, the child is unable to form a secure base because the child cannot consistently rely on the caregiver's response. The child will be very distressed on separation from the caregiver but has a mixed response when the caregiver returns, often displaying ambivalence. This is sometimes referred to as **anxious–ambivalent attachment** because the child is always anxious about the reliability of the caregiver.

Disorganized Attachment

Children with **disorganized attachment** show no clear pattern of behavior in response to the caregiver's absence or presence, but instead can show a mix of different behaviors. These can include avoidance or resistance; seeming dazed, frozen, or confused; or repetitive behaviors like rocking. Disorganized attachment is often associated with erratic behavior and social withdrawal by the caregiver. It may also be a red flag for abuse.

REAL WORLD

As a physician, you will be a mandated reporter. This means that you are required by law to report suspected cases of child abuse. Remember: It is better to report and be incorrect than to miss a potentially fatal scenario.

BEHAVIORAL SCIENCES GUIDED EXAMPLE WITH EXPERT THINKING

The Strange Situation is an experimental procedure used to assess attachment in infants who are between 9 and 18 months old. The procedure occurs over the course of twenty-one minutes and progresses as follows:

1. The parent and child are introduced to the experimental room.
2. The parent and child are left alone in the room. The infant is allowed to explore the room. The parent does not participate.
3. A stranger enters the room, converses with the parent, and attempts to engage with the child.
4. The parent conspicuously leaves the room, and the child is left with the stranger.
5. The parent returns and comforts the child.
6. Both the parent and the stranger leave, and the child is left alone.
7. The stranger returns and attempts to engage with the child.
8. The parent returns and comforts the child, and the stranger conspicuously leaves.

Researchers observe the amount of exploration the child engages in, reactions to the stranger, reactions to the absence of the parent, and the child's behavior upon reunion with the parent.

Experimental procedure. I'll take a brief moment to picture each step, but I'll have to return to this list if a question asks for details.

Not a traditional experiment so no real IV, the differing input is the child and the relationship to the parent, and the output is the child's behavior/attachment style.

This procedure has been replicated with nonhuman animals. In particular, it has been shown that dogs tend to form secure attachments with their owners. This procedure has been attempted with cats, but it appears that cats' behaviors may not serve the same function as behaviors of humans, and as such it is difficult to use procedures developed for humans to assess cat attachment. Furthermore, individual cats' behaviors are inconsistent throughout the procedure, possibly because the experimental procedure is a scenario that cats do not frequently encounter outside the laboratory.

Attachment in nonhuman animals.

Dogs are secure and cats are complicated.

Adapted from: Potter A, Mills DS (2015) Domestic Cats (*Felis silvestris catus*) Do Not Show Signs of Secure Attachment to Their Owners. *PLoS ONE* 10(9): e0135109. https://doi.org/10.1371/journal.pone.0135109

A child exposed to the Strange Situation was uninterested in exploring the room in the presence of either parent or stranger, and did not emotionally engage with either individual. In what ways do dogs' attachment behaviors and style differ from those of the child? What experimental limitations do the researchers point out regarding assessing attachment in cats?

The beginning of the question stem presents a scenario in which a child was reluctant to explore the surroundings, regardless of who was present. This description will have to be matched to our outside content knowledge regarding attachment. The different attachment styles are content we should know for the MCAT, and avoidant attachment is characterized by children who show no substantial behavioral change based on the presence or absence of a parent. Behaviorally, we can conclude the child described is displaying an avoidant attachment style. Dogs were described in the article as demonstrating secure attachment, so we'll want to recall the characteristics of a secure attachment and apply them to this situation. In secure attachment, the subject may or may not show distress when left with a stranger, but will definitely prefer the caregiver. The dogs must have displayed distress when left alone but were easily comforted when the caregiver returned. Specifically, then, the difference between the child and a typical dog should be twofold: the dogs should show greater engagement and positive response with their caregiver as compared to this child, and the dogs should show more distress when left alone without the caregiver as compared to the child.

The answer to the second question requires that we approach the problem in a similar way, this time applied to experimental design. We'll need to consider the description we were given and attempt to match the description to a vocabulary word or concept we've studied. Here, the researchers noted two issues. The first was that the behaviors measured in cats did not serve a function with respect to attachment. Since cats aren't social animals, trying to evaluate their behavior in a social context might be inappropriate. The experiment might not actually be measuring what it seeks to measure, which is a problem with internal validity. The author also mentions that the scenario presented in the experiment might not have a real-world analogue. When an experimental procedure might not actually apply to situations outside the lab, the experiment can be said to lack external validity.

In sum, the dog displayed secure attachment, while the infant displayed avoidant attachment. The application of this methodology to cats potentially lacks both internal (not measuring the targeted variable) and external (not bearing relevance to real-world analogues) validity.

Social Support

In psychology, **social support** is the perception or reality that one is cared for by a social network. Social support can be divided into many different categories: emotional, esteem, material, informational, and network support. While social support is present at all times, it is often most pronounced—and necessary—when someone suffers a personal or family tragedy.

Emotional support is listening, affirming, and empathizing with someone's feelings. It's the *I'm sorry for your loss* condolence card or a trip to the hospital to visit a sick relative. Many people equate social support with emotional support, but other forms of support exist as well.

Esteem support is similar, but touches more directly on affirming the qualities and skills of a person. Reminding others of the skills they possess to tackle a problem can bolster their confidence. For example, consider a friend who has missed a significant amount of school due to illness. Calling that friend a smart and efficient worker who should have no problem making up the work would be providing esteem support.

Material support, also called **tangible support**, is any type of financial or material contribution to another person. It can come in the form of making a meal for friends after they have lost a loved one or donating money to a person in need.

Informational support refers to providing information that will help someone. You will spend much of your career providing informational support to patients as you explain their diagnoses, potential treatment options, and risks and benefits of those treatment options.

Network support is the type of social support that gives a person a sense of belonging. This can be shown physically, as demonstrated in Figure 10.2, or can be accomplished through gestures, group activities, and shared experiences.

Figure 10.2 Network Support
A group hug creates a sense of belonging.

No matter the form, all of these social supports offer many different types of health benefits. Social support helps reduce psychological distress such as anxiety and depression. People with low social support show higher levels of major mental disorders, alcohol and drug use, and suicidal ideation. Beyond these intuitive improvements in mental health, there are also improvements to our physical health. Studies have found that people with low social support have a higher mortality risk from many different diseases, including diabetes, cardiovascular disease, and cancer. Strong social support appears to correlate with immunological health, too: those with higher social support are less likely to get colds and recover faster when they do.

Social Behaviors and Evolutionary Fitness

Many behaviors have neurological corollaries. Here, we will look at some specific behaviors and the brain regions that are implicated in causing them.

Foraging

The behavior of **foraging**, or seeking out and eating food, is driven by biological, psychological, and social influences. Biologically, hunger is driven by a complex pathway involving both neurotransmitters and hormones. The sensation of hunger is controlled by the **hypothalamus**. Specifically, the lateral hypothalamus promotes hunger, while the ventromedial hypothalamus responds to cues that we are full and promotes satiety. Thus, damage to the lateral hypothalamus will cause a person to lose all interest in eating; meanwhile, damage to the ventromedial hypothalamus will result in obesity because the individual never feels satiated. Foraging is also impacted by genetics. Certain genes play a role in the onset of foraging behavior and the division of tasks between members of the same group. Some species forage together while others engage in solitary foraging.

Cognitive skills play a role in the success of both solitary and group foraging. These skills include spatial awareness, memory, and decision making. In species that forage as a group, foraging is primarily a learned behavior. Young individuals learn through observing how to find and consume food and how to determine what is safe to eat, as shown in Figure 10.3. Animals also learn how to hunt by watching others. Some animals, such as wolves, hunt in packs that have strict rules regarding the order in which individuals are allowed to eat after a successful hunt.

Figure 10.3 Foraging Is Learned through Observational Learning

Mating and Mate Choice

A **mating system** describes the organization of a group's sexual behavior. Mating systems seen among animals include monogamy, polygamy, and promiscuity. **Monogamy** refers to an exclusive mating relationship. **Polygamy** means having exclusive relationships with multiple partners. Having exclusive relationships with multiple females is called **polygyny** and with multiple males is called **polyandry**. **Promiscuity** refers to a member of one sex mating with others without exclusivity. In most animal species, there is one dominant mating system; however, humans exhibit more flexibility. In humans, mating behavior is highly influenced by both biological and social factors. Humans also differ from animals by having formal relationships to correspond with mate choice. Mating may or may not be associated with these social relationships, such as marriage or dating.

Mate choice, or **intersexual selection**, is the selection of a mate based on attraction. **Mate bias** refers to how choosy members of the species are while choosing a mate. This bias is an evolutionary mechanism aimed at increasing the fitness of the species. It may carry **direct benefits** by providing material advantages, protection, or emotional support, or **indirect benefits** by promoting better survival in offspring.

KEY CONCEPT

Direct benefits provide advantages to the mate. Indirect benefits provide advantages to offspring.

There are five recognized mechanisms of mate choice:

- **Phenotypic benefits:** observable traits that make a potential mate more attractive to the opposite sex. Usually, these traits indicate increased production and survival of offspring. For example, males that appear more nurturing are more likely to care for, and promote the survival of, their offspring.
- **Sensory bias:** development of a trait to match a preexisting preference that exists in the population. For example, fiddler crabs are naturally attracted to structures that break up the level horizon because they may indicate a food source; male crabs take advantage of this fact by building pillars around their territory to attract mates.

- **Fisherian** or **runaway selection:** a positive feedback mechanism in which a particular trait that has no effect or a negative effect on survival becomes more and more exaggerated over time. In this model, a trait is deemed sexually desirable and thus is more likely to be passed on. This increases the attractiveness of the trait, which in turn increases the likelihood that it continues to be passed on. The bright plumage of the peacock, shown in Figure 10.4, is the prototypical example of Fisherian selection.

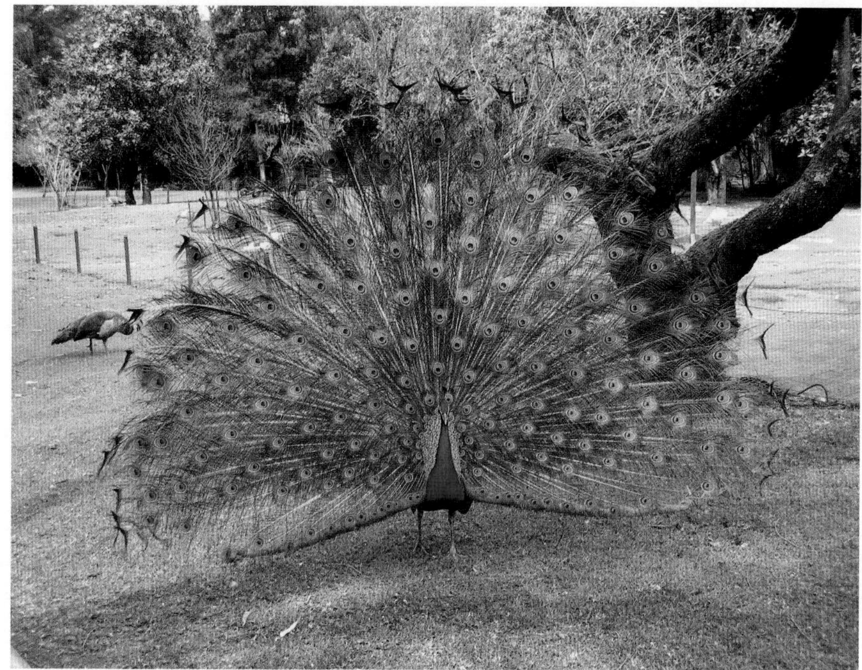

Figure 10.4 Fisherian Selection

The exaggerated plumage of the peacock is the prototypical example of Fisherian selection, in which the attractiveness of a trait that imparts a survival disadvantage leads to its continuation and exaggeration within the species.

- **Indicator traits:** traits that signify overall good health and well-being of an organism, increasing its attractiveness to mates. Notably, these traits may or may not be genetic in origin. For example, female cats are more attracted to male cats with clean and shiny coats; a dirty and dull coat may be related to an underlying genetic problem, or to malnutrition or infection.
- **Genetic compatibility:** the creation of mate pairs that, when combined, have complementary genetics. This theory provides a mechanism for the reduced frequency of recessive genetic disorders in the population: attraction to others who have starkly different genetic makeups reduces the probability of offspring being homozygotic for a disease-carrying allele.

Altruism

REAL WORLD

Altruism creates a bit of a problem for the traditional Darwinist model of evolution. Why would an organism sacrifice its own fitness for the fitness of another? Evolutionary biologists still wrestle with this question, but inclusive fitness offers at least one potential solution.

Altruism is a form of helping behavior in which the individual's intent is to benefit another at some cost to the self. Helping behavior can be motivated by selflessness, but can also be motivated by egoism or ulterior motives, such as public recognition. **Empathy** is the ability to vicariously experience the emotions of another, and it is thought by some social psychologists to be a strong influence on helping behavior. The **empathy–altruism hypothesis** is one explanation for the relationship between empathy and helping behavior. According to this theory, one individual helps another person when feeling empathy for the other person, regardless of the cost. This theory has been heavily debated, and more recent conceptions of altruism posit that an individual will help another person only when the benefits outweigh the costs for the individual.

Game Theory

Game theory attempts to explain decision-making behavior. The theory was originally used in economics and mathematics to predict interaction based on game characteristics, including strategy, winning and losing, rewards and punishments, and profits and cost. A game is defined by its players, the information and actions available to each player at decision points, and the payoffs associated with each outcome.

In the context of biology, game payoffs refer to fitness. Game theorists studying sex ratios in various species developed the concept of the **evolutionary stable strategy (ESS)**. When an ESS is adopted by a given population in a specific environment, natural selection will prevent alternative strategies from arising. The strategies are thus inherited traits passed along with the population, with the object of the game being becoming more fit than competitors.

One of the classic evolutionary games is the Hawk–Dove game. The game focuses on access to shared food resources. In each round, a player chooses one of two strategies: hawk or dove. The hawk exhibits a fighter strategy, displaying aggression and fighting until he wins or is injured. The dove exhibits a fight avoidance strategy, displaying aggression at first but retreating if the fight escalates. If not faced with a fight, the dove will attempt to share the food resources. There are three potential outcomes. If two hawks compete, one will win and one will lose. If a hawk and a dove compete, the hawk will invariably win. If two doves compete, they will share the food resources. The payoff in this case is based on both the value of the reward and the cost of fighting: If the reward is significantly larger than the cost of fighting, then hawks have an advantage. If the cost of fighting is significantly larger, doves have an advantage. There thus exists an equilibrium point where, based on the magnitude of the reward and the cost of fighting, the hawk and dove strategies can coexist as evolutionary stable strategies.

The Hawk–Dove game represents pure competition between individuals. However, social influences apply in nature and can result in four possible alternatives for competitors when dealing with strategic interactions. The four alternatives are shown in Figure 10.5 and are:

- **Altruism:** the donor provides a benefit to the recipient at a cost to the donor
- **Cooperation:** both the donor and recipient benefit by cooperating
- **Spite:** both the donor and recipient are negatively impacted
- **Selfishness:** the donor benefits while the recipient is negatively impacted

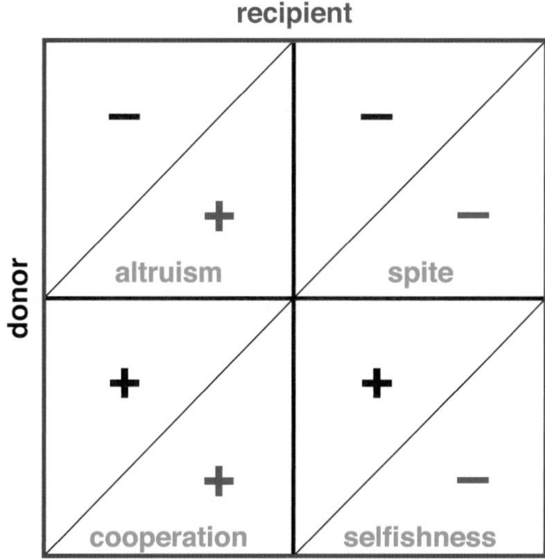

Figure 10.5 Strategic Alternatives for Socially Influenced Competitors

Other common strategy games, like rock–paper–scissors and chicken, can also be explained by game theory.

Inclusive Fitness

In evolutionary psychology, **inclusive fitness** is a measure of an organism's success in the population. This is based on the number of offspring, success in supporting offspring, and the ability of the offspring to then support others. Early descriptions of evolutionary success were based solely on the number of viable offspring of an organism. However, contemporary theories take into account the benefits of certain behaviors on the population at large. For example, the existence of altruism could be supported by the observation that close relatives of an individual will share many of the same genes; thus, promoting the reproduction and survival of related or similar individuals can also lead to genetic success. Other species show examples of inclusive fitness by protecting the offspring of the group at large. By sacrificing themselves to protect the young, these organisms ensure the passing of genes to future generations. Inclusive fitness therefore promotes the idea that altruistic behavior can improve the fitness and success of a species as a whole.

MCAT CONCEPT CHECK 10.1

Before you move on, assess your understanding of the material with these questions.

1. What is interpersonal attraction, and what are three factors that influence this attraction?

    -
    -
    -

2. For a behavior to be considered aggressive, is it necessary to have the intent to do harm? Does the behavior have to be physical in nature?

3. What are the four types of attachment? How does a child with each form of attachment act with regard to the caregiver?

Type of Attachment	Response to Caregiver

4. What is social support? List three of the common types of social support.

    -
    -
    -

5. What is altruism?

____

____

10.2 Social Perception and Behavior

High-Yield

LEARNING OBJECTIVES

After Chapter 10.2, you will be able to:

- Describe the primacy effect, recency effect, halo effect, fundamental attribution error, attribute substitution, just-world hypothesis, and self-serving bias
- Contrast dispositional and situational attributions, and what factors can make each one more likely
- Identify examples of attribution and attribution biases in real-world examples

Social perception is the name social psychologists give to how we form impressions about the characteristics of individuals and groups of people. We form impressions of others through observation of their behavior, past experiences, and personal beliefs and attitudes. We also feel the need to be able to explain and understand the behavior of others, a process we perform through attribution.

BRIDGE

Social perception is highly linked to attitudes; social perception focuses on how we form attitudes about specific characteristics of individuals and groups. Attitudes are discussed in detail in Chapter 8 of *MCAT Behavioral Sciences Review.*

Social Perception

Social perception is also referred to as **social cognition**, and provides the tools to make judgments and impressions regarding other people. These judgments and impressions include assessments of social roles, relationships, characteristics such as trustworthiness or friendliness, and **attributions**, which are explanations for the causes of a person's actions.

Components of Social Perception

There are three primary components of social perception: the perceiver, the target, and the situation. The **perceiver** is influenced by experience, motives, and emotional state. Past experiences affect our attitudes toward current and future experiences and can lead to particular expectations of events. Our motives influence what information we deem important and what we choose to ignore. Finally, emotional state can flavor our interpretation of an event. The **target** refers to the person about which the perception is made. Knowledge of the target can include past experiences or specific information that affect perception. When little information is available, there is a need for greater observation and interpretation by the perceiver. Finally, the **situation** is also important in developing perception. A given social context can determine what information is available to the perceiver.

Impression Bias

One model of social perception focuses on our selection of cues to form interpretations of others that are consistent over time. When coming into contact with an unfamiliar target, a perceiver takes in all cues from the target and environment, unfiltered. After becoming more familiar with a given target, the perceiver uses these cues to categorize the target: friend *vs.* enemy, caring *vs.* standoffish, open-minded *vs.* bigoted, and so on. Additional time spent with the target in the situational context will lead the perceiver to confirm the categorization. After this point, the perception of additional cues becomes selective in order to paint a picture of the target that is consistent with the perceptions the perceiver has already made. This theory is consistent with the **primacy effect**, which is the idea that first impressions are often more important than subsequent impressions. Sometimes, however, it is actually the most recent information we have about an individual that is the most important in forming our impressions; this is called the **recency effect**.

Individuals tend to organize the perception of others based on traits and personal characteristics of the target that are most relevant to the perceiver. This idea is referred to as the **reliance on central traits**. People may also project their own beliefs, opinions, ideas, and actions onto others. The categories we place others in during impression formation is based on **implicit personality theory**. This theory states that there are sets of assumptions people make about how different types of people, their traits, and their behavior are related. Making assumptions about people based on the category in which they are placed is known as **stereotyping**, and will be discussed in detail in the next section.

Halo Effect

The **halo effect** is a cognitive bias in which judgments about a specific aspect of an individual can be affected by one's overall impression of the individual. It is the tendency to allow a general impression about a person (*I like Jin*) to influence other, more specific evaluations about a person (*Jin is a good person, Jin is trustworthy, Jin can do no wrong*). The halo effect explains why people are often inaccurate when evaluating people that they either believe to be generally good or those that they believe to be generally bad. An individual's attractiveness has also been seen to produce the halo effect. As described earlier, attractiveness can be determined by a variety of traits, and the perception of these traits can impact the view of an individual's personality. It has been shown that people who are perceived as attractive are also more likely to be perceived as trustworthy and friendly.

Just-World Hypothesis

Another cognitive bias during impression formation is the **just-world hypothesis**. In a so-called just world, good things happen to good people and bad things happen to bad people; noble actions are rewarded and evil actions are punished. Consequences may be attributed to a universal restoring force; in some religions and cultures, this force is referred to as *karma*. A strong belief in a just world increases the likelihood of "blaming the victim" or stating that victims *get what they deserve* because such a worldview denies the possibility of innocent victims.

Self-Serving Bias

Self-identity and perception can be skewed through **self-serving bias**, also known as **self-serving attributional bias**, which refers to the fact that individuals credit their own successes to internal factors and blame their failures on external factors. The tendency to attribute good outcomes to our good traits or behaviors and to attribute bad outcomes to situational factors is used to protect our self-esteem. For example, students who earn a good grade on a test may attribute their success to their intelligence or to how intensely they studied. However, if they received a bad grade, they might attribute this outcome to poor teaching by the professor, unfair questions, or too long a test for the allotted time. These types of attributions have been found to occur in many settings including the workplace, school, interpersonal relationships, and athletics. Self-serving bias is influenced by motivational processes, like self-enhancement and self-verification. **Self-enhancement** focuses on the need to maintain self-worth, which can be accomplished in part by the self-serving bias. **Self-verification** suggests people will seek the companionship of others who see them as they see themselves, thereby validating a person's self-serving bias. Self-serving bias is also influenced by cognitive processes. For example, emotion is a factor in self-serving bias because emotion can impact self-esteem, which influences the need to protect one's self-identity. Individuals with higher self-esteem are more likely to protect this image and thus more likely to exhibit self-serving bias. Relationships to others also determine the likelihood of the bias: Individuals who have close relationships are less likely to attribute failures to one another, and instead will make joint attributions. On the other hand, strangers are much more likely to self-serve by placing blame for a failure on each other.

REAL WORLD

People with depression often have a reversed attributional bias, viewing their successes as caused by external factors (*I got lucky this time*) and failures as caused by internal factors (*It was all my fault*).

In-group *vs.* Out-group Bias

As suggested in Chapter 9 of *MCAT Behavioral Sciences Review* on social interaction, humans naturally come together to form groups, which results in the subjective categorization of in-group and out-group. In-group refers to other members of one's social group, while out-group refers to those who are not in the group. Given the propensity for humans to form groups, it's understandable that how an individual perceives members within their group (in-group) versus people outside their group (out-group) is heavily biased. Specifically, **in-group bias** refers to the inclination to view members in one's group more favorably, while **out-group bias** refers to the inclination to view individuals outside one's group harshly.

Attribution Theory

Another aspect of social cognition is explaining the behavior of others. It is human nature to observe and try to understand why others act the way they do. **Attribution theory** describes how individuals infer the causes of other people's behavior.

Dispositional and Situational Causes

Fritz Heider, one of the founders of attribution theory, divided the causes for attribution into two main categories: dispositional (internal) and situational (external). **Dispositional (internal)** attributions are those that relate to the person whose behavior is being considered, including beliefs, attitudes, and personality characteristics. **Situational (external)** attributions are those that relate to features of the surroundings, such as threats, money, social norms, and peer pressure. For instance, suppose you hear that a friend has been nominated for an academic award. Believing that the friend has been nominated because of hard work and personal effort would be a dispositional attribution. Contrarily, chalking up the nomination to luck would be a situational attribution. Situational attributions, therefore, consider the characteristics of the social context rather than the characteristics of the individual as the primary cause.

Cues

KEY CONCEPT

- Consistency cues—has consistent behavior over time
- Consensus cues—matches others' behavior
- Distinctiveness cues—uses similar behavior in similar situations

In order to understand the behavior of others, a variety of cues are used. These include consistency cues, consensus cues, and distinctiveness cues. **Consistency cues** refer to the behavior of a person over time. The more regular the behavior, the more we associate that behavior with the motives of the person. **Consensus cues** relate to the extent to which a person's behavior differs from others. If a person deviates from socially expected behavior, we are likely to form a dispositional attribution about the person's behavior. **Distinctiveness cues** refer to the extent to which a person engages in similar behavior across a series of scenarios. If a person's behavior varies in different scenarios, we are more likely to form a situational attribution to explain it.

MCAT EXPERTISE

Actor-observer asymmetry (or **bias**) results from the self-serving bias (by the actor) and the fundamental attribution error (by the observer). Actor-observer bias holds that, due to our unique knowledge about our own actions, we are more likely to make situational attributions for the self as compared to others. If you, rather than another person, failed to complete an assignment, for example, you would be far more likely to consider the situational factors involved in your behavior because you are intimately aware of them. This effect is most commonly seen with negative behaviors.

The **correspondent inference theory** takes this concept one step further by focusing on the intentionality of others' behavior. When an individual unexpectedly performs a behavior that helps or hurts us, we tend to explain the behavior by dispositional attribution. Thus, we may correlate these unexpected actions with the person's personality.

Fundamental Attribution Error

The **fundamental attribution error** posits that we are generally biased toward making dispositional attributions rather than situational attributions when judging the actions of others. For example, suppose that on a team project, other team members were unable to complete their assignments. According to the fundamental attribution error, our immediate response might be to assume that these team members are lazy or unreliable—both of which are dispositional attributions. We may ignore the possibility that the team members got ill, had too many concurrent assignments, or suffered a personal tragedy—all of which are situational attributions. The fundamental attribution error can present itself in positive contexts as well. Imagine if you observed someone getting out of their car to help an older adult across the road. According to the fundamental attribution error, you would likely make a dispositional attribution like, "What a kind stranger!" rather than a situational attribution like, "Oh, maybe that's their grandparent." Notice that in these examples, the dispositional attributions often provide simpler explanations than the situational attributions. This difference in complexity is actually the source of the fundamental attribution error: Assuming that a person's behaviors accurately portray who they are as a person is easier than speculating about what circumstances might have caused the observed behavior.

Attribute Substitution

Attribute substitution occurs when individuals must make judgments that are complex, but instead they substitute a simpler solution or apply a heuristic. When making automatic or intuitive judgments on difficult questions or scenarios, an individual may address a different question or scenario without even realizing a substitution has been made. In one study, individuals were asked to envision a sphere that could just fit inside a cube. They were then asked what percentage of the volume of the cube would be taken up by the sphere. This is challenging to envision, so most individuals likely simplified the problem in their minds to imagine a circle inside a square. The answers given in this study averaged around 74 percent, which is approximately the area of a square taken up by a circumscribed circle (79%), but significantly higher than the volume of a cube taken up by a circumscribed sphere (52%).

Attribute substitution can take place in far simpler setups as well. A classic example used in many psychology classes is the following question: *A pencil and an eraser cost $1.10 together. If the pencil costs one dollar more than the eraser, how much does the eraser cost?* Most individuals respond instinctively with the answer *ten cents*. It is easy to recognize that the pencil costs more, and to integrate the information given in the question stem (*$1.10* and *one dollar*) incorrectly.

This process is also common when dealing with size and color in optical illusions. For instance, when judging the size of figures in an image with perspective, the apparent sizes shown in the image can be distorted by three-dimensional context, as shown in Figure 10.6. The expected three-dimensional size of the figure, based on perspective cues, substitutes for the actual two-dimensional size of the birds within the image. It is interesting to note that painters and photographers with experience in two-dimensional images are less likely to substitute due to the fact that two-dimensional size is more understandable to their perception.

Figure 10.6 Attribute Substitution for Size in Optical Illusions
The birds are of identical size, but three-dimensional cues affect our interpretation of the image.

Shadows, patterns, the position of the sun, and other visual cues can also cause attribute substitution for color, as shown in Figure 10.7.

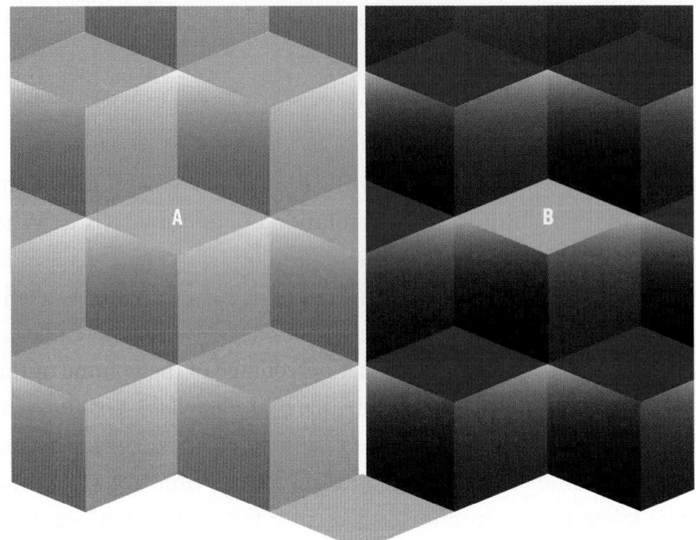

Figure 10.7 Attribute Substitution for Color in Optical Illusions
The central boxes labeled A and B are of identical color, but shadow cues affect our interpretation of the image.

Cultural Attribution

Another important factor in attribution is culture. The type of culture an individual belongs to plays a major role in the types of attributions the individual makes. Individualist cultures put high value on the individual, personal goals, and independence. Collectivist cultures view individuals as members of a group and place high value on conformity and interdependence. Individualists tend to make more fundamental attribution errors than those in collectivist cultures. Individualists are also more likely to attribute behavior to dispositional factors, whereas collectivists are more likely to attribute behavior to situational factors.

MCAT CONCEPT CHECK 10.2

Before you move on, assess your understanding of the material with these questions.

1. For each of the social cognitive biases below, provide a brief description:
   - Primacy effect:
   - Recency effect:
   - Halo effect:
   - Just-world hypothesis:
   - Self-serving bias:

2. What is attribution theory? What are the two types of attribution?
   -
   -

3. What is the fundamental attribution error?

4. What is attribute substitution?

10.3 Stereotypes, Prejudice, and Discrimination

LEARNING OBJECTIVES

After Chapter 10.3, you will be able to:

- Distinguish between stereotypes, prejudice, and discrimination
- List the types of social inequality that can influence prejudice
- Compare and contrast ethnocentrism and cultural relativism

KEY CONCEPT

- Stereotypes are cognitive
- Prejudices are affective
- Discrimination is behavioral

While stereotypes, prejudice, and discrimination are terms that are related and often used together, they are very different concepts. Stereotypes are viewed as cognitive, prejudice as affective, and discrimination as behavioral. **Stereotypes** refer to the expectations, impressions, and opinions about the characteristics of members of a group. Prejudice reflects the overall attitude and emotional response to a group. Discrimination refers to differences in actions toward different groups.

Note: Kaplan Test Prep does not endorse or encourage any of the stereotypes mentioned in this chapter; they are included only as examples.

Stereotypes

Despite their negative connotations, stereotypes often appear in everyday life, and can be beneficial when not directed toward other human beings. In a psychological sense, the purpose of a stereotype is to make sense of a complex world by categorizing and systematizing information in order to better identify items, predict their behavior, and react. In the context of stereotyping what different items of furniture look like, how different types of stores operate, or how different cuisines taste, stereotypes are extremely useful in defining categories and determining what does or does not fit into that category. However, when stereotypes are used to develop prejudices toward others and to discriminate, they are being appropriated for negative uses.

In the context of sociology, **stereotypes** occur when attitudes and impressions are based on limited and superficial information about a person or a group of individuals. The content of stereotypes are the attributes that people believe define and characterize a group. The **stereotype content model** attempts to classify stereotypes with respect to a hypothetical in-group using two dimensions: warmth and competence. Warm groups are those that are not in direct competition with the in-group for resources; competent groups are those that have high status within society. The four possible combinations of warmth and competence are shown in Figure 10.8 and are associated with distinct emotions.

		Competence	
		low	high
Warmth	high	**paternalistic stereotype** low status, not competitive	**admiration stereotype** high status, not competitive
	low	**contemptuous stereotype** low status, competitive	**envious stereotype** high status, competitive

Figure 10.8 Classifications of Stereotypes in the Stereotype Content Model
Adapted from Fiske et al. (2002)

Paternalistic stereotypes are those in which the group is looked down upon as inferior, dismissed, or ignored. **Contemptuous** stereotypes are those in which the group is viewed with resentment, annoyance, or anger. **Envious** stereotypes are those in which the group is viewed with jealousy, bitterness, or distrust. **Admiration** stereotypes are those in which the group is viewed with pride and other positive feelings.

Self-Fulfilling Prophecy

Stereotypes can lead to expectations of certain groups of individuals. These expectations can create conditions that then cause the expectations to become reality, a process referred to as **self-fulfilling prophecy**. For example, some medical students experience a self-fulfilling prophecy during their first days of surgery clerkship in medical school: During their first year in the wards, new students are stereotyped as being unable to quickly and efficiently throw knots during a surgery. With this knowledge in mind, these students are nervous to suture for the first time and may struggle with every step of the knot-tying process. This struggle validates the stereotype and thus completes the self-fulfilling prophecy.

Stereotype Threat

In some social situations, a person might be concerned or anxious about inadvertently confirming a negative stereotype about their social group. This concern is known as **stereotype threat**. Unfortunately, the feeling of stereotype threat often results in a self-fulfilling prophesy: People experiencing stereotype threat often exhibit stress arousal and are preoccupied by monitoring their own performance on a task, and these distractions can then lead to reduced performance on the task. An example of a well-studied group that often experiences stereotype threat is women

KEY CONCEPT

Stereotype threat is concern or anxiety about confirming a negative stereotype about one's group. This may hinder performance, which may actually create a self-fulfilling prophecy.

in mathematics. A study showed that women taking a math exam scored lower when the only other test takers in the room were men. The researchers concluded that when taking an exam with only men present, women test subjects were more concerned about stereotype threat, and performed more poorly as a result of their concerns. Researchers theorize that stereotype threat may be a contributing factor to long-standing racial and gender gaps in certain careers and in academic performance.

Prejudice

From a social psychology approach, **prejudice** is defined as an irrational positive or negative attitude toward a person, group, or thing, prior to an actual experience with that entity. The process of socialization results in the formation of attitudes regarding our own groups and a sense of identity as an individual and a group member. Prejudice can form in response to dissimilarities among groups, races, ethnicities, or even environments. While racial and ethnic prejudices against individuals are at the forefront of most people's minds, prejudices exist against objects and places as well. For instance, people have attitudes toward different regions of the country based on culture, weather, and history; which car manufacturers are the most reliable; what types of food are considered unhealthy; and even what types of animals make good pets. Prejudicial attitudes can run the gamut from hate to love, contempt to admiration, and indifference to loyalty.

Prejudices may be kept internally or shared with the larger community. **Propaganda** is a common way by which large organizations and political groups attempt to create prejudices in others. Propaganda posters often invoke messages of fear, and depictions of the target group are often exaggerated to an absurd degree.

Power, Prestige, and Class

There are a variety of social factors that influence prejudice. Three of the most important are power, prestige, and class. **Power** refers to the ability of people or groups to achieve their goals despite any obstacles, and their ability to control resources. **Prestige** is the level of respect shown to a person by others. **Class** refers to socioeconomic status. Social inequality, or the unequal distribution of power, resources, money, or prestige, can result in the grouping of *haves* and *have-nots*. *Have-nots* may develop a negative attitude toward *haves* based on envy. *Haves* may develop a negative attitude toward *have-nots* as a defense mechanism to justify the fact that they have more.

Ethnocentrism

Ethnocentrism refers to the practice of making judgments about other cultures based on the values and beliefs of one's own culture, especially when it comes to language, customs, and religion. Ethnocentrism can manifest in many ways, from innocent displays of ethnic pride to violent supremacy groups. Because of this, ethnocentrism is closely tied to the previously discussed concepts of in-group *vs.* out-group biases and group conflict.

Cultural Relativism

In order to avoid ethnocentrism, the concept of cultural relativism has been employed by sociologists to compare and understand other cultures. **Cultural relativism** is the recognition that social groups and cultures should be studied on their own terms. When studying a culture, social relativism acknowledges that the values, mores, and rules make sense in the context of that culture, and should not be judged against the norms of another culture. In other words, while one group may follow a given set of rules (say, the dietary rules of *kashrut* or *halal*), cultural relativism holds that those rules should not be perceived as superior or inferior to those of other cultures—just different.

Discrimination

Discrimination occurs when prejudicial attitudes cause individuals of a particular group to be treated differently from others. While prejudice is an attitude, discrimination is a behavior. As prejudice typically refers to a negative attitude, discrimination typically refers to a negative behavior. It is also important to note that prejudice does not always result in discrimination. For instance, a person might have strong feelings against a particular race (prejudice), but may not express those feelings or act on them. As social inequality influences prejudice, the same idea applies to discrimination. The unequal distribution of power, prestige, and class influence discrimination.

Individual *vs.* Institutional Discrimination

Discrimination can be either individual or institutional. **Individual discrimination** refers to one person discriminating against a particular person or group, whereas **institutional discrimination** refers to the discrimination against a particular person or group by an entire institution. Individual discrimination is considered to be conscious and obvious, and can be eliminated by removing the person who is displaying the behavior. Sociologists have begun to stress the need to focus on institutional discrimination, as it is discrimination built into the structure of society, so it is far more covert and harder to extricate. Because it is part of society, it is perpetuated by simply maintaining the status quo.

The United States has a long history of institutional discrimination against myriad groups. Perhaps the most overt example was that of racial segregation that existed in the early to mid-twentieth century. Even today, there are still concerns of institutional discrimination against individuals based on their race, ethnicity, gender identity, sexual orientation, religion, and other characteristics.

MCAT CONCEPT CHECK 10.3

Before you move on, assess your understanding of the material with these questions.

1. What are the distinctions between stereotypes, prejudice, and discrimination?
   - Stereotypes:

   ____

   - Prejudice:

   ____

   - Discrimination:

   ____

2. List three types of social inequality that can influence prejudice:
   -

   ____

   -

   ____

   -

   ____

3. What is the difference between ethnocentrism and cultural relativism?

   ____

Conclusion

Social psychology focuses on social behavior and the attitudes, perceptions, and influences of others that impact behavior. In this chapter, we first looked at social behaviors, including attraction, aggression, attachment, and social support. We also looked at the biological explanations of specific social behaviors, including foraging, mate choice, altruism, game theory, and inclusive fitness. We further defined the components of social perception and impression biases. The way we view ourselves also influences the way we view others and how we attribute behavior to others. Finally, we took a look at stereotypes, prejudice, and discrimination.

These last few topics demonstrate a negative side of classifying individuals. We can use classification to create hierarchies, inequities in opportunity and finances, as well as to silence or suppress communities. But classification can also serve a positive purpose. In social science, we often classify populations to study interactions between groups, changes in population makeup over time, and to track migration patterns. These classifications are considered in the field of demographics, which we will explore in the next chapter.

GO ONLINE

You've reviewed the content, now test your knowledge and critical thinking skills by completing a test-like passage set in your online resources!

CONCEPT SUMMARY

Social Behavior

- **Interpersonal attraction** is what makes people like each other and is influenced by multiple factors:
    - Physical attractiveness, which is increased with symmetry and proportions close to the **golden ratio**
    - Similarity of attitudes, intelligence, education, height, age, religion, appearance, and socioeconomic status
    - **Self-disclosure**, which includes sharing fears, thoughts, and goals with another person and being met with empathy and nonjudgment
    - **Reciprocity**, in which we like people who we think like us
    - **Proximity**, or being physically close to someone
- **Aggression** is a physical, verbal, or nonverbal behavior with the intention to cause harm or increase social dominance.
- **Attachment** is an emotional bond to another person, and usually refers to the bond between a child and a caregiver. There are four types of attachment:
    - **Secure attachment** requires a consistent caregiver so the child is able to go out and explore, knowing there is a secure base to return to; the child will show strong preference for the caregiver.
    - **Avoidant attachment** occurs when a caregiver has little or no response to a distressed, crying child; the child shows no preference for the caregiver compared to strangers.
    - **Ambivalent attachment** occurs when a caregiver has an inconsistent response to a child's distress, sometimes responding appropriately, sometimes neglectfully; the child will become distressed when the caregiver leaves and is ambivalent when the caregiver returns.
    - **Disorganized attachment** occurs when a caregiver is erratic or abusive; the child shows no clear pattern of behavior in response to the caregiver's absence or presence and may show repetitive behaviors.
- **Social support** is the perception or reality that one is cared for by a social network.
    - **Emotional support** includes listening to, affirming, and empathizing with someone's feelings.
    - **Esteem support** affirms the qualities and skills of the person.
    - **Material support** is providing physical or monetary resources to aid a person.
    - **Informational support** is providing useful information to a person.
    - **Network support** is providing a sense of belonging to a person.

- **Foraging** is searching for and exploiting food resources.
- A **mating system** describes the way in which a group is organized in terms of sexual behavior.
  - **Monogamy** consists of exclusive mating relationships.
  - **Polygamy** consists of multiple exclusive relationships, including **polygyny** (with multiple females) and **polyandry** (with multiple males).
  - **Promiscuity** means mating without exclusivity.
- **Mate choice**, or **intersexual selection**, is the selection of a mate based on attraction and traits.
- **Altruism** is a form of helping behavior in which people's intent is to benefit someone else at some cost to themselves.
- **Game theory** attempts to explain decision making between individuals as if they are participating in a game.
- **Inclusive fitness** is a measure of an organism's success in the population. This is based on the number of offspring, success in supporting offspring, and the ability of the offspring to then support others.

Social Perception and Behavior

- **Social perception** or **social cognition** is the way by which we generate impressions about people in our social environment. It contains a **perceiver**, a **target**, and the **situation** or social context of the scenario.
- **Implicit personality theory** states that people make assumptions about how different types of people, their traits, and their behavior are related.
- Certain cognitive biases impact our perceptions of others.
  - The **primacy effect** refers to when first impressions are more important than subsequent impressions.
  - The **recency effect** is when the most recent information we have about an individual is most important in forming our impressions.
  - A **reliance on central traits** is the tendency to organize the perception of others based on traits and personal characteristics that matter to the perceiver.
  - The **halo effect** is when judgments of an individual's character can be affected by the overall impression of the individual.
  - The **just-world hypothesis** is the tendency of individuals to believe that good things happen to good people and bad things happen to bad people.
  - **Self-serving bias** refers to the fact that individuals will view their own successes as being based on internal factors, while viewing failures as being based on external factors.

- **Attribution theory** focuses on the tendency for individuals to infer the causes of other people's behavior.
  - **Dispositional** (**internal**) causes are those that relate to the features of the person whose behavior is being considered.
  - **Situational** (**external**) causes are related to features of the surroundings or social context.
- **Correspondent inference theory** is used to describe attributions made by observing the intentional (especially unexpected) behaviors performed by another person.
- **Fundamental attribution error** is the bias toward making dispositional attributions rather than situational attributions in regard to the actions of others.
- **Attribute substitution** occurs when individuals must make judgments that are complex but instead substitute a simpler solution or heuristic.
- Attributions are highly influenced by the culture in which one resides.

Stereotypes, Prejudice, and Discrimination

- **Stereotypes** occur when attitudes and impressions are made based on limited and superficial information about a person or a group of individuals.
- Stereotypes can lead to expectations of certain groups, which can create conditions that lead to confirmation of the stereotype, a process referred to as **self-fulfilling prophecy**.
- **Stereotype threat** is concern or anxiety about confirming a negative stereotype about one's social group.
- **Prejudice** is defined as an irrational positive or negative attitude toward a person, group, or thing prior to an actual experience.
- **Ethnocentrism** refers to the practice of making judgments about other cultures based on the values and beliefs of one's own culture.
- **Cultural relativism** refers to the recognition that social groups and cultures should be studied on their own terms.
- **Discrimination** is when prejudicial attitudes cause individuals of a particular group to be treated differently from others.
  - **Individual discrimination** refers to one person discriminating against a particular person or group.
  - **Institutional discrimination** refers to the discrimination against a particular person or group by an entire institution.

ANSWERS TO CONCEPT CHECKS

10.1

1. Interpersonal attraction is what makes people like each other and is influenced by at least five factors discussed in the chapter: physical attractiveness, similarity, self-disclosure, reciprocity, and proximity.
2. No, in addition to behavior with the intent to cause harm, aggression can also be a behavior that increases relative social dominance. Aggression can be physical, verbal, or nonverbal.
3.

Type of Attachment	Response to Caregiver
Secure	Upset at departure of caregiver, comforted by return; trusts caregiver, who is viewed as a secure base
Avoidant	Shows no preference for a stranger or caregiver; shows little distress at departure and little relief by return of caregiver
Ambivalent	Distressed by departure of caregiver with mixed reactions at return
Disorganized	No clear pattern of behavior; sometimes exhibits repetitive behaviors or seems dazed, frozen, or confused

4. Social support is the perception or reality that one is cared for by a social network. There are five types discussed in this chapter: emotional support, esteem support, material support, informational support, and network support.
5. Altruism is a form of helping behavior in which people's intent is to benefit someone else at some cost to themselves.

10.2

1. The primacy effect is the power of first impressions over later impressions of an individual. The recency effect is weighing the most recent information of a person as the most important. The halo effect occurs when one applies general feelings about a person (usually, "good" or "bad") to specific characteristics of that person. The just-world hypothesis is the belief that good things happen to good people and bad things happen to bad people. Self-serving bias is the tendency to attribute our successes to internal factors and our failures to external factors.
2. Attribution theory focuses on the tendency of individuals to infer the causes of other people's behavior. Attributions are divided into two types: dispositional (internal) causes, which relate to the features of the target, and situational (external) causes, which relate to features of the surroundings or context.
3. Fundamental attribution error is the general bias toward making dispositional attributions rather than situational attributions about the behavior of others, especially in negative contexts.
4. Attribute substitution occurs when individuals must make judgments that are complex but instead substitute a simpler solution or heuristic.

10.3

1. Stereotypes occur when attitudes and impressions are made based on limited and superficial information about a person or a group of individuals and are cognitive. Prejudice is defined as an irrational negative, or occasionally positive, attitude toward a person, group, or thing, which is formed prior to an actual experience and is affective. Discrimination is when prejudicial attitudes cause individuals of a particular group to be treated differently than others and is behavioral.
2. Power, prestige, and class all influence prejudice through unequal distribution of wealth, influence, and resources.
3. Ethnocentrism refers to the practice of making judgments about other cultures based on the values and beliefs of one's own culture. Cultural relativism refers to the recognition that social groups and cultures must be studied on their own terms. In both cases, an individual perceives another group to which that individual does not belong; however, it is the reaction to that other group that determines which paradigm is being used.

SCIENCE MASTERY ASSESSMENT EXPLANATIONS

1. **A**

Each of the answer choices influences social attraction; however, proximity deals with the tendency to be attracted to those who are physically close by.

2. **B**

Aggression is influenced both by the amygdala and prefrontal cortex activity. Activity of the amygdala increases aggression. The prefrontal cortex should control aggression; decreased activity in the prefrontal cortex, therefore, is associated with increased aggression.

3. **D**

This attachment pattern is representative of secure attachment. Secure attachment is seen when a child has a consistent caregiver and is able to go out and explore, knowing there is a secure base to return to. The child will be upset at the departure of the caregiver and will be comforted and resume exploring upon the return of the caregiver.

4. **B**

Polygamy involves having exclusive relationships with several partners, several females (polygyny), or several males (polyandry), **(A)**. Monogamy, **(C)**, consists of exclusive mating relationships. Promiscuity, **(D)**, refers to mating with others without exclusivity.

5. **C**

A person with a ventromedial hypothalamus injury will never feel satiated when eating and will therefore never feel the sensation to stop eating. A person with a lateral hypothalamus injury will never feel hunger and will have decreased food intake, **(D)**.

6. **D**

Phenotypic benefits refer to observed traits in an individual that make them more attractive to other members of their species. Benefits associated with increased fitness through direct material advantages are direct benefits, **(C)**, while indirect benefits involve increased genetic fitness for offspring.

7. **A**

In evolutionary psychology, inclusive fitness is a measure of the number of offspring an individual has, how they support their offspring, and how their offspring can support others. Inclusive fitness promotes the idea that altruistic behavior can improve the fitness and success of a species; the behavior in this scenario can be described as altruism: benefiting another at one's own expense.

8. **D**

There are three primary components of perception: the perceiver, the target, and the situation.

9. **A**

The impressions we form when meeting others are influenced by a number of perceptual biases. The primacy effect refers to those occasions when first impressions are more important than subsequent impressions.

10. **C**

The halo effect is a cognitive bias in which judgments of an individual's character can be affected by the overall impression of the individual.

11. **C**

Self-serving bias refers to the fact that individuals will view their own successes as being based on internal factors, while viewing their failures as being based on external factors.

12. **C**

Types of attribution fall into two main categories: dispositional (internal) and situational (external). Dispositional (internal) causes are related to the features of the person whose behavior is being considered. Situational (external) causes are related to features of the surroundings.

13. **B**

Stereotype threat refers to the phenomenon of people being concerned or anxious about confirming a negative stereotype of their social group. Stereotype threat can hinder performance, creating a self-fulfilling prophecy.

14. **D**

Discrimination is when prejudicial attitudes cause individuals of a particular group to be treated differently than others. While prejudice is an attitude, discrimination is a behavior.

15. **C**

Game theory was originally designed to study decision-making behavior in economics and mathematics; it has since been used to describe decision making in politics, biology, philosophy, and other fields.

Consult your online resources for additional practice.

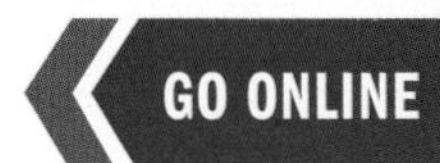

SHARED CONCEPTS

Behavioral Sciences Chapter 5
Motivation, Emotion, and Stress

Behavioral Sciences Chapter 6
Identity and Personality

Behavioral Sciences Chapter 8
Social Processes, Attitudes, and Behavior

Behavioral Sciences Chapter 9
Social Interaction

Behavioral Sciences Chapter 11
Social Structure and Demographics

Behavioral Sciences Chapter 12
Social Stratification

CHAPTER 11

SOCIAL STRUCTURE AND DEMOGRAPHICS

SCIENCE MASTERY ASSESSMENT

Every pre-med knows this feeling: there is so much content I have to know for the MCAT! How do I know what to do first or what's important?

While the high-yield badges throughout this book will help you identify the most important topics, this Science Mastery Assessment is another tool in your MCAT prep arsenal. This quiz (which can also be taken in your online resources) and the guidance below will help ensure that you are spending the appropriate amount of time on this chapter based on your personal strengths and weaknesses. Don't worry though—skipping something now does not mean you'll never study it. Later on in your prep, as you complete full-length tests, you'll uncover specific pieces of content that you need to review and can come back to these chapters as appropriate.

How to Use This Assessment

If you answer 0–7 questions correctly:

Spend about 1 hour to read this chapter in full and take limited notes throughout. Follow up by reviewing **all** quiz questions to ensure that you now understand how to solve each one.

If you answer 8–11 questions correctly:

Spend 20–40 minutes reviewing the quiz questions. Beginning with the questions you missed, read and take notes on the corresponding subchapters. For questions you answered correctly, ensure your thinking matches that of the explanation and you understand why each choice was correct or incorrect.

If you answer 12–15 questions correctly:

Spend less than 20 minutes reviewing all questions from the quiz. If you missed any, then include a quick read-through of the corresponding subchapters, or even just the relevant content within a subchapter, as part of your question review. For questions you answered correctly, ensure your thinking matches that of the explanation and review the Concept Summary at the end of the chapter.

1. Which of the following best describes a manifest function?

A. An intended positive effect on a system
B. An intended negative effect on a system
C. An unintended positive effect on a system
D. An unintended negative effect on a system

2. Studying why a nod means "yes" in many cultures is most representative of which of the following sociological concepts?

A. Demographic transition
B. Expectancy theory
C. Symbolic interactionism
D. Demographic shift

3. Which of the following ethical principles states that physicians should avoid using treatments with greater potential for harm than benefit?

A. Autonomy
B. Beneficence
C. Justice
D. Nonmaleficence

4. A Cuban-American man living in the United States has the dominant physical features of a Black man. He speaks Spanish, prefers Latin foods, and listens to Latin music. His preferences are best defined through which of the following attributes?

I. Race
II. Ethnicity
III. Culture

A. I only
B. II only
C. II and III only
D. I, II, and III

5. A patient who resides in the United States says, "I love you," and hugs the doctor after every routine visit. This behavior violates:

A. personal beliefs.
B. patient autonomy.
C. social values.
D. social norms.

6. Which of the following demographics can be measured in events per 1000 people per year?

I. Birth rate
II. Fertility rate
III. Mortality rate

A. I only
B. I and III only
C. II and III only
D. I, II, and III

7. Because there are more than 500 American Indian tribes, there are several different healing practices among them. Some tribes may have ceremonies that include chanting, singing, body painting, dancing, and even use of mind-altering substances to persuade the spirits to heal the sick person. These ceremonies are examples of:

A. latent functions.
B. rituals.
C. cultural barriers.
D. social movements.

8. Over the last few decades, the United States population has become:

A. bigger, older, and more diverse.
B. bigger, younger, and more diverse.
C. smaller, older, and less diverse.
D. smaller, older, and more diverse.

9. Which of the following is NOT an example of material culture?
- **A.** Traditional Kenyan clothing
- **B.** Japanese cuisine
- **C.** American values
- **D.** American Indian sand paintings

10. During which stage of demographic transition are both birth rates and mortality rates low?
- **A.** Stage 1
- **B.** Stage 2
- **C.** Stage 3
- **D.** Stage 4

11. Shortly after a state legalizes gambling in casinos, a formal coalition forms to oppose the building of any casinos in the major cities of the state. This scenario includes:
- **I.** conflict theory.
- **II.** social institutions.
- **III.** a social movement.

- **A.** I only
- **B.** I and III only
- **C.** II and III only
- **D.** I, II, and III

12. A young adult male claims to have had sexual relationships mostly with other men, although he has been attracted to women at times. What would be his most likely score on the Kinsey scale?
- **A.** 0
- **B.** 1
- **C.** 5
- **D.** 6

13. Which of the following demographic variables is known to be biologically determined?
- **A.** Gender
- **B.** Sex
- **C.** Ethnicity
- **D.** Sexual orientation

14. Which of the following would contribute to increasing population growth over time?
- **A.** A fertility rate less than 2
- **B.** An immigration rate larger than emigration rate
- **C.** An increase in mortality rate
- **D.** A decrease in birth rate

15. Urbanization can cause all of the following negative effects EXCEPT:
- **A.** decreased opportunity for social interaction.
- **B.** increased transmission of infectious disease.
- **C.** decreased air quality and sanitation.
- **D.** increased rates of violent crime.

Answer Key

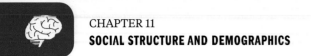

1. **A**
2. **C**
3. **D**
4. **C**
5. **D**
6. **B**
7. **B**
8. **A**
9. **C**
10. **D**
11. **D**
12. **C**
13. **B**
14. **B**
15. **A**

Detailed explanations can be found at the end of the chapter.

CHAPTER 11

SOCIAL STRUCTURE AND DEMOGRAPHICS

In This Chapter

CHAPTER PROFILE

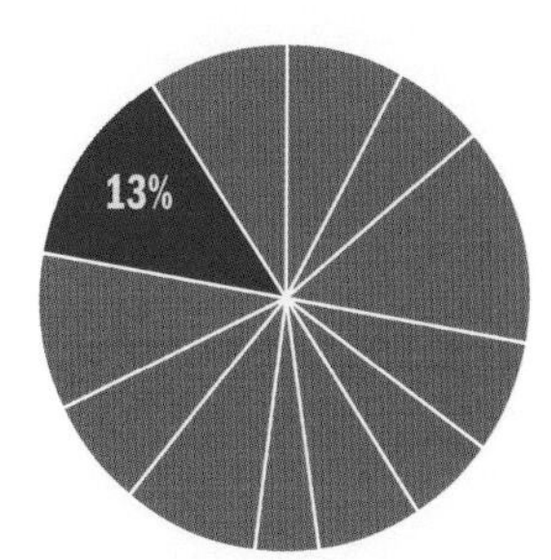

The content in this chapter should be relevant to about 13% of all questions about the behavioral sciences on the MCAT.

This chapter covers material from the following AAMC content categories:

7B: Social processes that influence human behavior

9A: Understanding social structure

9B: Demographic characteristics and processes

Introduction

A frail, older man is admitted to the Intensive Care Unit at a local hospital. He is gaunt, weighing just over 100 pounds, in severe respiratory distress, and nearing circulatory collapse. The intensive care team works to stabilize the patient by starting intravenous lines and pumping fluids. The patient screams statements in a language the team doesn't speak and is eventually sedated so they can intubate and ventilate him. They begin to run tests and discover that the man has widespread metastatic lung cancer that is unlikely to be cured or even controlled through chemotherapy or radiation. The patient's wife and children visit the next morning and are told of the bad news. After crying for some time, they turn to the intensive care team and ask them not to inform the patient of this grave diagnosis. They explain that in their culture family members are expected to make healthcare decisions for the ill to avoid burdening them with such matters. The members of the medical team, however, feel that they must adhere to the tenets of local medical ethics—and their country's laws—and allow the patient to make his own healthcare decisions. As the head of the hospital ethics committee, you get a call from the team to help them make this decision. What would you do?

Ethicists and sociologists alike wrestle with medical dilemmas like these. As a clinician in an ever-diversifying society, you will certainly run into challenging situations like these, where you must try to balance a patient's cultural beliefs with the processes and protocols of the healthcare system in which you work. In this chapter, we'll explore some of the sociological topics on the MCAT, focusing on theoretical models and the key institutions on which you are likely to be tested. We'll then explore culture itself. Finally, we'll describe demographics, the mathematical and statistical modeling of sociological concepts.

11.1 Sociology: Theories and Institutions

High-Yield

LEARNING OBJECTIVES

After Chapter 11.1, you will be able to:

- Recall the primary theses of major sociological theories, including functionalism, conflict theory, symbolic interactionism, social constructionism, rational choice theory, and feminist theory
- Contrast manifest and latent functions of given elements of a sociological system
- List the four key tenets of American medical ethics

Sociology is the study of society: how we create society, how we interact within and change society, and how we define what is normal and abnormal in society. Whereas psychologists focus on the behaviors of individuals, sociologists focus on the way groups organize and interact. Of course, it would be difficult to describe society as a whole since human society is not completely uniform and is instead made up of distinct cultures, subcultures, groups, and institutions, each with its own structure, patterns, and interests. As a result, sociologists study the subject at different levels: The **micro** level consists of family groups and local communities. The **meso** level consists of organizations, institutions, and ethnic subcultures. And the **macro** level consists of national and international systems.

Theoretical Approaches

Because human interaction is so complex, sociologists have proposed many models, called **theoretical approaches to sociology**, to help analyze and explain aspects of human social behavior. Sociologists don't necessarily disagree about which model is "best." Rather, each model was created to explain certain aspects of sociology, and is inadequate for explaining other aspects. For example, one theoretical approach we will examine is symbolic interactionism, which studies how humans interact and communicate using language, writing, and other symbols. By contrast, another approach, called conflict theory, examines how groups with little power in society can rise up and create more equality. These two models are not necessarily competing explanations. Rather, they examine such vastly different aspects of human interaction that both models are necessary!

As you read this section, keep track of which aspects of human behavior each model explains, and what the weaknesses of each model are. For example, approaches such as symbolic interactionism, social constructionism, and rational choice (exchange) theory tend to focus on society at the micro- and meso-levels. On the other hand, conflict theory, structural-functionalism, and feminist theory are more applicable at the macro-level.

Symbolic Interactionism

Communication among humans stretches far beyond just language, spoken, signed, or written. Our bodies are tools of communication through our posture, facial expressions, and informal hand gestures. We recognize and put meaning into images like the smiley face, traffic signs, or corporate logos. Sacred emblems like the cross, the Star of David, or the star and crescent communicate deep religious significance. Even our clothing can communicate meaning. Any object, image, sound, or action that carries meaning to humans is a **symbol**. Importantly, according to this definition, while many physical objects are symbols, a symbol need not be a physical object. For example, spoken words are symbols, and so are meaningful grunts. Body language and hand gestures are symbols. Anything that carries meaning beyond its own existence is a symbol.

Symbolic interactionism is an approach to sociology pioneered by influential sociologist George Herbert Mead that attempts to understand human action and interaction by studying the symbols we use to communicate. This theoretical approach makes three main assumptions about human behavior:

- Humans act toward symbols based on the meanings that these symbols carry.
- The meanings symbols carry come from social interaction.
- Humans interpret the meaning of symbols, and this interpretation influences action.

Therefore, according to symbolic interactionism, humans are different from lower animals in that lower animals simply respond to stimuli, while humans have the capability to interpret the stimulus first, then react. For example, a dog might bare its teeth, which is a gesture that other dogs always interpret as a symbol of aggression. Lower animals are not able to conceive of alternative meanings to gestures. By contrast, the thumbs up gesture in American culture generally signifies approval. In a different context, however, it could be used sarcastically to communicate disapproval. And in some Middle Eastern countries, it is an offensive gesture. The meaning of the thumbs up symbol therefore depends not only on social and cultural understanding, but also on a person's interpretation.

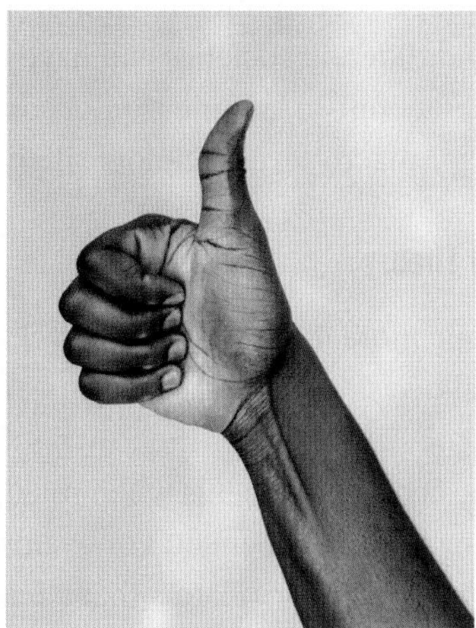

Figure 11.1 Symbolic Interactionism
Hand gestures do not always carry the same meaning across cultures. The thumbs-up is a sign of approval in American culture; in some Middle Eastern countries, it is an offensive gesture.

Not only do hand gestures carry different meanings in different cultures, but different cultures have unique languages and slang, alphabets and number systems, traffic signs, and so on. In fact, the meaning of a symbol might differ even from group to group within a culture. The shared understanding of symbols is therefore a micro- or meso-level phenomenon. One major limitation to the symbolic interactionist approach is that it overlooks macro-level structures, like cultural norms or class interactions.

Social Constructionism

KEY CONCEPT

Symbolic interactionism reflects on how we use symbols to interact with each other. Social constructionism reflects on how we, as a society, construct concepts and principles.

The agreed-upon meaning of symbols can change. For example, the peace sign, which today is a symbol signifying peace and freedom in general, originally was used to indicate support for nuclear disarmament specifically. Society's interpretation of the peace sign has therefore broadened over time. This example shows that the meaning of a symbol does not come from the symbol itself, but from implicit social agreement. Symbols are therefore examples of social constructs. A **social construct** is any idea that has been created and accepted by the people in a society. **Social constructionism** is the attempt to understand a society through the study of the society's social constructs.

Symbols are not the only type of social construct. Abstract ideas might also be social constructs. For example, the definitions of concepts like honor and justice rely on group agreement among individuals within a given society, and these concepts are therefore social constructs. Sociologists also apply social constructionism to physical objects, such as money. Paper money and coinage do not inherently have significant value; it is only because we, as a society, imbue these objects with value that they can be used to trade for goods and services. Other examples of social constructs include work ethic, acceptable dress, and gender roles.

As with symbolic interactionism, social constructionism is useful for explaining micro- and meso-level sociological phenomena. Any social construct depends on the society being studied, and different societies will have different social constructs. Therefore, like symbolic interactionism, a limitation of social constructionism is that this theory cannot account for macro-level sociological phenomena.

Rational Choice (Exchange) Theory

Rational choice theory is one more micro- to meso-level approach to sociology. This theory focuses on individual decision making. According to rational choice theorists, humans will make rational choices to further their own self-interests. This theory claims that people weigh the costs and benefits when making choices, ranking their options based on maximizing perceived benefit. For example, when deciding to purchase a new laptop, rational choice theory claims that a person will gather information about characteristics such as cost, performance, brand recognition, and so on, and then will rank their options and choose the best one.

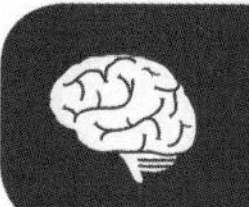

This laptop example illustrates that rational choice theory is influenced by the study of economics. Rational choice theory becomes a model of sociology when this idea of rational transactions is applied to the social interaction. The rational choice perspective views all social interactions as transactions that take into consideration the benefits and harms to the individual. Every outcome in a given social interaction can be associated with particular social rewards (such as accolades, honor, prestige, or social approval) and with particular punishments (such as embarrassment, humiliation, sanctions, or stigmatization). From the rational choice perspective, an individual carefully considers all of the possible rewards and punishments of each social action and chooses the option that results in the greatest social benefit.

Moreover, according to rational choice theorists, people evaluate whether there is reciprocity and balance in social relationships: People stay in relationships because they get something from the exchange, and they leave relationships when there are more social costs than benefits. Due to this view of relationships as exchanges of social value, rational choice theory is sometimes called **social exchange theory**, or just **exchange theory** for short.

Rational choice (exchange) theory is acceptable for explaining some micro- and meso-level sociological phenomena. However, rational choice (exchange) theory does not easily explain charitable, illogical, unselfish, or altruistic behavior.

Conflict Theory

Conflict theory is a macro theory that attempts to understand society by examining the inevitable conflicts between groups in society. Conflict theory has its origins in the writings of Karl Marx, a 19th century social philosopher who examined the influence of capitalism on 19th century society. **Capitalism** is an economic system in which individuals and corporations, rather than governments, own and control what Marx called the means of production, meaning property, machinery, factories, or any other means of creating a saleable good or service. According to Marx, such private ownership naturally leads to a small, wealthy **capitalist (bourgeoisie) class**, who control the means of production. In Marx's model, the rest of society is relegated to a lower **worker (proletariat) class** that performs manual labor. Because the capitalist class owns the means of production, this class has power over the worker class, and the disparity in power and resources between these two groups leads to conflict. According to Marx, the conflict in such a society would be a physical one: Eventually, the worker class would rise up and overthrow the capitalist class and form a new, classless society.

REAL WORLD

Conflict theory can be applied to healthcare and medicine. Conflict theorists would not deny that modern healthcare can help people maintain or restore their health; however, they may ask who holds the power in the healthcare system. Is it the patient? The doctor? Hospitals? Pharmaceutical companies? Insurance companies? The government? This is an issue the United States continues to grapple with.

Where Marx focused specifically on conflict between capitalists and workers, modern conflict theory expands this idea to examine any conflict between groups with more power and those with less. For example, conflict might exist between people of different generations, different religions, or different regions of a country. According to conflict theory, in any such conflict, individuals in the group with more power attempt to preserve their power by shaping the structure of society itself. The group with more power uses their influence to dictate the laws, customs, and cultural norms of the society. However, according to conflict theorists, if people in lower-status positions recognize this power differential and see that others share a common dissatisfaction, then these individuals can organize to form **interest groups**, through which they can use tools such as protesting or voting to enact change and equalize power. In short, according to conflict theory, for the more powerful in society, maintenance of the status quo is usually desirable, and for the less powerful, change comes through disruption and revolution.

While conflict theory is a useful model for describing many large-scale changes and other macro-level societal phenomena, it is not very effective for explaining the choices of individuals in society. Also, conflict theorists tend to focus on social stress and disharmony, so conflict theory is less effective than some other models at explaining social cohesion, cooperation, and altruism.

Structural-Functionalism

In some ways, **structural-functionalism** is the inverse of conflict theory. The founder of structural-functionalism, Émile Durkheim, was interested in how large societies survive over long time periods, and was therefore concerned with social cohesion and stability. Durkheim compared society to an organism and proposed that each group in society has a role to play in the overall health and operation of society. These roles might be very different, in the same way that different organs or even different cells have very different functions within an organism, but each is important. Durkheim called each social group's role its **function**: the contribution made by that group to the system. According to structural-functionalist theory, the different groups of society work together in an unconscious, almost automatic way toward maintenance of equilibrium.

In structural-functionalism, functions can either be manifest or latent. A **manifest function** is an intended consequence of the actions of a group within a society. When an organization or institution has unintended but beneficial consequences, these are called **latent functions**. For example, annual meetings of medical societies have the manifest function of educating a group of physicians, sharing research findings, and setting goals for the next year. Such meetings also create stronger interpersonal bonds between physicians and provide a sense of identity for the group, both of which are latent functions. On the other hand, while both manifest and latent functions provide a benefit to society, **dysfunctions** are negative consequences of the existence of an institution, organization, or interaction.

Because structural-functionalism focuses on social cohesion and equilibrium, this approach is not well suited for explaining social change. Additionally, while structural-functionalism attempts to explain how groups interact with other groups, it does not explain how individuals interact *within* a group. To understand those kinds of micro-level interactions, other models are better suited.

Feminist Theory

Feminist theory critiques the institutional power structures that disadvantage women in society. Feminist theory was originally an offshoot of conflict theory. From a conflict theory perspective, feminist theory describes society as inherently **patriarchal**, with men seeking to preserve their position of power over women through societal privilege and institutional discrimination. For example, some feminist theorists argue that the study of sociology itself has been historically dominated by a male perspective.

In developed countries, gender stratification and inequality typically lessens, often as a result of the activism of feminist interest groups. However, imbalances of power still exist. In the workplace, for example, the term **glass ceiling** refers to processes that limit the progress of women to the highest job positions because of invisible social barriers to promotion. In contrast, even in cases where men do not seek to climb the job ladder, invisible social forces sometimes push men up to higher positions, a phenomenon called the **glass escalator**. The glass escalator is especially prevalent for men working in traditionally female occupations.

BEHAVIORAL SCIENCES GUIDED EXAMPLE WITH EXPERT THINKING

In societies facing increasing amounts of migration and ethnic integration, concerns about intergroup conflicts have prompted research into the effects of diversity on a society as a whole. Previous studies suggest that ethnic diversity negatively impacts social cohesion, but critics have pointed out that historic measures of social cohesion have focused on single indicators such as trust or volunteering.

Older studies demonstrated a negative relationship between diversity and social cohesion, but those are in doubt.

The current study sought to operationalize social cohesion across a more comprehensive set of indicators. Ethnic diversity was found to be associated with lower social cohesion with respect to some indicators, particularly volunteering and feelings of safety, but this trend could not be generalized when including a larger scope of indicators of social cohesion, such as neighborhood social capital or belonging. When heterogeneous groups showed increased contact with one another, subject assessment via survey showed a significant increase in feelings of trust, safety, and social capital.

Several results here. Not only were previous studies possibly wrong, but new results show that integration increases *cohesion. Not an experiment so no experimenter-controlled IV, looking at association only.*

Adapted from: McKenna S, Lee E, Klik KA, Markus A, Hewstone M, Reynolds KJ (2018) Are diverse societies less cohesive? Testing contact and mediated contact theories. *PLoS ONE* 13(3): e0193337. https://doi.org/10.1371/journal.pone.0193337

How does conflict theory relate to the results of this study?

This question at first might seem tough, as the results of the study seem to suggest that there is less conflict than one would expect in more diverse societies. When faced with a question that asks us to apply a concept to a new situation, it's useful to think critically about the definition of that concept. According to conflict theory, institutions arise as a result of power differentials between individuals, and those institutions seek to maintain that power differential.

So where is the conflict in this situation? According to the study, there can be a great deal of diversity in an area that has low interaction between different groups. In those situations, some measures of social cohesion decrease, including feelings of safety. So when people are in close quarters with diverse others, particularly when not interacting with those others regularly, perceived threat to the establishment and maintenance of the power differential will increase, which increases the conflict between groups.

However, the study also found that when people in close quarters with a diverse population interact more with diverse others, they report increased feelings of trust, safety, and social capital, indicating a reduction in perceived threat from other groups. We could infer that this increased interaction may counter the conflict theory effects noted earlier: increased trust and safety reduces fear of others, generally, and fear of loss of power, specifically. With a reduction in perceived threat and increased social cohesion, motivation for conflict should decrease, as power differentials are less threatened or groups are actively integrated into the existing power structure.

So, in summary, conflict theory provides some explanation as to why those who are not interacting with their diverse society may experience more perceived threat and conflict motivation. The study further indicates the perceived threat effect, and possibly societal conflict overall, can be countered by increased social interaction.

Social Institutions

Social institutions are well-established social structures that dictate certain patterns of behavior or relationships and are accepted as a fundamental part of culture. Social institutions regulate the behavior of individuals in core areas of society. For example, family is a social institution that encourages learning of acceptable behavior, socialization, and bonding.

Institutions exist at the meso-level of sociological analysis because they are a part of society, but are not dependent upon the individuals involved—in the United States, the idea of "government" stays basically the same even as the president, senators, representatives, and other officials change. The exact nature of each institution differs from culture to culture, although each institution performs a similar role regardless of culture. Often, institutions are dependent upon and support one another, though competition over resources can bring institutions into conflict. A summary of six of the major social institutions is provided in Table 11.1 at the end of this section.

Family

The definition of **family** differs greatly from culture to culture. For some, the term means "those people to whom I maintain close ties and who are related to me by blood." For others, family is simply "the people who live in my house." In fact, even terms for different family members (such as *sister*, *father*, *cousin*, and so on) are not conserved across time and culture; different **patterns of kinship** may be reflected by these terms. For example, it is common in Hawaiian culture to refer to all family members as *cousins*, while this term would not be used by many other Americans to describe one's mother's brother (the term *uncle* being preferred by many continental Americans). Different patterns of kinship between societies have bearing on responsibility for child rearing, familial loyalty, and even the boundaries of what is considered incest.

Regardless of the definition, family is the most basic of institutions. It is the institution most closely tied to the individual and helps to meet many of our most basic needs, especially when we are young, providing food, shelter, emotional and physical security, and intimacy. Additionally, many life rituals and rites of passage, such as marriage, funerals, and graduations occur in the context of family.

BRIDGE

Families help to meet many of the basic needs at the base of Maslow's hierarchy pyramid. Maslow's hierarchy of needs is discussed in chapter 5 of *MCAT Behavioral Science Review*.

Sociologists studying family relationships may examine the stages of coupling (courtship, cohabitation, engagement, and marriage), changes in relationships between spouses through time, or parenting. Parenting is a complex topic that involves socialization of children; varied definitions of the role of father, mother, and child; and single parenting, same-sex parenting, adoption, and foster parenting. Not all families are composed of a mother, a father, and children. Alternative forms exist, including single-parent families; families that cohabitate with other family members beyond the nuclear family, such as grandparents, aunts, uncles, cousins, godparents, and surrogate kin; and families with marital disunions (divorce). A number of different family structures are illustrated in Figure 11.2.

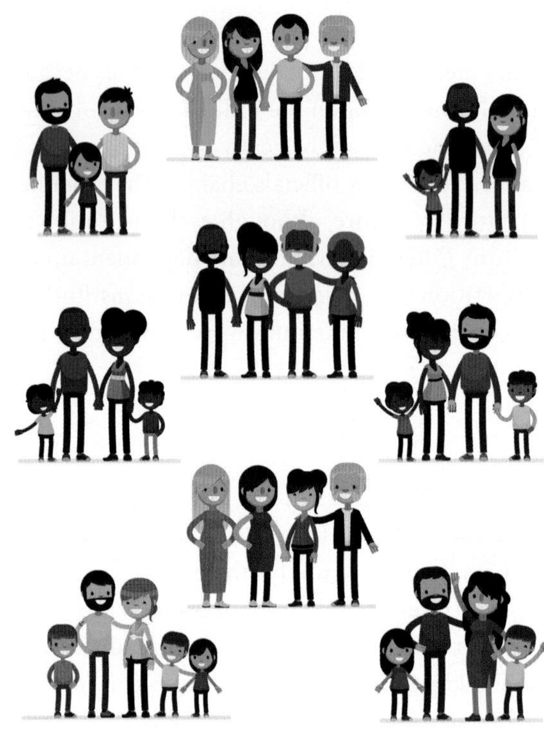

Figure 11.2 Various Family Structures

MCAT EXPERTISE

The MCAT will not expect you to know any specific demographic numbers for the exam, but you should be familiar with some of the recent trends seen in the United States population.

Divorce rates in the United States rose significantly in the second half of the twentieth century but have started to drop over the last two decades.

While the family can be a source of joy and support, it can also be a source of violence. Spousal abuse (**domestic violence**) is seen across all social classes and genders and can include not only physical violence, but sexual abuse, emotional abuse, and financial abuse. Domestic violence is the #1 cause of injury to American women, and is most common in families with drug abuse, especially alcoholism. Victims of domestic violence may find it challenging to leave the abusive relationship for a variety of reasons, including lack of a safe haven to escape to, financial restrictions, and psychological disorders (consider the connection to learned helplessness, described in Chapter 6 of *MCAT Behavioral Sciences Review*). **Elder abuse** is also seen across all socioeconomic classes, and most commonly manifests as neglect of an older relative—although physical, psychological, and financial abuse may occur as well. The caretaker of the individual is most commonly the source of abuse. Finally, **child abuse** also most commonly manifests as neglect, although physical, sexual, and psychological abuse are also common. During medical school, you will be trained to recognize certain signs suggestive of nonaccidental trauma, such as a broken femur in a child who is too young to have begun walking or burn marks on the buttocks from placing a child in scalding water. As a physician, you will be considered a **mandated reporter**, which means that you are legally required to report suspected cases of elder or child abuse. Domestic abuse does not fall under mandated reporting laws, but counseling and information about shelters for victims of intimate partner violence should be provided.

Education

Education, as an institution, aims to provide a population with a set of skills that will be useful to them or to society. In many societies, education is formal—in other words, it takes place in a setting designated for educational purposes using a prescribed curriculum. The function of formal education is to teach skills, facts, and mental processes, but the system also has a social latent function, providing opportunities for peer socialization and reinforcing social stratification, both within individual schools and through comparisons between schools. Education, therefore, includes not only the information and cognitive skills students learn but also the **hidden curriculum** of transmitting social norms, attitudes, and beliefs to students. Sociological investigations into education may focus on the ethics, morals, practices, political influence, finances, and values of an education system. Sociologists also explore educational trends, including grade inflation and deflation, adult education, online education, and accessibility of education.

Performance in the education system depends not only on a student's intrinsic abilities, but also on the education system itself. **Teacher expectancy** refers to the idea that teachers tend to get what they expect from students. Thus, a teacher who places high demands on students—but who also believes that students can rise to the challenge—will more often see students succeed than a teacher who places the same demands but doubts that the students can achieve them. This is an example of a self-fulfilling prophecy, discussed in Chapter 10 of *MCAT Behavioral Sciences Review*, and may be due to differences in how teachers motivate, interact, and offer feedback to their students.

Education is susceptible to inequalities across socioeconomic class. Lower socioeconomic status is associated with decreased accessibility to and quality of education. This is not an easy trend to reverse. Low funding, deprioritization of education, and poor historical performance can make it challenging for a failing school district to acquire resources and improve education to its students.

As mentioned earlier, institutions are intentionally or unintentionally connected, and there is a well-known, persistent association between education and medicine. Health disparities between more educated and less educated individuals are significant, and lack of education may be a hurdle to accessing or trusting healthcare providers.

Religion

As an institution, **religion** is a pattern of social activities organized around a set of beliefs and practices that seek to address the meaning of existence. **Religiosity** refers to how religious one considers oneself to be, and includes strength of religious beliefs, engagement in religious practices, and attitudes about religion itself.

More than 80% of the global population identifies with a religion. Most follow either Christianity or Islam; other world religions include Hinduism, Buddhism, Sikhism, and Judaism. These religions are generally divided into multiple **denominations** or **sects** that may share certain beliefs and practices but not others. A denomination is simply a part of a **church**, a term which can refer both to a large, universal religious group and to the building in which the congregation of such a group meets. Orthodox and Reform are examples of Jewish denominations, and the Sunni and Shia are examples of Islamic denominations. Denominations often coexist, but can come into conflict when their religious beliefs promote opposing values. Whereas the word *sect* was historically a pejorative term, it now refers more properly to a religious group that has chosen to break off from the parent religion. The Amish would be considered a sect of Christianity. In rare cases, a religious sect may take on extreme or deviant philosophies and transform into a **cult**.

As the twenty-first century continues, religious groups—many of which have existed in more or less the same form for thousands of years—grapple with finding a place in contemporary society. For many religious groups, this is reflected by a shift toward modernization within the religion and relaxing historical practices. For others, there is a shift away from religion as society **secularizes**, or moves from a world dominated by religion toward rationality and scientific thinking. For other groups, maintenance of strict adherence to religious code, or **fundamentalism**, predominates.

While spirituality and religion are not equivalent terms, they are often linked to each other because they both seek to understand the meaning of existence and to identify what is sacred. Spirituality and religion may play a role in a patient's understanding of disease, may impact healthcare decisions, and can be an essential component of the patient's coping mechanisms.

Government and Economy

Government and **economy** can be defined as systematic arrangements of political and capital relationships, activities, and social structures that affect rule making, representation of the individual in society, rights and privileges, division of labor, and production of goods and services. Notably, political and economic institutions impact all other institutions to some extent. That is, the government may sanction or define specific family structures, may finance and regulate education, may recognize some religions but not others, and may play a key role in funding and certifying healthcare and medicine. The effects of the economy on institutions can also be viewed from the individual level. For example, when the economy takes a downturn, large swaths of the population may have trouble supporting their families and paying for health insurance. Note that this institutional influence is bidirectional: because of the economic downturn and changes in family, education, or health, an individual may choose to vote a new political candidate into office, or to support or oppose a particular piece of legislation.

While an in-depth exploration of the American (or any other) government is outside the scope of the MCAT, you could be asked to distinguish between different types of government. A **democracy** allows every citizen a political voice, usually through electing representatives to office (i.e., a representative democracy). **Monarchies** include a royal ruler (a king or queen), although the ruler's powers may be significantly limited by the presence of a constitution, a parliamentary system, or some other legislative body. A **dictatorship** is a system where a single person holds power, and usually includes mechanisms to quell threats to this power. A **theocracy** is a system where power is held by religious leaders. Many of these systems of leadership are based around a **charismatic authority** (a leader with a compelling personality).

In comparative economics, the largest division is between capitalist and socialist economies. **Capitalist** economies focus on free market trade and laissez-faire policies, where success or failure in business is primarily driven by consumerism with as little intervention from central governing bodies as possible. In capitalism, a private owner or corporation maintains and profits from the success of the business. Capitalist societies encourage **division of labor**, where specific components of a larger task (say, developing, manufacturing, quality testing, and marketing goods) are separated and assigned to skilled and trained individuals. This promotes specialization and efficiency. **Socialist** economies, on the other hand, treats large industries as collective, shared businesses, and compensation is provided based on the work contribution of each individual into the system. Profit, then, is distributed equally to the workforce. There are many other forms of government and economy, but these would be defined and explained if necessary on Test Day.

Healthcare and Medicine

The institutions of **healthcare** and **medicine** are aimed at maintaining or improving the health status of the individual, family, community, and society as a whole. Healthcare is an ever-changing field, but some of the key goals in American healthcare over the past few decades include:

- Increased access to care
- Decreased costs of healthcare
- Prevention of disease before it occurs
- Association of patients with a primary care physician or a patient-centered medical home
- Increased education for the public with public health outreach
- Decreased paternalism (*doctor knows best* mentality)
- Reduced economic conflicts of interest for physicians
- **Life course approach to health** (maintaining and considering a comprehensive view of the patient's history beyond the immediate presenting symptoms)

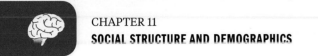

There has also been a shift in the role of the patient in the medical system. In the mid-twentieth century, sociologists believed that a patient who is ill takes on what was called the **sick role**. These sociologists believed that such patients are not responsible for their illness and are exempt from normal social roles. But on the other hand, these sociologists also believed that a patient had the obligation to want to become well and to seek out competent help. While this paradigm still exists, patients are now expected to take more ownership of their health through diet, exercise, seeking help before it is needed (through annual primary care visits and screenings), and so on. Modern sociologists also recognize that the notion that patients should feel an obligation to want to become better does not fit patients with chronic, long-term, or permanent conditions.

Even our understanding of disease has shifted over time. On the one hand, many ordinary parts of the human condition, including the full range of sexual orientations and gender identities, are no longer defined as medical conditions, as they misguidedly were in earlier editions of the DSM. On the other hand, other types of behaviors have become **medicalized**, meaning that they are now defined and treated as medical conditions.This latter shift can be seen in the addition of a number of diagnostic entities to the DSM-5, such as hoarding disorder and binge eating disorder.

The medical community also now recognizes that health is not only characterized by physiological wellness or disease, but also by a person's subjective experience of their health. The phrase **illness experience** refers to the ways in which people, rather than doctors, define and adjust to changes in their health. For example, people can experience disease without illness, as is the case for someone living a full and productive life with a manageable condition such as HIV. People can also experience illness without disease, such as when a person experiences somatic symptoms as a result of psychological stress.

Sociologists studying healthcare are often interested in **social epidemiology**. Epidemiology is the study of health and its determinants within a society (discussed in detail in Chapter 12 of *MCAT Behavioral Sciences Review*); social epidemiology is specifically the study of the effects of institutions, social structures, and relationships on health. Social epidemiologists study the effects of racial and economic inequality or government safety net legislation on health and access to healthcare. They might also be interested in how social conditions early in life affect healthcare outcomes later.

In addition, many sociologists investigate medical ethics. In the United States, physicians are expected to adhere to four key tenets of medical ethics:

- **Beneficence:** the physician has a responsibility to act in the patient's best interest.
- **Nonmaleficence:** *do no harm*; the physician has a responsibility to avoid treatments or interventions in which the potential for harm outweighs the potential for benefit.
- **Respect for patient autonomy:** the physician has a responsibility to respect patients' decisions and choices about their own healthcare. While there are exceptions to this rule (significant psychiatric illness interfering with decision-making capacity, children, public health threats), patients do have the right to refuse life-saving therapies.
- **Justice:** the physician has a responsibility to treat similar patients with similar care, and to distribute healthcare resources fairly.

SOCIAL INSTITUTION	NEEDS MET BY INSTITUTION	STATUSES	VALUES	NORMS
Education	Transmit knowledge and skills across generations	Teacher, student, dean, principal	Academic honesty, good grades	Doing homework, preparing for lectures, being kind to other students
Family	Regulate reproduction, socialize and protect children	Father, mother, son, daughter, brother, sister, uncle, aunt, grandparent	Sexual fidelity, providing for children, keeping a clean home, respect for caregivers	Having as many children as one desires, being faithful to one's spouse
Religion	Concerns about life and death, the meaning of suffering and loss, desire to connect with a creator	Clergy, worshipper, teacher, disciple, missionary, prophet, convert	God and holy texts should be honored	Going to services, following teachings of the religion, applying beliefs outside of worship
Government	Maintain social order, enforce laws	President, senator, lobbyist, voter, candidate	Transparency, accountability, professionalism	Acting in the best interest of constituents, debating political issues
Economy	Organize money, goods, and services	Worker, boss, buyer, seller, creditor, debtor, advertiser	Making money, paying bills on time, producing efficiently	Maximizing profits, *the customer is always right*, working hard
Medicine	Heal the sick and injured, care for the dying	Doctor, nurse, pharmacist, insurer, patient	Hippocratic oath, staying in good health, following care providers' recommendations	Beneficence, nonmaleficence, respect for autonomy, justice

Table 11.1 Social Institutions

MCAT CONCEPT CHECK 11.1

Before you move on, assess your understanding of the material with these questions.

1. What are manifest and latent functions?

   - Manifest functions:

   - Latent functions:

   ---

2. For each of the theoretical approaches listed below, what is the primary thesis or idea of the theory?

Theoretical Approach	Primary Thesis or Idea
Functionalism	
Conflict theory	
Symbolic interactionism	
Social constructionism	
Rational choice-exchange theory	
Feminist theory	

3. What are the four key tenets of American medical ethics? Provide a short description of each.

Ethical Principle	Description

11.2 Culture

LEARNING OBJECTIVES

After Chapter 11.2, you will be able to:

- Distinguish between material and symbolic culture
- Recall the definitions of, and the differences between, the values, beliefs, norms, and rituals of a culture

The study of culture is likely the most diverse and complex dimension within sociology. **Culture** can be defined as encompassing the entire lifestyle for a given group. It binds our nation-states, political institutions, marketplaces, religions, and ideologies. Culture flavors our interpretations of the world, and is generally passed through familial lines. In short, culture is what makes human societies unique from one another.

Material and Symbolic Culture

Sociologists view culture according to two different categories: material culture and symbolic culture.

Material Culture

One can discern a lot about people by looking at their **artifacts**: material items that they make, possess, and value. This examination surrounds **material culture**, which includes the physical items one associates with a given group, such as artwork, emblems, clothing, jewelry, foods, buildings, and tools. Sociologists explore the meaning of these objects to a given society.

An example of material culture in the United States is the American flag. This item is used to reinforce a sense of belonging via shared American citizenship. Other symbols that are considered traditionally American include barbecue, baseball, apple pie, and rock and roll.

Material culture is often most visible during ceremonies, such as birthdays, weddings, and funerals. Some artifacts of traditional Indian material culture are shown in Figure 11.3.

BRIDGE

A description of culture in the context of group processes is described in Chapter 8 of *MCAT Behavioral Sciences Review*

MCAT EXPERTISE

Ethnography is the study of cultures and customs, and **ethnographic methods** are experimental methods used to study the ethnicity or culture of a group.

BRIDGE

Symbols are also discussed in Chapter 6 of *MCAT Behavioral Sciences Review*

Figure 11.3 Material Culture

Material culture includes objects important to a group, including clothing, jewelry, cuisine, ceremonial objects, and so on.

Symbolic Culture

KEY CONCEPT

Material culture is associated with artifacts (objects). Symbolic culture is associated with ideas.

Symbolic culture, also called **nonmaterial culture**, focuses on the ideas that represent a group of people. These may be encoded in mottos, songs, or catchphrases, or may simply be themes that are pervasive in the culture. Phrases like *free enterprise* and *life, liberty, and the pursuit of happiness* are examples of American symbolic culture. Material culture is often the tangible embodiment of the underlying ideas of symbolic culture.

For any social group to remain connected over time, there must be a culture that binds its members together. In times of war and crisis, governments often draw upon symbolic culture to rally people to action, using songs, parades, discussion of heroes past, and so on, as shown in Figure 11.4. It is not a coincidence that most high schools have a school mascot, school colors, and a school song. Such cultural artifacts are in place to help create a shared sense of identity, loyalty, and belonging. Symbolic culture includes both cognitive and behavioral components; that is, it informs cultural values and beliefs, as well as cultural norms and communication styles.

Figure 11.4 Symbolic Culture
Symbolic culture includes ideas that identify a culture; it may be drawn upon to encourage loyalty or patriotism, as shown here.

Symbolic culture is usually slower to change than material culture, which can lead to the phenomenon of **culture lag**. The expansion of devices and technology in contemporary times are prototypical examples of culture lag: whereas American culture still prizes individuality and privacy, the development of smartphones and social media push toward a more community-oriented and less private world. Still, there is evidence that symbolic culture is beginning to change in response to these technological (material) innovations: younger generations appear to be less concerned about what personal information is publicly accessible than older generations.

Language

Language is the most highly developed and complex symbol system used by most cultures. Language consists of spoken, written, or signed symbols, which are regulated according to certain rules of grammar and syntax. Language enables us to share our ideas, thoughts, experiences, discoveries, fears, plans, and desires with others. Written language extends our capacity to communicate across both spatial and temporal boundaries. Without language, it would be difficult to transmit culture. Understanding a group's language is critical to understanding its culture.

BRIDGE

Language is critically important in the transmission of culture. It requires a complex interplay of multiple brain circuits, which are discussed in Chapter 4 of *MCAT Behavioral Sciences Review*.

Values, Beliefs, Norms, and Rituals

An important component of culture are the rules that structure society. **Values** are what a person deems important in life, which dictates one's ethical principles and standards of behavior. A **belief** is something that an individual accepts to be truth. Every culture has its own beliefs and value systems. This will be important in your future career, as patients tend to carry their beliefs into the healthcare system. For example, as described in the chapter introduction, some Asian cultures believe that healthcare decisions should be the responsibility of a patient's family, which avoids burdening the patient (who is already ill) with having to make such a decision. This belief is in direct contrast to the American belief that patient autonomy should be prized and that healthcare decisions should be made by a patient whenever possible. These conflicts can prove challenging to healthcare professionals, and there is not always one correct answer to such dilemmas. Such situations—when a cultural difference impedes interaction with others—are called **cultural barriers**.

REAL WORLD

Many health systems have an ethics board to deal with conflicts that may arise from differences in belief systems between patient and practitioner, among other ethical issues. These committees tend to facilitate discussion, rather than simply issuing a decision.

As described in Chapter 8 of *MCAT Behavioral Sciences Review*, **norms** are societal rules that define the boundaries of acceptable behavior. While norms are not laws, they do govern the behavior of many individuals in society and provide a sense of social control. Norms are what provide us with a sense of what is appropriate, what we should do, and what we should not do. Norms exist for behavior, speech, dress, home life, and more.

A **ritual** is a formalized ceremony that usually involves specific material objects, symbolism, and additional mandates on acceptable behavior. Rituals tend to have a prescribed order of events or routine. These rituals can be associated with specific milestones, such as a baby-naming, graduation ceremony, wedding, or funeral; with holidays, such as a Thanksgiving dinner, trick-or-treating on Halloween, or a Passover seder, shown in Figure 11.5; or with regular activities, such as a Catholic mass, a pregame pep rally, or even just getting ready in the morning (showering, brushing teeth, eating breakfast, and so on).

Figure 11.5 A Passover Seder Is an Example of a Ritual
Seder means "order" in Hebrew; most rituals have a specific order of events.

Evolution and Human Culture

Evolution may have selected for the development of culture. Culture serves as a method of passing down information from generation to generation; in prehistoric times, the transmission of information through culture served to teach future generations how to create tools, hunt, domesticate animals, and grow crops. Culture also creates a sense of loyalty and allegiance, which, as described in Chapter 10 of *MCAT Behavioral Sciences Review*, may help explain altruistic behavior. Finally, culture creates a sense of *us* vs. *them*, which presumably served a role in the dispersion of populations across the globe in different environmental niches.

Culture, in turn, may also influence evolution. There is evidence that some genetic traits may have been favored because of cultural values and beliefs. For example, human beings—at least those who are not lactose intolerant—are the only animals that are able to digest milk after adolescence; they are also the only animals that ingest another animal's milk. These evolutionary adaptations may have arisen out of Northern European cultures, which relied heavily on cattle farming for sustenance. A mutation permitting digestion of milk into adulthood presumably imparted a nutritional and survival advantage to certain individuals, and would thus be retained within the population.

MCAT CONCEPT CHECK 11.2:

Before you move on, assess your understanding of the material with these questions.

1. What are material and symbolic culture?
   - Material culture:

   ____

   - Symbolic culture:

   ____

2. What is the difference between a value and a belief?

   ____

11.3 Demographics

LEARNING OBJECTIVES

After Chapter 11.3, you will be able to:

- Distinguish between race and ethnicity
- Describe symbolic ethnicity
- Describe fertility rate, birth rate, and mortality rate and how they shift during a demographic transition
- Recall examples of proactive and reactive social movements and how the two types of movements differ

Demographics refer to the statistics of populations and are the mathematical applications of sociology. Demographics can be gathered informally, such as a professor asking how many freshmen, sophomores, juniors, and seniors are in a given course, or may be gathered formally. For example, the United States Census Bureau gathers demographic data about every individual in the country every ten years.

Common Demographic Categories

Demographers can classify individuals based on hundreds of different criteria. The MCAT will not expect you to know advanced topics within demographics, but familiarity with some of the common demographic categories as well as their implications on society and healthcare are important. In this section, we'll explore age, gender, race and ethnicity, sexual orientation, and immigration status.

Age

Aging is a inevitable process experienced by all people around the world. In this section we will explore the implications of an individual's age on healthcare, then dive deeper into its implications on society.

Considering an individual's age and cumulative life experiences when analyzing their personality, social status, health, and other social metrics is known as the **life course perspective** (sometimes referred to as the **life course approach**). For example, in healthcare, a psychiatrist may consider a patient's early life events and how those events continue to impact a patient's condition. Or perhaps a general practitioner chooses to conduct additional lung screenings on a patient who previously worked in a coal mine. In both these examples, the physician is incorporating the life course perspective into the treatment of the patient.

With the potentially large difference in the experiences between age cohorts, prejudice or discrimination based on a person's age can arise. This is known as **ageism** and can be seen at all ages. For example, young professionals entering the workplace are often viewed as being inexperienced, and their opinions and ideas may therefore be ignored or downplayed. Older individuals may be perceived as frail, vulnerable, or less intelligent, and may thus be treated with less respect.

In order to understand and analyze age-related differences, researchers can group individuals based on their age or birth year; these groupings are known as **age cohorts**

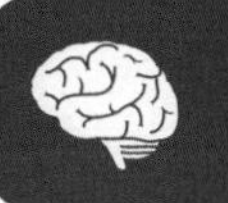

(sometimes called generational cohorts). The utility of age cohorts goes beyond understanding the differences of an individual's life course, as they allow researchers to look at a population at a macroscopic level. An analysis of a population's distribution among its age cohorts can predict demographic shifts, such as an aging population, the shift from a developing to developed economy, or a stable population. We'll explore examples of each below.

In the United States, many sociologists document a "graying of America" as the Baby Boomer generation ages. The term Baby Boomer stems from the large spike in fertility rates (birth rates) after World War II, or in other words, a "boom of babies." Due to the baby boom spanning from the 1940s to 1960s, over 70 million Americans will be 65 or older by 2030, representing nearly 20 percent of the population. Thus, the fastest-growing age cohort in the United States is the 85-or-older group. This has profound effects on healthcare: more than 40 percent of adult patients in acute care hospital beds are 65 or older. Considering this shift in demographics, government programs such as Medicare and Social Security will experience increased demand, which may pose challenges for these programs.

This situation is an application of the **dependency ratio**, which is the ratio of the number of members of a population that are not in the workforce to the number of members that are in the workforce. This ratio depends on two components, the youth ratio and age dependency ratio. The **youth ratio** is defined by the number of people under the age of 15 divided by the number of people aged 15–65. The **age dependency ratio** is defined by the number of people over 65 divided by the number of people aged 15–65. Applied to societies, the dependency ratio quantifies the economic burden felt by the working age population (15–65) in order to support the portion of the population outside of the workforce (under 15 and over 65).

In contrast to the United States, developing countries, such as Uganda, see the reverse trend with 48% of the population being under the age of 15 years old, resulting in a dependency ratio of about 1 (2% of the population is 65 years and over, which means the dependent population roughly equals the working age population). This can be explained by the country's steadily dropping infant mortality rate over the last several decades. Forecasting the next 10 to 20 years for Uganda would predict a large proportion of the population entering the working-age age cohort. Although this may lead to an increase in the country's economic productivity, this demographic shift must be matched with job growth. If that does not occur, unemployment rates will increase, which could lead to civil unrest or other negative outcomes. This shift from developing to developed country is explained by demographic transition theory, covered later in this chapter.

Finally, when a population's fertility rate and mortality rate remain relatively consistent over a long period of time, the distribution of the population among the age cohorts remains fairly constant. This is known as a **stable population**.

Gender

Sex and gender are not synonymous terms. **Sex** is a biological category. In many species, including humans, a phenotypical female is an individual that produces the larger gamete (the ovum) and carries offspring. **Gender** refers to a society's notions of femininity, masculinity, and other sexual identities. Gender is therefore a socially

REAL WORLD

The demographic tool of age cohorts has woven its way into our everyday language. If you've heard terms like the Silent Generation, Baby Boomers, Generation X, Millennial (Generation Y), and Generation Z, then you've encountered age cohorts. These generational terms group individuals by their birth year. However, age cohorts can be used to group individuals based on their current age, such as 0-14, 15-30, etc.

constructed set of ideas about what it means to be male, female, or otherwise in a given culture. A culture's ideas about gender usually include expected behavioral traits associated with a particular sex. These expected behavioral traits are known as **gender roles**. As such, gender roles are also social constructs. Once an individual understands these socially constructed behavioral expectations, an individual can adopt behaviors that project the gender that individual wishes to portray, which is known as the individual's **gender identity**.

Gender segregation is the separation of individuals based on perceived gender. Such segregation includes divisions of male, female, and gender-neutral bathrooms, or separating male and female sports teams. Differences between genders and the phenomenon of gender segregation do not necessarily imply inequality, although inequality can occur. **Gender inequality** is the intentional or unintentional empowerment of one gender to the detriment of others. In the presence of gender inequality, gender stratification may occur. **Gender stratification** is defined as any inequality in access to social resources that is based on gender, and is an example of social stratification in general, which will be studied in Chapter 12 of *MCAT Behavioral Sciences Review*. To illustrate the difference between gender segregation and gender stratification: Single-gender schools are an example of gender segregation. Children enrolled in such schools do not necessarily receive unequal qualities of education. However, if there is a systemic difference in resource allocation between single-gender schools, the result is uneven access to resources, leading to gender stratification.

Race and Ethnicity

REAL WORLD

Certain racial and ethnic groups have a higher incidence of specific health problems. For example, the Chinese population accounts for a disproportionate number of chronic hepatitis B infections and liver cancer. Mediterranean and African populations have a significantly higher rate of hemoglobinopathies (diseases related to hemoglobin). The Ashkenazi Jewish population has a higher rate of autoimmune diseases. Some American Indian populations are associated with gallbladder and biliary tree diseases. Being of a particular race or ethnicity is not necessary for the development of any disease, but may certainly be associated with increased risk.

The definition of race has changed through recent history, and continues to change. The term originally referred to speakers of a common language, and later indicated national origin. However, the term has also historically been used to denote certain shared phenotypic similarities between people. The five racial categories currently recognized by the U.S. Census exemplify these shifting definitions of race. These categories are: White, Black, Asian, American Indian or Alaskan native, and Native Hawaiian/Pacific Islander. Observe that some of these racial categories, like White and Black, describe phenotypic similarities, while other categories, like American Indian and Pacific Islander, are based on national origin. Furthermore, no other country uses these same five racial labels, and in fact the officially recognized races differ in each country. So, there is no uniform agreement about racial categories; rather each society generates its own racial labels, making race a social construct.

If race is not consistently defined, then why do sociologists concern themselves with this concept? The answer is that racial labels, though socially constructed, do materially affect the lives of people through institutionalized practices of preference and discrimination. In order to define race more scientifically, sociologists specify that the term **race** refers to socially constructed groupings of people based specifically on inherited phenotypic characteristics. Note that the human history of migration and mixing of populations means that there are few if any genetically isolated people left on earth. So sociologists recognize that scientifically categorizing people by genetic differences is not possible. Nevertheless, societies continue to generate racial labels based on perceived phenotypic differences, and so sociologists study how each society treats its socially defined racial groups.

An important takeaway from the above discussion is that sociologists narrow the definition of the term race to refer specifically to attempts to group people by phenotypic difference. However, sociologists recognize that societies also group people by shared language, cultural heritage, religion, and/or national origin. The term sociologists use for these types of groupings is **ethnicity**. While certain ethnicities are often associated with certain racial labels, race and ethnicity are distinct. Here is an illustration of the difference: African American individuals, African immigrants, and West Indian immigrants speak different languages and express different cultural norms. These three groups represent three different ethnicities. However, due to some phenotypic similarities shared by some members of these groups, individuals in these groups would generally be given the same racial label. Like race, ethnicity is also a social construct, in that ethnic labels and the criteria for inclusion in a certain ethnic group change from society to society and change over time.

Symbolic ethnicity describes a specific connection to one's ethnicity in which ethnic symbols and identity remain important, even when ethnic identity does not play a significant role in everyday life. For example, many Irish Americans in the United States celebrate their heritage only one day per year: St. Patrick's Day. In all other facets of life, these individuals' Irish-American ethnicity does not play a significant role. Other examples include attending folk festivals, visiting specific cultural locales for holidays, or participating in an ethnic celebration.

It is important to consider how race and ethnicity may affect one's ability to receive proper healthcare. The Agency for Healthcare Research and Quality (AHRQ), a government agency, reports that race and ethnicity influence a patient's chance of receiving many specific procedures and treatments. Whether due to conscious or unconscious bias, there is evidence that different races are not always offered the same level of care escalation in a medical emergency.

On the other hand, there are a number of public health outreach projects that target at-risk racial or ethnic populations through education, screening, and treatment. These specific strategies are geared to close gaps in health disparities. Many large university health systems run free clinics in local neighborhoods and may target specific populations; for example, some of these clinics will staff Spanish-speaking doctors and medical students to cater to the Hispanic immigrant population.

BRIDGE

Many public health outreach efforts are aimed at closing the gap in health disparities between populations. Health and healthcare disparities are discussed in Chapter 12 of *MCAT Behavioral Sciences Review*.

Sexual Orientation

Sexual orientation can be defined as the direction of one's sexual interest. In scientific and healthcare communities, sexual orientation has historically been divided into three categories:

- **Heterosexual**: attraction to individuals of a different sex
- **Bisexual** (or pansexual): attraction to members of multiple sexes
- **Homosexual**: attraction to individuals of the same sex

Sexual orientation involves a person's sexual feelings and may or may not be a significant contributor to that person's sense of identity. It may or may not be evident in the person's appearance or behavior. Disclosure of minority sexual orientations, sometimes called *coming out of the closet*, can be a major milestone in the absorption of sexuality into one's identity. This disclosure has also been shown sometimes to have therapeutic effects: coming out is associated with decreases in depressive and anxious symptoms linked to cortisol levels and stress.

Human sexuality continues to be an important area of research for psychologists, sociologists, and biologists alike, but evidence shows that sexuality is likely more fluid than previously believed. Alfred Kinsey was a pioneer in this area, and—in addition to a number of other models and publications—described sexuality on a zero to six scale, with zero representing exclusive heterosexuality and six representing exclusive homosexuality. When ranked on this **Kinsey scale**, few people actually fell into the categories of zero and six, with a significant proportion of the population falling somewhere between the two.

Sexual and gender identity minorities are often grouped together under the umbrella term **LGBTQ** (lesbian, gay, bisexual, transgender, and queer or questioning). In some cases, this acronym has been expanded to include other self-definitions of sexuality and sexual identity, including I (intersex) or A (asexual).

Several health disparities have been recognized within the LGBTQ community. The most significant historical disparity is HIV, which disproportionately affected individuals who are gay in urban environments during the early 1980s. While the prevalence of HIV is still slightly higher in men who have sex with men (MSM), it exists in all populations. Efforts to encourage safe sex and increase screening have helped to slow the epidemic of HIV, as has increased awareness of those with HIV/AIDS with projects like the AIDS Memorial Quilt, shown in Figure 11.6. Within the healthcare system, individuals who are lesbian receive less screening for cervical cancer and may not be screened for other sexually transmitted infections. Individuals who are transgender have multiple areas of increased risk, including off-label or unsupervised use of "street hormones" without proper counseling on their side effects.

Figure 11.6 The AIDS Memorial Quilt

Mental health disparities are also common in the LGBTQ community. LGBTQ youth are at significantly higher risk for bullying, victimization, and violence, and have higher rates of suicide. In adults, the LGBTQ population has a higher prevalence of depression and anxiety than their heterosexual and cisgender counterparts. In addition, men who are gay have an increased rate of eating disorders as compared to men who are heterosexual. A host of campaigns and outreach efforts have begun to target these disparities.

Immigration Status

According to the US Census Bureau, the nation's total recent immigrant population is growing rapidly; it was quantified at 40.4 million in 2011 and is expected to increase by roughly 20 million in the next two decades. This tells us that immigrants, whether documented or undocumented, are interwoven into every social structure and institution in the United States. The nativity of immigrant populations changes over time; in the most recent census, the largest proportions of immigrants had emigrated from Mexico, the Caribbean, and India. **Generational status** refers to the place of birth of a specific person or that person's parents. For instance, first generation refers to someone who is born outside of their place of residence. Second generation refers to a person that has at least one parent that is foreign-born.

Considering the number of immigrant communities, there are often barriers that affect interactions with social structures and institutions. The complex organization of the United States healthcare system is starkly different from those of most other nations, and this may present a barrier to use for immigrants. Language barriers may also make it difficult for immigrants to access healthcare or to take control of their healthcare decisions; telephone translation services have been created to help facilitate the conversation between clinician and patient. Racial and ethnic identity may be more apparent in first-generation immigrants, and the biases and prejudices against certain racial and ethnic groups might be compounded by the individual's immigrant status; this interplay between multiple demographic factors—especially when it leads to discrimination or oppression—is termed **intersectionality**. Finally, undocumented status presents a major barrier for many immigrants to access healthcare for fear of reporting and deportation.

Demographic Shifts and Social Change

Since 1950, the United States population has roughly doubled. In addition to increasing in size, the makeup of the American population has changed significantly. The average age in the United States has increased, and the population is continuing to become more racially and ethnically diverse. These are examples of **demographic shifts**: changes in the makeup of a population over time. These shifts can be measured by considering the **population density**, which counts the number of people per square kilometer of land area.

Population projections attempt to predict changes in population size over time, and can be assisted by historical measures of growth, understanding of changes in social structure, and analysis of other demographic information. To aid in the construction of population projections, **population pyramids** provide a histogram of the population size of various age cohorts, as shown in Figure 11.7.

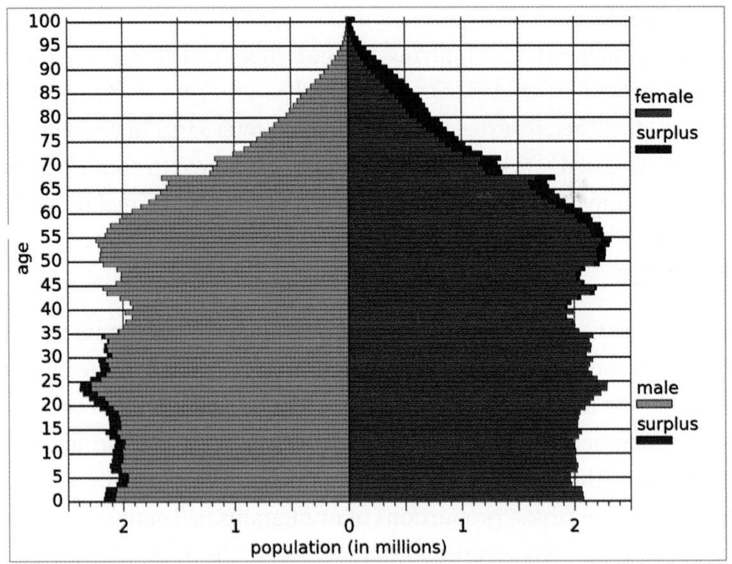

Figure 11.7 U.S. Population Pyramid, 2014

Surplus occurs when one gender has a larger population than another.

Fertility, Mortality, and Migration

KEY CONCEPT

Demographic statistics:

- Fertility rate = children per woman per lifetime
- Birth rate = children per 1000 people per year
- Mortality rate = deaths per 1000 people per year
- Migration rate = immigration rate minus emigration rate

The increased population of the United States is due to a number of factors that center around fertility, mortality, and migration. **Fertility rate** refers to the average number of children born to a woman over a lifetime in a population. In many parts of the world, fertility rate is the primary driver of population expansion; for example, in some parts of Africa, the average fertility rate is between four and eight children per woman, as seen in Figure 11.8. In the United States, fertility rates have trended downward over time; in 2013, the rate was still above two, indicating that fertility rates were still contributing to population growth.

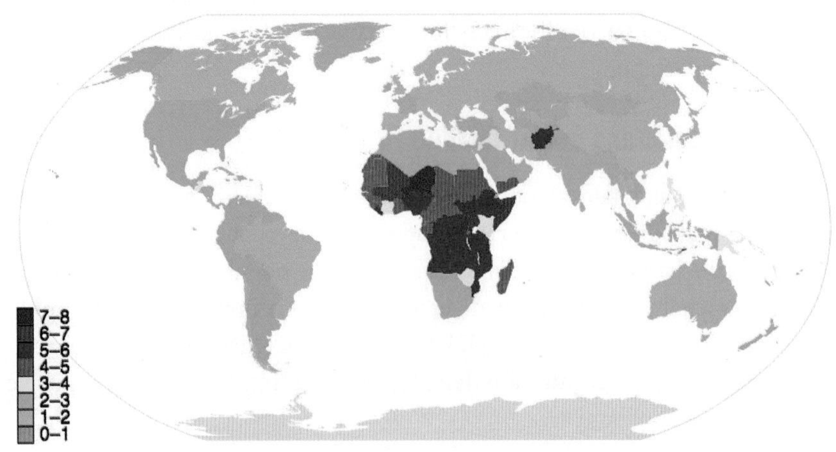

Figure 11.8 Fertility Rates around the World, 2013

Based on data from the CIA World Factbook; measured in children born per woman in the population

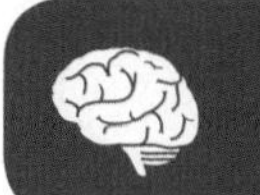

Mortality rates refer to the number of deaths in a population per unit time. Usually, this rate is measured in deaths per 1000 people per year. With advancements in healthcare and access, the mortality rate in the United States has dropped significantly over the past century. However, mortality rates are a significant brake on population growth in many parts of the world, as demonstrated in Figure 11.9. The decreased mortality rate in the United States is one contributor to the increase in average age of the population, as is a decreased fertility rate. In addition, the aging of the Baby Boomer generation, one of the largest generations in United States history, increases this average age. Both birth and mortality rates can be reported in multiple forms: the total rate for a population, the **crude rate** (adjusted to a certain population size over a specific period of time and multiplied by a constant to give a whole number), or age-specific rates.

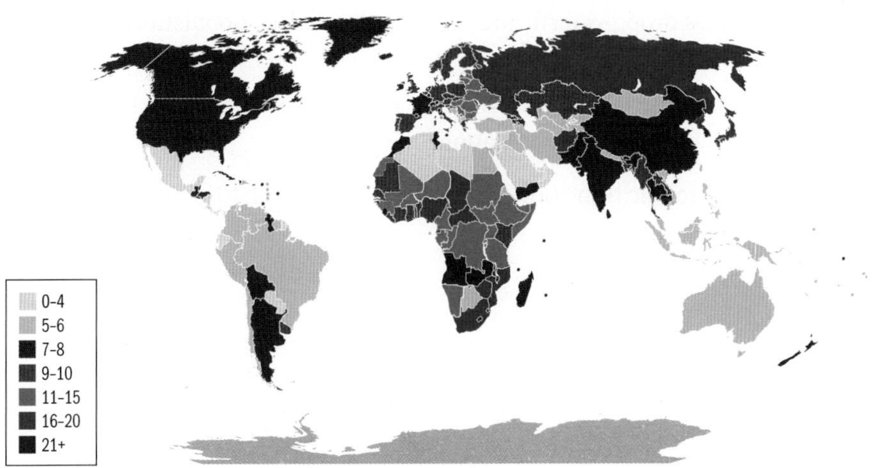

Figure 11.9 Mortality Rates around the World, 2009
Based on data from the CIA World Factbook; measured in deaths per 1000 individuals per year

Finally, **migration** is a contributor to population growth. **Immigration** is defined as movement into a new geographic space, whereas **emigration** is movement away from a geographic space. As described earlier, the United States continues to have larger net immigration than emigration, driving an increase in the population size. Immigration also increases the racial and ethnic diversity of the United States, as do increased mobility within the country and increases in intermarriage between different races and ethnicities. Migration can be motivated by both **pull factors**, which are positive attributes of the new location that attract new residents, and **push factors**, which are negative attributes of the old location that encourage existing residents to leave.

KEY CONCEPT

The United States population is getting bigger, older (average age has increased), and more diverse (through immigration, mobility, and intermarriage).

Demographic Transition Theory

While demographic shift refers to general changes in population makeup over time, **demographic transition** is a specific example of a demographic shift that occurs as a country develops from a preindustrial to an industrial economic system. Demographic transition has been seen in the United States since the Industrial Revolution and is currently occurring in many developing countries.

Demographic transition theory explains this link between economic development and demographic shift in four stages:

- **Stage 1**: Preindustrial society; birth and death rates are both high, resulting in a stable population.
- **Stage 2**: Economic progress leads to improvements to healthcare, nutrition, sanitation, and wages, causing a decrease in death rates. Thus, total population increases.
- **Stage 3**: Improvements in contraception, women's rights, and a shift from an agricultural to an industrial economy cause birth rates to drop. For example, with an industrializing society, children must go to school for many years to be productive in society and may need to be supported by parents for a longer period of time than was formerly the case. Thus families have fewer children, and birth rates drop. As birth and death rates equalize, population growth hits an inflection point and begins to level off.
- **Stage 4**: An industrialized society; birth and death rates are both low, resulting in a relative constant total population.

A model of demographic transition can be seen in Figure 11.10.

KEY CONCEPT

During demographic transition, mortality rate drops before birth rate. Therefore, the population grows at first while mortality rate is dropping, and then plateaus as the birth rate decreases as well.

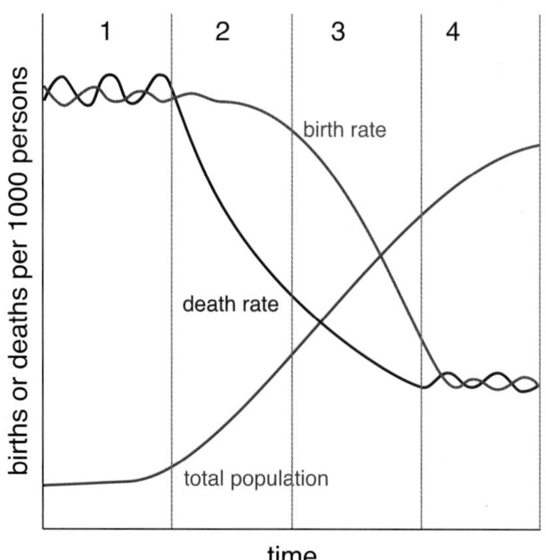

Figure 11.10 Demographic Transition

Recently, sociologists have described a fifth stage of demographic transition theory in which birth rates continue to drop and fall below the death rate, resulting in a decline of total population. Japan and Germany are currently experiencing demographic shifts expected in this theorized fifth stage.

Malthusian theory focuses on how the exponential growth of a population can outpace growth of the food supply and lead to social degradation and disorder. For example, some theorists predict that demographic transition among developing nations might cause growth in the world's population to outpace the world's ability to generate food. The resulting hypothetical mass starvation is called the Malthusian catastrophe. This result is similar to the death phase of bacterial growth, when resources in the environment have been depleted, as described in Chapter 1 of *MCAT Biology Review*.

Social Movements

Social movements are organized either to promote or to resist social change. These movements are often motivated by a group's perceived **relative deprivation**, which is a decrease in resources, representation, or agency relative to the whole of society, or relative to what the group is accustomed to from the past. Social movements that promote social change are termed **proactive**; those that resist social change are **reactive**. Members of social movements work to correct what they perceive as social injustices. Some examples of proactive movements include the civil rights movement, women's rights movement, gay rights movement, animal rights movement, and environmentalism movement. Some examples of reactive movements include the White supremacist movement, counterculture movement, antiglobalization movement, and anti-immigration movement. To further their goals, social movements may establish coordinated organizations. For example, some organizations associated with the proactive movements above include the National Association for the Advancement of Colored People (NAACP), American Civil Liberties Union (ACLU), Human Rights Campaign (HRC), Humane Society, and Greenpeace. Social movements may also seek to share their message through the media and demonstrations. Political involvement is also common through lobbying and donations.

Globalization

Globalization is the process of merging of the separate nations of the world into a single sociocultural entity, and is a relatively recent phenomenon spurred on by improvements in global communication technology and economic interdependence. Globalization leads to a decrease in the geographical constraints on social and cultural exchanges and can lead to both positive and negative effects. For example, the availability of foods (especially produce) from around the world during the entire calendar year can only be accomplished through trade with an extremely large number of world markets. However, significant worldwide unemployment, rising prices, increased pollution, civil unrest (particularly in unindustrialized or undemocratic nations), and global terrorism are negative effects of globalization.

Traditionally, the health sector has been organized at the national, state, or local level, but this is beginning to change. Groups such as the World Health Organization (WHO), the American Red Cross, and Doctors Without Borders supply aid to populations in need around the globe. Many medical schools are also increasing opportunities for medical students to complete rotations in other countries.

Urbanization

Urbanization refers to dense areas of population creating a pull for migration. In other words, cities are formed as individuals move into and establish residency in these new urban centers. Urbanization is not a new phenomenon; ancient populations established cities in Jerusalem, Athens, Timbuktu, and other locations. The economic opportunities offered in cities and creation of a large number of "world cities" has fueled an increase in urbanization during the last few decades. Currently, more than half of the world's populations live in what are considered urban areas. Sociologists and other professionals have found links between urban societies and health challenges related to water sanitation, air quality, environmental hazards, violence and injuries, infectious diseases, unhealthy diets, and physical inactivity.

Cities are rarely homogenous with respect to their population makeup. Most cities have areas that are more socioeconomically well-off and others that are more impoverished. Sometimes, members of specific racial, ethnic, or religious groups are concentrated in particular areas, usually due to social or economic inequities. In the most extreme cases, as shown in Figure 11.11, there arises an extremely densely populated area of a city with low-quality, often informal housing and poor sanitation.

Figure 11.11 A Disadvantaged Area in Jakarta, Indonesia

MCAT CONCEPT CHECK 11.3

Before you move on, assess your understanding of the material with these questions.

1. What is the difference between race and ethnicity?

____

2. What is symbolic ethnicity?

____

3. Define the following demographic statistics:

   - Fertility rate:

____

   - Birth rate:

____

   - Mortality rate:

____

4. During demographic transition, what happens to the mortality rate? To the birth rate?

   - Mortality rate:

____

   - Birth rate:

____

5. What are the two types of social movements? How do they differ?

   - ____
   - ____

Conclusion

There are three major trends that are changing our nation's healthcare needs and our patient population. First, the increased diversity in the American population as a whole (from immigration, increased social and academic mobility, and interconnectedness through technology) puts us in front of patients whose thoughts and beliefs about health and well-being may be starkly different from our own. Second, increased access to healthcare through reform legislation has allowed millions of Americans to reach providers for the first time. Finally, our successes in medicine and public health have increased survival rates of many formerly fatal conditions and have enabled us to live longer. This leaves us with an aging population, in which individuals may be coping with multiple illnesses simultaneously.

To arm physicians of the future with the skills needed to take care of this population, many medical schools are increasing their coursework in interpersonal skills (*Doctor–Patient Relationship*; *Doctoring*; or *Physician, Patient, and Society* are such courses at various schools), as well as **cultural sensitivity**, the recognition and respect of differences between cultures, and research ethics. This is part of the biopsychosocial model of medicine described in Chapter 7 of *MCAT Behavioral Sciences Review*. Knowledge of the structure of society and how it shifts over time, as explained in this chapter, will enhance your ability to communicate with patients. Unlike the old model of *doctor knows best* (often referred to as the paternalistic approach to medicine), today's doctors must work together with patients to find solutions to their health problems. By working with patients on their own terms, you will be able to help maintain and improve their health status and begin to correct the health inequities that exist in today's population. In the next chapter—the last of *MCAT Behavioral Sciences Review*—we will explore these inequities in resources, health status, and healthcare.

You've reviewed the content, now test your knowledge and critical thinking skills by completing a test-like passage set in your online resources!

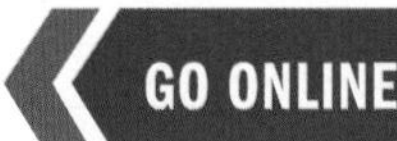

CONCEPT SUMMARY

Sociology: Theories and Institutions

- Theoretical approaches provide frameworks for the interactions we observe within society.
  - **Functionalism** focuses on the function of each component of society and how those components fit together. **Manifest functions** are deliberate actions that serve to help a given system; **latent functions** are unexpected, unintended, or unrecognized positive consequences of manifest functions.
  - **Conflict theory** focuses on how power differentials are created and how these differentials contribute to the maintenance of social order.
  - **Symbolic interactionism** is the study of the ways individuals interact through a shared understanding of words, gestures, and other symbols.
  - **Social constructionism** explores the ways in which individuals and groups make decisions to agree upon a given social reality.
  - **Rational choice theory** states that individuals will make decisions that maximize potential benefit and minimize potential harm; **exchange theory** applies rational choice theory within social groups.
  - **Feminist theory** critiques the institutional power structures that disadvantage women in society.
- **Social institutions** are well-established social structures that dictate certain patterns of behavior or relationships and are accepted as a fundamental part of culture. Common social institutions include the family, education, religion, government and the economy, and health and medicine.
- There are four key ethical tenets of American medicine.
  - **Beneficence** refers to acting in the patient's best interest.
  - **Nonmaleficence** refers to avoiding treatments for which risk is larger than benefit.
  - **Respect for autonomy** refers to respecting patients' rights to make decisions about their own healthcare.
  - **Justice** refers to treating similar patients similarly and distributing healthcare resources fairly.

Culture

- **Culture** encompasses the lifestyle of a group of people and includes both material and symbolic elements.
  - **Material culture** includes the physical items one associates with a given group, such as artwork, emblems, clothing, jewelry, foods, buildings, and tools.
  - **Symbolic culture** includes the ideas associated with a cultural group.
- **Cultural lag** refers to the idea that material culture changes more quickly than symbolic culture.
- A **cultural barrier** is a social difference that impedes interaction.
- **Language** consists of spoken, signed, or written symbols combined into a system and governed by rules.
- A **value** is what a person deems important in life.
- A **belief** is something a person considers to be true.
- A **ritual** is a formalized ceremonial behavior in which members of a group or community regularly engage. It is governed by specific rules, including appropriate behavior and a predetermined order of events.
- **Norms** are societal rules that define the boundaries of acceptable behavior.
- There is evidence that culture flows from evolutionary principles, and that culture can also influence evolution.

Demographics

- **Demographics** refer to the statistics of populations and are the mathematical applications of sociology. One can analyze hundreds of demographic variables; some of the most common are age, gender, race and ethnicity, sexual orientation, and immigration status.
  - **Ageism** is prejudice or discrimination on the basis of a person's age.
  - **Gender** is the set of behavioral, cultural, or psychological traits typically associated with a biological sex. **Gender inequality** is the intentional or unintentional empowerment of one gender to the detriment of the other.
  - **Race** is a social construct based on phenotypic differences between groups of people; these may be either real or perceived differences.
  - **Ethnicity** is also a social construct that sorts people by cultural factors, including language, nationality, religion, and other factors. **Symbolic ethnicity** is recognition of an ethnic identity that is only relevant on special occasions or in specific circumstances and does not specifically impact everyday life.
  - **Sexual orientation** can be defined by one's sexual interest toward members of same or different genders.
  - **Immigration** is the movement into a new geographic area. **Emigration** is the movement away from a geographic area.

- A **fertility rate** is the average number of children born to a woman over a lifetime in a population. A **birth rate** is relative to a population size over time, usually measured as the number of births per 1000 people per year.
- A **mortality rate** is the average number of deaths per population size over time, usually measured as the number of deaths per 1000 people per year.
- **Migration** refers to the movement of people from one geographic location to another.
- **Demographic transition** is a model used to represent drops in birth and death rates as a result of industrialization.
- **Social movements** are organized to either promote (**proactive**) or resist (**reactive**) social change.
- **Globalization** is the process of integrating a global economy with free trade and tapping of foreign labor markets.
- **Urbanization** refers to the process of dense areas of population creating a pull for migration or, in other words, creating cities.

ANSWERS TO CONCEPT CHECKS

11.1

1. Manifest functions are actions that are intended to help some part of a system. Latent functions are unintended, unstated, or unrecognized positive consequences of these actions on society.

2.

Theoretical Approach	Primary Thesis or Idea
Functionalism	Each part of society serves a function; when these functions work together correctly, society overall can function normally
Conflict theory	Power differentials are created when groups compete for economic, social, and political resources; these differentials contribute to the maintenance of social order
Symbolic interactionism	Humans communicate through words, gestures, and other symbols to which we attach meaning
Social constructionism	Individuals and groups make decisions to agree upon a given social reality
Rational choice theory	Individuals will make decisions that maximize potential benefit and minimize potential harm
Feminist theory	Critiques the institutional power structures that disadvantage women in society

3.

Ethical Principle	Description
Beneficence	Act in the patient's best interest
Nonmaleficence	*Do no harm*; avoid interventions where the potential for harm outweighs the potential for benefit
Respect for autonomy	Respect patients' decisions and choices about their own healthcare
Justice	Treat similar patients with similar care; distribute healthcare resources fairly

11.2

1. Material culture focuses on the artifacts associated with a group: the physical objects, such as artwork, emblems, clothing, jewelry, foods, buildings, and tools. Symbolic culture focuses on the ideas and principles that belong to a particular group.
2. A value is what a person deems to be important; a belief is what a person deems to be true. While these terms are often used interchangeably in everyday life, they have specific definitions in the social sciences.

11.3

1. Race is based on phenotypic differences between groups of people. Ethnicity is based on common language, religion, nationality, or other cultural factors.
2. Symbolic ethnicity is recognition of an ethnic identity on special occasions or in specific circumstances, but not during everyday life.
3. Fertility rate is the average number of children a woman has over a lifetime in a population. Birth rate is the number of births in a population per unit time, usually measured as births per 1000 people per year. Mortality rate is the number of deaths in a population per unit time, usually measured as deaths per 1000 people per year.
4. During demographic transition, both the mortality and birth rates decrease.
5. Proactive social movements are in favor of a specific social change. Reactive social movements run against a specific social change.

SCIENCE MASTERY ASSESSMENT EXPLANATIONS

1. **A**

A manifest function is an intended positive effect on a system. A latent function is an unintended positive effect on a system, **(C)**. A negative effect on a system, **(B)** and **(D)**, is termed a dysfunction.

2. **C**

Symbolic interactionism studies how individuals interact through a shared understanding of words, gestures, and other symbols. A nod is thus a symbol in many cultures that signifies "yes."

3. **D**

The principle of nonmaleficence states that physicians must not only act in their patient's best interest (beneficence, **(B)**), but must also avoid treatments where the potential for harm outweighs the potential for benefit.

4. **C**

Although one's dominant physical features are associated with race, this man's preferences are those of Latino ethnicity and culture. Ethnicity is a social construct that considers language, religion, nationality, and cultural factors. Culture relates to a group's way of life; the preferences listed in the question are examples of material and symbolic culture.

5. **D**

In the United States, it is not a common practice to say, "I love you," and hug one's physician after each routine checkup; therefore, this behavior could be considered deviant, going against the social norm. Values, **(C)**, are what an individual deems to be important.

6. **B**

Both birth rate and mortality rate can be measured per 1000 people per year. Fertility rate is measured in number of children per woman over a lifetime.

7. **B**

Rituals are formalized ceremonial behaviors in which members of a group or community regularly engage. Therefore, these activities are examples of rituals performed for healing.

8. **A**

With a decrease in mortality rate and a higher immigration rate than emigration rate, the United States population continues to grow, with an increasing average age and increasing racial and ethnic diversity.

9. **C**

Material culture includes any cultural artifact—an object to which meaning is assigned. Values are ideas, which are associated with symbolic culture.

10. **D**

During demographic transition, both birth rates and mortality rates are high in stage 1, **(A)**. Mortality rates drop during stage 2, **(B)**, and then birth rates drop during stage 3, **(C)**. In stage 4, both birth rates and mortality rates are low.

11. **D**

In this scenario, a group is fighting for social power, which is an aspect of conflict theory. Further, this group is an example of a reactive social movement because it is running counter to social change. The fact that gambling had been legalized implies the involvement of the government, a social institution.

12. **C**

This man is describing his sexuality as mostly homosexual, although he has also had some heterosexual attractions. The Kinsey scale scores a 6, **(D)**, as exclusively homosexual. A score of 3 would equate to bisexuality. Thus, this man would likely score a 4 or 5.

13. **B**

Sex is determined by one's genotype, and therefore is biologically determined. Gender, **(A)**, may or may not match biological sex and therefore is not biologically determined. Ethnicity, **(C)**, is a social construct that sorts people by cultural factors, and therefore is not biologically determined. Sexual orientation, **(D)**, may have some biological component, but the relative role of biology and environment is not yet known.

14. **B**

If the immigration rate in a geographic area is larger than the emigration rate, then there is a larger influx than efflux of people. This will increase the population of that area.

15. **A**

Urbanization is the migration of people into urban centers to create cities. The increased population density should provide additional opportunities for social interaction, not decreased opportunities.

Consult your online resources for additional practice.

SHARED CONCEPTS

Behavioral Sciences Chapter 6
Identity and Personality

Behavioral Sciences Chapter 8
Social Processes, Attitudes, and Behavior

Behavioral Sciences Chapter 9
Social Interaction

Behavioral Sciences Chapter 10
Social Thinking

Behavioral Sciences Chapter 12
Social Stratification

Biology Chapter 12
Genetics and Evolution

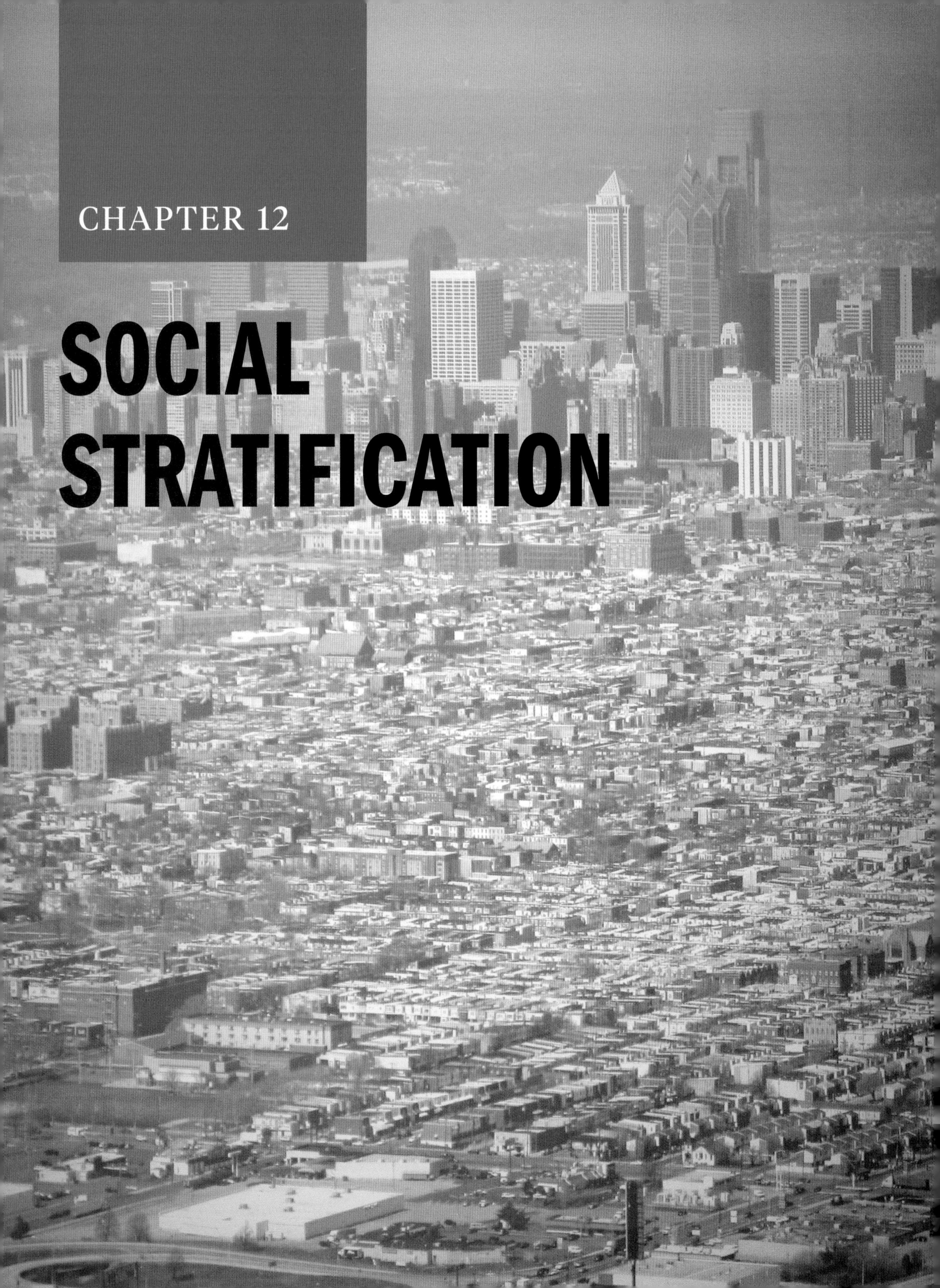

CHAPTER 12

SOCIAL STRATIFICATION

SCIENCE MASTERY ASSESSMENT

Every pre-med knows this feeling: there is so much content I have to know for the MCAT! How do I know what to do first or what's important?

While the high-yield badges throughout this book will help you identify the most important topics, this Science Mastery Assessment is another tool in your MCAT prep arsenal. This quiz (which can also be taken in your online resources) and the guidance below will help ensure that you are spending the appropriate amount of time on this chapter based on your personal strengths and weaknesses. Don't worry though—skipping something now does not mean you'll never study it. Later on in your prep, as you complete full-length tests, you'll uncover specific pieces of content that you need to review and can come back to these chapters as appropriate.

How to Use This Assessment

If you answer 0–7 questions correctly:

Spend about 1 hour to read this chapter in full and take limited notes throughout. Follow up by reviewing **all** quiz questions to ensure that you now understand how to solve each one.

If you answer 8–11 questions correctly:

Spend 20–40 minutes reviewing the quiz questions. Beginning with the questions you missed, read and take notes on the corresponding subchapters. For questions you answered correctly, ensure your thinking matches that of the explanation and you understand why each choice was correct or incorrect.

If you answer 12–15 questions correctly:

Spend less than 20 minutes reviewing all questions from the quiz. If you missed any, then include a quick read-through of the corresponding subchapters, or even just the relevant content within a subchapter, as part of your question review. For questions you answered correctly, ensure your thinking matches that of the explanation and review the Concept Summary at the end of the chapter.

1. Which of the following best describes the component of socioeconomic status attributable to direct individual efforts?
   - **A.** Ascribed status
   - **B.** Meritocratic competition
   - **C.** Anomic condition
   - **D.** Achieved status

2. Which of the following displays a correct association?
   - **A.** High social networking and low social capital
   - **B.** High social mobility and low social capital
   - **C.** Low social class and low social capital
   - **D.** Low social networking and high social capital

3. Which of the following phenomena are LEAST likely to coincide?
   - **A.** Hazardous waste facilities and low-income neighborhoods
   - **B.** Tuberculosis and poor living conditions
   - **C.** Environmental pollution and populations of underrepresented groups
   - **D.** Globalization and global equality

4. Which of the following trends is most likely FALSE?
   - **A.** Mortality rates are increased in groups with low income.
   - **B.** Life expectancy is decreased in groups with high income.
   - **C.** Birth weights are decreased in children of low-income women.
   - **D.** Rates of lung cancer are increased in low-income groups.

Questions 5–6 refer to the scenario described below.

A small town has 1000 residents, including 500 males and 500 females. In this town, 20 of the males have prostate cancer. During a calendar year, 10 more males are diagnosed with prostate cancer. Assume none of the males are cured or die during the year.

5. What is the prevalence of prostate cancer in the population that can develop the condition at the end of the year?
   - **A.** $10 \div 480$
   - **B.** $10 \div 1000$
   - **C.** $20 \div 500$
   - **D.** $30 \div 500$

6. What is the incidence of prostate cancer in this population during the year?
   - **A.** $10 \div 480$
   - **B.** $10 \div 1000$
   - **C.** $20 \div 500$
   - **D.** $30 \div 1000$

7. A low-income single parent works a part-time job and lives in a small apartment in the city. When the single parent's children grow up, they take similar jobs and live in similar housing. This is an example of:
   - **A.** upward social mobility.
   - **B.** downward social mobility.
   - **C.** social exclusion.
   - **D.** social reproduction.

8. Which of the following is true with regard to relative poverty?
   - A. Individuals in relative poverty have incomes below the poverty line.
   - B. Individuals in relative poverty exhibit downward social mobility.
   - C. Individuals in relative poverty may be in the upper class.
   - D. Individuals in relative poverty exhibit upward social mobility.

9. In comparison to urban centers, suburbs tend to have:
   - A. larger racially and ethnically underrepresented populations.
   - B. higher rates of poverty.
   - C. larger upper- and middle-class populations.
   - D. higher rates of crime and homicide.

10. Which of the following terms refers to the burden or degree of disease associated with a given illness?
    - A. Morbidity
    - B. Mortality
    - C. Second sickness
    - D. Chronicity

11. Compared to White Americans, which of the following racial or ethnic groups tends to have a better overall health profile?
    - A. African Americans
    - B. Asian Americans
    - C. Hispanic Americans
    - D. American Indians

12. Which of the following best describes the populations targeted by Medicare and Medicaid, respectively?
    - A. Medicare: mostly patients without employer-guaranteed healthcare; Medicaid: mostly patients who have recently immigrated
    - B. Medicare: mostly patients who have recently immigrated; Medicaid: mostly patients without employer-guaranteed healthcare
    - C. Medicare: mostly patients in older age groups; Medicaid: mostly patients with low socioeconomic status
    - D. Medicare: mostly patients with low socioeconomic status; Medicaid: mostly patients in older age groups

13. Morbidity is increased in low-income groups because of all of the following EXCEPT:
    - A. higher rates of obesity.
    - B. less access to healthcare.
    - C. higher rates of homicide.
    - D. lower rates of physical activity.

14. Hypertension (high blood pressure) can be diagnosed by having two or more blood pressure readings higher than 140/90 on two different occasions, separated by a week. Suppose that the criteria were changed to include anyone with a reading higher than 130/80 on at least one occasion. How would this change the prevalence of diagnosed hypertension in the population?
    - A. The prevalence would increase.
    - B. The prevalence would decrease.
    - C. The prevalence would remain the same.
    - D. There is not enough information to determine the change in prevalence.

15. Which of the following trends regarding healthcare disparities has NOT been documented?
    - A. Females are more likely to be insured than males.
    - B. Primary care use is more likely among males than females.
    - C. Low-income individuals have more difficulty accessing care than high-income individuals.
    - D. LGBTQ individuals have more barriers to healthcare than other individuals.

Answer Key

1. **D**
2. **C**
3. **D**
4. **B**
5. **D**
6. **A**
7. **D**
8. **C**
9. **C**
10. **A**
11. **B**
12. **C**
13. **C**
14. **A**
15. **B**

Detailed explanations can be found at the end of the chapter.

CHAPTER 12

SOCIAL STRATIFICATION

In This Chapter

CHAPTER PROFILE

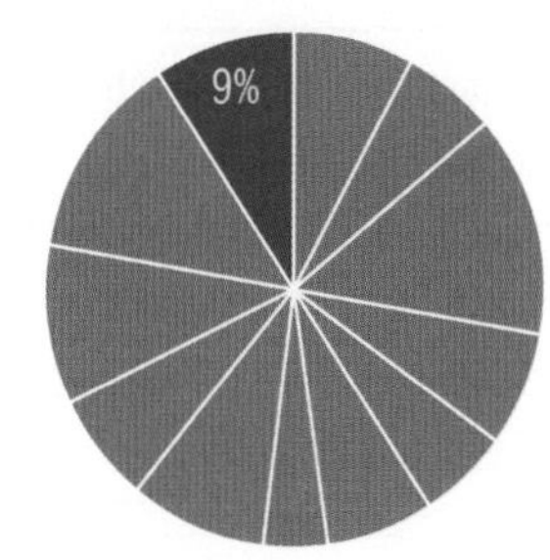

The content in this chapter should be relevant to about 9% of all questions about the behavioral sciences on the MCAT.

This chapter covers material from the following AAMC content categories:

8B: Social thinking

8C: Social interactions

10A: Social inequality

Introduction

The wonderfully witty Oscar Wilde once said, "Work is the curse of the drinking classes." While this quote is intended to be humorous, it does speak to the stereotypical characteristics associated with socioeconomic class. Some Americans think that class and social stratification are nonissues in our society. Unlike earlier feudal societies, most Americans are not royals or gentry, possessing inherited titles, land, or palaces; we're often considered to be a much more equality-oriented society, in keeping with our constitutional ideals. Yet how do we explain such differences in wealth, power, and privilege as a Manhattan lawyer driving a shiny Porsche past a homeless person rooting through a trash can? Such scenes make it hard to ignore the uneven distributions of material wealth and the overall social inequality in the United States.

To understand social inequalities in America and how such disparities impact health and healthcare services, we will examine several aspects of social stratification in terms of class, status, and social capital and how these intersect with race, gender, and age. We will also focus on patterns of social mobility and how poverty and location play major roles in health and illness. Later, we will connect how race, gender, and socioeconomic inequalities impact health profiles and access to quality healthcare.

12.1 Social Class

LEARNING OBJECTIVES

After Chapter 12.1, you will be able to:

- Explain how socioeconomic status of an individual or a community is determined
- Define social capital and social cohesion
- Identify groups that suffer disproportionate social inequality
- Describe the relationship between social mobility and merit

KEY CONCEPT

It's important to keep in mind that socioeconomic status is not only determined by merit (achieved status), but also external characteristics or outward appearances like skin color and gender (ascribed status).

BRIDGE

There are three major types of status: ascribed, achieved, and master status. While ascribed and achieved statuses are described here, remember that a master status is one that pervades all aspects of an individual's life. The role of statuses in social interaction is discussed in Chapter 9 of *MCAT Behavioral Sciences Review*.

A **social class** is defined as a category of people who share a similar socioeconomic position in society, and can be identified by looking at the economic opportunities, job positions, lifestyles, attitudes, and behaviors of a given slice of society. **Social cohesion**, or social integration, refers to the solidarity and sense of connectedness among different social groups and social classes in society.

Aspects of Social Stratification

Social stratification focuses on social inequalities and studies the basic question of who gets what and why. Social stratification is thus related to one's **socioeconomic status** (**SES**), which may depend on ascribed or achieved status, and causes the emergence of status hierarchies. **Ascribed status** derives from clearly identifiable characteristics, such as age, gender, and skin color; **achieved status** is acquired via direct, individual efforts. In other words, ascribed status is involuntary, while achieved status is obtained through hard work or merit. An important factor in achieved status is **educational attainment**, which is the highest degree obtained, or number of years of education completed. Caste and estate systems stratify by ascribed SES, while class systems stratify by achieved SES. After breaking free from British colonial rule, the United States moved toward a class-based system of social stratification.

Class, Status, and Power

There are three major classes—upper, middle, and lower—although these vary to different degrees in different locations. The upper class consists of those who have great wealth, along with recognized reputations and lifestyles, and have a larger influence on society's political and economic systems. In other words, the upper class has a high concentration of prestige and power. The middle class can be further divided into three levels: upper-middle, middle-middle, and lower-middle class. The middle class includes successful business and professional people (upper-middle), those who have been unable to achieve the upper-middle lifestyle because of educational and economic shortcomings (middle-middle), and those who are skilled and semiskilled workers with fewer luxuries (lower-middle). The lower class includes people who have lower incomes, and has a greatly reduced amount of sociopolitical power. The proportional improvement in healthcare as one moves up in socioeconomic status is called the **socioeconomic gradient** in health and development.

Prestige refers to the amount of positive regard society has for a given person or idea. Certain occupations, such as physicians, are broadly viewed with high levels of status, respect, and importance. Particular educational institutions, organizations, awards, and accolades may also be considered prestigious.

Power can be described as the ability to affect others' behavior through real or perceived rewards and punishments, and is based on the unequal distribution of valued resources. At its core, power defines the relationship between individuals, groups, and social institutions. Power relationships function to maintain order, organize economic systems, conduct warfare, and rule over and exploit people. As a result, power creates worldwide social inequalities as people tend to fall somewhere within the *haves* and the *have-nots*.

You may recall from Chapter 11 of *MCAT Behavioral Sciences Review* that Marxist theory, also called conflict theory, proposes that the *have-nots*, called the proletariat, could overthrow the *haves*, called the bourgeoisie, as well as the entire capitalist economy by developing class consciousness. **Class consciousness** refers to the organization of the working class around shared goals and recognition of a need for collective political action. By working together as one unit, the proletariat could revolt and take control of the political and economic system, laying the groundwork for a socialist state. The one major barrier to class consciousness, however, is **false consciousness**, a misperception of one's actual position within society. Members of the proletariat either do not see just how bad conditions are, do not recognize the commonalities between their own experiences and others, or otherwise are too clouded to assemble into the revolutionaries Marx envisioned.

Even in developed countries of the modern, globalized world, social inequality persists. Early sociologists explained that social inequality is further accelerated by what is called **anomie**, which refers to a lack of widely accepted social norms and the breakdown of social bonds between an individual and society. **Strain theory** focuses on how anomic conditions can lead to deviance. Anomic conditions include excessive individualism, social inequality, and isolation; these all erode **social solidarity**, which is the sense of community and social cohesion. Other sociologists have focused on the importance of social trust in the proper functioning of civil society. Social trust comes from two primary sources: social norms of reciprocity (*I'll scratch your back if you scratch mine*) and social networks. In the past several decades, as some societies have become more urbanized, self-oriented, and materialistic, associational ties have diminished and consequently have led to a decline in social capital. But what is the relationship between social stratification, social capital, and power?

KEY CONCEPT

Anomic conditions in postindustrial modern life have accelerated the decline of social inclusion and, as a result, have further obstructed opportunities to acquire social capital.

Social Capital

Essentially, **social capital** is the investment people make in their society in return for economic or collective rewards; the greater the investment, the higher the level of **social integration**, which is the movement of new or underrepresented populations into a larger culture while maintaining their ethnic identities. One of the main forms of social capital is the social network. Social networks can create two types of social inequality: situational (socioeconomic advantage) and positional (based on how connected one is within a network and one's centrality within that network). Inequality in networks creates and reinforces **privilege**, which is inequality in opportunity. Moreover, low social capital leads to greater social inequality. As social capital refers to the benefits one receives from group association, **cultural capital** refers to the benefits one receives from knowledge, abilities, and skills.

REAL WORLD

People who experience poor mental health are one of the largest disadvantaged groups to lack both strong and weak ties. Due to repercussions of social exclusion, these individuals may find that social capital is out of their reach. Consequently, this group is personally and socially disempowered, further propelling a cycle of exclusion. Social exclusion has huge financial repercussions on healthcare, with greater morbidity rates.

Communities are joined together through what are called strong and weak ties. **Strong ties** refer to peer group and kinship contacts, which are quantitatively small but qualitatively powerful. **Weak ties** refer to social connections that are personally superficial, such as associates, but that are large in number and provide connections to a wide range of other individuals. Social networking websites—especially those focusing on professional relationships—are examples of groups of weak ties. People without multiple weak ties, such as disadvantaged groups, may find accessing and contributing to social capital extremely difficult.

Intersections with Race, Gender, and Age

Social stratification, or, more properly, social inequality, remains higher among certain disadvantaged groups than others, including groups who are racially and ethnically underrepresented (especially Hispanic and African American people), households headed by women, and older adults. Think about who you know that may fall into these underprivileged or underserved groups. Your friends? Your relatives? Yourself? Socioeconomic inequalities remain in the United States. As described in Chapter 11 of *MCAT Behavioral Sciences Review*, this continued socioeconomic inequality is partially due to **intersectionality**—the compounding of disadvantage seen in individuals who belong to more than one underserved group.

Patterns of Social Mobility

Unlike a caste-based or estate-based system of social stratification, people in North America generally have the ability to move up or down from one class to another. In a class system, **social mobility**, also known as **structural mobility**, is typically the result of an economic and occupational structure that allows one to acquire higher-level employment opportunities given proper credentials and experience requirements. In the United States, the class system encourages this type of ambition through dedication and hard work, an ethos embodied in the phrase *The American Dream*.

Intergenerational and Intragenerational Mobility

Social mobility can either occur within a generation or across generations. **Intragenerational mobility** refers to changes in social status that happen within a person's lifetime, while **intergenerational mobility** refers to changes in social status from parents to children. Many people consider the United States to be *the land of opportunity*, where intragenerational and intergenerational mobility can easily occur. However, others argue that opportunities for social mobility are diminishing because the gap between the upper class and the middle and lower classes continues to widen.

MCAT EXPERTISE

For the guided example that follows, we were able to answer part of the question very quickly. On Test Day, all questions are multiple choice, so as soon as you have even part of the answer to a question, you can look at the answers and eliminate any that aren't a match.

BEHAVIORAL SCIENCES GUIDED EXAMPLE WITH EXPERT THINKING

Social class differences are related to health and mortality outcomes as a result of differences in stress, financial strain, and working conditions, among others. The present study focused on incidence of psychiatric disorder and its relation to social class, and social mobility as compared to parental social status in predicting mental health outcomes.

Purpose: evaluate relationship between psych disorders and socioeconomic status/social mobility

Researchers conducted a longitudinal study to target this relationship. Participants were adults born between 1949 and 1959. Participants' adult social class between 1980 and 1990 was coded by occupation type and compared to parental social class at participants' birth. Researchers then noted participants' psychiatric admissions between 1990 and 2005. Results are summarized in Figures 1 and 2.

Standard longitudinal study: collected outcome data over 15 years.

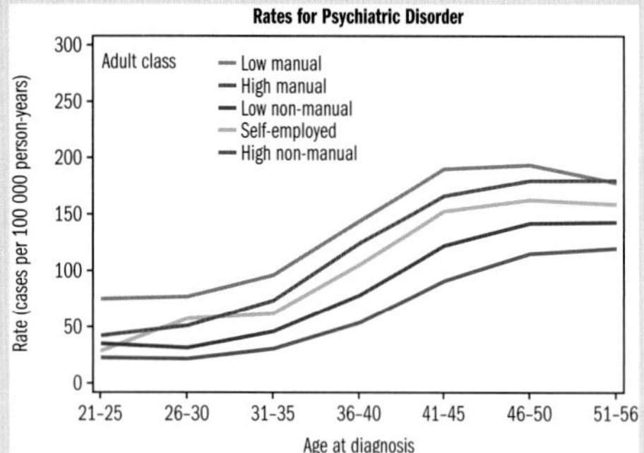

Figure 1 Incidence of psychiatric disorder by social class

IV: age, class

DV: rate of psychiatric cases

Results agree with the trend from the introduction, though not sure how self-employed fits.

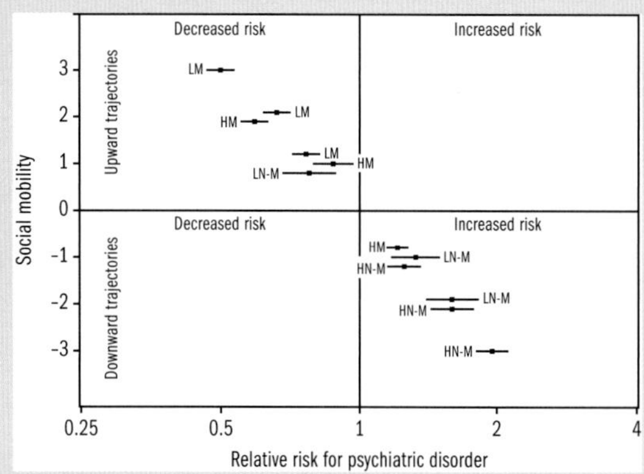

This figure is challenging, I will have to make sure to use the variable descriptions from the figure label to interpret. It looks like the LM (low manual) line in the top right corner means that this person's parents were LM and their own mobility is +3, which would put them at HN-M (high non-manual). These individuals have a 0.5 relative risk compared to LM individuals whose parents were LM, meaning that they're half as likely to develop mental health issues. The overall trend is linear and seems strong: higher upward mobility means lower risk.

Figure 2 Risk for psychiatric disorder resulting from social mobility compared to that of subjects who were socially stable. Social mobility scale and labels for confidence intervals are based on parental status, in order from high to low: high non-manual, low non-manual, high manual, low manual.

Figure label defines the variables, but I'll have to identify the relationship between them. The first letter of each word is the label in the graph.

Adapted from: Tiikkaja S, Sandin S, Malki N, Modin B, Sparén P, Hultman CM (2013) Social Class, Social Mobility and Risk of Psychiatric Disorder—A Population-Based Longitudinal Study. *PLoS ONE* 8(11): e77975. https://doi.org/10.1371/journal.pone.0077975

Researchers examined the records of three individuals: Person A (HM, parental class HN-M), Person B (LM, parental class LN-M), and Person C (HM, parental class HM). What can be predicted about the comparative risk of psychiatric disorder for A, B, and C?

This question is asking for a direct application of the results of the study, so the first thing to do to answer the question is to take a deeper look at the figures and determine what can be learned about these three individuals. Figure 1 gives overall incidence of mental health diagnoses for each class, and it looks like risk is inversely proportional to social class: people in lower social classes have a higher risk for psychiatric disorder. We can use this information alone to answer part of the question: Because the social class of person A is HM, i.e. "high manual," and the social class of person B is LM, i.e. "low manual," person B is in a lower social class and is therefore overall more likely to develop a psychiatric disorder.

Figure 2 is a little tougher to interpret, but we can start by finding the region of the graph where persons A and B would be located. Person A has parents in the HN-M class, but is in the HM class, for a mobility of −2. Person B has the same mobility, −2, from LN-M to LM. Looking at the data for individuals at −2 mobility, both persons A and B are about 1.75 times more likely to develop a psychiatric disorder than members of their respective social class with stable mobility. So, although person A will have a lower overall risk compared to person B due to social class, both persons A and B will have an elevated risk compared to socially stable members of their respective social class.

By contrast, person C shares a social class with their parents, meaning no additional risk is conferred to person C by downward social mobility. Given that persons A and C are in the same social class, but person C did not experience intergenerational downward mobility, person C would be expected to have a lower risk of disorder than person A.

Therefore, in order of risk, person C is at the lowest risk for psychiatric disorder based on the results of this study, followed by person A, with person B at the highest risk for psychiatric disorder based on both social status and downward social mobility.

Meritocracy

One of the largest factors driving American social mobility has been meritocratic competition or a merit-based system of social mobility. **Meritocracy** is a social structure in which intellectual talent and achievement are means for a person to advance up the social ladder. Given the rising levels of social inequality and concentration of wealth in the United States, some argue that motivation, a strong work ethic, a conscientious drive, and mastery of skills no longer offer the same opportunities for advancement. Some fear that the US meritocratic system is quickly becoming a **plutocracy**, or a rule by the upper classes. Nonetheless, merit still plays a key role in many segments of society, but merit does not always guarantee positive social mobility.

Upward and Downward Mobility

Social mobility usually occurs in one of two directions: up or down. Upward and downward mobility both refer to patterns of **vertical mobility**, or movement from one social class to another. **Upward mobility** is a positive change in a person's social status, resulting in a higher position. **Downward mobility** is the opposite: a negative change in a person's social status, wherein they fall to a lower position. Social mobility is often directly correlated with education, although other factors can contribute to upward mobility as well. Some of the best examples of upward mobility are seen with professional athletes, professional musicians, and entrepreneurs. So, in addition to education, athletics, music, and greater opportunities for small businesses may offer disadvantaged individuals more potential ladders to a higher social status.

Horizontal Mobility

Horizontal mobility is a change in occupation or lifestyle by an individual that keeps that individual within the same social class. For example, a construction worker who switches jobs to work in custodial services or mechanical maintenance has made a shift in occupation but typically remains in the lower-middle class.

Poverty

Poverty is defined by low socioeconomic status and a lack of possessions or financial resources. Poverty can be handed down from generation to generation, and can be defined on its own terms or in comparison to the rest of the population.

Social Reproduction

Social inequality, especially poverty (but also inherited wealth), can be reproduced or passed on from one generation to the next. This idea is referred to as **social reproduction**. Some consider social reproduction to be a cycle-of-poverty explanation for social inequality. In other words, the lifestyle of poverty, powerlessness, isolation, and even apathy is handed down from one generation to another as a feature of the society. However, there are many other factors that contribute to poverty, including where one lives and an emphasis on present orientation, in which people do not plan for the future. One theory, that of **structural poverty**, is based on the concept of "holes" in the structure of society being more responsible for poverty than the actions of any individual. Proponents of structural poverty argue that the same individuals do not by necessity occupy these "holes" from year to year, but the percentage of a society that falls under the poverty line stays relatively constant due to their existence. To understand how poverty is reproduced, it may be helpful to examine what types of poverty exist.

Absolute and Relative

On an **absolute** level, poverty is a socioeconomic condition in which people do not have enough money or resources to maintain a quality of living that includes basic life necessities such as shelter, food, clothing, and water. This absolute poverty view applies across locations, countries, and cultures. Poverty can also be defined as **relative**, in which people have less income and wealth in comparison to the larger population in which they live. For example, surviving on a low teaching salary while living in the Upper East Side of Manhattan may very well be considered poor relative to the other, far wealthier residents of that neighborhood.

In the United States, the official definition of the **poverty line** is derived from the government's calculation of the minimum income requirements for families to acquire the minimum necessities of life. Poverty is highly related to geography, as can be seen in Figure 12.1. One of the main problems with the official poverty line is that it is not contextualized according to geographic location and, as a result, does not take into account the cost of living in different communities. For example, the price of renting an apartment in a major urban center is much higher than the cost of rent in a rural small town. Some conceive of poverty as a form of powerlessness or a sociological and psychological condition of hopelessness, indifference, and distrust. In other words, poverty can be the result of the inability to control events that shape a person's life, often leading to a large degree of dependency on others.

KEY CONCEPT

In the United States, poverty is determined by the government's estimation of the minimum income requirements for families to acquire their minimum needs, such as shelter, food, water, and clothing. The problem with this official definition is that it fails to take into account geographical variables that impact the value of money in different locations. Certain areas are more costly to live in than others.

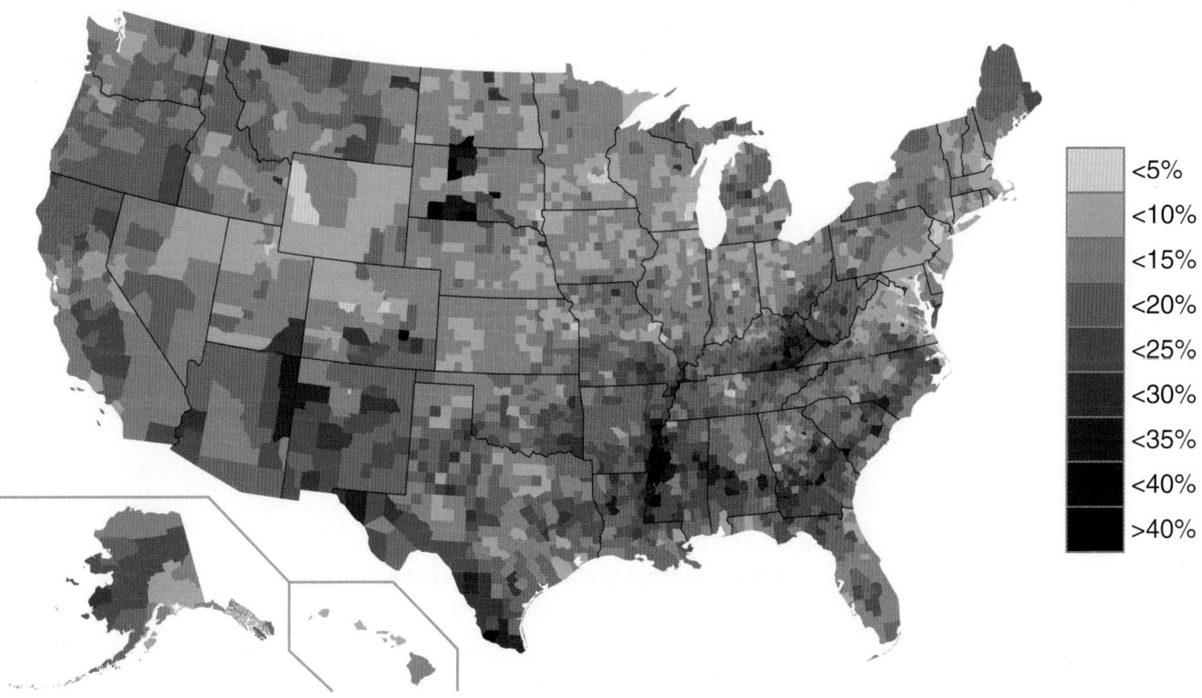

Figure 12.1 Percentage of Population Living below the Poverty Line by County, 2009

Social Exclusion

Social exclusion can arise from a sense of powerlessness when individuals who are poor or otherwise disadvantaged feel segregated and isolated from society. This feeling can create further obstacles to achieving self-help, independence, and self-respect. Disadvantaged groups, such as the racially and ethnically underrepresented, can experience magnified feelings of alienation and powerlessness when living in an affluent community. These feelings are similar to anomic conditions, which tend to further accelerate social inequality.

Spatial Inequality

Another important factor influencing poverty is the spatial setting of one's social life. In other words, where one lives plays a major role in the distribution of valuable resources. Does a person living in a low-income neighborhood in Mumbai have the same access to clean water and electricity as someone living in the posh neighborhood of London's West End? **Spatial inequality** focuses on social stratification across territories and their populations. Examining space helps to illuminate social inequalities because it attends to how geography influences social processes. Social categories such as gender, ethnicity and race, and class are distributed across spaces differently, as shown in Figure 12.2. In turn, these groups use spaces differently. For example, some cultures consider the home the center of family life, culture, and entertainment, while other cultures may view the home as merely a stop-off point for eating and sleeping while spending most of their time outside of the home.

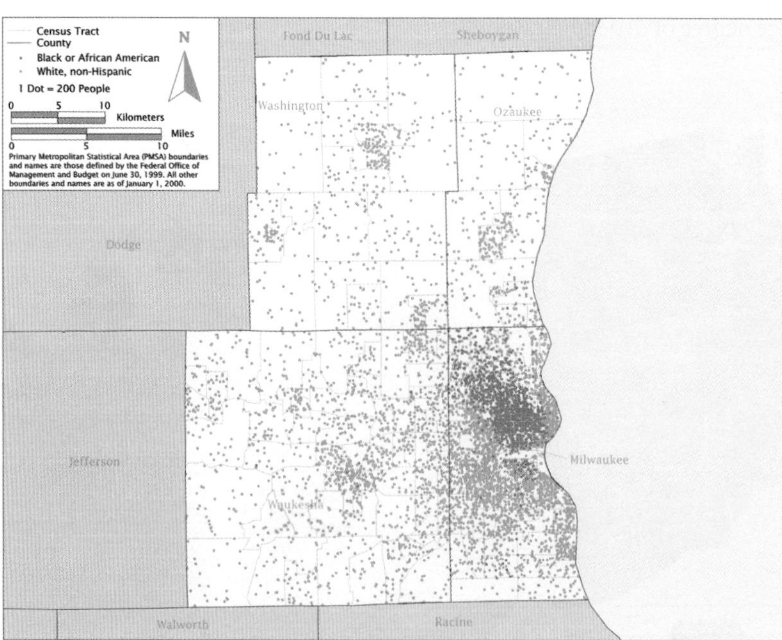

Figure 12.2 White and Black Population Distribution in Milwaukee County, WI

Space can be used to reinforce existing inequalities and can even amplify their effects, particularly poverty, leading to population segregation and the formation or expansion of destitute neighborhoods. Space can also be used to create social inequalities. In other words, social relationships between different agents, such as capitalists, laborers, the government, and citizens, result in spatially-varied social structures, built environments, and unequal regional development. For example, poorer neighborhoods tend to have less political and social influence than more affluent neighborhoods; as a result, "undesirable" buildings, like water refineries, trash-smoldering plants, and chemical manufacturers, tend to be placed in low-income areas, as shown in Figure 12.3. Citizens of these areas may lack the social resources to fight government and industry. To further understand spatial inequality, we must explore this idea on three levels: residential, environmental, and global.

Figure 12.3 Industrial Park in a Low Socioeconomic Area

Residential Segregation

Where one resides—an urban, suburban, or rural environment, and which neighborhood in that environment—has a substantial effect on how people interact, cooperate, and advance. The cultural diversity and anonymity of urban neighborhoods offer a person a greater range of opportunities than normally found in rural areas. For example, in urban environments, people are less likely to fall into their occupations and social positions because of familial ties. In rural environments, this is more likely to occur: *My parents were farmers, my grandparents were farmers, my great-grandparents were farmers; therefore, I will take up the family business when it's handed down to me.* People in urban areas tend to have more career options to choose from and can more easily improve their SES through such avenues as education, career choice, and marriage. In rural environments, these choices exist, but are less universally available. Such opportunities also do not always extend across urban environments. The neighborhood in which one lives plays a major role: affluent neighborhoods tend to have more homeowners, professionals and managers, college graduates, and higher-quality schools. Low-income neighborhoods tend to have greater poverty, unemployment rates, lower-quality schools, and higher rates of homelessness. Low-income neighborhoods are also less safe, with higher rates of violent crime, organized crime, and gang activity. These key attributes stratify neighborhoods and create unequal chances for people who live in these communities. The overall greater concentration of individuals who are poor in urban centers helps to explain **suburbanization**, or the migration pattern of the middle classes to suburban communities, as shown in Figure 12.4. The suburbs have become more attractive as they are generally cleaner and less crowded, have lower crime rates, and often have better school systems. Unlike the middle or upper classes, members of the lower class are often less able to relocate to areas that might offer them better opportunities. Many disadvantaged groups therefore remain in urban centers under poor living conditions. To make matters worse, this type of environment can easily expose low-income groups to illness and disease. Suburbanization can also lead to **urban decay**, in which a previously functional portion of a city deteriorates and becomes decrepit over time. Interestingly, this process can spontaneously reverse in the process of **urban renewal**, in which city land is reclaimed and renovated for public or private use. Urban renewal is often fueled by **gentrification**, when upper- and middle-class populations begin to purchase and renovate neighborhoods in deteriorated areas, displacing the low-SES population.

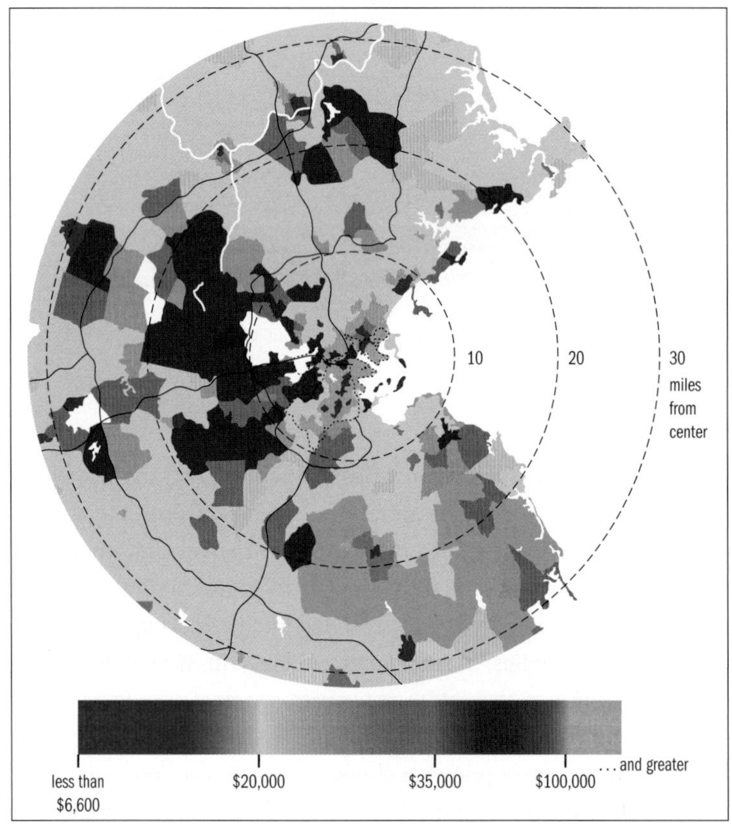

Figure 12.4 The Boston "Doughnut" of Income and Socioeconomic Status
Urban centers tend to contain lower-SES communities, surrounded by a ring of middle- to upper-SES suburbs. Further out are lower-SES exurbs and rural areas.

Environmental Justice

Poor living conditions and dangerous environmental conditions can result in an increase in illness and disease. Many low-income and underrepresented groups tend to reside closer to sites of environmental pollution because these areas are usually cheaper housing markets. Environmental risks, such as hazardous waste-producing plants and toxic waste dumps, tend to be located in low-income areas with a high concentration of groups who are racially and ethnically underrepresented. It is no surprise that inadequate housing, heating, and sanitation, in concert with toxin exposure, can contribute to acute medical problems. Illnesses such as influenza, pneumonia, substance use disorders, tuberculosis, and whooping cough are much more common among people living in poor-quality conditions. As mentioned earlier, these low-income areas also may lack the social and political power to prevent environmental risks from encroaching on their communities.

Global Inequalities

Poverty and social inequalities are not limited to hierarchies within a country. The **world system theory** categorizes countries and emphasizes the inequalities of the division of labor at the global level. **Core nations** focus on higher skills and higher paying productions while exploiting **peripheral nations** for their lower-skilled productions. **Semi-peripheral nations** are midway between the two—these nations work toward becoming core nations, while having many characteristics of peripheral nations. Much of the world, especially within semi-peripheral and peripheral nations, lives on less than the equivalent of $1.25 per day. Such rates of poverty are especially prevalent in parts of India, sub-Saharan Africa, and South Central Asia, as shown in Figure 12.5.

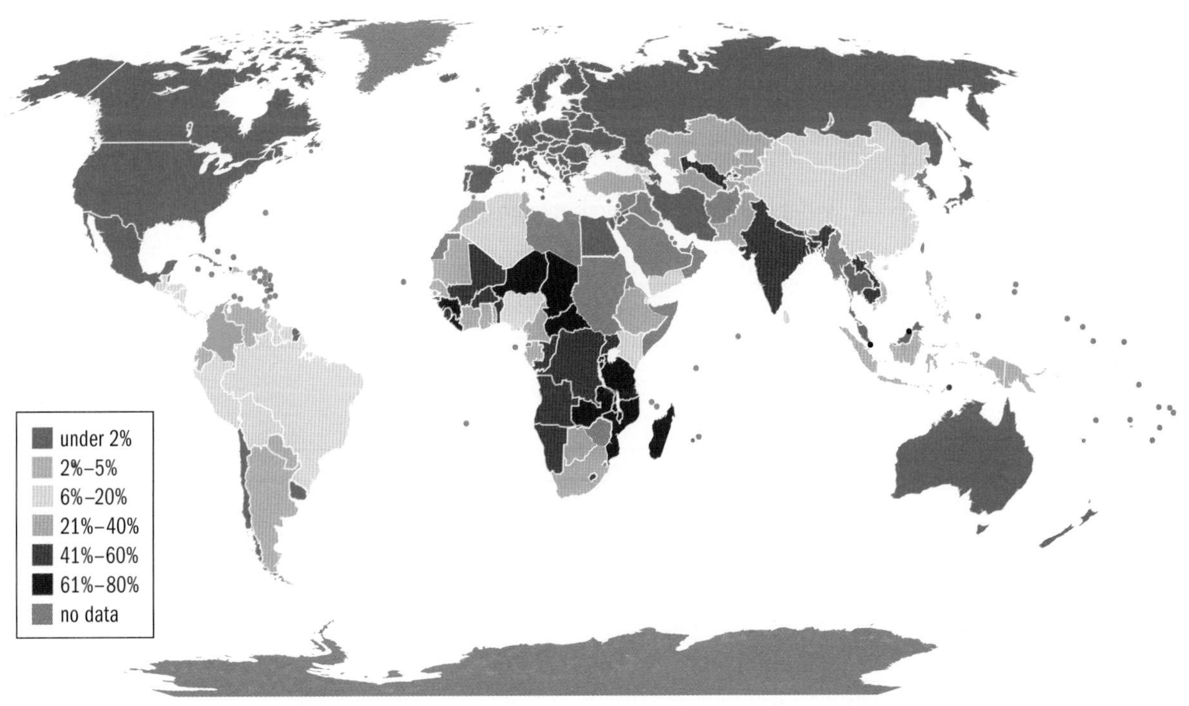

Figure 12.5 Percentage of Population Living on Less than $1.25 per Day, 2008

Largely due to the effects of globalization, massive restructuring of industry and trade patterns have had a major impact on local communities, specifically because of the production of cheap goods at suppressed rates for the global market. This reconfiguration obstructs or limits access to power and resources as the production of goods constantly shifts from location to location. Social inequalities have increased on a worldwide level as local communities become more and more subject to the ebb and flow of the global market. Since the advent of globalization, with the development of world cities, international communication chains, and global immigration, interaction between industrialized and developing nations has had more of an impact on peoples and regions within the state, and has thereby led to further inequalities in space, food and water, energy, housing, and education. Global inequality has been further exacerbated by an unprecedented large population spike, placing strain on the world's resources. The majority of the world also has limited access to healthcare. Consequently, many people around the globe suffer from malnutrition and parasitic and infectious diseases, and have higher rates of morbidity and mortality.

MCAT CONCEPT CHECK 12.1

Before you move on, assess your understanding of the material with these questions.

1. How is socioeconomic status (SES) determined?

   ____

2. How does social capital affect social cohesion?

   ____

3. What are some groups that suffer disproportionate social inequality?

   ____

4. What is the relationship between merit and social mobility?

   ____

5. With regard to health, which groups are most often affected by environmental hazards?

   ____

12.2 Epidemiology and Disparities

LEARNING OBJECTIVES

After Chapter 12.2, you will be able to:

- Identify the causes of Waitzkin's "second sickness"
- Explain why women are more likely to have better health profiles than men
- Describe the links between class, ethnicity, and healthcare disparities

An old saying intones that *Your health is your wealth.* This same correlation certainly holds true in reverse: the wealthier tend to have better health and better access to healthcare. Wherever there is low social capital, high urban degradation, interpersonal violence, and low social trust, the social environment is poor, and there is less protection against disease. As a result, class gradients often increase. Poor health conditions and lower life expectancy, as shown in Figure 12.6, are some of the many consequences of social stratification. Low-income groups are significantly worse off than the middle or upper classes when it comes to health disparities, meaning they tend to be sicker than others. **Social epidemiology** is a branch of epidemiology that studies the ways in which health and disease correlate to social advantages and disadvantages.

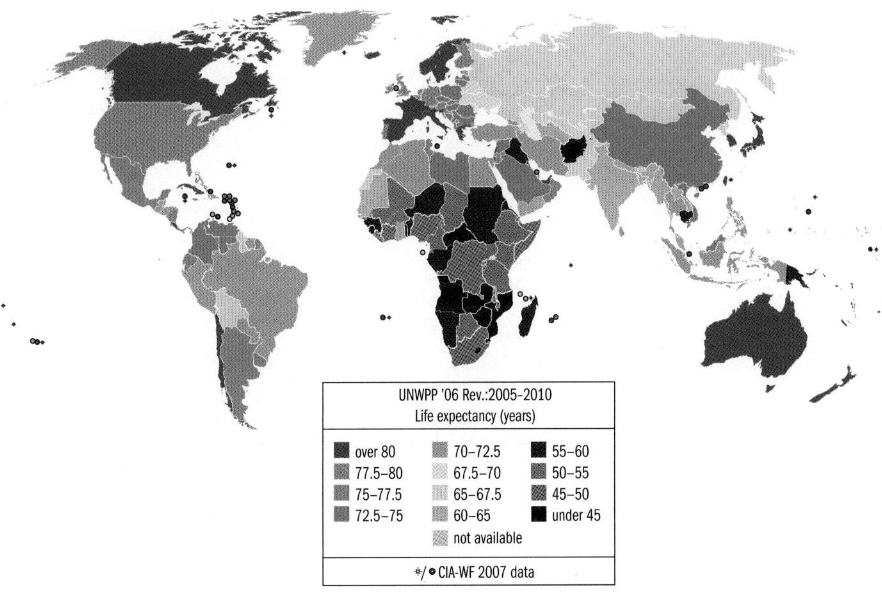

Figure 12.6 Life Expectancy at Birth, 2005-2010

To understand health statistics, it is important to define a few epidemiological terms. **Incidence** is defined as the number of new cases of an illness per population at risk in a given amount of time; for example, the number of new cases of lung cancer per 1000 at-risk people per year. **Prevalence** is a measure of the number of cases of an illness overall—whether new or chronic—per population in a given amount of time; for example, the number of people with new or chronic

KEY CONCEPT

Epidemiology statistics:

- Incidence = new cases/population at risk/time
- Prevalence = total cases/total population/time

Note that incidence is relative to the population at risk, not the total population; if you already have the illness, you are no longer at risk!

lung cancer per 1000 people per year. Health statistics are also given in terms of morbidity and mortality. **Morbidity** is the burden or degree of illness associated with a given disease, while **mortality** refers to deaths caused by a given disease.

Inequities in Health

Science has clearly demonstrated that poor environmental and social factors negatively impact health. This correlation was first demonstrated in the nineteenth century through public health efforts. One landmark example linking geography with disease was John Snow's investigation of a cholera outbreak in London in 1854. Snow tracked cases of cholera on a map, as shown in Figure 12.7, and was able to deduce that a water pump in the neighborhood was causing the spread of the infectious agent (a bacterium called *Vibrio cholerae*).

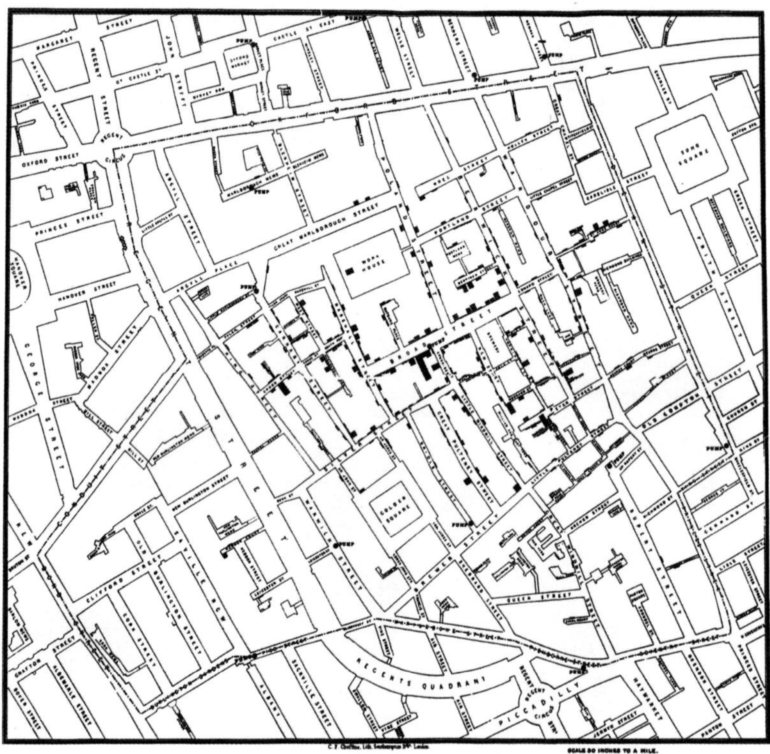

Figure 12.7 John Snow's Map of Disease Cases during the 1854 Cholera Outbreak in London

By tracing the geography of the disease, Snow deduced that a water pump was responsible for cholera transmission.

Health is dependent not only on geography, but also on social and economic factors. Over time, socioeconomic improvements lead to greater general health in the population, and the best health outcomes are generally seen in egalitarian societies. However, despite the ambitions of the modern **welfare state** (the system of government that protects the health and well-being of its citizens), the Black Report of 1980 showed that class differences in health still exist, with professional groups having longer life expectancies than working-class people. Howard Waitzkin described this outcome as the **second sickness**, which is an exacerbation of health outcomes caused by social injustice. As the Centers for Disease Control and Prevention (CDC) have shown, low-income groups are more likely to have poorer health, be uninsured, and die younger than middle- or upper-class adults. Poverty, in combination with a culture of inequality, leads to worse health outcomes, and this effect runs across gender, age, and racial and ethnic boundaries. For example, low-income women are more likely to deliver babies with low birth weights, thereby placing these babies at risk for numerous physical and cognitive problems in life. Similarly, impoverished members of racially and ethnically underrepresented groups have lower life expectancies. Members of the lower class, overall, are four times more likely to view themselves in worse health compared with affluent groups. Low-income groups are much more likely to develop life-shortening diseases such as lung cancer, diabetes, heart disease, and other degenerative illnesses. These groups are also more likely to commit suicide and die from homicide in comparison to wealthier adults. The infant mortality rate among the poor is also much higher; in some populations of the United States, the infant mortality rate can approximate that of developing countries. However, because of the correlation between poverty and racially and ethnically underrepresented groups, many of these characteristics apply to particular ethnic groups more than others.

When it comes to health and illness among racially and ethnically underrepresented groups, Asian Americans and Pacific Islanders have some of the best health profiles. Reports illustrate that, in comparison to White Americans, these groups have a lower rate of death associated with cancer, heart disease, diabetes, and infant mortality. African Americans appear to have a worse health profile in comparison to White Americans, showing higher rates of death linked to cancer, heart disease, diabetes, drug and alcohol use, infant mortality, and HIV/AIDS. African American infants have twice the infant mortality rate of White infants. Specifically, African American males have the lowest life expectancy of any racial or gender category. Hispanic Americans have a mixed profile in comparison to White Americans, in that they have lower mortality rates attributable to cancer, heart disease, and infant mortality, but higher mortality rates attributable to diabetes, alcohol and drug use, and HIV/AIDS. Hispanic Americans also have a high mortality rate from influenza, pneumonia, and accidents. American Indians also have a mixed profile in this regard, showing higher rates of death from diabetes, alcohol and drug use, and infant mortality, but lower mortality rates compared to White Americans from cancer, heart disease, and HIV/AIDS. American Indians also show some of the highest rates of death by suicide in comparison to the general population. This group also has some of the highest mortality rates linked to diabetes compared to any racial category.

When it comes to gender-related health disparities, most statistical information shows that women have better health profiles than men. This trend is true throughout the world. Female life expectancy has been consistently higher than male life expectancy since records began. While the gap in life expectancy is beginning to narrow in the United States, most countries still have higher life expectancies for female citizens than male citizens, as shown in Figure 12.8.

MCAT EXPERTISE

The MCAT will not expect you to be able to rattle off the relative rates of these illnesses across racial groups, but a sensitivity to these differences between groups may be important in passages related to sociology and public health.

KEY CONCEPT

Low-income groups, especially those who are racially and ethnically underrepresented, have an overall worse health profile in terms of morbidity and mortality rates.

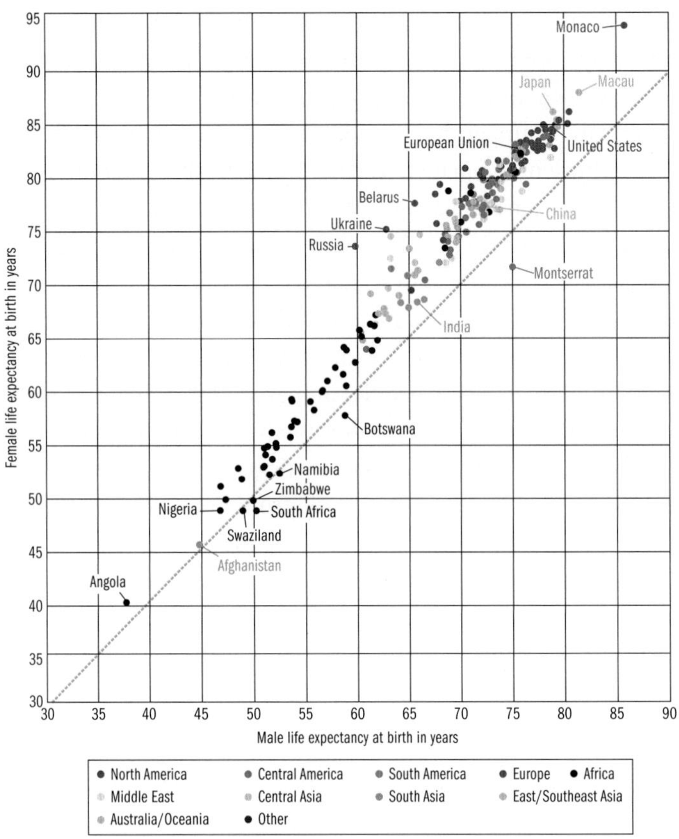

Figure 12.8 Male and Female Life Expectancies by Region and Country *Note that female life expectancies are almost universally higher than male life expectancies. Based on data from the CIA World Factbook.*

Mortality rates from heart disease, cancer, chronic lower respiratory diseases, and diabetes are higher for men than women. Men are also three times more likely than women to die from accidents, suicide, and homicide. Men are far less likely to seek medical attention than women. Men, especially those raised to have hypermasculine behaviors, may try to "tough it out" rather than go to a doctor. When men do seek healthcare, they are less likely to comply with medical instructions or adhere to medical advice.

KEY CONCEPT

In comparison to women, men tend to have worse mortality rates. However, women have higher morbidity rates. Differences in male and female health profiles are both biologically and sociologically determined, the latter being the result of factors like risk-taking behavior, hypermasculinity, and dangerous employment.

While women show better mortality rates, this is not the case when it comes to morbidity rates for certain acute and chronic diseases. More women than men suffer from infectious and parasitic diseases, digestive problems, respiratory conditions, high blood pressure (hypertension), arthritis, diabetes, and inflammatory bowel diseases (colitis). Women tend to suffer more from illnesses and disabilities than men, but their conditions are less often life threatening.

The reasons that men tend to have lower life expectancy rates are both sociological and biological. Sociologically, men are considered to be bigger risk takers, and therefore more likely to expose themselves to accidents and unintentional injuries. This is especially true of young men. Men are also more likely to be employed in dangerous jobs, such as the police force, steel industry, and coal mining. Men also have higher rates of alcohol use, speeding, and participation in potentially violent sports. Biologically speaking, men are at a disadvantage from infancy onward. Throughout life, men are more likely to come down with diseases that are life threatening.

Inequities in Healthcare

While the United States has one of the most advanced healthcare systems in the world, quality healthcare and services are not always extended to all. Like many institutions, quality healthcare favors those in higher social classes. Many people are frustrated by the way healthcare is delivered in the United States, especially with regard to health insurance. Further, the healthcare system of the United States is one of the few among industrialized nations that is not organized and planned by a central (governmental) system.

Passage of the Affordable Care Act (ACA) in 2010 was an attempt in the United States to increase coverage and affordability of insurance for all Americans, and also to reduce the overall costs of healthcare. Medicare and Medicaid are also programs that attempt to increase access to healthcare in the United States. **Medicare** covers patients over the age of 65, those with end-stage renal disease, and those with amyotrophic lateral sclerosis (ALS). **Medicaid** covers patients who are in significant financial need. However, disadvantaged groups, especially those who are poor, are still affected by disparities in healthcare both in terms of access and quality. Even those individuals who have Medicare or Medicaid may lack access, as many physicians will not accept such public insurance programs. Additionally, some doctors will not open practices in low-income neighborhoods, making access even harder for populations with low socioeconomic status. Consequently, individuals in the lower class are less likely to seek medical assistance until they are seriously ill; by then, intervention may be too late.

KEY CONCEPT

Medicare covers patients over 65, those with end-stage renal disease, and those with amyotrophic lateral sclerosis (ALS). Medicaid covers patients in significant financial need.

Some of the primary reasons low-income groups have higher mortality rates include poor access to quality medical care, poor nutrition, and feeling less in control of life circumstances. Those who are poor are more likely to smoke and be overweight or obese; they are less likely to engage in physical activity. In addition to socioeconomic status, race and ethnicity can create barriers to care. In race-concordant patient-physician relationships, the patient and physician are of the same race, whereas they are of different races in race-discordant relationships. Sometimes, culture and non-native language are viewed as contributors to pathology because they can act as obstacles to diagnosis and treatment. Also, despite efforts to systemically address unequal treatment of underrepresented populations by physicians, there are still inequalities and disparities in treatment relative to race and ethnicity over a wide range of medical specialties. In other words, underrepresented and low-income groups tend to face greater barriers to care, and poorer quality of care when they receive it. To be specific, it has been demonstrated that African Americans, Asian Americans, American Indians, and Hispanic Americans receive worse care than White Americans.

Quality of preventative care, acute treatment, and chronic disease management also differ regionally. States in New England and the Mid-Atlantic are shown to be in the top quartile of healthcare quality while states in the South are in the bottom quartile.

Outside of race and ethnicity, other identifiable characteristics may have a large impact on how patients are treated by their providers. One of the most common biases is discrimination against overweight and obese patients. This bias can apply to any individual who is overweight, regardless of socioeconomic status, gender, age, and racial and ethnic background, although there is a higher prevalence of obesity in low-income groups. Doctors may assume that being overweight or obese is the only cause of a patient's health problems, which can sometimes lead to misdiagnoses and ineffective treatments. This, in turn, damages the trust necessary to form a strong doctor–patient relationship. As a result, patients who are overweight or obese are more likely than other patients to switch doctors repeatedly. When one does not have a consistent primary care doctor, continuity of care is nearly nonexistent. Additionally, patients who are overweight or obese are less likely to have quality preventative care and screenings, including screenings for breast and colon cancer.

In terms of gender, women tend to be favored by the healthcare system. As a whole, women tend to fare better when it comes to accessing healthcare, largely because women are more likely to be insured. Women tend to utilize healthcare services more than men, with more examinations, blood pressure checks, lab tests, drug prescriptions, and physician visits per year. Women also receive more services per visit than men do. Healthcare use is likely more common among women due to higher morbidity rates among women for many illnesses, thereby facilitating the need to seek medical attention. On the other hand, women are more likely to be delayed or unable to obtain necessary medical care, dental care, and prescription medicines. Also, for LGBTQ individuals, discrimination and decreased access to healthcare are quite common, often due to homophobia and other prejudices.

MCAT CONCEPT CHECK 12.2

Before you move on, assess your understanding of the material with these questions.

1. What is Waitzkin's second sickness?

____

2. What is the relationship between class, ethnicity, and health?

____

3. Why are women more likely to have better health profiles than men?

____

4. What are some of the factors that contribute to healthcare disparities between classes?

____

Conclusion

So what have we learned? We do not live in a perfect world where valuable materials and resources are unlimited and evenly distributed. In the United States and on a worldwide level, social stratification is an unfortunate reality. In a class-based economic system such as ours, status and power are inextricably linked, which can either facilitate or hinder access to social capital and its associated rewards. This is especially the case for certain disadvantaged groups based on categories of class, race, gender, and age. While not perfect, our class system does tend to allow for upward social mobility either in one's lifetime or across generations. At the heart of America's socioeconomic values is the principle of meritocracy, which means that through hard work, credentials, and dedication, one can move up in society. However, many people in the United States remain impoverished. Social inequality and social exclusion make it increasingly difficult for low-income groups to improve their socioeconomic condition. Some hold that social inequalities such as poverty remain because these inequalities are passed down from one generation to another. While there is some truth to the social reproduction of poverty, one's situational context also plays a role through spatial inequality. Where one lives in this world has an impact on one's position in life, especially in terms of accessing key resources and prosperous opportunities. Spatial inequality remains at the residential, environmental, and global level.

But how do such social inequalities influence health and healthcare disparities? Well, as is the case with many facets of life, wealth matters. Those with greater income typically have access to better quality healthcare. This is especially the case in the U.S. healthcare system because of uneven levels of coverage and high healthcare costs. Socioeconomic status or class greatly impacts one's ability to navigate the system and procure healthcare in the United States. Low-income racially and ethnically underrepresented groups tend to be worse off, having both poorer health and poorer access to healthcare. Women, despite being more prone to chronic and degenerative diseases, tend to fare better than men when it comes to overall health and accessing and utilizing healthcare resources.

While the U.S. healthcare system is undergoing a significant reorientation and taking on a more preventative approach, it is more important now than ever to place a greater emphasis on sociological issues to understand the relationship between social stratification and health and healthcare disparities. Illness and disease are a product of social as well as psychological and physiological issues. This chapter is the last chapter in *MCAT Behavioral Sciences Review*; you have therefore covered all of the psychology and sociology content required for the MCAT. This is a fitting chapter to finish this discussion, and we leave you with a charge: as you prepare for the MCAT, medical school, and life as the physician you deserve to be, think about the changes that are needed in the U.S. healthcare system. Serve your local, national, and international community and improve the health status of those around you, while contributing to a future where everyone can access quality healthcare.

GO ONLINE

You've reviewed the content, now test your knowledge and critical thinking skills by completing a test-like passage set in your online resources!

CONCEPT SUMMARY

Social Class

- Social stratification is based on **socioeconomic status** (**SES**). Socioeconomic status depends on ascribed status and achieved status.
  - **Ascribed status** is involuntary and derives from clearly identifiable characteristics, such as age, gender, and skin color.
  - **Achieved status** is acquired through direct, individual efforts.
- A **social class** is a category of people with shared socioeconomic characteristics. The three main social classes are upper, middle, and lower class. These groups also have similar lifestyles, job opportunities, attitudes, and behaviors.
- **Prestige** is the respect and importance tied to specific occupations or associations.
- **Power** is the capacity to influence people through real or perceived rewards and punishments. It often depends on the unequal distribution of valued resources. Power differentials create social inequality.
- **Anomie** is a state of normlessness. Anomic conditions erode social solidarity by means of excessive individualism, social inequality, and isolation.
- **Social capital** is the investment people make in their society in return for economic or collective rewards. Social networks, either situational or positional, are one of the most powerful forms of social capital and can be achieved through establishing strong and weak social ties.
- **Meritocracy** refers to a society in which advancement up the social ladder is based on intellectual talent and achievement.
- **Social mobility** allows one to acquire higher-level employment opportunities by achieving required credentials and experience. Social mobility can either occur in a positive upward direction or a negative downward direction depending on whether one is promoted or demoted in status.
- **Poverty** is a socioeconomic condition. In the United States, the poverty line is determined by the government's calculation of the minimum income requirements for families to acquire the minimum necessities of life.
- **Social reproduction** refers to the passing on of social inequality, especially poverty, from one generation to the next.
- Poverty can either be absolute or relative.
  - **Absolute poverty** is when people do not have enough resources to acquire basic life necessities, such as shelter, food, clothing, and water.
  - **Relative poverty** is when one is poor in comparison to a larger population.
- **Social exclusion** is a sense of powerlessness when individuals feel alienated from society.

- **Spatial inequality** is a form of social stratification across territories and their populations, and can occur along residential, environmental, and global lines.
  - Urban areas tend to have more diverse economic opportunities and more ability for social mobility than rural areas. Urban areas also tend to have more neighborhoods that are low-income and racially and ethnically underrepresented than do rural areas.
  - Formation of higher-income suburbs is a common occurrence, and is due in part to the limited mobility of lower-income groups in urban centers.
  - **Environmental injustice** refers to an uneven distribution of environmental hazards in communities. Lower-income neighborhoods may lack the social and political power to prevent the placement of environmental hazards in their neighborhoods.
- **Globalization** has led to further inequalities in space, food and water, energy, housing, and education as the production of goods shifts to cheaper and cheaper labor markets. This has led to significant economic hardship in industrializing nations.

Epidemiology and Disparities

- **Incidence** is calculated as the number of new cases of a disease per population at risk in a given period of time: for example, new cases per 1000 at-risk people per year.
- **Prevalence** is calculated as the number of cases of a disease per population in a given period of time: for example, cases per 1000 people per year.
- **Morbidity** is the burden or degree of illness associated with a given disease.
- **Mortality** refers to deaths caused by a given disease.
- Health is dependent on geographic, social, and economic factors.
  - The **second sickness** refers to an exacerbation of health outcomes caused by social injustice.
  - Poverty is associated with worse health outcomes, including decreased life expectancy, higher rates of life-shortening diseases, higher rates of suicide and homicide, and higher infant mortality rates.
  - Certain racially and ethnically underrepresented groups have worse health profiles than others. African Americans have the worst health profiles; White Americans, American Indians, and Hispanic Americans have health profiles in the middle; and Asian Americans and Pacific Islanders have the best health profiles.
  - Females have better health profiles than males, including higher life expectancy, lower rates of life-threatening illnesses, and higher rates of accessing and utilizing health resources. However, females have higher rates of chronic diseases and higher morbidity rates.
- Efforts to improve healthcare for underserved populations include the **Affordable Care Act** (ACA) and the **Medicare** and **Medicaid** programs.

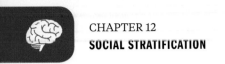

- Healthcare access and quality differ across the population.
  - Low-income groups and racially and ethnically underrepresented groups (specifically, African Americans, Asian Americans, American Indians, and Hispanic Americans) receive worse care than high-income groups and White Americans, respectively.
  - Biases against overweight or obese patients are associated with lower-quality treatment, including less preventative care and fewer screenings.
  - Women tend to have better access to healthcare and utilize more healthcare resources than men.
  - LGBTQ individuals may have barriers to care due to prejudices, discrimination, homophobia, and transphobia.

ANSWERS TO CONCEPT CHECKS

12.1

1. SES is determined by two factors: it can be ascribed according to physical or external characteristics such as age, gender, or skin color, or acquired through direct efforts such as hard work or merit.
2. The less social capital a person has (reduced network equality and equality of opportunity), the more social inequality. This, in turn, decreases social cohesion.
3. Social inequality is highest among racially and ethnically underrepresented groups (especially African Americans and Hispanic Americans), households headed by women, and older adults. It is also most prevalent among those living in poverty.
4. Social mobility can be dependent on intellectual talent and achievement (meritocratic competition) but can also be obstructed by concentrated power as well as discrimination based on ethnicity, gender, age, or other identifiable characteristics.
5. Environmental hazards tend to be located in low-income areas with a higher population of racially and ethnically underrepresented groups. Poor living conditions can result in increased illness and disease among these groups of people.

12.2

1. The second sickness refers to the fact that health outcomes are exacerbated by social inequalities and social injustice. As a result, higher-income groups have longer life expectancies than lower-income groups.
2. Low-income racially and ethnically underrepresented groups have higher morbidity rates and overall worse health compared to the middle and upper classes. The lower class also has higher infant mortality rates, homicide rates, and suicide rates in comparison to wealthier classes.
3. Women typically have longer life expectancies and an overall better health profile in comparison to men. This can be attributed to both biological and sociological causes: women are less likely to have life-threatening conditions, although they do have higher morbidity rates. Women are also more likely to seek care and to utilize healthcare services than men.
4. Low-income groups have less access to healthcare services and often experience lower quality healthcare. Poor Americans are less likely to be insured and consequently are less likely to seek medical attention until conditions have become extremely serious, thereby limiting available interventions. As a result, morbidity and mortality rates are highest among low-income groups.

SCIENCE MASTERY ASSESSMENT EXPLANATIONS

1. **D**

Social stratification based on direct efforts, such as merit, is a form of achieved socioeconomic status. Ascribed socioeconomic status, **(A)**, is based on identifiable external characteristics. Achieved status may be due to meritocratic competition, **(B)**, but other individual efforts can also be associated with achieved status. Anomic conditions, **(C)**, are those that cause a breakdown between the individual and society and erode social solidarity.

2. **C**

Low social class may lead to low social capital. Members of the lower class often have smaller numbers of weak ties in social networks, and therefore have less opportunity to invest in society and reap its benefits.

3. **D**

Globalization does not typically lead to global equality; rather, globalization tends to create further global inequalities. In regard to environmental justice, higher numbers of hazardous waste facilities tend to be found in low-income neighborhoods, **(A)**. Poor living conditions tend to be associated with greater health problems, including tuberculosis, **(B)**. Finally, environmental pollution is more prevalent in areas with larger populations of underrepresented groups, especially when low-income, **(C)**.

4. **B**

Groups with high income tend to have increased life expectancy rates, not decreased. People with low income have higher mortality rates than those with high income, **(A)**. Low-income women tend to have children with lower birth weights, **(C)**. Finally, rates of various diseases, including lung cancer, are increased among low-income groups, **(D)**.

5. **D**

Prevalence is defined as the total number of cases divided by the possibly affected population during a period of time. Here, the period of time is defined as one point: the end of the year. At the end of the year, there are 30 total cases in a population of 1000 individuals, but only 500 of those individuals can develop prostate cancer, meaning the prevalence is $30 \div 500$.

6. **A**

Incidence is defined as the total number of new cases divided by the at-risk population during a period of time. Here, the period of time is one year. There were 10 new cases in this year, and the at-risk population will be only the males who do not already have prostate cancer; the 20 males already diagnosed and the 500 females should not be included in the at-risk population. Therefore, the incidence in this population is $10 \div 480$.

7. **D**

In this scenario, the children remain in the same socioeconomic class as their parent, indicating a lack of social mobility, **(A)** and **(B)**. Rather, this is an example of social reproduction, in which social inequality, especially poverty, is passed from one generation to the next.

8. **C**

Relative poverty is a comparative term: it describes being poorer than those in the surrounding population. Members of the upper class can live in relative poverty compared to others in their neighborhood if they are not as well-off as their neighbors. Relative poverty is not directly associated with upward or downward social mobility, eliminating **(B)** and **(D)**; individuals living in relative poverty could exhibit mobility in either direction or no social mobility at all.

9. **C**

Suburbs tend to have larger upper- and middle-class populations than urban centers; urban centers tend to have larger low-socioeconomic status populations than suburbs. This is due, in part, to the increased mobility seen in upper- and middle-class populations, which permits their migration into the suburbs.

10. **A**

Morbidity refers to the burden of illness, or the severity or degree of illness. Mortality, **(B)**, refers to deaths caused by a given illness. Second sickness, **(C)**, is a term used to describe the exacerbation of health outcomes due to social injustice. Chronicity, **(D)**, refers to the duration of a disease, not its severity or significance for the patient.

11. **B**

In comparison to White Americans, Asian Americans tend to have better overall health profiles. African Americans, **(A)**, tend to have worse overall health profiles. Hispanic Americans and American Indians, **(C)** and **(D)**, both have mixed health profiles in comparison to White Americans: they are better off in some categories and worse off in others. However, Hispanic Americans and American Indians do not have better overall health profiles than White Americans.

12. **C**

Medicare covers patients over the age of 65 (older age groups), those with end-stage renal disease, and those with amyotrophic lateral sclerosis (ALS). Medicaid covers patients below a certain socioeconomic level.

13. **C**

Morbidity refers to the burden or severity of disease. All of the factors listed are true with regard to low-socioeconomic status populations; however, high homicide rates cause increases in mortality, not morbidity.

14. **A**

If the threshold for hypertension (high blood pressure) were lowered, more individuals would be fit the criteria for the disease. If the number of individuals with the disease increases and the population stays the same overall, there will be an increased prevalence of the disease.

15. **B**

In comparison to females, males visit primary care doctors less frequently. All of the other trends listed here have been documented.

Consult your online resources for additional practice. GO ONLINE

SHARED CONCEPTS

Behavioral Sciences Chapter 5
Motivation, Emotion, and Stress

Behavioral Sciences Chapter 6
Identity and Personality

Behavioral Sciences Chapter 7
Psychological Disorders

Behavioral Sciences Chapter 8
Social Processes, Attitudes, and Behavior

Behavioral Sciences Chapter 10
Social Thinking

Behavioral Sciences Chapter 11
Social Structure and Demographics

GLOSSARY

Absolute poverty–Poverty wherein people do not have enough resources to acquire basic life necessities such as shelter, food, clothing, and water.

Absolute threshold–The minimum stimulus energy needed to activate a sensory system.

Accommodation–Process by which existing schemata are modified to encompass new information.

Acetylcholine–A neurotransmitter associated with voluntary muscle control.

Achieved status–A status gained as a result of direct, individual action.

Acquisition–In classical conditioning, the process of taking advantage of reflexive responses to turn a neutral stimulus into a conditioned stimulus.

Actor-observer bias–The tendency to make situational attributions about the self, but dispositional attributions about others, regarding similar behaviors.

Adaptation–In perception, a decrease in stimulus perception after a long duration of exposure; in learning, the process by which new information is processed; consists of assimilation and accommodation.

Adaptive value–The extent to which a trait benefits a species by influencing the evolutionary fitness of the species.

Affect–The experience and display of emotion. Can be described as a scale, with both positive and negative affect having separate scales.

Afferent neuron–Sensory neurons which transmit information to the brain from the body in response to sensory input.

Ageism–Prejudice or discrimination on the basis of a person's age.

Agent of socialization–Any part of society that is important when learning social norms and values.

Aggression–A behavior with the intention to cause harm or increase relative social dominance; can be physical or verbal.

Agnosia–The loss of the ability to recognize objects, people, or sounds, though typically just one of the three.

Alcohol myopia–The inability to think about consequences and possible outcomes of one's actions due to alcohol intoxication.

Alertness–State of consciousness in which one is aware, able to think, and able to respond to the environment; nearly synonymous with arousal.

Algorithm–A formula or procedure for solving a certain type of problem.

Aligning actions–An impression management strategy in which one makes questionable behavior acceptable through excuses.

Alter-casting–An impression management strategy in which one imposes an identity onto another person.

Altruism–A form of helping behavior in which people's intent is to benefit someone else at a cost to themselves.

Alzheimer's disease–Degenerative brain disorder that is characterized by dementia and memory loss. Neurofibrillary tangles and β-amyloid plaques are phenomena found in the brains of Alzheimer's patients.

Amphetamine–A central nervous system stimulant that increases activity of both dopamine and norepinephrine in the brain.

Amygdala–A portion of the limbic system that is important for memory and emotion, especially fear.

Anomie–A state of normlessness; anomic conditions erode social solidarity by means of excessive individualism, social inequality, and isolation.

Anterograde amnesia–Form of memory loss in which new long-term memories cannot be established.

Anxiety disorders–Disorders that involve worry, unease, fear, and apprehension about future uncertainties based on real or imagined events that can impair physical and psychological health.

Aphasia–Deficit of language production or comprehension.

Appraisal model–A similar theory to the basic model, accepting that there are biologically predetermined expressions once an emotion is experienced; accepts that there is a cognitive antecedent to emotional expression.

Archetype–In Jungian psychoanalysis, a thought or image that has an emotional element and is a part of the collective unconsciousness.

Arcuate fasciculus–A bundle of axons that connects Wernicke's area (language comprehension) with Broca's area (motor function of speech).

Arousal–A psychological and physiological state of being awake or reactive to stimuli; nearly synonymous with alertness.

Arousal theory–A theory of motivation that states there is a particular level of arousal required in order to perform actions optimally; summarized by the Yerkes–Dodson law.

Ascribed status–A status that one is given at birth, such as race, ethnicity, or sex.

Assimilation–In psychology, the process by which new information is interpreted in terms of existing schemata; in sociology, the process by which the behavior and culture of a group or an individual begins to merge with that of another group.

Associative learning–The process by which a connection is made between two stimuli or a stimulus and a response; examples include classical conditioning and operant conditioning.

Attachment–An emotional bond to another person, particularly a parent or caregiver. The four main attachment styles are: secure, avoidant, ambivalent, and disorganized.

Attitude–A tendency toward expression of positive or negative feelings or evaluations of a person, place, thing, or situation.

Attribute substitution–A phenomenon observed when individuals must make judgments that are complex but instead substitute a simpler solution or perception.

Attribution theory–A theory that focuses on the tendency for individuals to infer the causes of other people's behavior.

Auditory cortex–Region of the temporal lobe devoted to sound processing.

Auditory pathway–After entering the brain, sound is processed by several regions, including the MGN, auditory cortex, superior olive, and inferior colliculus.

Authentic self–Who someone actually is, including both positive and negative attributes.

Automatic processing–The brain process most closely resembling autopilot, enabling performance of multiple activities at the same time.

Autonomic nervous system–The involuntary branch of the peripheral nervous system that controls involuntary functions such as heart rate, bronchial dilation, temperature, and digestion.

Autonomy–The ethical tenet that the physician has the responsibility to respect patients' choices about their own healthcare.

Availability heuristic–A shortcut in decision making that relies on the information that is most readily available, rather than the total body of information on a subject.

Avoidance learning–A form of negative reinforcement in which one avoids the unpleasantness of something that has yet to happen.

Babbling–Precursor to language known to spontaneously occur in children.

Back stage–In the dramaturgical approach, the setting where players are free from their role requirements and not in front of the audience; back stage behaviors may not be deemed appropriate or acceptable and are thus kept invisible from the audience.

Barbiturate–A drug that acts as a central nervous system depressant; often used for anxiety, for insomnia, and as an antiseizure medication.

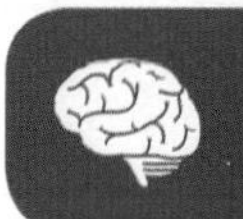

Basal ganglia–A portion of the forebrain that coordinates muscle movement and routes information from the cortex to the brain and spinal cord.

Base-rate fallacy–Using prototypical or stereotypical factors while ignoring actual numerical information when making a decision.

Basic model–First established by Charles Darwin, a theory that states that emotional expression involves a number of systems: facial expression as well as behavioral and physical responses; claims that emotions are universal and should be similar across cultures.

Behaviorism–B. F. Skinner's theory that all behaviors are conditioned. Behaviorism can be applied across many bodies of psychological thought, including theories of development, of identity, and of personality.

Belief–An acceptance that a statement is true or that something exists.

Belief perseverance–The inability to reject a particular belief despite clear evidence to the contrary.

Beneficence–The ethical tenet that the physician has a responsibility to act in the patient's best interest.

Benzodiazepine–A central nervous system depressant that is often used to reduce anxiety or promote sleep.

Biomedical approach–An approach to psychological disorders that considers only pathophysiological causes and offers pharmaceutical and medical solutions for symptom alleviation.

Biopsychosocial approach–An approach to psychological disorders that considers conditions and treatments to be dependent on biological, psychological, and social causes. Treatment under this approach includes both direct and indirect therapy.

Bipolar disorders–Class of mood disorders characterized by both depression and mania.

Birth rate–The number of births per population in a period of time; usually the number of births per 1000 people per year.

Bisexual–A sexual orientation wherein individuals are attracted to members of multiple sexes.

Bottom-up processing–Object recognition by parallel processing and feature detection in response to sensory stimuli.

Brainstem–The most primitive portion of the brain, which includes the midbrain and hindbrain; controls the autonomic nervous system and communication between the spinal cord, cranial nerves, and brain.

Broca's aphasia–Loss of the motor function of speech, resulting in intact understanding with an inability to correctly produce spoken language.

Broca's area–A brain region located in the inferior frontal gyrus of the frontal lobe (usually in the left hemisphere); largely responsible for the motor function of speech.

Bureaucracy–A formal organization with the goal of performing complex tasks as efficiently as possible by dividing work among a number of bureaus.

Bystander effect–The observation that, when in a group, individuals are less likely to respond to a person in need.

Cannon–Bard theory–A theory of emotion that states that a stimulus is first received and is then simultaneously processed physiologically and cognitively, allowing for the conscious emotion to be experienced.

Cataplexy–Loss of muscle control with intrusion of REM sleep during waking hours, usually caused by an emotional trigger.

Catatonia–Disorganized motor behavior characterized by various unusual physical movements or stillness.

Central nervous system (CNS)–The portion of the nervous system composed of the brain and spinal cord.

Cerebellum–A portion of the hindbrain that maintains posture and balance and coordinates body movements.

Cerebral cortex–The outermost layer of the cerebrum, responsible for complex perceptual, behavioral, and cognitive processes.

Cerebrospinal fluid (CSF)–An aqueous solution in which the brain and spinal cord rest; produced by cells lining the ventricles of the brain.

Cerebrum–A portion of the brain that contains the cerebral cortex, limbic system, and basal ganglia.

Characteristic institution–The social structure or institution about which societies are organized.

Chemoreceptors–Sensory neurons that respond to chemical stimuli.

Choice shift–This term is analogous to group polarization, but describes the behavior change of the group as a whole rather than the individual.

Circadian rhythm–The alignment of physiological processes with the 24-hour day, including sleep–wake cycles and some elements of the endocrine system.

Circular reaction–A repetitive action that achieves a desired response; seen during Piaget's sensorimotor stage.

Class consciousness–In Marxist theory, the organization of the working class around shared goals and recognition of a need for collective political action.

Classical conditioning–A form of associative learning in which a neutral stimulus becomes associated with an unconditioned stimulus such that the neutral stimulus alone produces the same response as the unconditioned stimulus; the neutral stimulus thus becomes a conditioned stimulus.

Cocaine–Drug that decreases reuptake of dopamine, norepinephrine, and serotonin, with effects similar to amphetamines.

Cognition–The process of acquiring knowledge and understanding through thought, experiences, and the senses; how we think and respond to the world.

Cognitive appraisal–The subjective evaluation of a situation that induces stress, consisting of both an initial primary appraisal and a potential secondary appraisal if a threat is revealed during primary appraisal.

Cognitive development–The development of one's ability to think and solve problems across the life span.

Cognitive dissonance–The simultaneous presence of two opposing thoughts or opinions.

Cognitive reassociation model–A model of aggression which states that we are more likely to respond aggressively when experiencing negative emotions.

Collective unconscious–In Jungian psychoanalysis, the part of the unconscious mind that is shared among all humans and is a result of our common ancestry.

Colliculi–Two structures in the midbrain involved in sensorimotor reflexes; the superior colliculus receives visual sensory input, and the inferior colliculus receives auditory sensory input.

Compliance–A change of behavior of an individual at the request of another.

Concordance rates–In twin studies, the presence of a trait in both twins.

Conditioned response–In classical conditioning paradigms, the reflexive response caused by a conditioned stimulus.

Conditioned stimulus–In classical conditioning paradigms, this is an initially neutral stimulus that is paired with an unconditioned stimulus to train a behavioral response, rendering the previously neutral stimulus a conditioned stimulus.

Conduction aphasia–A speech disorder characterized by the inability to repeat words with intact spontaneous speech production and comprehension; usually due to injury to the arcuate fasciculus.

Confirmation bias–A cognitive bias in which one focuses on information that supports a given solution, belief, or hypothesis and ignores evidence against it.

Conflict theory–A theoretical framework that emphasizes the role of power differentials in producing social order.

Conformity–The changing of beliefs or behaviors in order to fit into a group or society.

Consciousness–Awareness of oneself; can be used to describe varying levels of awareness that occur with wakefulness, sleep, dreaming, and drug-induced states.

Conservation–Concept seen in quantitative analysis performed by a child; develops when a child is able to identify the difference between quantity by number and actual amount, especially when faced with identical quantities separated into varying pieces.

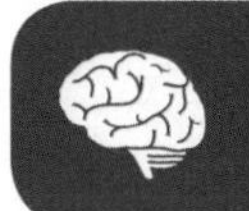

Constancy–In sensory perception, perceiving certain characteristics of object to remain the same despite differences in the environment.

Context effect–A retrieval cue by which memory is aided when a person is in the location where encoding took place.

Contralateral–On the opposite side of the body, relative to something else (usually a side of the brain).

Controlled (conscious) processing–Processing method used when a task requires complete attention.

Correspondent inference theory–A theory that states that people pay closer attention to intentional behavior than accidental behavior when making attributions, especially if the behavior is unexpected.

Cortical homunculus–A "map" that relates regions of the brain to the anatomical regions of the body.

Critical period–A time during development during which exposure to language is essential for eventual development of the effective use of language; occurs between two years of age and puberty.

Crystallized intelligence–Cognitive capacity to understand relationships or solve problems using information acquired during schooling and other experiences.

Cues–In understanding the behavior of others, indicators of the underlying cause of a behavior. This includes consistency cues, consensus cues, and difference cues.

Cultural capital–The benefits one receives from knowledge, abilities, and skills.

Cultural diffusion–The spread of norms, cultures, and beliefs throughout a culture.

Cultural relativism–The theory that social groups and cultures must be studied on their own terms to be understood.

Cultural sensitivity–Recognizing and respecting the differences between cultures.

Cultural syndrome–A shared set of beliefs, attitudes, norms, values, and behaviors organized around a central theme and found among people who speak the same language and share a geographic region.

Cultural transmission–The means by which a society socializes its members.

Culture–The beliefs, behaviors, actions, and characteristics of a group or society of people.

Culture shock–Cultural differences that are seen as quite dramatic when travelling outside of one's own society.

Deductive reasoning–A form of cognition that starts with general information and narrows down that information to create a conclusion.

Defense mechanism–In Freudian psychoanalysis, a technique used by the ego that denies, falsifies, or distorts reality in order to resolve anxiety caused by undesirable urges of the id and superego.

Deindividuation–The idea that people will lose a sense of self-awareness and can act dramatically differently based on the influence of a group.

Delirium–Rapid fluctuation in cognitive function that is reversible and has a nonpsychological cause.

Delusions–Fixed, false beliefs that are discordant with reality and not shared by one's culture, and are maintained in spite of strong evidence to the contrary.

Dementia–Intellectual decline starting with impaired memory and progressing to impaired judgment and confusion.

Demographic shift–A change in the makeup of a population over time.

Demographic transition–The transition from high birth and mortality rates to lower birth and mortality rates, seen as a country develops from a preindustrial to an industrialized economic system.

Demographics–The statistical arm of sociology, which attempts to characterize and explain populations by quantitative analysis.

Depressant–Any substance that reduces nervous system function.

Depressive disorder–Sadness meeting certain conditions of severity and duration such that a diagnosis of a mental health issue is warranted. Depressive disorders include, among others: major depression, dysthymic disorder, and seasonal affective disorder.

Depressive episode–A period of at least two weeks in which there is a prominent and persistent depressed mood or lack of interest and at least four other depressive symptoms.

Deviance–The violation of norms, rules, or expectations within a society.

Diagnostic and Statistical Manual of Mental Disorders **(DSM)**–The guide by which most psychological disorders are characterized, described, and diagnosed; currently in its fifth edition (DSM-5, published May 2013).

Diencephalon–A portion of the prosencephalon that becomes the thalamus, hypothalamus, posterior pituitary gland, and pineal gland.

Differential association theory–Theory that deviance can be learned through interactions with others who engage in deviant behavior, provided those interactions outnumber interactions with those who conform to social norms in number and/or importance.

Disconfirmation principle–The idea that states that if evidence obtained during testing does not confirm a hypothesis, then the hypothesis is discarded or revised.

Discrimination–In classical conditioning, the process by which two similar but distinct conditioned stimuli produce different responses; in sociology, when individuals of a particular group are treated differently than others based on their group.

Discriminative stimulus–In behavioral conditioning, a stimulus whose presence indicates the opportunity for reward.

Dishabituation–A sudden increase in response to a stimulus, usually due to a change in the stimulus or addition of another stimulus; sometimes called resensitization.

Displacement–A defense mechanism by which undesired urges are transferred from one target to another, more acceptable one.

Display rules–Cultural expectations of how emotions can be expressed.

Dispositional (internal) attributions–Attributions that relate to the decisions or personality of the person whose behavior is being considered.

Dissociative disorders–Disorders that involve a perceived separation from identity or the environment.

Distal stimulus–Part of the outside world that serves as a source for stimuli that reach the sensory neurons.

Distant networks–Networks that are looser and composed of weaker ties.

Distress–The stress response to unpleasant stressors.

Divided attention–The ability to attend to multiple stimuli simultaneously and to perform multiple tasks at the same time.

Dizygotic twins–Fraternal twins who share approximately 50% of their genes, as with most siblings.

Dominant hemisphere–The side of the brain that provides analytic, language, logic, and math skills; in most individuals, the left hemisphere.

Dopamine–A neurotransmitter associated with smooth movements, steady posture, the reward pathway, and psychosis.

Dramaturgical approach–An impression management theory that represents the world as a stage and individuals as actors performing to an audience.

Dreaming–Phenomenon which mostly occurs during REM sleep. Theories proposed to explain this phenomenon include activation-synthesis theory, problem-solving dream theory, and cognitive process dream theory.

Drive reduction theory–A theory that explains motivation as being based on the goal of eliminating uncomfortable internal states.

Drives–Deficiencies that activate particular behaviors focused on a goal, which can be further subdivided into either primary (body-sustaining) or secondary (not biologically necessary) drives.

Dual-coding theory–A cognitive theory that states that both visual and verbal associations are used to encode and retrieve information.

Duplicity theory of vision–A theory which holds that the retina contains two types of specialized photoreceptors: rods specialized for light and dark perception and cones specialized for color perception.

Dyssomnia–A sleep disorder in which one has difficulty falling asleep, staying asleep, or avoiding sleep.

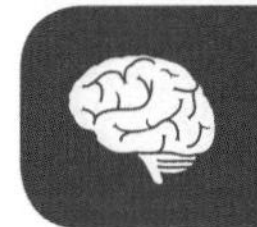

Ecstasy–Common name for MDMA (3,4-methylenedioxy-*N*-methylamphetamine); a central nervous system stimulant with effects similar to both amphetamines and hallucinogens.

Efferent neurons–Motor neurons that transmit information from the central nervous system to the periphery.

Ego–In Freudian psychoanalysis, the part of the unconscious mind that mediates the urges of the id and superego; operates under the reality principle.

Egocentrism–Self-centered view of the world in which one is not necessarily able to understand the experience of another person; seen in Piaget's preoperational stage.

Elaboration likelihood model–A theory in which attitudes are formed and changed through different routes of information processing based on the degree of deep thought given to persuasive information. There are two possible processing routes within this model: central route processing (deep thinking or elaborative) and peripheral route processing (non-elaborative).

Elaborative rehearsal–The association of information in short-term memory to information already stored in long-term memory; aids in long-term storage.

Electroencephalography (EEG)–A test used to study the electrical patterns of the brain under varying conditions; consists of multiple electrodes placed on the scalp. Characteristic EEG patterns include beta, alpha, theta, and delta waves, as well as patterns associated with REM sleep.

Emotion–A feeling and state of mind derived from circumstances, mood, or relationships.

Emotional support–Listening to, affirming, and empathizing with someone's feelings as part of social support.

Empathy–The ability to vicariously experience the emotions of another.

Empathy-altruism hypothesis–Theory that one individual helps another when they feel empathy for the other person.

Encoding–The process of receiving information and preparing it for storage; can be automatic or effortful.

Endorphins–Natural painkillers produced by the brain.

Epinephrine–A neurotransmitter associated with the fight-or-flight response.

Errors of growth–Misuse of grammar characterized by universal application of a rule, regardless of exceptions; seen in children during language development.

Escape learning–A form of negative reinforcement in which one reduces the unpleasantness of something that already exists.

Esteem support–Affirming qualities and skills of the person as part of social support.

Ethnic enclave–Locations with a high concentration of one specific ethnicity that can often slow assimilation.

Ethnicity–A social construct that sorts people by cultural factors, including language, nationality, religion, and other factors.

Ethnocentrism–The practice of making judgments about other cultures based on the values and beliefs of one's own culture.

Eustress–The stress response to positive conditions.

Evolutionary stable strategy–A strategy that, once adopted, will use natural selective pressure to prevent alternate strategies from arising.

Exchange theory–In social structure, an extension of rational choice theory that focuses on interactions in groups. Exchange theory holds that behavior is engaged in based on expectancy of future rewards and/or punishments.

Expectancy-value theory–The amount of motivation needed to reach a goal is the result of both expectation of success in reaching the goal and degree to which reaching the goal is valued.

Explicit memory–Memory that requires conscious recall, divided into facts (semantic memory) and experiences (episodic memory); also known as declarative memory.

Extinction–In classical conditioning, the decrease in response resulting from repeated presentation of the conditioned stimulus without the presence of the unconditioned stimulus.

Extrapyramidal system–Part of the basal ganglia that modulates motor activity.

Extraversion–In trait theory, the degree to which an individual is able to tolerate social interaction and stimulation.

Extrinsic motivation–Motivation that is external, or outside the self, including rewards and punishments.

False consciousness–In Marxist theory, a misperception of one's actual position within society.

Family group–A group determined by birth, adoption, and marriage rather than self-selection (as in a peer group).

Fertility rate–The average number of children born to a woman during her lifetime in a population.

Fisherian selection–Also called runaway selection, this is a positive feedback mechanism in which a trait with no impact (or a negative impact) on survival becomes more and more exaggerated over time, especially if the trait is deemed sexually desirable.

Fixation–In Freudian psychoanalysis, the result of overindulgence or frustration during a psychosexual stage causing a neurotic pattern of personality based on that stage.

Flat affect–Behavior characterized by showing virtually no signs of emotion or affective expression.

Fluid intelligence–Ability to quickly identify relationships and connections, and then use those relationships and connections to make correct deductions.

Foraging–The act of searching for and exploiting food resources.

Forebrain–A portion of the brain that is associated with complex perceptual, cognitive, and behavioral processes such as emotion and memory.

Fornix–A long projection from the hippocampus that connects to other nuclei in the limbic system.

Front stage–In the dramaturgical approach, the setting where players are in front of an audience and perform roles that are in keeping with the image they hope to project about themselves.

Frontal lobe–A portion of the cerebral cortex that includes the prefrontal cortex and the motor cortex; it controls motor processing, executive function, and the integration of cognitive and behavioral processes.

Functional attitudes theory–Theory that attitudes serve four functions: knowledge, ego expression, adaptation, and ego defense.

Functional fixedness–The inability to identify uses for an object beyond its usual purpose.

Functionalism–A theoretical framework that explains how parts of society fit together to create a cohesive whole, via both manifest (intended to help some part of the system) and latent (unintended positive) functions.

Fundamental attribution error–The general bias toward making dispositional attributions rather than situational attributions when analyzing another person's behavior.

Game theory–A model that explains social interaction and decision making as a game, including strategies, incentives, and punishments.

γ-aminobutyric acid (GABA)–A neurotransmitter associated with stabilizing and quelling brain activity.

Ganglia–Collections of neuron cell bodies found outside the central nervous system.

Gemeinschaft und Gesellschaft–Theory that distinguishes between two major types of groups: communities (*Gemeinschaften*), which share beliefs, ancestry, or geography; and societies (*Gesellschaften*), which work together toward a common goal.

Gender–The set of behavioral, cultural, or psychological traits typically associated with a biological sex.

Gender inequality–The intentional or unintentional empowerment of one gender to the detriment of others.

Gender segregation–The separation of individuals based on perceived gender.

General adaptation syndrome–Sequence of physiological responses developed by Selye in response to stress, initiating with alarm, followed by resistance, and finally exhaustion.

Generalization–In classical conditioning, the process by which two distinct but similar stimuli come to produce the same response.

Genotype–The genetic makeup of an individual.

Gentrification–The process of renewal of low income areas by upper-class

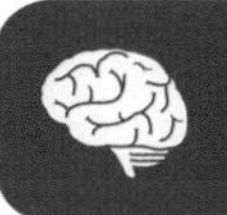

populations, ultimately displacing the lower income residents.

Gestalt principles–Goverened by the law of prägnanz, ways for the brain to infer missing parts of a picture when a picture is incomplete.

Globalization–The process of integrating the global economy with free trade and tapping of foreign labor markets.

Glutamate–An excitatory neurotransmitter in the central nervous system.

Glycine–An inhibitory neurotransmitter in the central nervous system.

Group–A social entity that involves at least two people, usually those sharing common characteristics.

Group conformity–Compliance with a group's goals, even when the group's goals may be in direct contrast to an individual's goals.

Group polarization–The tendency toward decisions that are more extreme than the individual inclinations of the group members.

Groupthink–The tendency for groups to make decisions based on ideas and solutions that arise within the group without considering outside ideas and ethics; based on pressure to conform and remain loyal to the group.

Gyrus–A ridge of the cerebral cortex.

Habituation–A decrease in response caused by repeated exposure to a stimulus.

Hallucinations–Perceptions that are not due to external stimuli but have a compelling sense of reality.

Hallucinogens–A group of drugs that cause distortions of reality in users, including lysergic acid diethylamide (LSD) and psilocybin-containing mushrooms.

Halo effect–A cognitive bias in which judgments of an individual's character can be affected by the overall impression of the individual.

Heterosexual–A sexual orientation wherein individuals are attracted to members of a different sex.

Heuristic–A rule of thumb or shortcut that is used to make decisions.

Hidden curriculum–In education, the transmission to students of social norms, attitudes, and beliefs.

Hierarchy of salience–Theory of identity organization that posits that we let situations dictate which identity holds the most importance at any given moment.

Hindbrain–A portion of the brain that controls balance, motor coordination, breathing, digestion, and general arousal processes.

Hippocampus–A portion of the limbic system that is important for memory and learning.

Homosexual–A sexual orientation wherein individuals are attracted to members of the same sex.

Humanistic theory–The set of theories that hold that personality is the result of the conscious feelings we have for ourselves as we attempt to attain our needs and goals. The theories of Kelly, Maslow, Lewin, and others fall into this category.

Hypnagogic hallucinations–Hallucinations that occur when going to sleep; seen in narcolepsy.

Hypnopompic hallucinations–Hallucinations that occur when awakening from sleep; seen in narcolepsy.

Hypnosis–An altered state of consciousness in which a person appears to be awake but is, in fact, in a highly suggestible state in which another person or event may trigger action by the person.

Hypothalamus–A portion of the forebrain that controls homeostatic and endocrine functions by controlling the release of pituitary hormones.

Id–In Freudian psychoanalysis, the part of the unconscious resulting from basic, instinctual urges for sexuality and survival; operates under the pleasure principle and seeks instant gratification.

Ideal self–The person one would optimally like to be.

Identity–A piece of an individual's self-concept based on the groups to which that person belongs and relationships to others.

Identity shift effect–When an individual's state of harmony is disrupted by a threat of social rejection, the individual will often conform to the norms of the group, followed by a corresponding identity shift to reduce cognitive dissonance.

Immediate networks–Networks that are dense with strong ties; generally overlap with distant networks.

Implicit memory–Memory that does not require conscious recall; consists of skills and conditioned behaviors.

Implicit personality theory–A theory that states that people tend to associate traits and behavior in others, and that people have the tendency to attribute their own beliefs, opinions, and ideas onto others.

Impression management–Behaviors that are intended to influence the perceptions of other people about a person, object, or event.

Incentive–A reward intended to motivate particular behaviors.

Incentive theory–Theory that behavior is motivated by the desire to pursue rewards and avoid punishments.

Incidence–The number of new cases of a disease per population at risk in a given period of time; usually, new cases per 1000 at-risk people per year.

Inclusive fitness–A measure of reproductive success; depends on the number of offspring an individual has, how well they support their offspring, and how well their offspring can support others.

Individual discrimination–One person discriminating against a particular person or group.

Inductive reasoning–A form of cognition that utilizes generalizations to develop a theory.

Inferior colliculus–Region of the midbrain that receives and integrates sensory input from the auditory system, and is involved in reflexive reactions to auditory input.

Information processing model–Model of human cognition containing four key components: information intake, information analysis, situational modification, and content/complexity of problem.

Informational support–Support given by providing information to help another person.

Ingratiation–An impression management strategy that uses flattery to increase social acceptance.

In-group–A social group to which people experience a sense of belonging or one in which they identify as a member.

Innate behavior–A behavior that is genetically programmed or instinctive.

Insomnia–Sleep disorder characterized by either an inability to fall asleep or difficulty staying asleep.

Instinct–An innate behavioral response to stimuli.

Instinct theory–In motivation, the theory that people are driven to engage in behaviors based on evolutionarily preprogrammed instincts.

Instinctive drift–The tendency of animals to resist learning when a conditioned behavior conflicts with the animal's instinctive behaviors.

Institutional discrimination–Discrimination against a particular person or group by an entire institution.

Intelligence quotient–Numerical measurement of intelligence, usually accomplished by some form of standardized testing.

Interaction process analysis–A technique of observing and immediately classifying the activities of small groups.

Interference–A retrieval error caused by the learning of information; can be proactive (old information causes difficulty learning new information) or retroactive (new information interferes with older learning).

Internalization–Changing one's behavior to fit with a group while also privately agreeing with the ideas of the group.

Interneuron–A neuron found between sensory and motor neurons; involved in the reflex arc.

Interpersonal attraction–The force that makes people like each other.

Intersectionality–The interconnected nature of social categorizations as they apply to a given individual/group, especially when they lead to discrimination or oppression.

Intrinsic motivation–Motivation that is internal or that comes from within.

Intuition–Perceptions about a situation that may or may not be

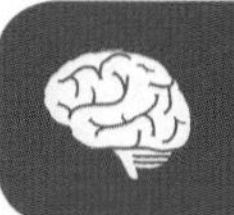

supported by available evidence, but are nonetheless perceived as information that may be used to make a decision.

Ipsilateral–On the same side of the body, relative to something else (usually a side of the brain).

Iron law of oligarchy–Democratic or bureaucratic systems naturally shift to being ruled by an elite group.

James–Lange theory–A theory of emotion that states that a stimulus results in physiological arousal, which then leads to a secondary response in which emotion is consciously experienced.

Justice–In medical ethics, the tenet that the physician has a responsibility to treat similar patients with similar care and to distribute healthcare resources fairly.

Just-noticeable difference (jnd)–The minimum difference in magnitude between two stimuli before one can perceive this difference; also called a difference threshold.

Just-world hypothesis–The cognitive bias that good things happen to good people, and bad things happen to bad people.

Labeling theory–Theory that labels given to people affect not only how others respond to that person, but also the person's self-image.

Language–Spoken or written symbols (verbal and nonverbal symbols), which are regulated according to certain rules of conduct or social norms and used for communication.

Language acquisition device (LAD)–An innate capacity for language acquisition that is triggered by exposure to language; part of the nativist (biological) perspective of language acquisition.

Latent learning–Learning that occurs without a reward but that is spontaneously demonstrated once a reward is introduced.

Learned helplessness–A state of hopelessness and resignation resulting from being unable to avoid repeated negative stimuli; often used as a model of depression.

Learning–In psychology, the way in which new behaviors are acquired.

Learning (behaviorist) theory–A theory that attitudes are developed through forms of learning (direct contact, direct interaction, direct instruction, and conditioning).

Libido–In Freudian psychoanalysis, the sex or life drive.

Life course approach to health–An analysis of health and probable outcomes that includes consideration of the patient's entire history.

Limbic system–A portion of the cerebrum that is associated with emotion and memory and includes the amygdala and hippocampus.

Linguistic relativity hypothesis–A hypothesis suggesting that one's perception of reality is largely determined by the content, form, and structure of language; also known as the Whorfian hypothesis.

Locus of control–The characterization of the source of influences on the events in one's life; can be internal or external.

Long-term memory–The relatively limitless form of memory reserved for information that is sufficiently rehearsed or of sufficient impact. There are both implicit and explicit forms of long-term memory.

Long-term potentiation–The strengthening of neural connections due to rehearsal or relearning; thought to be the neurophysiological basis of long-term memory.

Looking-glass self–Social psychological construct stating that the self is developed through interpersonal reactions, specifically through a person's understanding of the perception others have of them.

Magnocellular cells–In vision processing, cells that have high temporal resolution and detect motion.

Maintenance rehearsal–Repetition of a piece of information either to keep it within working memory or to store it.

Malthusian theory–Theory of demographic transition that focuses on how population growth can outpace food supply growth and lead to social degradation and disorder.

Managing appearances–An impression management strategy in which one uses props, appearance, emotional expression, or associations with others to create a positive image.

Manic episode–A period of at least one week with prominent and persistent elevated or expansive mood and at least two other manic symptoms.

Maslow's hierarchy of needs–Abraham Maslow's theory that certain needs will yield a greater influence on motivation. Maslow's hierarchy consists of five "levels" of need.

Mass hysteria–A shared, intense concern about the threats to society.

Master status–A status with which a person is most identified.

Mate bias–A measure of how choosy members of a species are in choosing a mate, based upon both direct and indirect benefits of mate selection.

Mate choice–The selection of a mate based on attraction and traits.

Material culture–The physical items one associates with a given cultural group.

Material support–Providing economic or other physical resources to aid a person as part of social support.

Mating system–The way in which a group organizes its sexual behavior and sexual relationships.

McDonaldization–A shift in focus toward efficiency, predictability, calculability, and control in societies.

Meditation–A state of consciousness entered voluntarily, characterized by a decreased level of physiological arousal and a quieting of the mind.

Medulla oblongata–A portion of the brainstem that regulates vital functions, including breathing, heart rate, and blood pressure.

Melatonin–A serotonin derivative secreted by the pineal gland that is associated with sleepiness.

Meninges–A thick layer of connective tissue that covers and protects the brain; composed of the dura mater, arachnoid mater, and pia mater.

Mental set–A tendency to repeat solutions that have yielded positive results at some time in the past.

Mere exposure effect–An explanation of attraction, also called the familiarity effect, which holds that people prefer stimuli that they have been exposed to more frequently.

Meritocracy–A society in which advancement up the social ladder is based on intellectual talent and achievement.

Mesencephalon–The embryonic portion of the brain that becomes the midbrain.

Mesolimbic reward pathway–Dopaminergic pathway in the brain including the nucleus accumbens, ventral tegmental area, and the medial forebrain bundle. This pathway is normally involved in motivation and emotional response, and is involved in drug addiction.

Metencephalon–The embryonic portion of the brain that becomes the pons and cerebellum.

Midbrain–A portion of the brainstem that manages sensorimotor reflexes to visual and auditory stimuli and gives rise to some cranial nerves.

Migration–The movement of people from one population to another, including immigration and emigration.

Mirror neurons–Neurons located in the frontal and parietal lobes and which fire both when an individual performs an action and when an individual sees that action performed.

Misinformation effect–A phenomenon in which memories are altered by misleading information provided at the point of encoding or recall.

Mnemonic–A technique that aids in memory recall.

Monogamy–An exclusive mating relationship.

Monozygotic twins–Identical twins, sharing the same genetic material.

Mood disorder–A mental health diagnosis category containing disorders primarily characterized by disturbance in mood. This includes depressive disorders, substance-induced, and bipolar disorders.

Moral reasoning–Kohlberg's theory of personality evelopment, which is focused on the development of moral thinking through preconventional, conventional, and postconventional stages.

Morbidity–The burden or degree of illness associated with a given disease.

Morphology–The structure of words, including their building blocks (prefixes, suffixes, and so on).

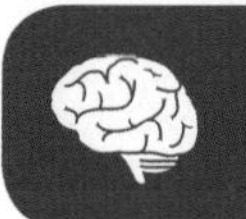

Mortality rate–The number of deaths in a population per unit time.

Motivation–The process of psychological and physical requirements, goals, or desires causing behavior.

Motor neuron–A neuron that transmits motor information from the spinal cord and brain to the periphery.

Multiculturalism–The encouragement of multiple cultures in a society to enhance diversity. Also referred to as cultural diversity.

Multiple intelligences–The idea that intelligence may exist in multiple areas, not just in the areas typically assessed by traditional intelligence quotient tests.

Myelencephalon–The embryonic portion of the brain that becomes the medulla oblongata.

Narcolepsy–A sleep disorder characterized by a lack of voluntary control over the onset of sleep; also involves cataplexy and hypnagogic and hypnopompic hallucinations.

Nativist theory (of language)–Theory credited to Noam Chomsky that posits the existence of an innate capacity for language, referred to as the language acquisition device.

Needs–Physiological and psychological requirements that motivate and influence behavior.

Negative symptoms–In mental illness, symptoms characterized by the absence of normal or desired behaviors.

Neologism–Coining a new word; seen in schizophrenia.

Network–A term used to describe the observable pattern of social relationships among individual units of analysis.

Network redundancy–Overlapping contact points within a social network.

Network support–Providing a sense of belonging as part of social support.

Neurocognitive models of dreaming–Models of dreaming that correlate subjective, cognitive experiences of dreaming with measurable physiological changes.

Neuroleptics (antipsychotics)–A class of drugs used to treat schizophrenia by blocking dopamine receptors.

Neuromodulator–Peptides that act as signaling molecules in the central nervous system; they are slower to act and longer lasting than neurotransmitters.

Neuroplasticity–Change in neural connections caused by learning or a response to injury.

Neuropsychology–The study of functions and behaviors associated with specific regions of the brain.

Neurosis–In Freudian theory, a disorder that occurs in response to the anxiety of a fixation during childhood that impacts personality development.

Neuroticism–In trait theory, the degree to which an individual is prone to emotional arousal in stressful situations.

Neurotransmitter–A chemical that transmits signals from a neuron to a target cell across a synapse.

Neurulation–Stage in development in which the ectoderm furrows over the notochord, forming the neural crest and neural tube.

Night terror–An experience of intense anxiety during sleep, causing the sleeper to scream in terror with no recall of the event in the morning; occurs during slow-wave sleep.

Nondominant hemisphere–The side of the brain associated with sensitivity to the emotional tone of language, intuition, creativity, music, and spatial processing; in most individuals, the right hemisphere.

Nonmaleficence–The ethical tenet that the physician has a responsibility to avoid interventions in which the potential for harm outweighs the potential for benefit.

Non-rapid eye movement (NREM) sleep–Stages 1 through 4 of sleep; contains ever-slowing brain waves as one gets deeper into sleep.

Nonverbal communication–How people communicate, intentionally or unintentionally, without using words; examples include body language, gestures, and facial expressions.

Norepinephrine–A neurotransmitter associated with wakefulness and alertness.

Norms–Societal rules that define the boundaries of acceptable behavior.

Obedience–The changing of behavior of an individual based on a command from someone seen as an authority figure.

Object permanence–Knowledge that an object does not cease to exist even when the object cannot be seen; a milestone in cognitive development.

Observational learning–A form of learning in which behavior is modified as a result of watching others.

Obsessive-compulsive disorders–This category, which also includes related disorders, describes the set of disorders where people feel the need to check things repeatedly or have certain thoughts repeatedly, without the ability to control these thoughts or activities.

Occipital lobe–A portion of the cerebral cortex that controls visual processing.

Operant conditioning–A form of associative learning in which the frequency of a behavior is modified using reinforcement or punishment.

Opiates–A drug family consisting of naturally occurring, highly addictive, pain-reducing drugs used in both medical and recreational settings; opioids are synthetic versions of these drugs.

Opponent-process theory–A theory that states that the body will adapt to counteract repeated exposure to stimuli, such as seeing afterimages or ramping up the sympathetic nervous system in response to a depressant.

Organization–A specific type of group characterized by five traits: formality, hierarchy of ranked positions, large size, complex division of labor, and continuity beyond its members.

Ought self–The representation of the way others think one should be.

Out-group–A social group with which an individual does not identify.

Overconfidence–A tendency to interpret one's decisions, knowledge, or beliefs as infallible.

Parallel play–Play style in which children can play alongside each other without interfering in each other's behavior.

Parallel processing–The ability to simultaneously analyze and combine information regarding multiple aspects of a stimulus, such as color, shape, and motion.

Parasomnia–A sleep disorder characterized by abnormal movements or behaviors during sleep.

Parasympathetic nervous system–A branch of the autonomic nervous system that promotes resting and digesting; associated with relaxed states, reductions in heart and respiration rates, and promotion of digestion.

Parietal lobe–A portion of the cerebral cortex that controls somatosensory and spatial processing.

Parkinson's disease–A disease characterized by slowness in movement, resting tremor, pill-rolling tremor, masklike facies, cogwheel rigidity, and a shuffling gait; caused by destruction of dopaminergic neurons in the substantia nigra.

Parvocellular cells–In visual processing, cells which have very high spatial resolution and detect shape.

Peer group–A group of self-selected equals that forms around common interests, ideas, preferences, and beliefs.

Peer pressure–The social influence placed on an individual by other individuals who are perceived as equals.

Perception–Processing of incoming information to comprehend and respond to the current incoming stimuli.

Peripheral nervous system (PNS)–The portion of the nervous system composed of nerve tissue and fibers outside the central nervous system.

Personality–The set of thoughts, feelings, traits, and behaviors that are characteristic of an individual across time and different locations.

Personality disorders–Disorders that involve patterns of behavior that are inflexible and maladaptive, causing distress or impaired function in at least two of the following: cognition, emotion, interpersonal functioning, or impulse control.

Phenotype–The expressed traits of an individual based on their genotype.

Phoneme–Individual speech sound associated with a language.

Phonology–The set of sounds that compose a language.

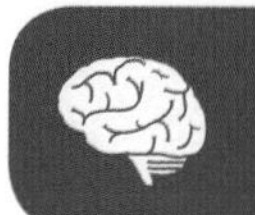

Piaget's theory–Piaget's theory of cognitive development divided the life span into sensorimotor, preoperational, concrete operational, and formal operational stages.

Pineal gland–A brain structure located near the thalamus that secretes melatonin.

Pituitary gland–The "master gland" of the endocrine system that triggers hormone release in other endocrine glands.

Place theory–Theory of sound conduction in the ear that holds that vibration on particular areas of the basilar membrane determines perception of pitch, also referred to as tonotopical organization.

Polyandry–A mating system in which an individual has exclusive relationships with several males.

Polygamy–A mating system in which an individual has multiple exclusive relationships.

Polygyny–A mating system in which an individual has exclusive relationships with several females.

Pons–A portion of the brainstem that relays information between the cortex and medulla, regulates sleep, and carries some motor and sensory information from the face and neck.

Positive symptoms–Behaviors, thoughts, or feelings added to normal behavior.

Poverty–A socioeconomic condition of low resource availability; in the United States, the poverty line is determined by the government's calculation of the minimum income requirements for families to acquire the minimum necessities of life.

Power–The capacity to influence people through the real or threatened use of rewards and punishments; often based on unequal distribution of valued resources.

Pragmatics–The ways in which use of language can be altered, depending on social context.

Prejudice–An irrational positive or negative attitude toward a person, group, or thing, formed prior to actual experience.

Prestige–In sociology, the amount of positive regard society has for a given person or idea.

Prevalence–The number of cases of a disease per population in a given period of time; usually, cases per 1000 people per year.

Primacy effect–The phenomenon of first impressions of a person being more important than subsequent impressions.

Primary group–A group wherein the interactions are direct, with close bonds, providing relationships to members that are very warm, personal, and intimate.

Primary stress appraisal–An initial evaluation of the environment to determine if there is an associated threat.

Priming–A retrieval cue by which recall is aided by a word or phrase that is semantically related to the desired memory.

Primitive reflexes–Reflexes present in infants that disappear with age.

Prodromal phase–A phase of poor adjustment that precedes the full onset of schizophrenia.

Projection–A defense mechanism by which individuals attribute their undesired feelings to others.

Projection area–A portion of the cerebral cortex that analyzes sensory input.

Promiscuity–A mating system in which an individual mates with others without exclusivity.

Proprioception–The ability to tell where one's body is in space.

Prosencephalon–The embryonic portion of the brain that becomes the forebrain.

Prosody–The rhythm, cadence, and inflection of speech.

Prospective memory–Remembering to perform a task at some point in a future.

Proximal stimulus–A stimulus that directly interacts with and affects sensory receptors.

Proximity–An aspect of interpersonal attraction based on being physically close to someone.

Psychoanalytic theory–In personality theory, the set of theories based on the assumption that unconscious internal states motivate overt actions and determine personality. The theories of several psychologists, including both Freud and Jung, fall into this category.

Psychological disorder–A set of thoughts, feelings, or actions that are considered deviant by the culture at hand and that cause noticeable distress to the sufferer.

Psychophysics–The study of the relationship between the physical nature of stimuli and the sensations/perceptions they evoke.

Psychosocial development–Erikson's theory of personality development, which is based in the concept that personality is developed based on a series of crises deriving from conflicts between needs and social demands.

Psychoticism–In trait theory, the measure of nonconformity or social deviance of an individual.

Punishment–In operant conditioning, the use of an aversive stimulus designed to decrease the frequency of an undesired behavior.

Race–A social construct based on phenotypic differences between groups of people; these may be either real or perceived differences.

Racial formation theory–Theory that racial identity is fluid and dependent on political, economic, and social factors.

Racialization–The definition or establishment of a group as a particular race.

Rapid eye movement (REM) sleep–Sleep stage in which the eyes move rapidly back and forth and physiological arousal levels are more similar to wakefulness than sleep; dreaming occurs during this stage.

Rational choice theory–In social structure, the theory that individuals consider benefits and harms to themselves in any given social interaction and choose the best possible action.

Rationalization–A defense mechanism by which individuals explain undesirable behaviors in a way that is self-justifying and socially acceptable.

Reaction formation–A defense mechanism by which individuals suppress urges by unconsciously converting them into their exact opposites.

Reappraisal–Process for ongoing monitoring of a continuing source of stress that cannot be dealt with via the normal 2-step appraisal method.

Recency effect–The phenomenon in which the most recent information we have about an individual is most important in forming our impressions.

Reciprocal determinism–In the social cognitive perspective, the notion that thoughts, feelings, behaviors, and environment interact to determine behavior in a given situation.

Reciprocal liking–The phenomenon whereby people like others better when they believe the other person likes them.

Reciprocity–An aspect of interpersonal attraction based on the idea that we like people who we think like us.

Recognition-primed decision model–A decision-making model in which experience and recognition of similar situations one has already experienced play a large role in decision making and actions; also one of the explanations for the experience of intuition.

Reference group–The group to which individuals compare themselves for a given identity.

Reflex–A behavior that occurs in response to a given stimulus without higher cognitive input.

Reflex arc–A neural pathway that controls reflex actions.

Regional cerebral blood flow (rCBF)–A technique used to record patterns of neural activity based on blood flow to different areas of the brain measured using detection of inhaled radioactive markers.

Regression–A defense mechanism by which an individual deals with stress by reverting to an earlier developmental state.

Reinforcement–In operant conditioning, the use of a stimulus designed to increase the frequency of a desired behavior.

Reinforcement schedule–The schedule by which reinforcement is administered for behavior in operant conditioning; reinforcement schedules can be fixed or variable, and can be based on a ratio or an interval between rewards.

Relative poverty–Poverty in comparison to the larger surrounding population.

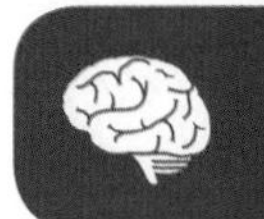

Reliance on central traits–The tendency to organize the perception of others based on traits and personal characteristics of the target that matter to the perceiver.

REM rebound–Phenomenon in which one spends an increased time in REM sleep following a period of sleep deprivation.

Representativeness heuristic–A shortcut in decision making that relies on categorizing items on the basis of whether they fit the prototypical, stereotypical, or representative image of the category.

Repression–A defense mechanism by which the ego forces undesired thoughts and urges to the unconscious mind.

Response bias–The tendency of subjects to respond systematically to a stimulus in a particular way due to nonsensory factors.

Reticular formation–A structure in the brainstem that is responsible for alertness.

Retrieval–The process of demonstrating that information has been retained in memory; includes recall, recognition, and relearning.

Retrograde amnesia–A form of memory loss that impacts long-term memories of events prior to the time of injury.

Rhombencephalon–The embryonic portion of the brain that becomes the hindbrain.

Ritual–A formalized ceremony that usually involves specific material objects, symbolism, and additional mandates on acceptable behavior.

Role–A set of beliefs, values, attitudes, and norms that define expectations for behavior associated with a particular status.

Role conflict–A difficulty in satisfying role requirements or expectations among various roles.

Role engulfment–Internalizing a label and assuming the role implied by the label leads to the assumed role taking over a person's identity.

Role partner–The person with whom one interacts while playing a particular role; each role partner provides a different set of behavioral expectations.

Role performance–Carrying out the behaviors associated with a given role.

Role set–A group of role partners relative to a given status.

Role strain–Difficulty in satisfying multiple requirements of the same role.

Role-taking–Roleplaying, by which children come to understand the perspectives of others and the ways in which these perspectives may differ from their own.

Sanction–A societally enforced punishment or reward for behavior. Formal sanctions are those enforced by social institutions (laws), and informal sanctions are enforced by social behaviors (ostracization, etc.).

Schachter–Singer theory–A theory of emotion that states that both physiological arousal and cognitive appraisal must occur before an emotion is consciously experienced.

Schema–An organized pattern of thought and behavior; one of the central concepts of Piaget's stages of cognitive development.

Schizophrenia–A psychotic disorder characterized by gross distortions of reality and disturbances in the content and form of thought, perception, and behavior.

Second sickness–The concept proposed by Howard Waitzkin that poor health outcomes are exacerbated by social injustice.

Secondary group–Groups wherein interactions are based on weaker, impersonal bonds.

Secondary stress appraisal–The interpretation of primary stress appraisal to determine emotional response to a given threat.

Selective attention–The ability to focus on a single stimulus even while other stimuli are occurring simultaneously.

Self-concept–The sum of the thoughts and feelings about oneself; includes self-schemata and appraisal of one's past and future self.

Self-determination theory–Need-based motivational theory that emphasizes the importance of autonomy, competence, and relatedness.

Self-disclosure–An aspect of interpersonal attraction or impression management in which one shares fears, thoughts, and goals with another person in the hopes of being met with empathy and nonjudgment.

Self-discrepancy theory–Theory that each of us has three selves: the actual self, the ideal self, and the ought self.

Self-efficacy–The degree to which individuals see themselves as being capable at a given skill or in a particular situation.

Self-enhancement–In self-serving bias, the need to maintain self worth through internal attribution of success and external attribution of failure.

Self-esteem–An individual's feelings of self-worth.

Self-fulfilling prophecy–The phenomenon of a stereotype creating an expectation of a particular group, which creates conditions that lead to confirmation of this stereotype.

Self-handicapping–An impression management strategy wherein people create obstacles to avoid self-blame when they do not meet expectations.

Self-presentation–The process of displaying oneself to society through culturally accepted actions and behaviors.

Self-reference effect–The tendency for individuals to best recall information that they can relate to their own experiences.

Self-schema–A self-given label that carries with it a set of qualities.

Self-serving bias–The idea that individuals will view their own success as being based on internal factors, while viewing failures as being based on external factors.

Semantic network–Organization of information in the brain by linking concepts with similar characteristics and meaning.

Semantics–The association of meaning with a word.

Sensation–Transduction of physical stimuli into neurologic signals.

Sensitive period–A time during which environmental input has a maximal impact on the development of a particular ability.

Sensory memory–Visual (iconic) and auditory (echoic) stimuli briefly stored in memory; fades very quickly unless attention is paid to the information.

Sensory neuron–A neuron that transmits information from sensory receptors to the central nervous system.

Septal nuclei–Part of the limbic system and one of the pleasure centers of the brain.

Serial position effect–The tendency to better remember items presented at the beginning or end of a list; related to the primacy and recency effects.

Serotonin–A neurotransmitter associated with mood, sleep, eating, and dreaming.

Sexual orientation–The direction of one's sexual interest.

Shadowing–An experimental technique in which participants recite speech immediately after hearing it.

Shaping–In operant conditioning, the process of conditioning a complicated behavior by rewarding successive approximations of the behavior.

Short-term memory–Memory which fades quickly, over about 30 seconds without rehearsal, and which is limited in capacity by the 7 ± 2 rule.

Sick role–Theory that a person who is ill enters a role of "sanctioned deviance," in which they are not responsible for the illness and are exempt from social norms.

Signal detection theory–A theory of perception in which internal (psychological) and external (environmental) context both play a role in the perception of stimuli.

Similarity–An aspect of interpersonal attraction based on being alike in attitudes, intelligence, education, height, age, religion, appearance, or socioeconomic status.

Situational (external) attributions–Attributions that relate to features of the surroundings, such as threats, money, social norms, and peer pressure, rather than to features of the individual.

Sleep apnea–Sleep disorder in which a person may cease to breathe while sleeping; may be due to obstruction or a central (neurological) cause.

Sleep cycle–A single complete progression through each stage of sleep.

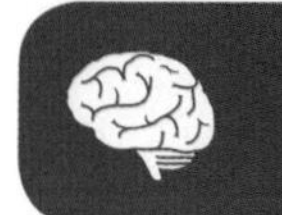

Slow-wave sleep–Consists of NREM sleep stages 3 and 4; also called delta-wave sleep.

Social action–Actions and behaviors that individuals are conscious of and performing because others are around.

Social capital–The investment people make in their society in return for economic or collective rewards.

Social class–A category of people with a shared socioeconomic background that exhibit similar lifestyles, job opportunities, attitudes, and behaviors.

Social cognitive theory–A theory that attitudes are formed through observation of behavior, cognition, and the environment.

Social construction model–A theory of emotional expression that assumes there are no biologically wired emotions; rather, they are based on experiences and situational context alone.

Social constructionism–A theoretical approach that uncovers the ways in which individuals and groups participate in the formation of their perceived social reality.

Social control–Regulating the behavior of individuals and groups within a society.

Social exclusion–The sense of being separated from and powerless in society when impoverished.

Social facilitation–The tendency of people to perform at a different level based on the fact that others are around.

Social institutions–Well-established, structured patterns of behavior or relationships that are accepted as a fundamental part of a culture.

Social interaction–The ways in which two or more individuals can shape each other's behavior.

Social interactionist theory–In language development, the theory that language acquisition is driven by the desire to communicate. This theory includes both biological and social processes.

Social loafing–The tendency of individuals to put in less effort in group settings as compared to an individual setting.

Social mobility–The movement of individuals in the social hierarchy through changes in income, education, or occupation.

Social movements–Philosophies that drive large numbers of people to organize to promote or resist social change.

Social perception–Understanding the thoughts and motives of other people present in the social world; also referred to as social cognition.

Social reproduction–The concept that social inequality, especially poverty, can be reproduced and passed on from one generation to the next.

Social stratification–Organization of societies into a hierarchical system, usually based on socioeconomic status and social class.

Social structure–A system of people within a society organized by a characteristic pattern of relationships.

Social support–The perception or reality that one is cared for by a social network.

Socialization–The process of developing and spreading norms, customs, and beliefs.

Socioeconomic status–Social standing or class of an individual or group, determined as a combination of education, income, and occupation.

Sociology–The study of society, including how it is created, interacted with, defined, and institutionalized.

Somatic nervous system–The voluntary branch of the peripheral nervous system, which consists of sensory and motor neurons used to control bodily movements.

Somatic symptom disorders–Mental health disorders marked by bodily symptoms that cause significant stress or impairment. This category of disorder also includes related disorders, such as illness anxiety and conversion disorders.

Somatosensation–The sense of touch, which contains multiple modalities: pressure, vibration, pain, and temperature.

Somatosensory cortex–Region of the parietal lobe located on the postcentral gyrus and involved in somatosensory information processing.

Somnambulism–Sleep disorder in which one carries out actions while asleep; also called sleepwalking.

Source-monitoring error–A memory error by which a person remembers the details of an event but confuses the context by which the details were gained; often causes people to remember events that happened to someone else as having happened to themselves.

Spacing effect–The phenomenon of retaining larger amounts of information when the amount of time between sessions of relearning is increased.

Spatial inequality–A form of social stratification across territories and their populations that can involve residential, environmental, or global components.

Spontaneous recovery–The reappearance of a conditioned response previously determined to be extinct.

Spreading activation–The unconscious activation of closely linked nodes of a semantic network.

State-dependent memory–A retrieval cue by which memory is aided when a person is in the same state of emotion or intoxication as when encoding took place.

Status–A position in society used to classify individuals.

Stereocilia–Structures on hair cells in the ear that sway with the movement of endolymph, causing receptor potential in the hair cells and ultimately leading to detection of incoming sound.

Stereotype content model–A model that classifies stereotypes using two dimensions: warmth and competence.

Stereotype threat–A feeling of anxiety about confirming a negative stereotype about one's social group.

Stereotypes–Attitudes and impressions that are made based on limited and superficial information about a person or a group of individuals.

Stigma–The extreme disapproval or dislike of a person or group based on perceived differences in social characteristics from the rest of society.

Stimulant–A drug that causes an increase in central nervous system arousal.

Stimulus–Any energy pattern that is sensed in some way by the body; includes visual, auditory, and physical sensations, among others.

Storage–The retention of encoded information; divided into sensory, short-term, and long-term memory.

Strain theory–Theory that explains deviance as a natural reaction to the disconnect between social goals and social structure.

Stress–The response to significant events, challenges, and decisions.

Stressors–Biological elements, external conditions, or events that lead to a stress response.

Structural poverty–Theory that poverty is due to inadequacies in societal and economic structure.

Subcultures–Groups of people within a culture that distinguish themselves from the primary culture to which they belong.

Sublimation–A defense mechanism by which unacceptable urges are transformed into socially acceptable behaviors.

Subliminal perception–Perception of a stimulus below a threshold (usually the threshold of conscious perception).

Substantia nigra–Part of the basal ganglia responsible for dopamine release that permits proper functioning of the rest of the basal ganglia.

Sulcus–A fold in the cerebral cortex.

Superego–In Freudian psychoanalysis, the part of the unconscious mind focused on idealism, perfectionism, and societal norms.

Superior colliculus–Structure in the midbrain that receives visual input and impacts eye movements and object oriented behaviors.

Symbolic culture–The nonmaterial culture that represents a group of people; expressed through ideas and concepts.

Symbolic ethnicity–An ethnic identity that is only relevant on special occasions or in specific circumstances and that does not impact everyday life.

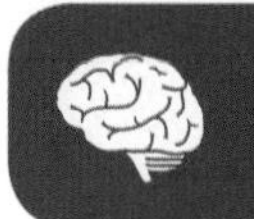

Symbolic interactionism–A theoretical framework that studies the way individuals interact through a shared understanding of words, gestures, and other symbols.

Sympathetic nervous system–The branch of the autonomic nervous system that controls the fight-or-flight response; associated with stressful situations that increase heart and respiration rates and decrease digestion.

Synaptic pruning–Adjustment of neural connections throughout life, involving breaking of weak neural connections and bolstering of strong neural connections.

Syntax–The way in which words are organized to create meaning.

System for multiple level observation of groups (SYMLOG)–A method of studying group dynamics; focuses on three fundamental dimensions of interaction: dominance *vs.* submission, friendliness *vs.* unfriendliness, and instrumentally controlled *vs.* emotionally expressive.

Tactical self–In impression management, the person one markets oneself to be when adhering to others' expectations.

Telencephalon–A portion of the prosencephalon that becomes the cerebrum.

Temporal lobe–A portion of the cerebral cortex that controls auditory processing, memory processing, emotional control, and language.

Tetrahydrocannabinol (THC)–The main active ingredient in marijuana.

Thalamus–A portion of the forebrain that serves as a relay and sorting station for sensory information, and then transmits the information to the cerebral cortex.

Theory of mind–The ability to sense how another's mind works.

Threshold–Also called limina; the minimum amount of a stimulus that renders a difference in perception.

Tolerance–Decreased response to a drug after physiological adaptation.

Top-down processing–Object recognition driven by memories and expectations that allow the brain to first recognize the whole object, and then recognize components based on existing expectations.

Trait theory–Personality theory that is focused on describing individual personalities as the sum of characteristic behaviors.

Transduction–Conversion of physical, electromagnetic, auditory, and other stimuli to electrical signals in the nervous system.

Transformational grammar–A linguistic theory that focuses on how changes in word order affect meaning.

Two-point threshold–The minimum distance necessary between two points of stimulation on the skin such that the points will be felt as two distinct stimuli.

Type theory–Theorifes of personality that are focused on creating taxonomies, or finite lists, of personality types.

Unconditioned response–In classical conditioning paradigms, the innate response brought about by an unconditioned stimulus prior to any conditioning.

Unconditioned stimulus–In classical conditioning paradigms, a stimulus that brings about an innate response.

Universal emotions–Emotions that are recognized by all cultures; include happiness, sadness, contempt, surprise, fear, disgust, and anger.

Urbanization–The process whereby large numbers of people migrate to and establish residence in relatively dense areas of population.

Value–What one deems important in life.

Ventricle–An internal cavity within the brain; cells lining the ventricles produce cerebrospinal fluid.

Verbal communication–The use of spoken or signed language.

Vestibular sense–One of the functions of the ear, the detection of rotational and linear acceleration to maintain awareness of body rotation and movement.

Visual cortex–Region of the occipital lobe devoted to processing visual information.

Visual pathways–Term which refers to both anatomical connections between eyes and brain and the flow of information from eyes to brain. The visual pathway contains several brain regions including the LGN, the visual cortex, and the superior colliculus.

Vygotsky theory–In cognitive development, the theory that the engine driving cognitive development is childhood internalization of culture.

Weber's law–A theory of perception that states that there is a constant ratio between the change in stimulus magnitude needed to produce a just noticeable difference and the magnitude of the original stimulus.

Wernicke–Korsakoff Syndrome–A condition resulting from chronic thiamine (vitamin B_1) deficiency, which is common in individuals who have alcoholism; characterized by severe memory impairment with changes in mental status and loss of coordination.

Wernicke's aphasia–Loss of language comprehension, resulting in fluid production of language without meaning.

Wernicke's area–A brain region located in the superior temporal gyrus of the temporal lobe (usually in the left hemisphere); largely responsible for language comprehension.

Working memory–Form of memory that allows limited amounts of information in short-term memory to be manipulated.

World system theory–World system theory argues there are global level inequalities in the division of labor between core, semi-peripheral, and peripheral nations.

Yerkes-Dodson law–A theory that there is a U shaped function relating arousal level and performance, dictating that performance is worst at the extreme high and low ends of arousal and optimal at intermediate arousal levels. This law is the basis of the Yerkes-Dodson law of social facilitation, a specific application of the Yerkes-Dodson law.

Zone of proximal development–Those skills that a child has not yet mastered but can accomplish with the help of a more knowledgeable other.

INDEX

Note: Material in figures or tables is indicated by italic *f* or *t* after the page number.

A

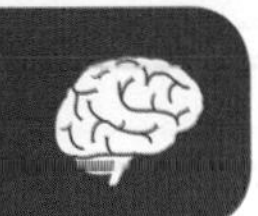

B

C

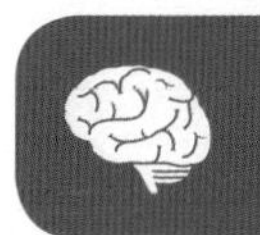

D

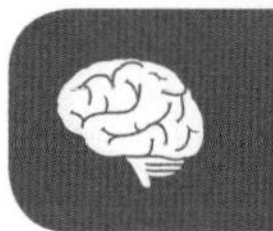

 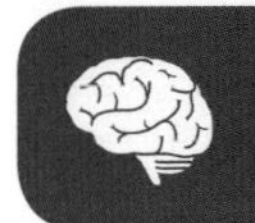

N

O

P

R

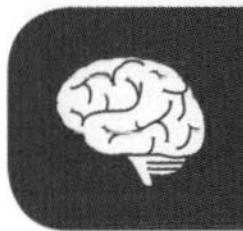

T

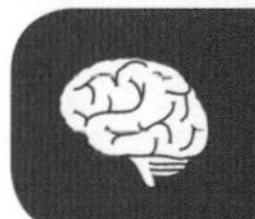

ART CREDITS

Chapter 1 Cover—Image credited to nobeastsofierce. From Shutterstock.

Figure 1.2—Image credited to Alila Medical Media. From Shutterstock.

Figure 1.3—Image credited to Alila Medical Media. From Shutterstock.

Figure 1.7—Image credited to User: _moep_. From Wikimedia Commons.

Figure 1.15—From Shutterstock.

Chapter 2 Cover—Image credited to Phatic Photography. From Shutterstock.

Chapter 3 Cover—Image credited to Fer Gregory. From Shutterstock.

Figure 3.4—Image credited to Public Library of Science. From Wikimedia Commons. Copyright © 2006. Used under license: CC-BY-2.5.

Chapter 4 Cover—Image credited to Igor Zh. From Shutterstock.

Figure 4.10—Image credited to User: Nitramas. From Wikimedia Commons.

Figure 4.13—Image credited to William Rafti. From Wikimedia Commons. Copyright © 2006. Used under license: CC-BY-2.5.

Figure 4.14—Image credited to User: Saxo. From Wikimedia Commons.

Chapter 5 Cover—Image credited to A.B.G. From Shutterstock.

Figure 5.1—Image credited to Mohd Fuad Salleh/EyeEm. From Getty Images.

Chapter 6 Cover—Image credited to Kanea. From Shutterstock.

Chapter 7 Cover—Image credited to hikrcn. From Shutterstock.

Figure 7.3—Image credited to User: Stepan Kapl. From Shutterstock.

Figure 7.4c—Image credited to Jose Luis Calvo. From Shutterstock.

Chapter 8 Cover—Image credited to file404. From Shutterstock.

Figure 8.2—Image credited to RonTech2000. From Getty Images.

Figure 8.5—Image credited to Newhoggy-commonswiki. From Wikimedia Commons.

Figure 8.6—Image credited to User: russavia. From Wikimedia Commons. Copyright © 2013. Used under license: CC-BY-2.0.

Figure 8.7—Image credited to User: kance. From Wikimedia Commons. Copyright © 2006. Used under license: CC-BY-2.0.

Chapter 9 Cover—Image credited to 10’000 Hours. From Getty Images.

Figure 9.2—Image credited to aldomurillo. From Getty Images.

Figure 9.3—Image credited to Tobertz Chavez. From Wikimedia Commons. Copyright © 2006. Used under license: CC-BY-2.0.

Figure 9.4—Image credited to Josh Plueger. From Wikimedia Commons.

Figure 9.5—Image credited to J. Tautz and M. Kleinheinz. From the Public Library of Science. Copyright © 2004. Used under license: CC-BY-2.5.

Chapter 10 Cover—Image credited to eyecrave productions. From Getty Images.

Figure 10.1—Image credited to Michael L. Baird. From Wikimedia Commons. Copyright © 2008. Used under license: CC-BY-2.0.

Figure 10.3—Image credited to AndreAnita. From Shutterstock.

Figure 10.4—Image credited to Alex Pronove. From Wikimedia Commons. Copyright © 2011. Used under license: CC-BY-2.5.

Chapter 11 Cover—Image credited to John M Lund Photography Inc. From Getty Images.

Figure 11.1—Image credited to Yana Iskayeva. From Getty Images.

Figure 11.2—Image credited to Baluchis. From Getty Images.

Figure 11.3—Image credited to arfabita. From Shutterstock.

Figure 11.10—Image credited to User: NikNaks93. From Wikimedia Commons.

Figure 11.11—Image credited to User: Jonathan McIntosh. From Wikimedia Commons.

Chapter 12 Cover—Image credited to Olivier Le Queinec. From Shutterstock.

Figure 12.1—Image credited to User: TastyCakes. From Wikimedia Commons. Copyright © 2009. Used under license: CC-BY-3.0.

Figure 12.3—Image credited to the National Park Service. From Wikimedia Commons.

Notes

Notes

Notes

Notes

Notes

Notes

Notes

Notes

Notes

Notes

Notes

Notes